Prudent Practices in the Laboratory

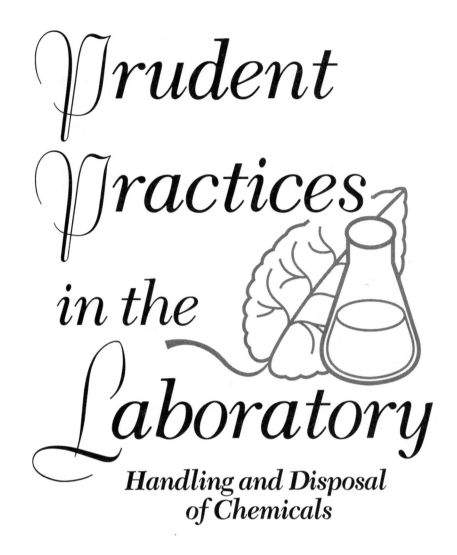

Handling and Disposal of Chemicals

Committee on Prudent Practices for Handling, Storage,
and Disposal of Chemicals in Laboratories

Board on Chemical Sciences and Technology
Commission on Physical Sciences, Mathematics, and Applications
National Research Council

NATIONAL ACADEMY PRESS
Washington, D.C. 1995

NATIONAL ACADEMY PRESS • 2101 Constitution Avenue, N.W. • Washington, D.C. 20418

NOTICE: The project that is the subject of this report was approved by the Governing Board of the National Research Council, whose members are drawn from the councils of the National Academy of Sciences, the National Academy of Engineering, and the Institute of Medicine. The members of the committee responsible for this report were chosen for their special competences and with regard for appropriate balance.

This report has been reviewed by a group other than the authors according to procedures approved by a Report Review Committee consisting of members of the National Academy of Sciences, the National Academy of Engineering, and the Institute of Medicine.

Support for this project was provided by the Department of Energy, American Chemical Society, National Science Foundation, National Institutes of Health, Camille and Henry Dreyfus Foundation, Howard Hughes Medical Institute, Chemical Manufacturers Association, National Institute of Standards and Technology, Occupational Safety and Health Administration, and Environmental Protection Agency, Office of Administration.

Library of Congress Cataloging-in-Publication Data

Prudent practices in the laboratory : handling and disposal of chemicals / Committee on Prudent Practices for Handling, Storage, and Disposal of Chemicals in Laboratories, Board on Chemical Sciences and Technology, Commission on Physical Sciences, Mathematics, and Applications, National Research Council.
    p. cm.
  Includes bibliographical references and index.
  ISBN 0-309-05229-7
  1. Hazardous substances. 2. Chemicals—Safety measures. 3. Hazardous wastes. I. National Research Council (U.S.). Committee on Prudent Practices for Handling, Storage and Disposal of Chemicals in Laboratories.
T55.3.H3P78 1995
660'.2804--dc20   95-32461

Printed in the United States of America

First Printing, August 1995
Second Printing, January 1999
Third Printing, January 2000
Fourth Printing, November 2001
Fifth Printing, December 2003

COMMITTEE ON PRUDENT PRACTICES FOR HANDLING, STORAGE, AND DISPOSAL OF CHEMICALS IN LABORATORIES

EDWARD M. ARNETT, Duke University, *Chair*
W. EMMETT BARKLEY, Howard Hughes Medical Institute
PETER BEAK, University of Illinois at Urbana-Champaign
EDWIN D. BECKER, National Institutes of Health
HENRY E. BRYNDZA, E.I. du Pont de Nemours & Co.
IMOGENE L. CHANG, Cheyney University
CAROL CREUTZ, Brookhaven National Laboratory
RICK L. DANHEISER, Massachusetts Institute of Technology
ERIC M. GORDON, Affymax Research Institute
ROBERT J. LACKMEYER, C/VS Inc.
LEE MAGID, University of Tennessee at Knoxville
THOMAS F. McBRIDE, U.S. Department of Energy
ANN M. NORBERG, 3M
EDWARD W. PETRILLO, Bristol-Myers Squibb
STANLEY H. PINE, California State University at Los Angeles
FAY M. THOMPSON, University of Minnesota

TAMAE MAEDA WONG, Study Director
KASANDRA GOWEN, Project Assistant
SARAH W. PLIMPTON, Editorial Assistant
JENNIFER F. BUTERA, Project Assistant

SUBCOMMITTEE ON ASSESSING CHEMICAL HAZARDS

RICK L. DANHEISER, Massachusetts Institute of Technology, *Chair*
W. EMMETT BARKLEY, Howard Hughes Medical Institute
PETER BEAK, University of Illinois at Urbana-Champaign
JAMES A. BOND, Chemical Industry Institute of Toxicology
WILLIAM M. HAYNES, Monsanto Co.
CURTIS D. KLAASSEN, University of Kansas
EDWARD W. PETRILLO, Bristol-Myers Squibb
CHARLES F. REINHARDT, E.I. du Pont de Nemours & Co.
PHILIP G. WATANABE, Dow Chemical Co.

SUBCOMMITTEE ON LABORATORY SPACE AND EQUIPMENT

ROBERT J. LACKMEYER, C/VS Inc., *Chair*
DALE T. HITCHINGS, Hitchings Associates, P.C.
JOHN S. NELSON, Affiliated Engineers Inc.

SUBCOMMITTEE ON MIXED WASTE

EDWIN D. BECKER, National Institutes of Health, *Chair*
PATRICIA A. BAISDEN, Lawrence Livermore National Laboratory
THOMAS F. CECICH, Glaxo Inc.
ERIC M. GORDON, Affymax Research Institute
PETER A. REINHARDT, University of Wisconsin at Madison

The National Academy of Sciences is a private, nonprofit, self-perpetuating society of distinguished scholars engaged in scientific and engineering research, dedicated to the furtherance of science and technology and to their use for the general welfare. Upon the authority of the charter granted to it by the Congress in 1863, the Academy has a mandate that requires it to advise the federal government on scientific and technical matters. Dr. Bruce M. Alberts is president of the National Academy of Sciences.

The National Academy of Engineering was established in 1964, under the charter of the National Academy of Sciences, as a parallel organization of outstanding engineers. It is autonomous in its administration and in the selection of its members, sharing with the National Academy of Sciences the responsibility for advising the federal government. The National Academy of Engineering also sponsors engineering programs aimed at meeting national needs, encourages education and research, and recognizes the superior achievements of engineers. Dr. Harold Liebowitz is president of the National Academy of Engineering.

The Institute of Medicine was established in 1970 by the National Academy of Sciences to secure the services of eminent members of appropriate professions in the examination of policy matters pertaining to the health of the public. The Institute acts under the responsibility given to the National Academy of Sciences by its congressional charter to be an adviser to the federal government and, upon its own initiative, to identify issues of medical care, research, and education. Dr. Kenneth I. Shine is president of the Institute of Medicine.

The National Research Council was organized by the National Academy of Sciences in 1916 to associate the broad community of science and technology with the Academy's purposes of furthering knowledge and advising the federal government. Functioning in accordance with general policies determined by the Academy, the Council has become the principal operating agency of both the National Academy of Sciences and the National Academy of Engineering in providing services to the government, the public, and the scientific and engineering communities. The Council is administered jointly by both Academies and the Institute of Medicine. Dr. Bruce M. Alberts and Dr. Harold Liebowitz are chairman and vice chairman, respectively, of the National Research Council.

Preface

In the early 1980s, the National Research Council (NRC) produced two major reports on laboratory safety and laboratory waste disposal: *Prudent Practices for Handling Hazardous Chemicals in Laboratories* (1981) and *Prudent Practices for Disposal of Chemicals from Laboratories* (1983). To provide safety and waste management guidance to laboratory workers, managers, and policymakers that would be responsive to knowledge and regulations in the 1990s, the NRC's Board on Chemical Sciences and Technology initiated an update and revision of the earlier studies.

After extensive consultation with members of the broad chemistry and laboratory communities, the full committee was appointed in September 1992. It first convened in November 1992 and held five additional meetings during the next two years. Several highly specialized areas were addressed by the appointment of several subcommittees, which met in conjunction with the full committee or independently as appropriate.

The Committee on Prudent Practices for Handling, Storage, and Disposal of Chemicals in Laboratories and its subcommittees were charged to:

- establish the scope of changes and new material required to update *Prudent Practices 1981* and *Prudent Practices 1983*,
- evaluate recent developments and trends in the scientific communities and regulatory areas,
- develop strategies for implementing safety programs, which include risk assessment methods in planning laboratory work with hazardous chemicals,
- develop a follow-up plan for training aids by obtaining consensus on the report and reviewing suggestions, and
- address such topics as procurement, storage, and disposal of chemicals; hazards of known chemicals; handling of chemicals; work practices; generation and classification of chemical waste; off-site transportation and landfills; and incinerators and small-scale combusters.

Prudent Practices 1981 and *Prudent Practices 1983* were conceived during the late 1970s in recognition of growing public expectations for health and safety in the workplace, protection of the environment, and the responsible use of hazardous chemicals. Since their original publication in the early 1980s, these reports have been distributed widely both nationally and internationally. In 1992, the International Union of Pure and Applied Chemistry and the World Health Organization published *Chemical Safety Matters*, a document based on *Prudent Practices 1981* and *Prudent Practices 1983*, for wide international use.

The original motivation for drafting *Prudent Practices 1981* and *Prudent Practices 1983* was to provide an authoritative reference on the handling and disposal of chemicals at the laboratory level. These volumes not only served as a guide to laboratory workers, but also offered prudent guidelines for the development of regulatory policy by government agencies concerned with safety in the workplace and protection of the environment. Pertinent health-related parts of *Prudent Practices 1981* are incorporated in a nonmandatory section of the OSHA Laboratory Standard (29 CFR 1910.1450; reprinted as Appendix A). OSHA's purpose was to provide guidance for developing and implementing its required Chemical Hygiene Plan.

Now, after nearly a decade and a half, the present volume (*Prudent Practices 1995*) responds to societal and technical developments that are driving significant change in the laboratory culture and laboratory operations relative to safety, health, and environmental protection. The major drivers for this new culture of laboratory safety include the following:

- The increasing regulatory compliance burden and associated time and financial penalties for noncompliance;
- The OSHA performance-based Laboratory Standard that places responsibility on individual laboratories to develop site-specific laboratory health programs, including certain elements such as written procedures, a designated coordinator for the written procedures, employee information and training, and compliance with OSHA-specified exposure limits;
- An increasingly litigious society and the growth of tort law;
- The increase in "public interest" groups and the realization by laboratory operators that operation of a laboratory is a privilege that carries a responsibility to go beyond mere compliance to "doing what is right" in the eyes of fellow workers and society;
- The myriad technical advances in our understanding of hazards and risk evaluation, improvements in chemical analysis, improvements in miniaturization and automation of laboratory operations, and the availability of vastly improved safety equipment, atmosphere-monitoring devices, and personal protective equipment; and
- A greater understanding and acceptance of the critical elements necessary for an effective culture of safety.

After careful consideration of these technical, regulatory, and societal changes, the committee chose to rewrite, rather than simply revise, much of the material in the previous two volumes and to condense them into a single one. In this 1995 revision, the committee has sought primarily to describe this new laboratory culture, identify its key elements, and provide certain information and procedures that have been developed within that culture. To ensure prudent handling in a coordinated manner from "cradle to grave," this new volume incorporates much material from the *Prudent Practices 1981* and *Prudent Practices 1983* volumes.

In addition, in response to users of *Prudent Practices 1981* who have emphasized the value of the information on how to handle compounds that pose special hazards, the committee has compiled Laboratory Chemical Safety Summaries (Appendix B) that provide chemical and toxicological information for 88 substances commonly found in laboratories. Although most of the information provided for these compounds will maintain its value, data on some properties, especially toxicological ones, should be updated frequently. Accordingly, the most recent Material Safety Data Sheets provided by the manufacturer or other updated sources should be consulted before work is done with hazardous compounds.

At every stage in the development of this book, the committee has maintained a close dialogue with the community of expected users through discussions with experts, participation of observers at committee meetings, and presentations to various professional organizations. In addition, subcommittees of experts were appointed to provide advice in several specialized areas. The goal in these discussions with authorities and with the general community of industrial and academic researchers and teachers has been to determine what are considered *prudent practices* for laboratory operations.

"Laboratory" means (following the OSHA Laboratory Standard) "a workplace where relatively small quantities of hazardous chemicals are used on a nonproduction basis." Through definition of the corollary terms "laboratory scale" and "laboratory use," OSHA expanded on this definition to encompass additional criteria: a laboratory is a place in which (1) "containers used for reactions, transfers, and other handling of substances are designed to be easily and safely manipulated by one person," (2) "multiple chemicals or chemical procedures are used," and (3) "protective laboratory practices and equipment are available and in common use to minimize the potential for employee exposure to hazardous chemicals." The definition excludes operations (1) in which the procedures involved are part of or in any way simulate a production process or (2) whose function is to produce commercial quantities of materials.

Dialogue with the chemical community has shown that there are many effective ways in which institutions can organize for safety in the laboratory when there is a sincere commitment to safe practice and institutional support. Accordingly, a single organizational model of institutional safety cannot be proposed as being typical. The aim throughout has been to offer generally useful guidelines rather than specific blueprints.

Public support for the laboratory use of chemicals depends on compliance with regulatory laws as a joint responsibility of everyone who handles or makes decisions about chemicals, from shipping and receiving clerks to laboratory workers and managers, environmental health and safety staff, and institutional administrators. This shared responsibility is now a fact of laboratory work as inexorable as the properties of the chemicals that are being handled. The use of chemicals, like the use of automobiles or electricity, involves some irreducible risks. However, all three of these servants to humankind have demonstrated benefits that enormously outweigh their costs if they are handled sensibly. The passage of time has demonstrated the value of *Prudent Practices 1981* and *Prudent Practices 1983* not only as guides to safe laboratory practice but also through their influence on the drafting of reasonable regulations. The committee hopes that its efforts will have a comparable beneficial impact as chemistry continues its central role in society.

Acknowledgments

Many technical experts, representing a wide variety of laboratories that use chemicals, provided input to this book. Their involvement through participation at workshops and committee meetings, submission of written materials, and review of technical material prepared by the committee has enhanced the book, and their efforts are greatly appreciated. The Committee on Prudent Practices for Handling, Storage, and Disposal of Chemicals in Laboratories thanks the following people both for their participation in the workshops and for contributions to the revision of *Prudent Practices 1995*.

Robert Alaimo, Proctor & Gamble; Bruce Backus, University of Minnesota; David Bammerlin, BP Warrensville; John Bartmess, University of Tennessee; John Beltz, Purdue University; Charles E. Billings, Massachusetts Institute of Technology; Daniel Brannegan, Pfizer; William H. Breazeale, Jr., Francis Marion University; Ronald Bresell, University of Wisconsin; Don G. Brown, University of Washington; Elise Ann B. Brown, USDA; Holmes C. Brown, Afton Associates, Inc.; Judy Brown, Edison Career Center; Rebecca Byrne, University of Illinois; Elna Clevenger, National Cathedral School; Robert G. Costello, W.R. Grace & Co.; Elizabeth Cotsworth, Environmental Protection Agency; Jeffrey L. Davidson, Environmental Protection Agency; Hugh Davis, Environmental Protection Agency; M. Sue Davis, Brookhaven National Laboratory; Gary Diamond, Syracuse Research Corporation; Howard Dobres, Drug Enforcement Administration; Laurence J. Doemeny, National Institute for Occupational Safety and Health; Greg L. Engstrom, 3M; Sandra A. Filippi, Prince George's Community College; Edward Gershey, Rockefeller University; Renae Goldman, 3M; Judith Gordon, Environmental Protection Agency; Carl Gottschall, E.G.&G.; Frederick D. Greene, Massachusetts Institute of Technology; Rolf Hahne, Dow Chemical Co.; Clayton E. Hathaway, Chesterfield, Missouri; Donna Heidel, R. W. Johnson; Jennifer Hernandez, Graham & James; Joseph Kanabrocki, University of Wisconsin; Glenn Ketcham, University of California; Robert Kohler, Monsanto; Po Yung Lu, Oak Ridge National Laboratory; Maureen Matkovich, American Chemical Society; Greg McCarney, 3M; Anne McCollister, Risk Communication International; Robert Meister, Eli Lilly; William G. Mikell, E.I. du Pont de Nemours & Co. (ret.); L. Jewel Nicholls, Sauk Rapids, Minnesota; Dan Pilipauskas, G. D. Searle; John E. Pingel, University of Illinois; Frank Priznar, Weston Inc.; Edward H. Rau, National Institutes of Health; Cynthia L. Salisbury, Compliance Solutions Inc.; David Schleicher, American Chemical Society; Eileen B. Segal, *Chemical Health and Safety*; William E. Shewbart, Dow Chemical Co.; Richard Shuman, Merck & Company; Reinhard Sidor, General Electric Co.; John Softy, Environmental Safety Office; Mary Ann Solstad, SOLSTAD Health and Safety Evaluations; Christine Springer, Scripps Research Institute; Ralph Stuart, University of Vermont; Martin J. Steindler, Argonne National Laboratory; Stephen A. Szabo, Conoco Inc./DuPont; Linda J. Tanner, 3M; David Vandenberg, University of California at Santa Barbara; George H. Wahl, North Carolina State University; Kenneth Williamson, Mount Holyoke College; Howard Wilson, Environmental Protection Agency; and Nola Woessner, University of Illinois.

Although the above list is extensive, it does not include all the individuals who have contributed their time, energy, and knowledge to this project. In full recognition that this report would not have been produced without the involvement of individuals not specifically mentioned here, the committee acknowledges their efforts by thanking the community at large.

Contents

Figures and Tables

Prudent Practices in the Laboratory

Handling and Disposal of Chemicals

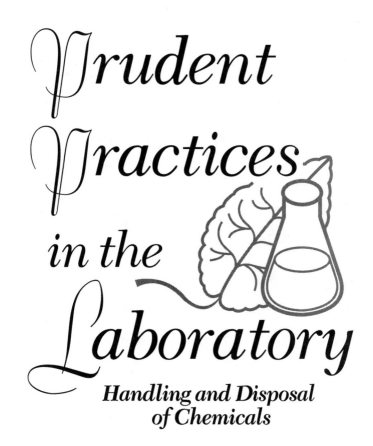

Overview and Recommendations

This book has been prepared by a National Research Council (NRC) committee in response to the growing recognition of the central place of chemistry in society, the special risks that are encountered by people who work with chemicals in the laboratory, and the potential hazards that are presented to the public by their use, transport, and disposal. Increased appreciation of the hazards related to certain chemicals has fostered a new "culture of safety" in many laboratories where chemicals are handled and chemical waste is generated and disposed of. Due in part to the publication of the NRC's *Prudent Practices for Handling Hazardous Chemicals in Laboratories* (hereinafter *Prudent Practices 1981*), and *Prudent Practices for Disposal of Chemicals from Laboratories* (hereinafter *Prudent Practices 1983*), there have been dramatic changes in attitudes toward shared responsibility by laboratory practitioners, management, and government at the federal, state, and local levels. These have been reflected in the OSHA Laboratory Standard (29 CFR 1910.1450; reprinted as Appendix A), which provides a legal, institutionalized, performance-based framework for safe, responsible laboratory work. There have also been important technical improvements that reduce the risks of handling chemicals in the laboratory and the cost of handling chemical waste.

Charged with the responsibility of evaluating the chemical, regulatory, and educational changes that have had an impact on the handling of chemicals in the laboratory since the previous reports were published, the committee has assessed the needs of all those who manage, handle, and dispose of chemicals in the laboratory workplace, where relatively small quantities of hazardous materials are used on a nonproduction basis. The committee was chosen for its breadth of expertise in chemistry, education, and environmental health and safety operations, and it has also called on a wider community of experts through the appointment of special subcommittees for assessing chemical hazards, design of laboratory space and equipment, pollution prevention, and the disposal of multihazardous waste. In addition, a number of meetings of the committee with different chemical user groups have helped to structure and clarify the committee's recommendations.

This volume was prepared primarily for those who use chemicals in laboratories, ranging from researchers and students to a broad array of technicians. Chemicals are handled not only by chemists, but also by biologists, physicists, geologists, materials engineers, and others. Accordingly, this volume is intended for general use but with the recognition that a basic familiarity with chemical nomenclature and its relation to molecular structure and chemical behavior is necessary in order to understand and use many parts of this work. A wider audience of administrative and chemical hygiene officers and environmental health and safety officers in educational, chemical, and regulatory institutions is also envisioned.

Although some readers may wish to become familiar with the entire book, others may be concerned with only one or two chapters, such as Chapter 3 (Evaluating Hazards and Assessing Risks in the Laboratory), Chapter 4 (Management of Chemicals), or Chapter 5 (Working with Chemicals). Others may be concerned with only Chapter 7 (Disposal of Waste) or Chapter 9 (Governmental Regulation of Laboratories). In deference to readers whose use of this book may be infrequent and specifically focused, or may perhaps occur under emergency conditions, the chapters are free-standing, even if this arrangement leads to repetition of some topics, albeit within different contexts.

THE CULTURE OF LABORATORY SAFETY (CHAPTER 1)

The new culture of laboratory safety implements the priority of "safety first" through a greatly increased emphasis on experiment planning, including habitual attention to risk assessment and consideration of hazards for oneself, one's fellow workers, and the public. So important is the formal framework for experiment planning that this volume has been structured around the sequence of steps described in Chapter 2. The key word "prudence" provides a middle pathway between the extremes of stultifying overregulation and a reckless rush to "get the job done" in the laboratory. A prudent attitude toward dealing with hazards in the laboratory is characterized by a determination to make every effort to be informed about risks and reduce them to a minimum while recognizing that the notion of "zero risk" in laboratory operations (or any other workplace) is an impossible ideal. However, an accident-free workplace can be approached by setting a goal of zero incidents and zero excuses. Continuous basic respect and care for the health and safety of laboratory workers and the greater society constitute the starting point for *Prudent Practices in the Laboratory: Handling and Disposal of Chemicals*.

Education for Safety and Pollution Control

Good attitudes toward rational risk assessment and safe habits, as well as awareness of the expectations of others who might be affected by laboratory work, should be instilled in laboratory workers from their earliest experiences with observing and performing laboratory operations. Early in primary school children should be involved in thinking through possible implications of and risks in experiments that they see or do, learning that this is part of the way science is done. If good habits are inculcated from the beginning, participation in the culture of safety will be natural and painless; if these lessons are neglected until college or graduate school, or until the first industrial job, reeducation can be a difficult, expensive, and perhaps even a dangerous initiation.

At the present time, there is a wide range of safety consciousness and safety preparation among individuals entering high school, college, graduate school, and industry. Teaching laboratories at all levels are faced with the problem of young personnel with diverse backgrounds and various levels of preparation. Some who are beginning their first college laboratory course work may have had no previous hands-on training in handling chemicals or equipment and may even carry a "chemophobic" prejudice to the workbench. Others may come well prepared to assume personal responsibility for risk assessment and safety planning in their experiments if their instructors in high school have trained them to share in every stage of experiment planning, with attention to suitable waste disposal as a routine component. If the research environment in their undergraduate and graduate schools has not emphasized shared responsibility and good working relationships with environmental health and safety personnel, even professional chemists entering well-run industrial or government laboratories may find it difficult to adjust to the environment of fully accountable experiment planning.

Factors Affecting Safety Practices in Laboratories

A wide variety of factors, both internal and external, have affected the conduct of laboratory work during the past 15 years. Public concern for safety in the workplace and protection of the environment through pollution prevention has resulted in a voluminous array of regulations designed to control every stage of the transportation of chemicals to and from laboratories, their handling within the laboratory workplace, and their final disposal. Safe practice by laboratory workers requires continuing attention and education; it cannot

be assumed to be optional. Accordingly, an infrastructure of professionals trained in environmental health and safety has developed who serve at the interface between federal, state, and local regulatory agencies and the educational and industrial laboratories where chemicals are handled. An increasing climate of litigation has also sharpened the awareness of everyone on the ladder of responsibility, from directors and trustees to maintenance personnel, about the price that may have to be paid if accidents occur as a result of the illegal or irresponsible handling of chemicals or chemical waste. "Down the sink" disposal, for example, is no longer routine.

Many steps have been taken to improve the safety of equipment for handling and experimenting with chemicals. An increasing trend toward miniaturizing chemical laboratory operations has reduced the volume and, therefore, the cost of acquiring chemicals and handling all aspects of waste, including the removal of toxic vapors from the laboratory. Concurrent with miniaturization has been the development of instruments with vastly increased sensitivity and speed of operation so that both the quantities of materials and the turnaround time required for obtaining answers to experimental questions have been reduced drastically. In some teaching and research programs, simulation by computer has replaced experiments that pose a particular danger. Waste minimization and pollution control techniques, such as recycling and the development of more efficient or nontoxic synthetic routes, have gained increasingly high priority as the cost of waste disposal has escalated.

Unquestionably, most laboratories are safer places to work now than they were 15 years ago. However, the ultimate key to maintaining a safe environment lies in the attitude and behavior of the individual worker.

PRUDENT PLANNING OF EXPERIMENTS (CHAPTER 2)

Preparation for running an experiment in the laboratory has always required forethought in order to assemble the necessary chemicals, purify them to an acceptable level, set up and test the required apparatus, and use precedents from the literature, or in-house reports, to consider the appropriate scale and conditions to be employed. Depending on the individual worker, the supervisors, and the requirements of the institution, the process may be formalized to different degrees. The new culture of laboratory safety recognizes that it is wise to formalize the process of experiment planning, both in the interest of safety and to ensure compliance with regulations for the handling of reactants required for the proposed processing and disposal of all waste generated. The sequence of steps for planning is so fundamental and repeatable that the committee has structured this book in accordance with the work-flow diagram, Figure 1.2, given in Chapter 1. Each chapter is organized around a key step in that protocol.

There is great diversity in the formality and means by which the planning structure applies. For example, laboratory manuals for students provide a complete package of planned experiments and detailed technical directions. Future manuals should include questions and assignments that involve the student actively in considering the risks, regulations, and waste disposal costs for alternative approaches to the problem under discussion. In research laboratories where important steps of the planning process have, over time, become standard operating procedure, mental planning may be sufficient to enable doing the next routine experiment safely and effectively. In contrast, a completely new type of experiment involving unfamiliar materials and unprecedented hazards may require formal planning for every stage, and discussion with experts outside the immediate research group and members of the environmental health and safety office should be considered carefully.

EVALUATING HAZARDS AND ASSESSING RISKS IN THE LABORATORY (CHAPTER 3)

A first step in planning a new experiment is consideration of the types of hazards that may be posed by toxic, flammable, reactive, and/or explosive materials encountered in the proposed work. Even if the experiment is part of a series that has been repeated so many times that every step has become routine, the change in conditions or compounds that makes this a new experiment, rather than just a repetition, may introduce a new hazard, unprecedented in the series; for example, a minor molecular modification may result in a sharp increase in toxicity or likelihood of explosion.

Thus, planning for the first experiments in a new field, as well as experiment planning by inexperienced workers, requires considerable investigation of published resources and discussion with competent experts on potential technical hazards and regulatory requirements. Advice from environmental health and safety or industrial hygiene personnel may be particularly valuable in the initial stage of planning.

A wide variety of published resources describe the dangerous properties of specific compounds. All of the common types of explosive, reactive, and flammable compounds are well documented. Because the functional groups that are associated with these properties are clearly identified, and their hazards well known, it

is usually possible to predict the potential for a serious accident. However, unprecedented accidents are reported occasionally, and new substances with unknown properties are continually being generated as products and by-products of chemical experiments.

At the present time, the least predictable dangerous chemical property is toxicity. In addition to discussing toxicity extensively, Chapter 3 provides general guidelines and references to specific toxicological information. Unless there is very well established evidence that a chemical is innocuous, the following warnings should always be borne in mind:

- Assume that all chemicals encountered in the laboratory are potentially toxic to some degree. Because the risk posed by a toxic chemical depends on the extent of exposure and the chemical's inherent toxicity, minimize exposure by avoiding skin contact and inhalation exposure through proper clothing and ventilation as habitual safe practice.
- Treat any mixture of chemicals as potentially more toxic than its most toxic component.
- Treat all new compounds, or those of unknown toxicity, as though they could be acutely toxic in the short run and chronically toxic in the long run. Because typical reactions produce a variety of by-products that are often unidentified or unknown, reaction products should be assumed to be toxic during work-up. Even though the likelihood is small that any given unknown chemical is very toxic, and the potential dose is usually low, laboratory researchers and workers may be exposed to thousands of chemicals during a professional lifetime, and there is a reasonable probability of eventual dangerous accidental exposure to a toxic substance. A habit of minimizing exposure should be cultivated.

The flammability, corrosiveness, and explosibility of chemicals and their combinations must also be considered along with toxicity in evaluating hazards and planning how to deal with them. In addition, wise risk management requires taking into account the amount of material to be used in an experiment.

Chapters 3, 5, and 6 are the heart of this volume because of their detailed guidelines regarding laboratory hazards and procedures. The content of these chapters inevitably overlaps because assessing the risks of a given procedure with a given chemical (Chapter 3) may also suggest the best ways to work with it (Chapters 5 and 6). Some of the most notable laboratory hazards arise from the interactions of equipment with chemicals (e.g., fires from sparking motors near flammable vapors) or from special nonchemical properties such as radioactivity or biological contamination.

Of the various resources that describe the properties of selected chemicals, two listed in Chapter 3 deserve particular comment: Material Safety Data Sheets and Laboratory Chemical Safety Summaries.

Federal law requires that Material Safety Data Sheets (MSDSs) be provided to users of chemicals by their manufacturers and distributors. MSDSs provide necessary information about precautions for protecting against known hazards associated with the subject product and often include useful information on chemical, physical, and toxicological properties, along with suggestions for storing, transporting, and disposing of chemicals. MSDSs are the best *general* source of information available, and they should be consulted as a first step in assessing the risk associated with doing an experiment. However, because there is currently no standard format for MSDSs, their quality varies widely, and the information that they contain may be inappropriate for laboratory use.

In consideration of the special problems of planning chemical experiments in the laboratory, the committee prepared Laboratory Chemical Safety Summaries (LCSSs) for 88 carefully chosen chemicals; these summaries are included as Appendix B of this book. Since many of these 88 chemicals are representative of a class of potentially hazardous compounds, the LCSSs can also be used as guides to handling many other compounds with related chemical structures. The LCSSs provide concise critical discussions, in a style readily understandable to laboratory workers, of the toxicity, flammability, reactivity, and explosibility of the subject chemicals. Directions for handling, storage, and disposal and special instructions for first aid and emergency response are given. The 88 LCSSs provide considerably greater coverage of specific and representative chemicals than was available in *Prudent Practices 1981* and *Prudent Practices 1983* and, unlike most MSDSs, are designed especially for laboratory workers.

MANAGEMENT OF CHEMICALS (CHAPTER 4)

Virtually every stage in the life cycle of a chemical has undergone dramatic change during the past 15 years as the new culture of laboratory safety has become established. Necessarily, the new ways that chemicals are acquired, tracked through an institution, stored, and delivered to the laboratory must be considered in contemporary experiment planning along with the detailed conduct of the experiment and the follow-up stages of handling all products and waste. Factors that once played at most a minor role in the handling of chemicals are now central. Almost all of these are related directly or indirectly to the legal, bureaucratic, and associated financial costs that have resulted from elevating the priority of safety in the workplace and protection of the public and the environment. The costs of documenting, handling, and disposing of all unwanted chemicals (i.e., waste) from completed experiments have increased enormously. Consequently, strategies that once were accepted as prudent, frugal, or, at worst, harmless are now no longer common practice. Fortunately, a number of technical advances, such as miniaturization and the large-scale management of information by computers, have allowed partial accommodation to the new forces affecting education and research in laboratories. Still, there is no doubt that the way science is done has changed enormously and that allowing adequate time and money for managing chemicals has become a major factor in planning experiments. Chapter 4 contains guidelines for the safe acquisition and storage of hazardous chemicals.

The prudent handling of chemicals now requires reducing the volume of every component to the minimum necessary to achieve the goals for which it was acquired. Any excess should be disposed of quickly and legally, unless there is a justifiable future use for it. In minimizing risks to laboratory, transport, and storeroom personnel, and to minimize the cost of waste disposal, source reduction is the first step. One should order and have on hand only what is necessary for currently planned experiments. No longer is it frugal to accept gifts of unneeded materials on the chance that they might be useful in the future or to buy the "large economy size" and store unused leftover chemicals for potential but unknown applications. The American Chemical Society booklet "Less Is Better" (1993) emphasizes the safety and financial reasons for buying chemicals in small packages: reduced risk of breakage, reduced risk of exposure following an accident, reduced storage cost, reduced waste from decomposition during prolonged storage in partially empty bottles, and reduced disposal cost for small containers of unused material. A well-planned experiment should reflect the "just in time" acquisition strategies used in modern manufacturing. If possible, the responsibility for storing and inventorying chemicals should remain with the supplier.

For chemicals likely to be used in the near future, storage is reasonable and can be a frugal component of a well-managed plan for handling chemicals. In general, though, if a chemical has not been used during the two years since it was placed in laboratory storage, the chance is small that it will ever be used again. Three years is a reasonable deadline for use, recognizing the value of shelf space, the deterioration of many chemicals, and the enormous price of disposal if the label decomposes or falls off and the compound becomes an "unknown." Unused remainders can be handled best by maintaining a comprehensive, reliably updated inventory, especially in a large institution. Maintaining a readily accessible inventory can be expedited if all chemicals are bar-coded and records of their status are continuously updated and made available through computer networking (see ''Recommendations'' below). Manufacturers and vendors of laboratory chemicals also can play an increasingly valuable role by responding to the needs of their customers to reduce the scale of experimentation, maintain continuously updated inventories accessible over networks, and reduce the cost of waste disposal.

The costs of acquisition, storage, and disposal can be minimized by conducting experiments on the smallest practical scale, a practice that also reduces the risk of hazards from exposure, fire, and explosion. Microscale reactions can now be run conveniently with less than 100 milligrams of solids or 100 microliters of liquids, compared with the traditional 10 to 50 grams and 100 to 500 milliliters.

Traditionally, the hazardous properties of chemicals have been regarded as a significant factor in planning experiments only if extreme toxicity or danger from explosion was apparent. The present emphasis on reducing risks and waste of all kinds may suggest the substitution of different solvents or less hazardous synthetic routes. Although it may not always be feasible to improve safety through the use of more benign materials, it is always appropriate to consider the possibility of reducing risks in this way, especially if precedents for the planned work were taken from the older literature, where safety and pollution problems were afforded less weight than is now given to them.

WORKING WITH CHEMICALS (CHAPTER 5)

Chapter 5 is a comprehensive manual for the safe handling of hazardous chemicals commonly used in laboratories. Like Chapters 3 and 6, Chapter 5 is intended to be used as a daily reference guide to appropriate standards of professional laboratory performance. In addition to discussing handling of chemicals in a variety of specific circumstances, it addresses issues such as proper protective clothing, good housekeeping, and necessary preparation for accidents and concludes with an alphabetical listing of especially hazardous materials.

WORKING WITH LABORATORY EQUIPMENT (CHAPTER 6)

Chapter 6 explains how to use the various kinds of equipment associated with handling hazardous chemicals. Although dangers such as electrical shock from bad wiring, falls on flooded floors, or cuts from broken glassware are not unique to the laboratory workplace, their consequences for laboratory personnel can be compounded because of the added hazards of toxicity, flammability, corrosiveness, and reactivity that characterize many chemicals. The accidental dropping of a glass container of a volatile poison or the fire hazard from sparking electrical equipment or switches in the presence of flammable fumes, for example, present potentially serious situations of a kind that must be kept in mind when laboratory experiments are planned and conducted.

The special hazards that accompany the use of electrical equipment (e.g., stirrers, pumps, and heating/cooling devices), the precautions necessary for handling gases in various containers and systems, and the equipment for dealing with and preventing many kinds of laboratory accidents are discussed in detail.

DISPOSAL OF WASTE (CHAPTER 7)

Concern about the fate of used or unwanted products of chemical reactions has not been a significant part of the traditional culture of laboratory workers. To emphasize the high priority that waste disposal has assumed in modern laboratory operations, the committee was charged to merge the subject matter of *Prudent Practices 1983* on the disposal of waste from laboratories with that of *Prudent Practices 1981* on the handling of hazardous chemicals in laboratories. Furthermore, it was asked to investigate the especially vexing problems of handling multihazardous chemical, biological, and radioactive waste and to propose recommendations for dealing with it. These are offered in Chapter 7. In view of the crucial role of regulations in dealing with laboratory waste, Chapter 9 on governmental regulation of laboratories should be referred to frequently as background for Chapter 7.

Waste is generally defined as excess, unneeded, or unwanted material. Because these terms are fairly subjective, regulatory agencies have attempted to provide more objective and specific definitions. However, the regulatory viewpoint that a material is waste if it is abandoned or ''inherently wastelike'' remains inescapably subjective. Although the residues from cleaned-up spills are obviously waste, the point at which a laboratory worker decides that a given chemical is no longer potentially useful may be difficult to define. Once the determination has been made, the waste must be handled within the constraints of legal guidelines that are usually defined according to the nature of the waste (chemical, radioactive, biological), the type and degree of hazard that it presents, and its quantity. Enlightened risk management also dictates that the amount of material be one factor in decisions on handling and disposal of waste.

The Environmental Protection Agency (EPA) has formulated most of the regulations for assessing risks from chemical waste and for dealing with them. Waste chemicals are characterized as ignitable, corrosive, reactive, and toxic. The responsibility for determining whether a waste is hazardous, and for characterizing the hazard, rests with the waste's generator, who may consult the appropriate LCSS, MSDS, or other published listing when dealing with a fairly common chemical. For other kinds of waste, workers who are well versed in reasoning by analogy from structural formulas should be able to make an educated guess by referring to related compounds whose molecules contain common structural units. If there is serious doubt, questions can be forwarded to an institution's environmental health and safety office or to the regional EPA office.

Hazardous Chemical Waste

Hazardous waste should be identified clearly so that its origin can be traced. Waste management facilities are prohibited from handling materials that are not identified and classified by hazard. Unidentified materials must be analyzed according to the following criteria before they will be accepted by most waste disposal firms: physical description; water reactivity; water solubility; pH; ignitability; and the presence of an oxidizer, sulfides or cyanides, halogens, radioactive materials, biohazardous materials, or toxic constituents. Chapter 7 provides detailed procedures for testing unknown materials.

Chemical waste should be accumulated at a central site where it can be sorted, stored temporarily, and prepared for disposal by commingling (according to regulation) or allowable on-site treatment for hazard reduction or, perhaps, recycling. During any waste-handling processes at a central site, or on the way to or from it, personnel should be protected: in particular, removal of toxic vapors, fire suppression, and spill control should be provided for. All personnel responsible for handling chemical waste must be trained.

Typically, chemical waste is sent for ultimate disposal to a landfill or incinerator in a 55-gallon Lab Pack or bulk solvent drum. Records must be maintained of the quantity, identity, and analyses (if necessary) of waste and of shipping and verification of disposal.

In many cases, the cost of waste handling and removal may be lowered sharply by using appropriate deactivation procedures for hazard reduction. Resource Conservation and Recovery Act (RCRA) regulations define the term "treatment" broadly, but the cost of being an EPA-permitted treatment facility is too high for most laboratories. However, small-scale treatments such as acid-base neutralization can be carried out as part of an experiment plan. Because illegal treatment can lead to fines of up to $25,000 per day of violation, it is important to check with an institution's environmental health and safety office or EPA before engaging in treatment procedures of any scale. Chapter 7 provides hazard reduction procedures for dealing with some common classes of chemicals. For disposal of small quantities of waste, approved procedures for identification, pickup, and delivery to a central site should be used to avoid risking citation for illegal treatment without a permit.

Incineration, addition to a landfill, release to the atmosphere, and discarding in the normal trash or the sanitary sewer are all options for disposal, depending on whether or not a waste is hazardous and on how it is regulated. Nonhazardous materials such as potassium chloride, sugars, amino acids, and noncontaminated chromatography resins or gels can usually be disposed of in the regular trash. Broken glass, needles, and sharp objects should be disposed of in special containers for the protection of custodial personnel, whose welfare must always be considered when dealing with laboratory waste. Spills present a wide range of hazards, depending on the nature and volume of the material, and should normally be dealt with by an institution's environmental health and safety office.

Multihazardous Waste

Multihazardous waste is a by-product of various kinds of critically important work in, for example, clinical and environmental laboratories. With the help of several experts as part of a special subcommittee, the committee studied the disposal of various combinations of chemical, radioactive, and biological waste. Few disposal facilities exist for multihazardous waste, and some waste materials are so unique and occur in such small quantities that there is no commercial incentive for developing special legal means for handling them.

Although there are no general federal regulations covering disposal of biohazardous or infectious waste, OSHA regulates the handling of some kinds of laboratory waste containing human body fluids, and local ordinances may apply to other types. Generally, biological waste may be disinfected, autoclaved, incinerated, or sent to the sanitary sewer.

For disposal of a multihazardous waste, the goal may be reduction to a waste that presents a single hazard, which can then be managed as a chemical, biological, or radioactive waste. Each option should be ranked, ordered, and prioritized according to the degree of risk posed. Any combination of methods that poses unacceptable risk to waste handlers should be rejected.

Chemical-Radioactive Waste

The management of mixed waste (chemical-radioactive) is often complicated by regulations whose application to a particular case is inconsistent with the relative risk posed by each component hazard. For example, chemical waste containing short-half-life radionuclides is managed best by being held for a period long enough to allow safe decay (e.g., 10 half-lives, but not to exceed 2 years). However, EPA regulations and state laws may limit storage of hazardous chemical waste to 90 days. Chemical-radioactive (mixed) waste is difficult to deal with, primarily because of EPA regulations that prevent on-site storage until *de minimis* levels of radioactivity can be reached and stringent U.S. NRC regulations for the management of low-activity radioactive waste that poses no significant risk to the public or the environment. Used flammable liquid scintillation cocktails, phenol/chloroform nucleic acid extractants from radioactive cells, neutralized radioactive trichloroacetic acid solutions, and some gel electrophoresis waste are examples of chemical-radioactive waste. Techniques for minimizing these types of waste include the use of nonhazardous chemical substitutes so that the waste can be handled simply as radioactive and treated by U.S. NRC "decay-in-storage" regulations. In some cases, EPA-approved chemical hazard reduction methods may be applied and the waste treated as radioactive.

Chemical-Biological Waste

Most, but not all, chemical-biological waste is best dealt with as chemical waste after due consideration is taken of special restrictions that may apply to the biological component if it is putrescible, infectious, or biohazardous. Incineration as a hazardous chemical is usually preferable because animal and medical waste incinerators are not licensed to burn regulated chemical waste. Many types of biological fluid waste containing chemical components can be disposed of in the sanitary sewer, but local approval may be required. Autoclaving can sterilize infectious waste, which then can be treated as chemical waste. However, autoclaving may volatilize chemicals, which could then pose hazards to personnel or could damage the autoclave. Waste and "sharps" of all kinds from laboratories working with hepatitis B or human immunodeficiency

viruses must be handled with special care under the OSHA Bloodborne Pathogen Standard.

Radioactive-Biological Waste

If short-half-life radionuclides are present, decay-in-storage until U.S. NRC regulations allow disposal as biological waste is the appropriate strategy for radioactive-biological waste. Preliminary disinfection or freezing should be used to protect personnel who handle putrescible waste during radionuclide decay. Appropriate options for ultimate disposal are incineration after the waste has reached U.S. NRC-approved levels of radioactivity or alkaline digestion and submission to the sanitary sewer in accordance with local regulations. Particular attention must be given to the handling or cleaning of radioactive laboratory ware, and to the proper disposal of needles, broken glassware, or sharps from biological or medical laboratories.

Chemical-Radioactive-Biological Waste

As indicated above, a combination of waste types may be very difficult to deal with and should always be considered case by case. Decay-in-storage to acceptable levels of radioactivity can reduce the problem to that of handling chemical-biological waste. Autoclaving or use of a disinfectant may be needed to reduce the hazard of biological waste during storage. Unlike the radioactive and biological component of a multihazardous waste, the chemical content does not usually vary significantly with time, although the possibility of treatment to convert hazardous chemical content to nonhazardous should be considered as part of the overall approach to waste management. Before initiating any experiment that might lead to chemical-biological-radioactive waste, researchers are advised to consult with their environmental health and safety office and/or waste removal contractor to avoid an intractable disposal situation. Growing recognition by regulatory agencies of the special problems of multihazardous waste management offers hope that disposal will become increasingly manageable.

LABORATORY FACILITIES (CHAPTER 8)

Chemical laboratories are the most common type of workplace where a wide variety of chemicals are handled on a routine basis. They have evolved into unique facilities designed to deal with many of the hazards described in this book. Chapter 8 discusses the modern laboratory environment as an essential component of the culture of safety and outlines the important role of safety inspection programs. Laboratory ventilation systems are described in considerable detail.

Although many other types of buildings in a large institution can be converted fairly readily from one use to another, the special demands of chemical laboratories require that they be dedicated to their unique purpose. Because of the great expense for a laboratory's construction and operation, the intricacy of its facilities, and its important role in protecting the public and the workers who use it, laboratory personnel should have a thorough understanding of specialized facilities designed for their safety. For the full value of a modern laboratory to be realized, it is important to maintain good working relationships between laboratory personnel and the facility engineering and maintenance staff as well as with environmental health and safety workers.

Perhaps the single most important safety system in chemical laboratories, and certainly the one most responsible for their unique design and cost, is the ventilation system, especially the fume hoods that remove toxic vapors from the workplace. During the past 15 years, virtually every aspect of air handling in the laboratory has been refined, and the issue of acceptable emissions to the outside environment has come under increasing scrutiny. As demands for "zero emissions" become more frequent, the hood is viewed increasingly as a necessary safety device that should be used for removing toxic vapors in case of an accident. Although this viewpoint may be unrealistic if carried to the extreme, it requires special consideration in the planning and execution of laboratory chemistry.

Many of the regulations that now impinge on the handling and disposal of hazardous chemicals in laboratories attempt to formalize their safe operation and to penalize noncompliance. Laboratory inspections are an important means for ensuring not only that laboratories are maintaining a safe operating standard, but also that they are being operated in an efficient manner that justifies their expense. Suitable laboratory inspections cover a wide range of formality and detail. At one end is the informal peer review by fellow workers or the laboratory supervisor giving collegial advice on safe procedures or identifying deteriorating equipment that needs maintenance or lapses in good housekeeping. At the other end are formal inspections by regulatory officers looking for noncompliance with local, state, or federal laws. Chapter 8 provides a variety of options and advice on conducting inspections, along with a basic checklist of common hazards that should be considered.

GOVERNMENTAL REGULATION OF LABORATORIES (CHAPTER 9)

In recognition of the enormous impact of federal, state, and local regulations on the planning and perfor-

mance of every stage of the handling of chemicals as they move to, in, and from laboratories, Chapter 9 outlines the vast and intricate legal framework by which organizations, laboratory workers, and supervisors are held accountable for compliance. Most institutions that deal with chemicals in laboratories, and dispose of them, have organized professional infrastructures to give advice and help implement the laws. The most essential regulations are promulgated under the OSHA Laboratory Standard and the Resource Conservation and Recovery Act, which were conceived to protect the public, the environment, and the individual laboratory worker. Noncompliance may expose workers to unnecessary risks, undermine the public's confidence in its institutions, and lead to fines of up to $25,000 per day of violation and severe criminal penalties. Prudent practice in the laboratory is mandated by law and enforceable through citations.

Although some regulations have not recognized the laboratory as a special environment for using chemicals, the OSHA Laboratory Standard specifies that each institution accountable for handling and disposal of chemicals must develop its own Chemical Hygiene Plan for implementing the requirements of the Laboratory Standard. Each employer is required to "furnish to each of his employees . . . a place of employment . . . free from recognized hazards that are likely to cause death or serious physical harm . . . ." The individual employee is required to "comply with occupational safety and health standards and all rules . . . which are applicable to his own actions and conduct." Although the position of students, in contrast to that of employees, is not covered explicitly, custodial and maintenance personnel who work in laboratories or handle chemicals or chemical waste are clearly protected by requirements for training and other safeguards. Provision is made for the development and enforcement of state OSHA regulations.

In addition to the Laboratory Standard, the Hazard Communication Standard (29 CFR 1910.1200) applies to all nonlaboratory businesses or operations "where chemicals are either used, distributed, or produced" and is more stringent than the Laboratory Standard in some respects. Other OSHA standards concerning level of exposure apply to hundreds of chemicals and are included in the LCSSs prepared for this report and in many MSDSs.

The Resource Conservation and Recovery Act (RCRA; 42 USC 6901 et seq.) applies to waste reduction and disposal of laboratory chemicals from "cradle to grave." Laboratory workers should be aware of RCRA definitions of "generators" of different amounts and types of hazardous waste as described in Chapter 7, and of the legal limitations on moving and disposing of hazardous chemical waste as defined by RCRA.

Although the sewer was the traditional dumping ground for many liquid laboratory wastes, such disposal is now out of the question for most waste chemicals and carries heavy penalties. Nevertheless, not all chemical waste is hazardous, and some may be disposed of in the sanitary sewer under carefully defined conditions. RCRA also controls and defines the in-laboratory treatment of hazardous wastes. Some methods for treatment, most of which may require a permit, are discussed in Chapter 7.

Several sets of wide-reaching regulations also directly affect laboratory operations. The Clean Air Act (42 USC 7401 et seq.) regulates emissions into the air and sets specific limits on the disposal of volatiles through the fume hood system. To protect both the community and the emergency response personnel that may be put at risk by a laboratory accident, the Superfund Amendments and Reauthorization Act (SARA; 42 USC 9601 et seq., 11000 et seq.) requires that inventories of hazardous chemicals be maintained and made available to the public. The Toxic Substances Control Act (TSCA; 15 USC 4601 et seq.) is concerned with the manufacture, distribution, and processing of new chemicals that are unusually dangerous to health and/or the environment.

Of necessity, the legislation and the activities of regulatory agencies referred to in this book represent only the most important and relevant of those concerned with chemicals in laboratories. It is essential that good communication be maintained among all laboratory workers and their colleagues in the institutional offices that are responsible for interfacing with regulatory agencies and keeping abreast of new laws.

RECOMMENDATIONS

In its entirety, this book constitutes a strong recommendation to workers in laboratories to exercise prudence in designing and carrying out their studies so as to maintain a safe workplace and safe operational procedures. In addition, the committee has identified a number of specific areas that need improvement, not only by laboratory workers themselves, but also by regulators at all levels and by chemical suppliers, in order to enhance the climate for laboratory safety. Summarized below are the committee's findings and specific recommendations for action.

Recommendations to the Environmental Protection Agency and Other Regulatory Agencies

In contrast to manufacturing facilities, laboratories generally use small amounts of a large variety of chemicals, often in frequently changing procedures; in addition, they are staffed by highly trained technical workers. As a consequence, overly prescriptive regulations for laboratories may not efficiently protect personnel and the environment. The OSHA Laboratory Standard, promulgated in 1990, formally recognized several unique aspects of laboratories and laboratory operations and established a performance-based system for regulating them. Such a performance-based system is often more effective, both for the laboratories being regulated and for those regulatory agencies concerned with health, safety, and the environment.

• The committee recommends that regulations directed to laboratories be *performance based* and be structured to take into account the unique aspects and professional expertise within the laboratory.

• The committee recommends that each federal regulatory agency establish a formal channel (e.g., through its science advisor or through a formal advisory committee) to ensure incorporation of research laboratory scientists' viewpoints in the preparation and implementation of regulations affecting research laboratories.

Pollution prevention and waste minimization have become major goals of U.S. industry and regulatory bodies. The technical expertise of well-trained laboratory workers offers opportunities for waste minimization through recycling, treatment of waste, and collection of laboratory quantities of waste. Such treatment of laboratory waste offers the possibility of substantially reducing the risks associated with certain hazardous waste before sending it off-site.

• The committee recommends that the Environmental Protection Agency extend its permit-by-rule provisions to allow scientifically sound treatment of small quantities of waste generated in laboratories.

• The committee recommends that the Environmental Protection Agency allow storage of small quantities of waste in laboratory facilities for periods longer than the current time limitation on storage of hazardous waste.

The laboratory community is trained to adhere to standard methods and protocols. But as students and career laboratory personnel advance from one stage to another, or change professional laboratory positions, relocation among geographical regions is frequent. Often, such job relocations require costly and time-consuming relearning of previously acquired safety and waste management habits and retraining in alternative methods because of the absence of a uniform regulatory environment. Worker safety and waste disposal procedures are seriously impeded when regulatory requirements vary among regions.

• The committee recommends that federal, state, and local lawmakers and regulators strive for conformity and consistency in the regulations that affect laboratories.

If disposal of multihazardous waste is to be accomplished safely and cost effectively, regulations affecting its individual components should not conflict, and the regulatory framework should be based on risk priority. For example, health and safety considerations may sometimes demand that greater emphasis be given to one component of the total chemical, biological, and radiological toxicity, while existing regulations would require equal emphasis for all components. This situation arises, for example, with mixtures containing trace amounts of radioactive material in which the major health and safety risk is associated with a chemical or biological component.

• The committee recommends that the U.S. Nuclear Regulatory Commission and the Environmental Protection Agency establish *de minimis* levels for radionuclides, below which laboratory waste can be disposed of without regard to radioactivity.
• The committee recommends that the Environmental Protection Agency encourage safe disposal of chemical-radioactive (mixed) waste materials with short half-lives by excluding the decay-in-storage period from the current 90-day limitation on storage of hazardous waste.

Requirements for multiple EPA identification numbers for a single campus create an unnecessary administrative burden.

• The committee recommends that the Environmental Protection Agency allow the use of one EPA identification number for all chemical waste generated on a single campus of an educational institution.

Complete safety procedures for an institution must incorporate emergency planning and must recognize the role of external agencies. It is essential that emergency procedures be developed that will minimize risk to personnel and allow emergency response workers to function effectively. Clear lines of communication must be maintained. In addition, local regulations affecting laboratories should be oriented toward reduction of risk. Federal law, for example, specifically exempts laboratories from detailed reporting of every chemical stored in a facility, enabling a reporting system focused on those chemicals that would pose the greatest risks in an emergency.

• The committee recommends that laboratory personnel, in cooperation with the institutional health and safety structure, establish ongoing relationships and clear lines of communication with emergency response teams.
• The committee recommends that emergency response regulations require inventory information only on those containers with chemicals in quantities large enough to pose a significant risk to personnel or the environment in the case of an emergency release or fire.

Small colleges and high schools often do not have an environmental health and safety office or the resources to manage laboratory waste. Teachers are thus left to shoulder the burden of waste disposal, and their attention can be diverted from core science teaching as a result.

• In order to support the teaching of laboratory courses in small colleges and high schools, the committee recommends a careful review of current record-keeping requirements to avoid excessive burdens on teachers at small institutions.

Recommendations to the Industrial Sector

Prudent management of chemicals is important for a variety of environmental, social, and economic reasons. Manufacturers and suppliers of chemicals play a central role in these efforts because they manage the commercial flow of chemicals and have responsibility for the procedures by which chemicals are packaged and shipped. Uniform identification of chemicals by manufacturers and suppliers could help to reduce risks in the storage, use, and disposal of chemicals, improve the management of chemicals in the laboratory, and enhance emergency response preparedness.

• The committee recommends that chemical suppliers adopt a uniform bar code identification system that would facilitate establishment and maintenance of laboratory chemical inventory and tracking systems.
• The committee recommends that all laboratory chemicals be labeled with the date of manufacture.
• The committee recommends that chemical suppliers adopt uniform color coding and bar coding for compressed gas cylinders.

The policies and practices of commercial manufacturers and suppliers of laboratory chemicals directly affect the management of chemicals in the laboratory, especially the ability to practice effective pollution-prevention techniques such as source reduction and recycling-reuse-recovery. The costs and risks associated with disposal of waste can be reduced greatly if materials never enter the waste stream.

- The committee recommends that chemical suppliers provide and promote to their customers the option of purchasing small quantities of chemicals.
- The committee recommends that chemical suppliers develop a mechanism whereby laboratories can return unopened containers of chemicals.

Recommendations to Chemical Laboratories

Academic departments and industrial laboratories that use chemicals are advised to develop internal safety groups that are complementary to their institution's environmental health and safety organization and include representation from all segments of the laboratory work force. Such internal safety groups should meet regularly to discuss departmental safety and waste disposal issues and to analyze any incidents that may have occurred. To achieve maximum effectiveness in working with an institution's environmental health and safety organization to minimize risks, a departmental safety group must have the full support of the institution's administration.

- The committee recommends that chemical laboratories establish their own safety groups or committees at the department level, composed of a cross section of laboratory workers, including students and support staff as well as faculty in academic laboratories.

The ability of any laboratory to operate in a manner that minimizes risks to personnel and the environment is dependent on laboratory workers who understand and carry out prudent practices for handling, storage, and disposal of chemicals. Training of laboratory personnel in safety and waste management is essential and must be followed up with an appropriate inspection system to ensure that safe practices are followed. Safety training must include discussion of chemical hazards, equipment hazards, laboratory safety and environmental systems, and the potential impact of laboratory work on these systems.

- The committee recommends that any laboratory using hazardous chemicals should provide appropriate training in safety and waste management for all laboratory workers, including students in laboratory classes.
- The committee recommends that laboratories using hazardous chemicals should incorporate institutionally supported laboratory and equipment inspection programs into their overall health and safety programs.

The Culture of Laboratory Safety

1.A INTRODUCTION

Over the past century, chemistry has made great contributions toward our understanding of and our ability to manipulate the physical and biological world. Most of the items we take for granted in our day-to-day life involve synthetic or natural chemical processing. Indeed, even our own bodies may be viewed as chemical machines, now that molecular biology has removed the traditional boundary between chemistry and biology. The chemical laboratory has become the center for acquiring knowledge and developing new materials for future use, as well as for monitoring and controlling those chemicals currently used routinely in thousands of commercial processes. Many of these chemicals are beneficial, but others have the potential to cause damage to human health and the environment, and therefore also to the public attitude toward the chemical enterprise on which we all so heavily depend.

Since the age of alchemy, some chemicals have demonstrated dramatic and dangerous properties, which have required the development of special techniques for handling them safely. We also know now that many more are insidious poisons. Until recently, the chemical hazards in many laboratories were not accepted and taken into account by those working in them, and, accordingly, the necessity of putting "safety first" was not fully appreciated. During the "heroic age" of chemistry the notion of martyrdom for the sake of science was actually accepted widely, according to an 1890 address by the great chemist August Kekulé: "If you want to become a chemist, so Liebig told me, when I worked in his laboratory, you have to ruin your health. Who does not ruin his health by his studies, nowadays will not get anywhere in Chemistry" (as quoted in Purchase, 1994). In sharp contrast, a growing recognition of moral responsibility and mounting public pressure have made institutions housing chemical laboratories accountable for providing safe working environments for those employed in them and complying with extensive regulation of the transport of chemicals to the laboratories and removal of waste from them. The "old days" of easygoing attitudes toward laboratory safety and down-the-sink disposal are over! Laboratories have become safe places to work.

1.B THE NEW CULTURE OF LABORATORY SAFETY

A new culture of safety consciousness, accountability, organization, and education has developed in the laboratories of the chemical industry, government, and academe. To a degree that could scarcely have been foreseen 25 years ago, programs have been implemented to train[1] laboratory personnel and to monitor the handling of chemicals from the moment they are ordered until their departure for ultimate treatment or disposal.

Workers in many hazardous fields[2] (e.g., seamen and construction workers) have developed traditions of working together for mutual protection and the maintenance of correct professional standards. In the same way, laboratory workers have come to realize that the welfare and safety of each individual depends on clearly defined attitudes of teamwork and personal responsibility. Learning to participate in this culture of habitual risk assessment, experiment planning, and consideration of worst-case possibilities for oneself and one's fellow workers is as much a part of a scientific education as learning the theoretical background of experiments or the step-by-step protocols for doing them in a professional and craftsmanlike manner.

Accordingly, a crucial component of chemical education at every level is to nurture basic attitudes and habits of prudent behavior in the laboratory so that safety is a valued and inseparable part of all laboratory activity. In this way, "safety first" becomes an internalized attitude, not just an external expectation driven by institutional rules. This process must be part and parcel of each person's chemical education throughout his or her scientific career. One aim of the present volume is to encourage academic institutions to address this responsibility effectively and cultivate their students' participation in the culture of laboratory safety as a solid basis for their careers as professional chemists.

1.C RESPONSIBILITY AND ACCOUNTABILITY FOR LABORATORY SAFETY

The culture of laboratory safety depends ultimately on the working habits of individual chemists and their

---

[1] Throughout this book, the committee uses the word *training* in its usual sense of "making proficient through specialized instruction" with no direct reference to regulatory language.

[2] With regard to safe use of chemicals, the committee distinguishes between hazard, which is an inherent danger in a material or system, and the risk that is assumed by using it in various ways. *Hazards* are dangers intrinsic to a substance or operation; *risk* refers to the probability of injury associated with working with a substance or carrying out a particular laboratory operation. For a given chemical, risk can be reduced; hazard cannot.

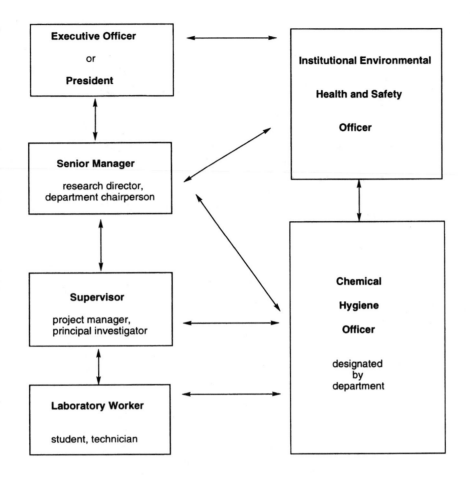

FIGURE 1.1 Pattern of interactions through which laboratory safety can be arranged within an institution.

sense of teamwork for protection of themselves, their neighbors, and the wider community and environment. However, safety in the laboratory also depends on well-developed administrative structures and supports that extend beyond the laboratory's walls within the institution. There are many ways that the detailed organization for laboratory safety can be arranged. Within a business, academic, or governmental institution, organizations often involve interactions such as those shown in Figure 1.1.

The protection of health and maintenance of safety constitute a moral obligation shared by everyone. Federal, state, and local laws and regulations make safety a legal requirement and an economic necessity as well. Laboratory safety, therefore, is not a purely voluntary function; it requires mandatory safety rules and programs and a commitment to them. A sound safety organization that is respected by all requires the wholehearted participation of laboratory administrators, employees, and students.

The ultimate responsibility for safety within any institution lies with its chief executive officer or president. That individual must provide the leadership to ensure that an effective safety program is in place so that all institutional officials will demonstrate a sincere and continuing interest in the program. Even a well-conceived safety program is apt to be treated casually by the workers if it is neglected by top management. Initiative and support for good safety programs, like most other institutional plans, usually come from the top down.

Although the responsibility for safety in a department or other administrative unit lies with its director or chairperson, the responsibility for the delineation of the appropriate safety procedures and the instruction of those who will carry out the operation lie with the project manager or principal investigator. The responsibility for safety during the execution of an operation lies with those technicians, students, and other workers who actually perform that operation. Nevertheless, the primary responsibility for maintaining safe behavior in a safe laboratory environment remains with the project manager or principal investigator. Each institution should develop policies that help to determine who has accountability for accidents or safety violations.

While the school principal or college president is ultimately responsible for the safety of students in courses that involve laboratory activity, the laboratory instructor carries direct responsibility for what actually takes place under his or her direction. The instructor is responsible for developing the positive attitudes and habits of the culture of laboratory safety as well as the necessary skills for handling chemicals safely.

The expanding system of federal, state, and local regulations for the handling and disposal of chemicals has resulted in institutional infrastructures to oversee compliance with safety laws. Most industrial, governmental, and academic institutions that maintain laboratory operations have an environmental health and safety office made up of individuals with appropriate professional credentials. These individuals may have expertise in chemical safety, industrial hygiene, engineering, biological safety, environmental health, occupational medicine, health physics, fire safety, or toxicology. Functions of environmental health and safety offices generally include technical consultation, hazardous waste management, accident reviews, inspections and audits, compliance monitoring, training, recordkeeping, and emergency response. These offices assist laboratory management in establishing safety policies and promoting high standards of laboratory safety. To be most effective, they should share in a genuine partnership with all department chairpersons or directors, principal investigators or managers, and laboratory workers in helping to design safety programs that provide technical guidance and training support that are relevant to the operations of the laboratory, are practical to carry out, and comply with the law. They should help technical and professional personnel to be aware of their legal responsibilities without being overwhelmed by a sense of unlimited liability from a mass of regulations.

In view of the importance of the environmental health and safety office to the whole safety enterprise, it should be directed by people who are truly knowledgeable about the operations. Safety directors should be given a high level of authority and responsibility for the development of a unified safety program. The safety director should also have direct access, when necessary, to people at the highest level in the institution who carry its ultimate accountability to the public through the media and the law. Department chairpersons need to deal directly with the safety officers, who are not only knowledgeable about safety regulations and consistent in enforcing them but also appreciate the unique problems of progressive training and pru-

dent operations in academic teaching and research institutions.

1.D SPECIAL SAFETY CONSIDERATIONS IN ACADEMIC LABORATORIES

Academic laboratories, like industrial and government ones, are concerned with meeting the fundamental safety goals of minimizing accidents and injuries, but there are differences that should be recognized when developing prudent and realistic safety programs for teaching institutions. Forming the foundation for a lifelong attitude of safety consciousness, risk assessment, and prudent laboratory practice should be an integral part of every stage of scientific education—in the classroom, in textbooks, and in the laboratory from the earliest exposures in primary or secondary school through graduate and postdoctoral training. Teaching and academic institutions have this essential and unique responsibility. They are also faced with the special problems that go with a rapid turnover of young people. The manifold requirements for recordkeeping and waste handling can be especially burdensome for overworked teachers in high school or college laboratories.

In addition to providing well-trained students, institutions with graduate programs also have the responsibility of discovering new knowledge through research programs, and these often involve unpredictable hazards. The safety goals and the allocation of resources to achieve them are sufficiently different for high school, undergraduate, and graduate teaching laboratories that they are discussed separately here. In research universities the goals for safety in teaching laboratories and in research laboratories usually overlap but may also compete for attention and funds.

1.D.1 High School Teaching Laboratories

Recognizing and evaluating hazards, assessing risks, selecting appropriate practices, and performing them proficiently are essential elements of laboratory safety. The training to lay the foundation for acquiring these skills begins with the student's first experience in the laboratory. Even the earliest chemical experiments should cover the proper approach to dealing with the principal hazardous properties of chemicals (e.g., flammability, reactivity, corrosiveness, and toxicity) as an introduction to laboratory safety, and should also begin to instill responsibility for sound environmental practice when managing chemical waste. Advanced high school chemistry courses should assume the same responsibilities for developing professional attitudes toward safety and pollution control as are expected of college and university courses.

1.D.2 Undergraduate Teaching Laboratories

Undergraduate chemistry courses are faced with the problem of introducing inexperienced people (frequently in enormous numbers) to the laboratory culture, including the handling of hazardous materials. Although many students come to their first undergraduate course with good preparation from their high school science courses, others may be "chemophobic," having a prejudice against chemicals of all kinds. They must learn to evaluate intelligently the wide range of hazards in laboratories and learn the techniques by which potential dangers can be controlled routinely with negligible risk.

In research universities the primary responsibility for undergraduate laboratory teaching is often assigned to teaching assistants, who may have widely different backgrounds and communication skills. At the same time that they are adapting to their first teaching experience, they are also trying to handle their first graduate courses. Some who are unfamiliar with safe laboratory practice and the proper disposal of chemical waste may have to learn new attitudes and habits at the same time that they are teaching them to the undergraduates. The supervision of teaching assistants and monitoring of their performance must be taken as a special departmental responsibility because the safe, meaningful operation of undergraduate laboratories depends so heavily on them.

In research universities the inherent problems of transmitting good laboratory training to undergraduates can be compounded by the conflicting demands for resources and attention between research and undergraduate teaching. Commitment of the entire faculty to laboratory safety and the responsible disposal of chemicals is a crucial factor in the initiation of all students into the laboratory culture at every level.

1.D.3 Academic Research Laboratories

Advanced training in safety is an important component of education through research. Unlike laboratory course work, where training comes primarily from repeating well-established procedures perfected through many years of experience, research often includes the production of new materials by unprecedented methods, which may involve unknown hazards. As a result, academic research laboratories may place an enormous range of processes and products in the hands of young investigators of widely varied scientific experience. Often the transition from undergraduate laboratory course work to the first experience of independent research is too abrupt. If high school

and college laboratory work has placed all of the responsibility for safety planning on the teachers and graduate research leaves it all to the student, neophyte researchers will be ill-prepared to face the many real hazards of the research laboratory. Given these circumstances, heavy responsibility rests with the faculty to provide the safest possible environment for research by careful oversight and example. Faculty should pay particular attention to the introduction of first-year graduate students to the research laboratory. To further enhance a good safety environment, many chemistry departments now present regular courses on laboratory safety for incoming graduate students and require that postdoctoral associates show proficiency on a regular safety test before starting laboratory work. Such preparations contribute greatly to the complete professional training required by most chemical companies.

Safety training must be a continuing process; it should become an integral part of the daily activities of laboratory workers and those who are accountable for them. It need not be an arduous task. As a student or laboratory worker learns a new protocol, safe practices relevant to it should also be emphasized in the normal setting of the laboratory, with the careful guidance of a mentor and the shared responsibility of colleagues. Opportunities to encourage and enhance informal safety training through collegial interactions should be pursued vigorously as a valuable way to exchange safety information, convey meaningful guidance, and sustain an atmosphere in which colleagues reinforce each others' good work habits.

Formal safety education for advanced students and laboratory workers should be made as relevant to their work activities as possible. Training that is conducted simply to satisfy regulatory requirements tends to subordinate the relevant safety issues to details associated with compliance. Such bureaucratic safety management has actually worked against fostering positive safety attitudes in many well-experienced laboratory workers and has undermined the credibility of warnings about bona fide hazards by emphasizing pro forma violation of rules.

Although principal investigators and project managers are legally accountable for the maintenance of safety in laboratories under their direction, this activity, like much of the research effort, is distributable. Well-organized academic research groups develop hierarchical structures of experienced postdoctoral associates, graduate students at different levels, undergraduates, and technicians, which can be highly effective in transmitting the importance of safe, prudent laboratory operation. When the principal investigator offers leadership that demonstrates a deep concern for safety, the university safety program thrives. However, if the principal investigator's attitude is laissez-faire or hostile to the university safety program, careless attitudes can take hold of the whole group and set the stage for accidents, costly litigation, and expensive reeducation for those who move on to a more responsible institution.

1.E THE SAFETY CULTURE IN INDUSTRY

Industrial laboratories that use chemicals engage in extremely varied activities. Some are at the heart of the chemical industry, with the complete complement of research, analytical, pilot plant, and production facilities engaged in making chemicals. In others the use of chemicals is more incidental to the production of special products.

Not surprisingly, the degree of commitment to environmental health and safety programs varies widely as well. Many chemical companies have recognized both their moral responsibility and their self-interest in developing the best possible safety programs. Others have done little more than is absolutely required by law and regulations, if that. Unfortunately, the bad publicity from a serious accident or violation in one carelessly operated laboratory or plant tarnishes the credibility of all those whose operations are above reproach. The public perception that chemical companies have "deep pockets" can place a high price on chemical accidents.

The industrial or government laboratory environment can provide strong corporate structure and discipline for maintaining a well-organized safety program where the safety culture is thoroughly understood, respected, and enforced from the highest level of management down. New employees coming from the more casual atmosphere of some academic research laboratories are often surprised to discover "a new world" of attention to the detailed planning and extensive checking that are required in preparation for running experiments. In return for their efforts, they enjoy the sense of security that goes with high professional standards.

Industrial safety programs can face several obstacles. Financial limitations to safety programs can prevent optimal development in business as well as academe. However, the short-sighted bottom-line thinking that can surface when management is separated from the laboratory by geography or by a lack of commitment to safety is a more common problem for industrial laboratories. Poor communication between laboratory workers and environmental health and safety officers can lead to adversarial relations when a perception develops that a bureaucracy is generating rules for

the sake of rules. Compliance with institutional safety regulations does not guarantee a real acceptance of the culture of safety.

1.F FACTORS THAT ARE CHANGING THE CULTURE OF SAFETY

Over the past 20 years, several trends have emerged that are changing the shape of the safety enterprise in the chemical laboratories of industry, government, and academe. These factors include advances in technology and changes in cultural values and in the legal and regulatory climate.

1.F.1 Advances in Technology

Several recent advances in technology have begun to change the safety requirements in chemical laboratories. For example, in response to the increasingly high cost of handling chemicals through the whole cycle from purchase to waste disposal, there has been a steady movement toward miniaturizing chemical operations in both teaching and research laboratories. This trend not only has had a significant effect on laboratory design but also has reduced costs of acquiring, handling, and disposing of chemicals. Another trend—motivated at least partially by safety concerns—is the simulation of laboratory experiments by computer. Such programs are a valuable conceptual adjunct to laboratory training but are by no means a substitute for it. As mentioned above, only students who have been educated carefully through a well-graded series of hands-on experiments in the laboratory will have the confidence and expertise needed to handle real laboratory procedures in a safe manner as they move on to advanced courses or research work.

1.F.2 The Culture of Pollution Prevention

A recent and widely accepted cultural change that affects laboratory work is the concept of pollution prevention. The idea is simplicity itself: if one makes less waste, there is less waste to dispose of, and therefore less impact on the environment. A frequent, but not universal, corollary is that costs are also reduced.

The terms "waste reduction," "waste minimization," and "source reduction" are often used interchangeably with "pollution prevention." In most cases the distinction is not important. However, the term "source reduction" may be used in a narrower sense than the other terms, and the limited definition has even been suggested as a regulatory approach that mandates pollution prevention. The narrow definition of source reduction includes only procedural and process changes that actually produce less waste. The definition does not include recycling or treatment to reduce the hazard of a waste. For example, changing to microscale techniques is considered source reduction, but recycling a solvent waste is not.

1.F.2.1 Waste Management Hierarchy for Pollution Prevention

The distinctions made above are the result of an approach to waste management that incorporates a hierarchy of pollution prevention techniques. At the top of the hierarchy is *source reduction*, which is always the preferred technique. Source reduction can be achieved by using a smaller quantity of material or a less hazardous material or by making a process more efficient. However, while source reduction is highly desirable, it is not always technically or economically feasible.

The second level of the hierarchy is *recycling/reuse/ recovery*. The distinction here is that the waste requires some input of energy (e.g., distillation) before it can be reused. Because of this additional waste-handling step, there is also an increased potential for spillage and other fugitive losses of material that would not occur had the waste not been generated in the first place. However, when source reduction techniques are not available or practical, recycling, reuse, or recovery can be important alternatives to disposal.

The third level is *treatment*. Generally, treatment of a waste renders it less hazardous or nonhazardous but does not allow reuse of the material. For materials that have no potential for recovery and are not amenable to source reduction, treatment—such as neutralization of acids, incineration of organic sludges, and oxidation of cyanides—becomes an important part of the waste management system.

The last level (and least desirable alternative) is *land disposal*. Certain hazardous wastes, particularly the heavy metals, cannot be rendered completely nonhazardous and cannot realistically be recovered. They can, however, be stabilized to reduce the likelihood of movement in the environment, and regulations require this procedure before land disposal is permitted. Whereas land disposal of laboratory waste (in "Lab Packs") was once the most common form of waste management, it is now rarely used, and then only in very specialized situations and locations.

1.F.2.2 Making Pollution Prevention Work

Many advantages can be gained by taking an active pollution prevention approach to laboratory work, and these are well documented throughout this book. Some

potential drawbacks do exist, and these are discussed as well and should be kept in mind when planning activities. For example, dramatically reducing the quantity of chemicals used in teaching laboratories may leave the student with an unrealistic appreciation of their behavior when used on a larger scale. Also, certain types of pollution prevention activities, such as solvent recycling, may cost far more in dollars and time than the potential value of recovered solvent. Certain waste treatment procedures may even have regulatory strictures placed against them.

Perhaps the most significant impediment to comprehensive waste reduction in laboratories is the element of scale. Techniques that are practical and cost-effective on a 55-gallon or tank car quantity of material may be highly unrealistic when applied to a 50-gram (or milligram) quantity. Evaluating the costs of both equipment and time becomes especially important when dealing with very small quantities.

1.F.3 Changes in the Legal and Regulatory Climate

Many important changes in the legal and regulatory climate over the last 20 years have added to the changing culture of safety. Because of increased regulation, the collection and disposal of laboratory waste now constitute a major budget item in the operation of every chemical laboratory. Also, it is now widely recognized that protection of students and research personnel from toxic materials is not only a moral obligation but also an economic necessity—the price of accidents in terms of time and money spent on fines for regulatory violations and on litigation can be very high.

In response to the heightened concern for safety in the workplace, the OSHA Laboratory Standard (29 CFR 1910.1450) requires every institution that handles chemicals to develop a Chemical Hygiene Plan. This requirement has generated a greater awareness of safety issues at all educational science and technology departments and research institutions. Although the priority assigned to safety varies widely among personnel within chemistry departments and divisions, increasing pressure is coming from several other directions in addition to the regulatory agencies and accident litigation. In some cases, significant fines to principal investigators who have received citations for safety violations have increased the faculty's concern for laboratory safety. Boards of trustees or regents of educational institutions often include prominent industrial leaders who are highly aware of the increasing national concern with safety and environmental issues and are particularly sensitive to the possibility of institutional liability as a result of laboratory accidents. Academic and government labo-

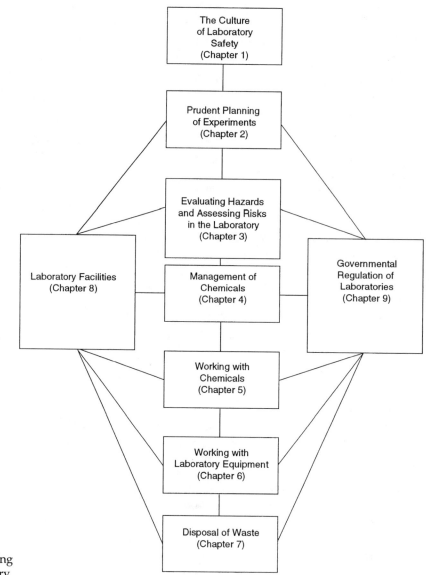

FIGURE 1.2 Protocol for planning and carrying out an experiment with chemicals in a laboratory.

ratories, like industrial ones, can be the targets of expensive lawsuits. The trustees can assist academic officers both by helping to develop an appropriate institutional safety system with an effective environmental health and safety office and by supporting departmental requests for modifications of facilities that are necessary for compliance with safety regulations.

Increasing concern for laboratory safety is also being engendered by federal granting agencies. The fact that negligent or cavalier treatment of laboratory safety regulations may jeopardize not only an individual investigator's but also a department's or an institution's ability to obtain funding may become a powerful incentive for improvement in this area.

1.G ORGANIZATION OF THIS BOOK

This book is organized around the protocol for planning and executing an experiment with chemicals in a laboratory. Figure 1.2 outlines the work flow. The chapters that follow represent the likely steps in this process, and guidelines on how to infuse the new culture of laboratory safety into each step are presented throughout the book.

# Prudent Planning of Experiments

2.A INTRODUCTION

The cornerstone of a sound program for prudent laboratory practices is a process designed to comprehensively review the operations and potential hazards associated with each experiment[1] over its life cycle. This review should take place before any work is conducted. The diverse nature of research and development activities makes it advisable to have such a process in place as part of the scientific method of experimentation. In laboratories where this preliminary survey is routinely practiced, it has proved to be useful in both the maintenance of safe laboratory operations and the minimization of chemical exposure and waste generation. Because of the diversity of types of researchers and laboratory work, such processes—both formal and informal—can help individuals associated with new, modified, or unfamiliar experiments or procedures to plan and work safely, responsibly, and productively. By first evaluating the work area, materials, equipment, and procedures in depth, hidden hazards may be identified and addressed. The pre-experiment review process can also help to ensure that every experiment and laboratory operation complies with all applicable laws, regulations, and other policies. Moreover, by addressing all relevant health, safety, and environmental issues when an experiment is first conceived, further research, scale-up, or development based on it can be made safer and more effective.

In this chapter the concept of experiment planning is addressed as the first step in ensuring that prudence is exercised in conducting laboratory operations. Because of the diversity in experiments, laboratory workers, facilities, and hazards, experiment planning can be a complex process for which it is impractical to structure rules. The committee's approach has been to consider the likely steps involved in conducting any chemical experiment. These range from developing a clear understanding of the goals and objectives of the project to providing for the acquisition and handling of materials and equipment all the way to the storage and ultimate disposal of all chemicals, both desired products and waste.

The steps to be considered in planning an experiment can be described by a flowchart such as the one depicted in Figure 1.2. (section 1.G), which shows the individual steps in a laboratory experiment. These steps, in turn, correspond to the different chapters of this book. For the experienced laboratory worker, these steps are understood intuitively, and the pre-experiment review is primarily a thought process, with perhaps a brief written description of the experiment plan. For others, the review of some or all steps on such a flowchart (and the corresponding chapters in this book), along with a more formal documentation process, is in order.

2.B LEVELS OF FORMALITY IN EXPERIMENT PLANNING

As mentioned above, the types of laboratory experiments are diverse and may be conducted by a wide range of practitioners whose skills and backgrounds may be enormously varied, even within a single discipline or institution. Thus, the degree of formality and documentation necessary for prudent planning is a matter of judgment. Decisions to be made in experiment planning will be affected by the knowledge and skills of the personnel, the scale of the experiments, the particular hazards of the materials or operations being contemplated, the institution's policies for planning and conducting experiments, and the regulatory environment in which the experiments are to be performed.

Because of this diversity among practitioners, what may pose a significant and unfamiliar challenge to one laboratory worker may be second nature to another. For example, the very limited skills of high school or undergraduate chemistry students may demand that extensive written planning of basic laboratory procedures take place before an experiment is attempted. For graduate students or seasoned research chemists, those same operations may be sufficiently familiar that pre-experiment review of an informal "mental checklist" and a single line entry in a laboratory notebook may, in such a circumstance, be entirely prudent in laboratory practice. Similarly, what may be a straightforward distillation of organic solvent to a Ph.D. graduate student in synthetic organic chemistry may be an unfamiliar and, thus, rather more hazardous procedure for even an experienced professor of theoretical chemistry.

To allow for these differences in level of experience, many institutions have established general guidelines for several different levels of formality, which are scaled according to the estimated risks. Depending on the situation, these might include (1) simple mental evaluation of hazards for straightforward experiments

---

[1]Throughout this book the term *experiment* includes the entire range of laboratory operations from science classes to large-scale experimental programs involving many people over a considerable period of time, all of which require some degree of forethought.

by experienced practitioners, (2) more formal discussion of the experiment and options with experienced peers for more hazardous operations, or (3) a formal hazard review process with complete written documentation of the procedures to be employed for new, unfamiliar, or intrinsically hazardous operations. For example, the following special areas of laboratory work will almost always require some approval steps: work with radioactive materials; experiments involving pathogens that cause serious or lethal infection; high- and low-pressure work; research involving especially hazardous materials; and experiments being scaled up.

Diversity in local, state, and federal regulatory issues and institutional policies also enters into the planning of an experiment. Simply "thinking things through" and recording a description of a procedure in a laboratory notebook might be a fully prudent practice for handling one chemical, while the identical procedure for another compound subject to specific regulations or for a large quantity of the same compound might demand a detailed written experiment plan, review by others, authorizing signatures, and accounting of material balances.

It is clear that no single, universally applicable description of "good" experiment planning exists and that the level of formality to be considered prudent in pre-experiment planning is a matter of judgment. In an area where potential hazards exist, more attention to planning is clearly better than less.

2.C INDIVIDUAL RESPONSIBILITIES FOR PLANNING EXPERIMENTS

Implementation of effective pre-experiment review programs must be initiated and backed by the highest level of leadership in an organization. Primary responsibility for day-to-day implementation of such programs should rest with individuals who supervise particular laboratory activities. While the experiments may be prepared and conducted by the laboratory workers, it remains the responsibility of the laboratory supervisor to determine what level of experiment planning is appropriate and to be accountable for necessary training, documentation, and compliance with regulations.

The laboratory workers involved with the experiment or procedure should participate actively and monitor the planning process carefully. When planning for new or unfamiliar procedures or experiments, the workers should review the literature and consult experts to assist with the review. These experts may be outside the regular chain of leadership in the organization or may even be outside the organization alto-

gether. They could include program leaders, co-workers, and safety, health, toxicology, and industrial hygiene personnel who are associated with chemical research. Experimenters should also consult appropriate sections of this book and any other available safety, toxicology, and industrial hygiene reference materials that might aid in planning the experiment. At the completion of the pre-experiment review process, the workers should have complete familiarity with the planned activities, their associated risks, all protective measures needed, and contingency plans to deal with unexpected events or accidents. The protection of the individual worker and the public is paramount. When conducting laboratory activities, workers not only must have the knowledge necessary to ensure their own safety and that of co-workers and society, but also must be willing to accept the responsibility for that safety.

2.D INSTITUTIONAL POLICIES AND EMERGENCY RESPONSE PLANNING

Just as those proposing an experiment have responsibilities for safety in laboratory work, the institution in which the experiment is to be conducted is also responsible for certain aspects of experiment planning. This is the case for academic departments in universities and national laboratories as well as for private corporations. Because of the scope of institutional responsibilities, it is generally less effective for an institutional bureaucracy than for the experienced professionals directly involved in the work to attempt to set guidelines for specific experiments. However, the institution shares the ethical, legal, and financial burden of ensuring that experimental work is carried out safely and responsibly; thus, the institution must establish general guidelines for what constitutes prudence in laboratory work practices. The institution is responsible for setting standards and keeping records of any necessary training of laboratory workers. Moreover, in specific circumstances the institution may spell out guidelines for working with specific hazards, as in the case of an especially toxic compound or a federally regulated drug intermediate.

In addition to setting the general tone for work practices, the institution is responsible for developing and implementing laboratory policies and standards for emergency response procedures and training. This responsibility is best handled at the institutional level by a central environmental health and safety office. Activities of such an office might include developing contingency plans for handling injuries, chemical spills, explosions, fires, natural disasters, the loss of power or water pressure, and other emergencies. Such a central office should also make all laboratory workers familiar with applicable laws as well as the institution's policies and plans so that considering emergency response becomes a natural aspect of experiment planning for all laboratory workers. Many institutions that have worked with their local response agencies (e.g., hospital emergency rooms and fire stations) have found that such planning helps all parties develop a better understanding of the requirements for effective emergency response and a clearer appreciation of the potential magnitude and likelihood (or frequency) of the services that might be needed. These policies should include the broad types of emergency response mentioned above, as well as the maintenance of, and training in the use of, fire extinguishers, first aid kits, spill cleanup kits, eyewashes and safety showers, self-contained breathing apparatus, and so forth.

It is also the responsibility of the institution to determine the level of documentation appropriate for different laboratory operations, including experiment planning, as mentioned above, as well as emergency response planning. In some instances, minimum standards have already been spelled out by regulatory agencies, and the institution must ensure compliance. In other circumstances, such as for emergency response planning, it may simply be a good idea to establish laboratory standards for documenting the location of flammable solvent storage, the best routes for laboratory and building evacuation, the decision trees specifying the contact person, and so forth. Prudence also requires that appropriate levels of documentation be provided to emergency responders. Although some regulatory environments may specify that all facilities that handle chemicals must provide local response agencies with complete lists of chemical storage by location (with accompanying Material Safety Data Sheets (MSDSs)), this type of detail is rarely appropriate for a laboratory where small quantities of thousands of different chemicals (many new and/or not yet fully characterized) are used. Even if a laboratory were able to comply with such a stipulation, the local response agency could easily be inundated with the enormous volume of paperwork and be unable to easily find the needed information in an emergency. A multipage inventory of complex names and structures of research chemicals, many of which never have been published in the literature, would be of little value to a public safety response team facing a laboratory fire. In planning for emergencies, therefore, the actual needs of the responders should be the highest priority, and the appropriate level of information should be provided to ensure effective emergency response.

Finally, some institutions, particularly those in private industry, require extensive documentation of the planning process for all experiments, while others do not. Similarly, undergraduate teaching laboratories often require detailed written experiment plans before work can commence, whereas this level of written experiment planning is much less common in graduate research programs. In the end, the extent of documentation is probably much less important than establishing a culture in which workers think through the potential hazards of experiments they plan to conduct and seek out the resources necessary to ensure that experiments are conducted safely. Documenting the experiment planning process is one tool that may help to build this proactive safety culture, but it will probably not be sufficient to ensure safe work if other drivers for prudent practices are absent.

2.E STEPS FOR PLANNING AN EXPERIMENT

One mechanism to facilitate effective planning is to consider the steps of an experiment in a flowchart. When the fundamental steps in the research process and the flow of work through each step are understood, the critical issues for laboratory work can be addressed in the sequence in which they are likely to be encountered. Once the goals and objectives of the experiment have been clearly formulated, the planning can begin. Consideration must be given, in turn, to risk assessment, acquisition and storage of chemicals, handling of chemicals and equipment, and disposal of waste. Other customized flowcharts, with more or less detail, should be considered by laboratories that employ different procedures or use only a few of these steps.

The *general* steps a laboratory worker must consider in planning an experiment are highlighted in the following paragraphs. The actual execution of each step is discussed in much greater detail in the following chapters.

Just as a clear understanding of research goals and objectives is an essential part of any scientific investigation, so also is a clear understanding of the goal of "safety first," and how it meshes with the research goals and objectives, an essential part of planning. The research goals and objectives should be stated clearly in order to generate unambiguous data and to facilitate consideration of such matters as source reduction and the substitution of benign alternatives to some reagents. Pollution prevention methods, in turn, can minimize exposure to hazards and the potential risk to the researchers while also minimizing the cost and waste disposal requirements associated with the experiment.

Following the philosophy described in Chapter 1, the development of a laboratory culture that emphasizes safe work practices and workplaces should be an important goal of the leadership of an institution. An inspection system to audit work practices is recommended to ensure that laboratory activities are conducted prudently and comply with all regulatory requirements and local policies. Owing to the diversity and number of activities conducted in a typical laboratory, inspections commonly focus on particular experiments or procedures that may contain special hazards. However, random checking of more routine operations may uncover laxity or hidden problems not envisaged when the experiment was first planned. The nature and structure of an inspection will vary with circumstances, but clearly inspections should not be conducted in circumstances where they may actually increase laboratory hazards. However, the basic concepts used by a financial auditor to define a scope, conduct an audit, document the findings, require a response to the findings, and ensure that these responses address the findings properly can apply quite readily to the audit of pre-experiment reviews and general laboratory safety inspections.

2.E.1 Chapter 3: Evaluating Hazards and Assessing Risks in the Laboratory

Complete assessment of hazards should be made for all materials and suspected products associated with the experiment or procedure. Chapter 3 provides the basis for interpreting and applying much of the available hazard information. Both literature resources and knowledgeable contacts inside the local research community and in other institutions can provide additional information. If risks are determined to be unacceptable, experiments can be redesigned to minimize the volumes of chemicals used or to employ less hazardous alternatives that might do the job equally well. Some important considerations include volumes and flow rates to be employed, amounts required, physical properties of materials to be used, potential for exposure, regulatory concerns, and emergency response for unexpected events.

The Material Safety Data Sheet (MSDS) for each hazardous chemical is one of the resources that should be incorporated into experiment planning. However, because of the inconsistent quality of information found in MSDSs, Laboratory Chemical Safety Summaries (LCSSs), which are compiled in Appendix B, should be consulted or developed for the materials involved. In any case, the experiment planner needs to be aware that the existing regulations do not necessarily represent the full complement of prudent practices for handling hazardous materials and that other input is therefore essential.

In many experiments, new materials are produced whose physical properties and toxicity are unknown. Product mixtures should, therefore be regarded with suspicion of hazard until their compositions can be determined and they can be proven safe. Some provision for protecting those involved in the analysis of these product compositions from potential hazards must also be considered in experiment planning. Moreover, not all experiments proceed in the expected manner. A critical analysis should also involve consideration of the accidents that could occur in even simple experiments.

2.E.2 Chapter 4: Management of Chemicals

The experiment plan should include provisions for acquiring and storing chemicals and equipment to be used in the procedures. Some considerations for management of materials include effective labeling; inventory maintenance and reagent tracking; source reduction and materials sharing; compound shelf life; monitoring of reactive chemicals; hazards associated with storage of incompatibles, flammables, reactive chemicals, and so on; and the regulations governing shipping and storage of chemicals.

2.E.3 Chapter 5: Working with Chemicals

While the subject of Chapter 5 is at the heart of every experiment, because of the diversity of possible laboratory procedures, it is impossible to anticipate all of the potential issues that should be specified for a "generic experiment." Instead, the worker who is planning the experiment needs to rely on judgment and consultation with the literature and fellow scientists in determining which factors require particular attention. In any case, the proposed experimental procedure should be considered in adequate detail before any laboratory operations begin. Certainly these preparations should include steps such as sample preparation, equipment assembly and commissioning, start-up and calibration of equipment, data acquisition, product isolation and characterization, and storage and disposal of materials after the work is completed. Special consideration should be given to planning for unattended operations, novel equipment that is to be purchased or fabricated, and experiments that are undergoing significant scale-up.

In experimental work, it is important to recognize that although accidents can be minimized, the nature of gaining new knowledge suggests that they can never be eliminated completely. Any good experimental design process should identify hazards and develop contingency plans to deal with the unexpected so that people are not hurt, facilities are not damaged, and the public and the environment are protected. But unexpected situations can develop despite the best experiment planning. With good contingency plans, the worker might be able to change reaction conditions or procedures to obtain valuable information from experiments that do not proceed as expected, while, at least, preventing them from becoming harmful accidents.

2.E.4 Chapter 6: Working with Laboratory Equipment

A complete assessment should be made of the equipment proposed for the experiment to highlight any associated hazards. The location of the equipment within the work space should also be noted. The equipment hazards to be considered include those associated with reactors, tubing, relief devices, pumps, refrigerators, glassware, heat sources, electrical devices, lasers, ultrasound generators, photochemical equipment, compressed gases, and equipment for working at temperature extremes. Consideration should be given to whether proper maintenance procedures have been followed and documented for all equipment. The proper use of personal protective equipment such as aprons, face shields, gloves, safety glasses, and respirators should also be planned. Certain equipment will require the use of warning signs, lights, barriers, equipment monitors, alarms, and safety interlocks, particularly when temperature extremes, pressurized gases, or extremely hazardous substances are involved. The use of certain materials might also require industrial hygiene monitoring and/or special occupational health reviews.

Various institutions and local, state, and federal agencies may require certain considerations, documentation, or training for some laboratory operations, particularly (but not restricted to) those involving especially hazardous materials, equipment, or procedures. The laboratory worker and supervisor are responsible for understanding and ensuring compliance with such mandates, but they need to be aware that not all hazards are regulated and not all regulations are sufficient and that other safety measures may be necessary.

2.E.5 Chapter 7: Disposal of Waste

Environmental and waste disposal issues for source reduction, waste minimization, and recycling of materials must be considered in any experiment plan. The chemical composition of all products and waste materials generated by the experiment should be considered, and appropriate handling and disposal procedures for each of these materials should be evaluated in advance. Careful attention to regulatory requirements is essential for waste disposal. Special issues to consider include the frequency and amount of waste generated, methods to minimize waste, steps to neutralize waste or render it nonhazardous, procedures for dealing with unstable waste or waste that requires special storage and handling, and the compatibility of materials being accumulated. During the planning stage, particular attention should be given to the minimization of multi-hazardous waste, such as waste that represents both a chemical and a biological hazard.

Precautions should be taken to minimize the release of hazardous chemicals to the environment. A fume hood is a safety device and not a waste disposal facility. Therefore, fume hoods should not be used to dispose of volatile hazardous materials—to do so could cause toxic materials to be released. Special ventilation and exhaust systems, scrubbers, filters, or some other control equipment for discharges to the air or chemical sewer systems may be required under some circumstances.

2.E.6 Chapter 8: Laboratory Facilities

The facilities proposed for an experiment should be assessed completely to identify any associated hazards and to determine if the facilities are adequate for the purposes of the experiment being planned. The location of the equipment in the work space relative to the location of emergency response facilities should be considered. Work with hazardous chemicals should be carried out with fume hoods, elephant trunks, and glove boxes for some operations. The use of certain materials might also require industrial hygiene monitoring and/or special occupational health reviews. General consideration of the type of work space, its layout, and infrastructure may be appropriate. Special needs for bench space, storage, ventilation, shielding, and so forth might also affect the planning of the experiment.

2.E.7 Chapter 9: Governmental Regulation of Laboratories

Regulations are an intrinsic part of modern laboratory work that cannot be separated easily from other matters and should be considered at each step of experiment planning. It is only prudent for laboratory workers and supervisors to ensure regulatory compliance in conducting laboratory experiments. However, the responsibility of leadership goes beyond compliance to the protection of individual laboratory workers, society, and the environment. While regulations are designed to protect these entities, not all regulations are sufficient or prudent in this regard. Therefore, it falls to each worker as well as the supervisor to work toward the goal of safe, responsible laboratory performance.

3 Evaluating Hazards and Assessing Risks in the Laboratory

3.A INTRODUCTION

A key element of planning an experiment involves assessing the hazards and potential risks associated with the chemicals and laboratory operations to be employed in a proposed experiment. This chapter provides a practical guide for the laboratory worker engaged in these activities. Section 3.B introduces the sources of information where laboratory workers can find data on toxic, flammable, reactive, and explosive chemical substances as well as physical, biological, and radioactive hazards. Section 3.C discusses the toxic effects of laboratory chemicals. The first part of this section presents the basic principles that form the foundation for evaluating hazards for toxic substances. The remainder of the section describes how the laboratory worker can use this understanding and the sources of information introduced above to assess the risks associated with potential hazards of chemical substances and then to select the appropriate level of laboratory practice as discussed in Chapter 5. Sections 3.D and 3.E present guidelines for evaluating hazards associated with the use of flammable, reactive, and explosive substances and physical hazards, respectively. Finally, there is a brief reference to biohazards and hazards from radioactivity in sections 3.F and 3.G, respectively.

Although the responsibility for carrying out the hazard evaluations and risk assessments described here generally lies primarily with the laboratory worker who will actually be conducting the proposed experiment, this activity often requires consultation with other colleagues and superiors. For example, depending on the level of training and experience of the laboratory worker, the involvement of the worker's immediate laboratory supervisor may be advisable and in some instances essential. In addition, many institutions have environmental health and safety offices, where industrial hygiene specialists are available to advise laboratory workers and their supervisors on issues involved in the assessment of risks of laboratory chemicals. Chemical hygiene officers, required by federal regulation, play similar departmental roles in many institutions.

3.B SOURCES OF INFORMATION

3.B.1 Chemical Hygiene Plan

Beginning in 1991, every laboratory in which hazardous chemicals are in use has been required by federal law to have a written Chemical Hygiene Plan (CHP), which includes provisions capable of protecting personnel from the "health hazards associated with the chemicals present in that laboratory." All laboratory workers should be familiar with and have ready access to their institution's CHP. In some laboratories, CHPs include standard operating procedures for work with specific chemical substances, and in these cases the CHP may be sufficient as the primary source of information used for risk assessment and experiment planning. However, most CHPs provide only general procedures for handling chemicals, and in these cases prudent experiment planning requires that the laboratory worker consult additional sources for information on the properties of the substances that will be encountered in the proposed experiment.

3.B.2 Material Safety Data Sheets

Federal law requires that manufacturers and distributors of chemicals provide users with Material Safety Data Sheets (MSDSs), which are designed to provide the information needed to protect users from any hazards that may be associated with the product. MSDSs have become the primary vehicle through which the potential hazards of materials obtained from commercial sources are communicated to the laboratory worker. Institutions are required by law to retain and make readily available to workers the MSDSs provided by chemical suppliers.

As the first step in a risk assessment, laboratory workers should examine their plan for a proposed experiment and identify the chemicals whose toxicological properties they are not already familiar with from previous experience. The MSDS for each unfamiliar chemical should then be examined. Procedures for accessing MSDS files vary from institution to institution. In some cases, MSDS files may be present in each laboratory, while in many cases complete files of MSDSs are maintained only in a central location, such as the institution's environmental health and safety office. Some laboratories now have the capability to access MSDSs electronically, either from CD-ROM disks or via computer networks. As a last resort, the laboratory worker can always contact the chemical supplier directly and request that an MSDS be sent by mail.

MSDSs are concise technical documents, generally two to five pages in length. An MSDS typically begins with a compilation of data on the physical, chemical, and toxicological properties of the substance and then provides generally concise suggestions for handling, storage, and disposal. Finally, emergency and first aid procedures are usually outlined. At present there is no required format for an MSDS; however, it is expected that the Occupational Safety and Health Administration (OSHA) will soon adopt a general 16-part format proposed by the American National Standards Institute (ANSI). The following is a guide to the information typically found in an MSDS:

1. *Name of supplier (with address and phone number) and date MSDS was prepared or revised.* Toxicity data and exposure limits sometimes undergo revision, and for this reason MSDSs should be reviewed periodically to check that they contain up-to-date information. Phone numbers are provided so that, if necessary, users can contact the supplier to obtain additional information on hazards and emergency procedures.

2. *Name of the chemical.* For products that are mixtures, this section may include the identity of most but not every ingredient. Common synonyms are usually listed.

3. *Physical and chemical properties.* Data such as melting point, boiling point, and molecular weight are included here.

4. *Physical hazards.* This section provides data related to flammability, reactivity, and explosibility hazards.

5. *Toxicity data.* OSHA and American Conference of Governmental Industrial Hygienists (ACGIH) exposure limits (as discussed below in section 3.C) are listed. Many MSDSs provide lengthy and comprehensive compilations of toxicity data and even references to applicable federal standards and regulations.

6. *Health hazards.* Acute and chronic health hazards are listed, together with the signs and symptoms of exposure. The primary routes of entry of the substance into the body must also be described. In addition, potential carcinogens are explicitly identified. In some MSDSs, this list of toxic effects is quite lengthy and may include every possible harmful effect the substance can have under the conditions of every conceivable use.

7. *Storage and handling procedures.* This section usually consists of a list of precautions to be taken in handling and storing the material. Particular attention is devoted to listing appropriate control measures, such as the use of engineering controls and personal protective equipment necessary to prevent harmful exposures. Because an MSDS is written to address the largest scale that the material could conceivably be used on, the procedures recommended may involve more stringent precautions than are necessary in the context of laboratory use.

8. *Emergency and first aid procedures.* This section usually includes recommendations for firefighting procedures, first aid treatment, and steps to be taken if the material is released or spilled. Again, the measures outlined here are chosen to encompass worst-case scenarios, including accidents on a larger scale than could conceivably occur in a laboratory.

9. *Disposal considerations.* Many MSDSs provide guidelines for the proper disposal of waste material.

10. *Transportation information.* It is important to remember that this chapter is concerned only with evaluating the hazards and assessing the risks associated with chemicals *in the context of laboratory use.* MSDSs, in contrast, must address the hazards associated with chemicals in all possible situations, including industrial manufacturing operations and large-scale transportation accidents. For this reason, some of the information in an MSDS may not be relevant to the handling and use of that chemical in a laboratory. For example, most MSDSs stipulate that self-contained breathing apparatus and heavy rubber gloves and boots be worn in cleaning up spills, even of relatively nontoxic materials such as acetone. Such precautions, however, might be unnecessary in the case of laboratory-scale spills of acetone and other substances of low toxicity.

Originally, the principal audience for MSDSs comprised health and safety professionals (who are responsible for formulating safe workplace practices), medical personnel (who direct medical surveillance programs and treat exposed workers), and emergency responders (e.g., fire department personnel). With the promulgation of federal laws such as the Hazard Communication Standard (29 CFR 1910.1200) and the OSHA Laboratory Standard (29 CFR 1910.1450), the audience for MSDSs has been expanded to include laboratory workers in industrial and academic laboratories. However, not all MSDSs are written to meet the requirements of this new audience effectively.

In summary, among the currently available resources, MSDSs remain the best single source of information for the purpose of evaluating the hazards and assessing the risks of chemical substances. However, laboratory workers should recognize the limitations of MSDSs as applied to laboratory-scale operations:

1. The quality of MSDSs produced by different chemical suppliers varies widely. The utility of some MSDSs is compromised by vague and unqualified generalizations and internal inconsistencies.

2. MSDSs must describe control measures and precautions for work on a variety of scales, ranging from microscale laboratory experiments to large manufacturing operations. Some procedures outlined in an MSDS may therefore be unnecessary or inappropriate for laboratory-scale work. An unfortunate consequence of this problem is that it tends to breed a lack of confidence in the relevance of the MSDS to laboratory-scale work.

3. Many MSDSs comprehensively list all conceivable health hazards associated with a substance without differentiating which are most significant and which are most likely to actually be encountered. This can make it difficult for laboratory workers to distinguish highly hazardous materials from moderately hazardous and relatively harmless ones.

3.B.3 Laboratory Chemical Safety Summaries

As discussed above, although MSDSs are invaluable resources, they suffer some limitations as applied to risk assessment in the specific context of the laboratory. Appendix B introduces the concept of the Laboratory Chemical Safety Summary (LCSS), which is specifically tailored to the needs of the laboratory worker. As indicated in their name, LCSSs provide information on chemicals in the context of laboratory use. These documents are summaries and are not intended to be comprehensive or to fulfill the needs of all conceivable users of a chemical. In conjunction with the guidelines described in this chapter, the LCSS provides essential information required to assess the risks associated with the use of a particular chemical in the laboratory.

The format, organization, and contents of LCSSs are discussed in detail in the introduction to Appendix B. Included in an LCSS are the key physical, chemical, and toxicological data necessary to evaluate the relative degree of hazard posed by a substance. LCSSs also include a concise critical discussion, presented in a style readily understandable to laboratory workers, of the toxicity, flammability, reactivity, and explosibility of the chemical; recommendations for the handling, storage, and disposal of the title substance; and first aid and emergency response procedures.

Appendix B contains LCSSs for 88 chemical substances. Several criteria were used in selecting these chemicals, the most important consideration being whether the substance is commonly used in laboratories. Preference was also given to materials that pose relatively serious hazards. Finally, an effort was also made to select chemicals representing a variety of different classes of substances, so as to provide models for the future development of additional LCSSs.

3.B.4 Labels

Commercial suppliers are required by law to provide their chemicals in containers affixed with precautionary labels. Labels usually present concise and nontechnical summaries of the principal hazards associated with their contents. Note that precautionary labels should not replace MSDSs and LCSSs as the primary source of information for risk assessment in the laboratory. However, labels can serve as valuable reminders of the key hazards associated with the substance.

3.B.5 Additional Sources of Information

The resources described above provide the foundation for risk assessment of chemicals in the laboratory. This section highlights the sources that should be consulted for additional information on specific harmful effects of chemical substances. Although MSDSs and LCSSs include considerable information on toxic effects, in some situations the laboratory worker should seek additional, more detailed information. This step is particularly important when the worker is planning to use chemicals that have a high degree of acute or chronic toxicity or when it is anticipated that work will be conducted with a particular toxic substance frequently or over an extended period of time. Section 3.B of this chapter provides explicit guidelines as to how laboratory workers can use the information in an MSDS or LCSS to recognize when it is necessary to seek such additional information.

The following annotated list provides references on the hazardous properties of chemicals in the approximate order of their utility in assessing risks in the laboratory. *The first six references are particularly valuable sources of information, and it is strongly recommended that copies of these be made readily accessible to laboratory workers at all times.* A compilation of related materials and recommended resources can be found in the bibliography.

1. *Occupational Health Guidelines for Chemical Hazards*, U.S. DHHS; F. W. Mackison, R. S. Stricoff, and L. J. Partridge, editors, DHHS (NIOSH) Publication Number 81-123, U.S. Government Printing Office, Washington, D.C., 1981, and a supplement published as DHHS (NIOSH) Publication No. 89-104, U.S. Government Printing Office, Washington, D.C., 1988. The guidelines currently cover almost 400 substances and are based on the information assembled under the Standards Completion Program, which served as the basis for the promulgation of federal occupational health regulations ("substance-specific standards"). Typically five pages in length and written clearly at a level that should be readily understood by laboratory workers, each set of guidelines includes information on physical, chemical, and toxicological properties, signs and symptoms of exposure, and considerable detail on control measures, medical surveillance practices, and emergency first aid procedures. However, some guidelines date back to 1978 and may not be current, particularly with regard to chronic toxic effects.

2. *Chemical Safety Data Sheets,* Royal Society of Chemistry, five volumes, Cambridge, United Kingdom, 1989–1992. This excellent collection of data sheets summarizes hazard information on more than 500 chemicals. These are more useful for the laboratory worker than most MSDSs and are similar in aim to the LCSSs. Sections include threshold limit values, physical properties, chemical hazards, biological hazards (e.g., vapor inhalation, eye contact, skin contact, swallowing), carcinogenicity, mutagenicity, reproductive hazards, first

aid, handling and storage, disposal, and fire precautions. Each summary includes a list of references.

3. *A Comprehensive Guide to the Hazardous Properties of Chemical Substances*, P. A. Patnaik, Van Nostrand Reinhold, New York, 1992. This particularly valuable guide is written at a level appropriate for the typical laboratory worker. It covers about 1,500 substances; sections in each entry include uses and exposure risk, physical properties, health hazards, exposure limits, fire and explosion hazards, and disposal/destruction. Entries are organized into chapters according to functional group classes, and each chapter begins with a general discussion of the properties and hazards of the class.

4. *Threshold Limit Values for Chemical Substances and Physical Agents and Biological Exposure Indices, 1994–1995*, American Conference of Governmental Industrial Hygienists (ACGIH), Cincinnati, Ohio, 1994. A handy booklet listing ACGIH threshold limit values (TLVs) and short-term exposure limits (STELs). These values are under continuous review, and this booklet is updated annually. The ACGIH's multivolume publication *Documentation of the Threshold Limit Values and Biological Exposure Indices* reviews the data (with reference to literature sources) that were used to establish the threshold limit values.

5. *Fire Protection for Laboratories Using Chemicals* (NFPA Standard Code No. 45), National Fire Protection Association, Quincy, Massachusetts, 1991. This is the national fire safety code pertaining to laboratory use of chemicals.

6. *Bretherick's Handbook of Reactive Chemical Hazards*, 4th edition, L. Bretherick, Butterworth, London, 1990. An extremely comprehensive compilation of examples of violent reactions, fires, and explosions due to unstable chemicals, as well as reports on known examples of incompatibility between reactive chemicals.

7. *Sax's Dangerous Properties of Industrial Materials*, 8th edition, three volumes, Richard J. Lewis, Sr., Van Nostrand Reinhold, New York, 1992. This compilation of data for 20,000 chemical substances contains much of the information found in a typical MSDS, including physical and chemical properties, data on toxicity, flammability, reactivity, and explosibility, and a concise safety profile describing symptoms of exposure. This is a useful reference for checking the accuracy of an MSDS and a valuable resource to assist workers in preparing their own LCSSs.

8. *Fire Protection Guide to Hazardous Materials*, 10th edition, National Fire Protection Association, Quincy, Massachusetts, 1991. This resource contains hazard data on more than 400 chemicals.

9. *Patty's Industrial Hygiene and Toxicology*, 4th edition, G. D. Clayton and F. E. Clayton, editors, Wiley-Interscience, New York, 1994, Volume 2, *Toxicology* (part C). A classic and authoritative reference on the toxicology of different classes of organic and inorganic compounds. The six parts of volume 2 consist of several thousand pages of information organized by functional group class. The focus in *Patty's* is on health effects; hazards due to flammability, reactivity, and explosibility are not covered.

10. *Proctor and Hughes' Chemical Hazards of the Workplace*, 3rd edition, G. J. Hathaway, N. H. Proctor, J. P. Hughes, and M. L. Fischman, editors, Van Nostrand Reinhold, New York, 1991. This resource provides an excellent summary of the toxicology of 542 chemicals. Most entries are one to two pages in length and include signs and symptoms of exposure with reference to specific clinical reports.

11. *Handbook of Toxic and Hazardous Chemicals and Carcinogens*, 3rd edition, two volumes, Marshall Sittig, Noyes Publications, Park Ridge, New Jersey, 1991. This very good reference, which is written with the industrial hygienist in mind, covers 800 substances.

12. *Sigma-Aldrich Library of Chemical Safety Data*, 2nd edition, Robert E. Lenga, editor, two volumes, Sigma-Aldrich, Milwaukee, Wisconsin, 1988. This compilation of safety data for approximately 14,500 chemicals is in tabular form. It presents considerably less information than is found in a typical MSDS or LCSS, but it is convenient as a single source of information for a very large number of substances.

13. *Clinical Toxicology of Commercial Products*, 5th edition, Robert E. Gosselin, Roger P. Smith, and Harold C. Hodge, Williams & Wilkins, Baltimore, Maryland, 1984. This reference is designed to assist the physician in dealing with cases of acute chemical poisoning. It contains trade names of products and their ingredients.

14. *Casarett and Doull's Toxicology: The Basic Science of Poisons*, 4th edition, M. O. Amdur, J. Doull, and C. D. Klaassen, editors, Pergamon Press, New York, 1991. This complete and readable overview of toxicology is a good textbook but is not arranged as a ready reference for handling laboratory emergencies.

15. *Catalog of Teratogenic Agents*, 7th edition, Thomas H. Shepard, Johns Hopkins University Press, Baltimore, Maryland, 1992. This catalog is one of the best references available on the subject of reproductive and developmental toxins.

16. *The Laboratory Environment*, R. Purchase, editor, Special Publication Number 136, Royal Society of Chemistry, Cambridge, United Kingdom, 1994.

3.B.6 Computer Services

In addition to computerized MSDSs, a number of computer databases are available that supply data for creating or supplementing MSDSs. The National Library of Medicine (NLM) and the Chemical Abstracts databases are examples. These and other such databases are accessible through various on-line computer data services; also, most of this information is available as CD-ROM and computer updates. Many of these services can be accessed for up-to-date toxicity information.

3.B.6.1 The National Library of Medicine Databases

The databases supplied by NLM are easy to use and relatively inexpensive. TOXLINE, the best source of information for most people, covers data published from 1981 to the present. For data published in the period from 1965 through 1980, TOXLINE65, a back file of TOXLINE, is also available. The telephone number to call for information and instructions on obtaining an NLM account is 1-800-638-8480.

Other databases supplied by NLM are the Hazardous Substance Data Base (HSDB), the Registry of Toxic Effects of Chemical Substances (RTECS), and the Medical Literature Analysis and Retrieval System (MEDLARS). NLM also supplies other specialized databases called CANCERLIT, DART, GENETOX, IRIS, CCRIS, and CHEMID.

3.B.6.2 Chemical Abstracts Databases

Another source of toxicity data is Chemical Abstracts (CA). In addition to the NLM, several services provide CA, including Knight-Ridder Information (formerly DIALOG), ORBIT, STN, and Ovid Technologies (formerly CD Plus). Searching procedures for CA depend on the various services supplying the database. Searching costs are considerably higher than for NLM databases because CA royalties must be paid. Telephone numbers for the above suppliers are as follows:

Knight-Ridder Information	1-800-334-2564
ORBIT	1-800-456-7248
STN	1-800-848-6533
Ovid Technologies	1-800-289-4277

Specialized databases are available from a vendor called Chemical Information Systems (CIS) for aquatic toxicity, dermal toxicity, EPA TSCA FYI, 8(d) and 8(e) studies, and so on. The CIS telephone number is 1-800-CIS-USER.

Searching any database is best done using the Chemical Abstracts Service (CAS) Registry Number for the particular chemical. Free text searching is available on most of the databases except MEDLINE, which has a controlled vocabulary. As mentioned above, a menu-driven format is available to aid the inexperienced user. Equipment needed to do a search includes a computer terminal, a modem for accessing the on-line database by telephone, and a printer. Results of the search can also be captured by using an electronic format (e.g., a floppy disk).

3.B.6.3 Informal Forum

The "Letters to the Editor" column of *Chemical & Engineering News*, published weekly by the American Chemical Society, has become an informal but widely accepted forum for the reporting of anecdotal information on chemical reactivity hazards and other safety-related information. This publication is accessible via full-text searching services provided by STN.

3.C TOXIC EFFECTS OF LABORATORY CHEMICALS

3.C.1 Basic Principles

The chemicals encountered in the laboratory have a broad spectrum of physical, chemical, and toxicological properties and physiological effects. The risks associated with the use of laboratory chemicals must be well understood prior to their use in an experiment. The risk of toxic effects is related to both the extent of exposure and the inherent toxicity of a chemical. As discussed in detail below, extent of exposure is determined by the dose, the duration and frequency of exposure, and the route of exposure. Exposure to even large doses of chemicals with little inherent toxicity, such as phosphate buffer, presents low risk. In contrast, even small quantities of chemicals with high inherent toxicity or corrosivity may cause significant adverse effects. The duration and frequency of exposure are also critical factors in determining whether a chemical will produce harmful effects. In some cases, a single exposure to a chemical is sufficient to produce poisoning. On the other hand, for many chemicals repeated exposure is required to produce toxic effects. For most substances, the route of exposure (through the skin, the eyes, the gastrointestinal tract, or the respiratory tract) is also an important consideration in risk assessment. In the case of chemicals that are systemic toxicants, the internal dose to the target organ is a critical factor.

When considering possible toxicity hazards while planning an experiment, it is important to recognize that *the combination of the toxic effects of two substances may be significantly greater than the toxic effect of either substance alone.* Because most chemical reactions are likely to produce mixtures of substances whose combined toxicities have never been evaluated, it is prudent to assume that mixtures of different substances (i.e., chemical reaction mixtures) will be more toxic than their most toxic ingredient. Furthermore, chemical reactions involving two or more substances may form reaction products that are significantly more toxic than the starting reactants. This possibility of generating toxic reaction products may not be anticipated by the laboratory worker in cases where the reactants are mixed unintentionally. For example, inadvertent mixing of formaldehyde (a common tissue fixative) and hydrogen chloride could result in the generation of bis(chloromethyl)ether, a potent human carcinogen.

It is essential that all laboratory workers understand certain basic principles of toxicology and learn to recognize the major classes of toxic and corrosive chemicals. The next sections of this chapter summarize the key concepts involved in assessing the risks associated with the use of toxic chemicals in the laboratory. (Also see Chapter 5, section 5.D.)

3.C.1.1 Dose-Response Relationships

Toxicology, the science of poisons, is the study of the adverse effects of chemicals on living systems. The basic tenet of toxicology is that no substance is entirely safe and that all chemicals result in some toxic effects if a high enough amount (dose) of the substance comes in contact with a living system. Paracelsus (1493–1541) elegantly articulated this simple concept five centuries ago when he noted, "All substances are poisons; there is none which is not a poison. The right dose differentiates a poison. . . ." This is perhaps the most important concept for all laboratory workers to be cognizant of. For example, ingestion of water, a vital substance for life, can result in death if a sufficiently large amount (i.e., gallons) is ingested at one time. On the other hand, sodium cyanide, a highly lethal chemical, will produce no permanent effects if a living system is exposed to a sufficiently low dose. The single most important factor that determines whether a substance will be harmful (or, conversely, safe) to an individual is the relationship between the amount (or concentration) of the chemical and the toxic effect it produces. For all chemicals, there is a range of concentrations that result in a graded effect between the extremes of no effect and death. In toxicology, this is referred to as the dose-response relationship for the chemical. The dose is the amount of the chemical and the response is the effect of the chemical. This relationship is unique for each chemical, although for many similar types of chemicals, the dose-response relationships are very similar. Among the thousands of laboratory chemicals, there is clearly a wide spectrum of doses that are required to produce toxic effects and, in some cases, even death. For most

chemicals, a threshold dose has been established (by rule or by consensus) below which a chemical is not considered to be harmful.

Some chemicals (e.g., dioxin) will produce death in laboratory animals upon exposure to microgram doses and therefore are obviously extremely toxic. Other substances, however, may have no harmful effects following doses in excess of several grams. One way to evaluate the acute toxicity (i.e., the toxicity occurring after a single exposure) of laboratory chemicals involves consideration of their lethal dose 50 (LD_{50}) or lethal concentration 50 (LC_{50}) value. The LD_{50} is defined as the amount of a chemical that when ingested, injected, or applied to the skin of a test animal under controlled laboratory conditions will kill one-half (50%) of the animals. The LD_{50} is usually expressed in units of milligrams or grams per kilogram of body weight. For volatile chemicals (i.e., chemicals with sufficient vapor pressure that inhalation is an important route of chemical entry into the body), the LC_{50} is often reported instead of the LD_{50}. The LC_{50} is the concentration of the chemical in air that will kill 50% of the test animals exposed to it. The LC_{50} is usually given in units of parts per million, milligrams per liter, or milligrams per cubic meter. Also reported are LC_{lo} and LD_{lo} values, which are defined as the lowest concentration or dose that causes the death of test animals. In general, the larger the value of the LD_{50} or LC_{50}, the more chemical it takes to kill the test animals and therefore the lower the toxicity of the chemical. Although lethal dose values may vary among animal species and between animals and humans, the relative toxicity of different substances is usually relatively constant, and chemicals that are highly toxic to animals are generally highly toxic to humans.

3.C.1.2 Duration and Frequency of Exposure

Toxic effects of chemicals can occur after single (acute), intermittent (repeated), or long-term, repeated (chronic) exposure. An acutely toxic substance can cause damage as the result of a single, short-duration exposure. Hydrogen cyanide, hydrogen sulfide, and nitrogen dioxide are examples of acute toxins. In contrast, a chronically toxic substance causes damage after repeated or long-duration exposure or causes damage that becomes evident only after a long latency period. Chronic toxins include all carcinogens, reproductive toxins, and certain heavy metals (e.g., mercury, lead) and their compounds. Many chronic toxins are extremely dangerous because of their long latency periods: the cumulative effect of low exposures to such substances may not become apparent for many years.

In a general sense, the longer the duration of exposure, that is, the longer the body (or tissues in the body) is in contact with a chemical, the greater the opportunity for toxic effects to occur. Frequency of exposure also has an important influence on the nature and extent of toxicity. The total amount of a chemical required to produce a toxic effect is generally less for a single exposure than for intermittent or repeated exposures. More total chemical is required to produce toxicity for intermittent or chronic exposure because many chemicals can be eliminated from the body, because tissue injuries can often be repaired, and because adaptation of tissues can occur over time. Some toxic effects occur only after chronic exposure; this is because sufficient amounts of chemical cannot be attained in the tissue by a single exposure. Sometimes a chemical has to be present in a tissue for a considerable time to produce injury. For example, the neurotoxic and carcinogenic effects from exposure to heavy metals usually require long-term repeated exposure.

The time between exposure to a chemical and onset of toxic effects varies depending on the chemical and the exposure. For example, the toxic effects of carbon monoxide, sodium cyanide, and carbon disulfide are evident within minutes. For many chemicals, the toxic effect is most severe between one and a few days after exposure. However, some chemicals produce ''delayed'' toxicity; in fact, the neurotoxicity produced by some chemicals is not observed until a few weeks after exposure. The most delayed toxic effect produced by chemicals is cancer: in humans, it usually takes 10 to 30 years between exposure to a known human carcinogen and the detection of a tumor.

3.C.1.3 Routes of Exposure

Exposure to chemicals in the laboratory can occur by several different routes: (1) inhalation, (2) contact with skin or eyes, (3) ingestion, and (4) injection. Important features of these different pathways are detailed below.

3.C.1.3.1 *Inhalation*

Toxic materials that can enter the body via inhalation include gases, the vapors of volatile liquids, mists and sprays of both volatile and nonvolatile liquid substances, and solid chemicals in the form of particles, fibers, and dusts. Inhalation of toxic gases and vapors can produce poisoning by absorption through the mucous membranes of the mouth, throat, and lungs and can also damage these tissues seriously by local action. Inhaled gases and vapors can pass into the capillaries of the lungs and be carried into the circulatory system. This absorption can be extremely rapid. Because of the large surface area of the lungs in humans (about 75 square meters (m^2)), this is the main site for absorption of many toxic materials.

The factors governing the absorption of gases and vapors from the respiratory tract differ significantly from those that govern the absorption of particulate substances. Factors controlling the absorption of inhaled gases and vapors include the solubility of the gas in body fluids and the reactivity of the gas with tissues and the fluid lining the respiratory tract. Gases or vapors that are highly water-soluble, such as methanol, acetone, hydrogen chloride, and ammonia, dissolve predominantly in the lining of the nose and windpipe (trachea) and therefore tend to be absorbed from those regions. These sites of absorption are also potential sites of toxicity. Formaldehyde is an example of a reactive, highly water-soluble vapor for which the nose is a major site of deposition. In contrast to water-soluble gases, reactive gases with low water-solubility, such as ozone, phosgene, and nitrogen dioxide, penetrate farther into the respiratory tract and thus come into contact with the smaller tubes of the airways. Gases and vapors that are not water-soluble but are more fat-soluble, such as benzene, methylene chloride, and trichloroethylene, are not completely removed by interaction with the surfaces of the nose, trachea, and small airways. As a result, these gases penetrate the airways down into the deep lung, where they can diffuse across the thin lung tissue into the blood. The more soluble a gas is in the blood, the more of it will be dissolved and transported to other organs.

In the case of inhaled solid chemicals, an important factor in determining if and where a particle will be deposited in the respiratory tract is its size. One generalization is that the largest particles ($\geq$5 microns (μm)) are deposited primarily in the nose, smaller particles (1 to 5 μm) in the trachea and small airways, and the smallest particles in the lungs. Thus, depending on the size of an inhaled particle, it will be deposited in different sections of the respiratory tract, and the location can affect the local toxicity and the absorption of the material. In general, particles that are water-soluble will dissolve within minutes or days, and chemicals that are not water-soluble but have a moderate degree of fat-solubility will also clear rapidly into the blood. Those that are not water-soluble or highly fat-soluble will not dissolve and will be retained in the lungs for long periods of time. Metal oxides, asbestos, and silica are examples of water-insoluble inorganic particles that might be retained in the lungs for years.

A number of factors can affect the airborne concentrations of chemicals. Vapor pressure (the tendency of molecules to escape from the liquid or solid phase into the gaseous phase) is the most important characteristic of a chemical to consider. The higher the vapor pressure, the greater the potential concentration of the chemical in the air. For example, acetone (with a vapor pressure of 180 millimeters of mercury (mmHg) at 20 °C) could reach an equilibrium concentration in air of 240,000 parts per million (ppm), or 24%. (This value is approximated by dividing the vapor pressure of the chemical by the atmospheric pressure—760 mmHg—and multiplying by 1,000,000 to convert to ppm.) Fortunately, the ventilation present in most laboratories prevents an equilibrium concentration from developing in the breathing zone of the laboratory worker.

Even very low vapor pressure chemicals can be dangerous if the material is highly toxic. A classic example is elemental mercury. Although the vapor pressure of mercury at room temperature is only 0.0012 mmHg, the resulting equilibrium concentration of mercury vapor is 1.58 ppm, or about 13 milligrams per cubic meter (mg/m^3). The TLV for mercury is 0.05 mg/m^3, more than 2 orders of magnitude lower.

The vapor pressure of a chemical increases with temperature; therefore, heating of solvents or reaction mixtures increases the potential for high airborne concentrations. Also, a spilled volatile chemical can evaporate very quickly because of its large surface area, creating a significant exposure potential. It is clear that careful handling of volatile chemicals is very important; keeping containers tightly closed or covered and using volatiles in fume hoods are techniques that should be used to avoid unnecessary exposure to inhaled chemicals.

Certain types of particulate materials can also present the potential for airborne exposures. If a material has a very low density or a very small particle size, it will tend to remain airborne for a considerable time. For example, the very fine dust cloud generated by emptying a low-density particulate (e.g., vermiculite) into a secondary container will take a long time to settle out, and these particles can be inhaled. Such operations should therefore be carried out in a fume hood.

Operations that generate aerosols (suspensions of microscopic droplets in air), such as vigorous boiling, high-speed blending, or bubbling gas through a liquid, increase the potential for exposure via inhalation. Consequently, these and other such operations on toxic chemicals should also be carried out in a hood.

3.C.1.3.2   Contact with Skin or Eyes

Contact with the skin is a frequent mode of chemical injury in the laboratory. Many chemicals can injure the skin directly. Skin irritation and allergic skin reactions are a common result of contact with certain types of chemicals. Corrosive chemicals can cause severe burns when they come in contact with the skin. In addition to causing local toxic effects, many chemicals are absorbed through the skin in sufficient quantity to produce systemic toxicity. The main avenues by which chemicals enter the body through the skin are the hair follicles, sebaceous glands, sweat glands, and cuts or abrasions of the outer layer. Absorption of chemicals through the skin depends on a number of factors, including chemical concentration, chemical reactivity, and the solubility of the chemical in fat and water. Absorption is also dependent on the condition of the skin, the part of the body exposed, and duration of contact. Differences in skin structure affect the degree to which chemicals can be absorbed. In general, toxicants cross thin skin (e.g., scrotum) much more easily than thick skin (e.g., palms). When skin is damaged, penetration of chemicals increases. Acids and alkalis can injure the skin and increase its permeability. Burns and skin diseases are the most common examples of skin damage that can increase penetration. Also, hydrated skin absorbs chemicals better than dehydrated skin. Some chemicals such as dimethyl sulfoxide can actually increase the penetration of chemicals through the skin by increasing its permeability.

Contact of chemicals with the eyes is of particular concern because these organs are so sensitive to irritants. Few substances are innocuous in contact with the eyes; most are painful and irritating, and a considerable number are capable of causing burns and loss of vision. Alkaline materials, phenols, and strong acids are particularly corrosive and can cause permanent loss of vision. Because the eyes contain many blood vessels, they also can be a route for the rapid absorption of many chemicals.

3.C.1.3.3   Ingestion

Many of the chemicals used in the laboratory are extremely hazardous if they enter the mouth and are swallowed. The gastrointestinal tract, which consists of the mouth, esophagus, stomach, and small and large intestines, can be thought of as a tube of variable diameter (about 5 m in length) with a large surface area (about 200 m^2) for absorption. Toxicants that enter the gastrointestinal tract must be absorbed into the blood to produce a systemic injury. Sometimes a chemical is caustic or irritating to the gastrointestinal tract tissue itself. Absorption of toxicants can take place along the entire gastrointestinal tract, even in the mouth, and depends on many factors, including the physical properties of the chemical and the speed at which it dissolves. Absorption increases with surface area, permeability, and residence time in various segments of the tract. Some chemicals increase intestinal permeability and thus increase the rate of absorption. More chemical will be absorbed if the chemical remains in the intestine for a long time. If a chemical is in a relatively insoluble, solid form, it will have limited contact with gastrointestinal tissue, and its rate of absorption will be low. If it is an organic acid or base, it will be absorbed in that part of the gastrointestinal tract where it is most fat-soluble. Fat-soluble chemicals are absorbed more rapidly and extensively than water-soluble chemicals.

3.C.1.3.4 Injection

Exposure to toxic chemicals by injection does not occur frequently in the chemical laboratory. However, it can occur inadvertently through mechanical injury from "sharps" such as glass or metal contaminated with chemicals or when chemicals are handled with syringes. The intravenous route of administration is especially dangerous because it introduces the toxicant directly into the bloodstream, eliminating the process of absorption. Nonlaboratory personnel, such as custodial workers or waste handlers, must be protected from this form of exposure by putting all "sharps" in special trash containers and never in the ordinary scrap baskets. Hypodermic needles with blunt ends are available for laboratory use.

3.C.2 Types of Toxins

Exposure to a harmful chemical can result in local toxic effects, systemic toxic effects, or both. Local effects involve injury at the site of first contact. The eyes, the skin, the nose and lungs, and the digestive tract are typical sites of local reactions. Examples of local effects include (1) ingestion of caustic substances causing burns and ulcers in the mouth, esophagus, stomach, and intestines, (2) inhalation of hazardous materials causing toxic effects in the nose and lungs, and (3) contact with harmful materials on the skin or eyes leading to effects ranging from mild irritation to severe tissue damage. Systemic effects, by contrast, occur after the toxicant has been absorbed from the site of contact into the bloodstream and distributed throughout the body. While some chemicals produce adverse effects on all tissues of the body, other chemicals tend to selectively injure a particular tissue or organ without affecting others. The affected organs (e.g., liver, lungs, kidney, central nervous system) are referred to as the target organs of toxicity. The target organ of toxicity is not necessarily the organ where the highest concentration of the chemical is achieved. Hundreds of different systemic toxic effects of chemicals are known. Systemic effects can result from single

(acute) exposures or from repeated or long-duration (chronic) exposures, becoming evident only after a long latency period.

Toxic effects can be further classified as reversible or irreversible. Reversible toxicity is possible because in some cases tissues have the capacity to repair toxic damage, so that the damage disappears following cessation of exposure. Irreversible damage, in contrast, persists even after cessation of exposure. Recovery from a burn is a good example of reversible toxicity; cancer is generally thought to be irreversible.

The chemicals used in the laboratory can be grouped among several different classes of toxic substances. Many chemicals display more than one type of toxicity. The following are the most common classes of toxic substances encountered in laboratories.

3.C.2.1 Irritants

Irritants are noncorrosive chemicals that cause reversible inflammatory effects (swelling and redness) on living tissue by chemical action at the site of contact. A wide variety of organic and inorganic chemicals are irritants, and consequently, skin and eye contact with all chemicals in the laboratory should be avoided.

3.C.2.2 Corrosive Substances

Corrosive substances cause destruction of living tissue by chemical action at the site of contact and can be solids, liquids, or gases. Corrosive effects can occur not only on the skin and eyes, but also in the respiratory tract and, in the case of ingestion, in the gastrointestinal tract as well. Corrosive materials are probably the most common toxic substances encountered in the laboratory. Corrosive liquids are especially dangerous because their effect on tissue generally takes place very rapidly. Bromine, sulfuric acid, aqueous sodium hydroxide solution, and hydrogen peroxide are examples of highly corrosive liquids. Corrosive gases are also frequently encountered. Gases such as chlorine, ammonia, and nitrogen dioxide can damage the lining of the lungs, leading, after a delay of several hours, to the fatal buildup of fluid known as pulmonary edema. Finally, a number of solid chemicals have corrosive effects on living tissue. Examples of common corrosive solids include sodium hydroxide, phosphorus, and phenol. Dust from corrosive solids can be inhaled and cause serious damage to the respiratory tract.

There are several major classes of corrosive substances. Strong acids such as nitric, sulfuric, and hydrochloric acid can cause serious damage to the skin and eyes. Hydrofluoric acid is particularly dangerous and produces slow-healing, painful burns. Strong bases, such as the metal hydroxides and ammonia, make up another class of corrosive chemicals. Strong dehydrating agents, such as phosphorus pentoxide and calcium oxide, have a powerful affinity for water and can cause serious burns upon contact with the skin. Finally, strong oxidizing agents, such as concentrated solutions of hydrogen peroxide, can also have serious corrosive effects and should never come into contact with the skin or eyes.

3.C.2.3   Allergens

A chemical allergy is an adverse reaction by the immune system to a chemical. Such allergic reactions result from previous sensitization to that chemical or a structurally similar chemical. Once sensitization occurs, allergic reactions can result from exposure to extremely low doses of the chemical. Allergic reactions can be immediate, occurring within a few minutes after exposure. Anaphylactic shock is a severe immediate allergic reaction that can result in death if not treated quickly. If this is likely to be a hazard for a planned experiment, advice on emergency response should be obtained. Allergic reactions can also be delayed, taking hours or even days to develop. The skin is usually the site of such delayed reactions, in which cases it becomes red, swollen, and itchy.

It is important to recognize that delayed chemical allergy can occur even some time after the chemical has been removed. Contact with poison ivy is a familiar example of an exposure that causes a delayed allergic reaction. Also, just as people vary widely in their susceptibility to sensitization by environmental allergens such as dust and pollen, individuals may also exhibit wide differences in their sensitivity to laboratory chemicals. Examples of substances that may cause allergic reactions in some individuals include diazomethane, dicyclohexylcarbodiimide, formaldehyde, various isocyanates, benzylic and allylic halides, and certain phenol derivatives.

3.C.2.4   Asphyxiants

Asphyxiants are substances that interfere with the transport of an adequate supply of oxygen to the vital organs of the body. The brain is the organ most easily affected by oxygen starvation, and exposure to asphyxiants can lead to rapid collapse and death. Simple asphyxiants are substances that displace oxygen from the air being breathed to such an extent that adverse effects result. Acetylene, carbon dioxide, argon, helium, ethane, nitrogen, and methane are common asphyxiants. It is thus important to recognize that even

chemically inert and biologically benign substances can be extremely dangerous under certain circumstances. Certain other chemicals have the ability to combine with hemoglobin, thus reducing the capacity of the blood to transport oxygen. Carbon monoxide, hydrogen cyanide, and certain organic and inorganic cyanides are examples of such substances.

3.C.2.5   Carcinogens

A carcinogen is a substance capable of causing cancer. Cancer, in the simplest sense, is the uncontrolled growth of cells, and it can occur in any organ. The mechanism by which cancer develops is not well understood, but the current thinking is that some chemicals interact directly with DNA, the genetic material in all cells, to result in permanent alterations. Other chemical carcinogens can modify DNA indirectly by changing the way the cells grow. Carcinogens are chronically toxic substances; that is, they cause damage after repeated or long-duration exposure, and their effects may become evident only after a long latency period. Carcinogens are particularly insidious toxins because they may have no immediate apparent harmful effects.

3.C.2.6   Reproductive and Developmental Toxins

Reproductive toxins are substances that have adverse effects on various aspects of reproduction, including fertility, gestation, lactation, and general reproductive performance. Developmental toxins are substances that act during pregnancy to cause adverse effects on the embryo or fetus. These effects can include lethality (death of the fertilized egg, the embryo, or the fetus), malformations (this class of substances is also called teratogens), retarded growth, and postnatal functional deficiencies. When a pregnant woman is exposed to a chemical, generally the fetus is exposed as well because the placenta is an extremely poor barrier to chemicals. Reproductive toxins can affect both men and women. Male reproductive toxins can in some cases lead to sterility. Two well-known male reproductive toxins are ethylene dibromide and dibromochloropropane.

3.C.2.7   Neurotoxins

Neurotoxic chemicals can induce an adverse effect on the structure or function of the central and/or peripheral nervous system, which can be permanent or reversible. In some cases the detection of neurotoxic effects may require specialized laboratory techniques, but often they can be inferred from behavior such as slurred speech and staggered gait. Many neurotoxins are chronically toxic substances whose adverse effects are not immediately apparent. At the present time, because of the limited data available in this area, significant uncertainties attend the assessment of risks associated with work with neurotoxic substances.

3.C.2.8   Toxins Affecting Other Organs

Target organs outside the reproductive and neurological systems can also be affected by toxic substances found in the laboratory. Most of the chlorinated hydrocarbons, benzene, other aromatic hydrocarbons, some metals, carbon monoxide, and cyanides, among others, can produce one or more effects in target organs. Such an effect may be the most probable result of exposure to the particular chemical. Although this chapter does not include specific sections on liver, kidney, lung, or blood toxins, many of the LCSSs mention those effects in the toxicology section.

3.C.3   Assessing Risks Due to the Toxic Effects of Laboratory Chemicals

The first step in assessing the risks associated with a planned laboratory experiment involves identifying which of the chemicals to be used in the proposed experiment are potentially hazardous substances. The OSHA Laboratory Standard (29 CFR 1910.1450) defines a hazardous substance as

> a chemical for which there is statistically significant evidence based on at least one study conducted in accordance with established scientific principles that acute or chronic health effects may occur in exposed employees. The term "health hazard" includes chemicals which are carcinogens, toxic or highly toxic agents, reproductive toxins, irritants, corrosives, sensitizers, hepatotoxins, nephrotoxins, neurotoxins, agents which act on the hematopoietic systems, and agents which damage the lungs, skin, eyes, or mucous membranes.

The OSHA Laboratory Standard further requires that certain chemicals be identified as "particularly hazardous substances" and handled using special additional procedures. Particularly hazardous substances include chemicals that are "select" carcinogens (those strongly implicated as a potential cause of cancer in humans), reproductive toxins, and compounds with a high degree of acute toxicity. Highly flammable and explosive substances make up another category of hazardous compounds, and the assessment of risk for these classes of chemicals is discussed in section 3.D. This section considers the assessment of risks associated with work with specific classes of toxic chemicals,

including those that pose hazards due to acute toxicity and chronic toxicity.

3.C.3.1   Acute Toxicants

Acute toxicity is the ability of a chemical to cause a harmful effect after a single exposure. Acutely toxic agents can cause local toxic effects, systemic toxic effects, or both, and this class of toxicants includes corrosive chemicals, irritants, and allergens (sensitizers). Among the most useful parameters for assessing the risk of acute toxicity of a chemical are its LD_{50} and LC_{50} values, selected with due regard for the possible routes of exposure. In interpreting these lethal dose and lethal concentration values, the following points should be considered. The LD_{50} is the *mean* dose causing death in animals, and it should be recognized that the *minimum* dose causing death in some proportion of the test population will be much lower, with significant illness or harm short of lethality probably occurring at even lower doses. Finally, it is assumed that the lethal dose for animals (usually rodents) is an appropriate predictor of the lethal dose in humans.

In assessing the risks associated with acute toxicants, it is useful to classify a substance according to the acute toxicity hazard level as shown in Table 3.1. LD_{50} values can be found in the Laboratory Chemical Safety Summary (LCSS) or MSDS for a given substance, and in references such as *Sax's Dangerous Properties of Industrial Materials* (Lewis, 1992), *Sigma-Aldrich Library of Chemical Safety Data* (Lenga, 1988), and *A Comprehensive Guide to the Hazardous Properties of Chemical Substances* (Patnaik, 1992). Table 3.2 relates test animal LD_{50} values expressed as milligrams or grams per kilogram of body weight to the probable human lethal dose, expressed in easily understood units, for a 70-kilogram (kg) person.

Special attention must be given to any substance classified according to the above criteria as having a high level of acute toxicity hazard. Chemicals with a high level of acute toxicity make up one of the categories of "particularly hazardous substances" defined by the OSHA Laboratory Standard. Any compound rated as highly toxic in Table 3.1 meets the OSHA criteria for handling as a particularly hazardous substance.

Table 3.3 lists some of the most common chemicals with a high level of acute toxicity that are encountered in the laboratory. These compounds must generally be handled using the additional procedures outlined in Chapter 5, section 5.D. In some circumstances, it may not be necessary to employ all of these special precautions, such as when the total amount of an acutely toxic substance to be handled is a small fraction of the harmful dose. It is an essential part of prudent experiment planning to determine whether a chemical with a high

TABLE 3.1 Acute Toxicity Hazard Level

Hazard Level	Toxicity Rating	Oral LD_{50} (Rats, per kg)	Skin Contact LD_{50} (Rabbits, per kg)	Inhalation LC_{50} (Rats, ppm for 1 h)	Inhalation LC_{50} (Rats, mg/m³ for 1 h)
High	Highly toxic	<50 mg	<200 mg	<200	<2,000
Medium	Moderately toxic	50 to 500 mg	200 mg to 1 g	200 to 2,000	2,000 to 20,000
Low	Slightly toxic	500 mg to 5 g	1 to 5 g	2,000 to 20,000	20,000 to 200,000

TABLE 3.2 Probable Lethal Dose for Humans

Toxicity Rating	Animal LD_{50} (per kg)	Lethal Dose When Ingested by 70-kg (150-lb) Human
Extremely toxic	Less than 5 mg	A taste (less than 7 drops)
Highly toxic	5 to 50 mg	Between 7 drops and 1 teaspoonful
Moderately toxic	50 to 500 mg	Between 1 teaspoonful and 1 ounce
Slightly toxic	500 mg to 5 g	Between 1 ounce and 1 pint
Practically nontoxic	Above 5 g	Above 1 pint

SOURCE: Modified, by permission, from Gosselin et al. (1984). Copyright 1984 by Williams & Wilkins, Baltimore.

degree of acute toxicity should be treated as a "particularly hazardous substance" in the context of a specific planned use. This determination not only will involve consideration of the total amount of the substance to be used, but also will require a review of the physical properties of the substance (e.g., is it volatile? does it tend to form dusts?), its potential routes of exposure (e.g., is it readily absorbed through the skin?), and the circumstances of its use in the proposed experiment (e.g., will the substance be heated? is there likelihood that aerosols may be generated?). Depending on the worker's level of experience and the degree of potential hazard, this determination may require consultation with supervisors and safety professionals.

Because the greatest risk of exposure to many laboratory chemicals is by inhalation, it is essential that laboratory workers understand the use of exposure limits that have been established by agencies such as ACGIH

TABLE 3.3 Examples of Compounds with a High Level of Acute Toxicity

Acrolein	Nickel carbonyl
Arsine	Nitrogen dioxide
Chlorine	Osmium tetroxide
Diazomethane	Ozone
Diborane (gas)	Phosgene
Hydrogen cyanide	Sodium azide
Hydrogen fluoride	Sodium cyanide
Methyl fluorosulfonate	(and other cyanide salts)

and OSHA. The threshold limit value (TLV), assigned by the ACGIH, defines the concentration of a chemical in air to which nearly all individuals can be exposed without adverse effects. The TLV-TWA (threshold limit value-time weighted average) refers to the concentration safe for exposure during an entire 8-h workday, while the TLV-STEL (threshold limit value-short term exposure limit) is a higher concentration to which workers may be exposed safely for a 15-min period. OSHA defines the permissible exposure limit (PEL) analogously to the ACGIH values, with corresponding TWA and STEL limits. TLV and PEL values allow the laboratory worker to quickly determine the relative inhalation hazards of chemicals. In general, substances with PELs or TLVs of less than 50 ppm should be handled in a fume hood. Comparison of these values to the odor threshold for a given substance will often indicate whether the odor of the chemical provides sufficient warning of possible hazard. However, individual differences in ability to detect some odors as well as anosmia, or "olfactory fatigue," can limit the usefulness of odors as warning signs of overexposure. LCSSs contain information on odor threshold ranges and whether a substance is known to cause olfactory fatigue. Finally, a variety of devices are available for measuring the concentration of chemicals in laboratory air, so that the degree of hazard associated with the use of a chemical can be assessed directly. The industrial hygiene offices of many institutions can assist laboratory workers in measuring the air concentrations of chemicals.

3.C.3.2 Corrosive Substances, Irritants, and Allergens

Lethal dose and other quantitative toxicological parameters generally provide little guidance in assessing the risks associated with corrosives, irritants, and allergens (sensitizers), because these toxic substances exert their harmful effects locally. When planning an experiment that will involve the use of corrosive substances, basic prudent handling practices should be reviewed to ensure that the skin, face, and eyes are protected adequately by the proper choice of corrosion-resistant gloves and protective clothing and eyewear, including, in some cases, face shields. Similarly, LD_{50} data are not an indicator of the irritant effects of chemicals, and therefore special attention should be paid to the identification of irritant chemicals by consulting LCSSs, MSDSs, and other sources of information. Allergens are another class of acute toxicants whose effects are not included in LD_{50} data. Individuals may differ widely in their tendency to become sensitized to allergens, so it is prudent to regard compounds with a proven ability to cause sensitization as highly toxic agents. Once a person has become sensitized to an allergen, subsequent contact can lead to immediate or delayed allergic reactions. Furthermore, sensitization to a specific substance can persist for many years. Because an allergic response can be triggered in a sensitized individual by an extremely small quantity of the allergen, it may occur despite personal protection measures that are adequate to protect against the acute effects of chemicals. Laboratory workers should be alert for signs of allergic responses to chemicals.

3.C.3.3 Carcinogens

Because cancer is such a widespread cause of human mortality, and because exposure to chemicals may play a significant role in the onset of cancer, a great deal of attention has been focused on evaluation of the carcinogenic potential of chemicals. However, the vast majority of the substances involved in research, especially in laboratories concerned primarily with the synthesis of novel compounds, have not been tested for carcinogenicity. Compounds that are known to pose the greatest carcinogenic hazard are referred to as "select carcinogens," and they constitute another category of substances that must be handled as "particularly hazardous substances" according to the OSHA Laboratory Standard. A select carcinogen is defined in the OSHA Laboratory Standard as a substance that meets one of the following criteria:

1. It is regulated by OSHA as a carcinogen.
2. It is listed as "known to be a carcinogen" in the latest *Annual Report on Carcinogens* issued by the National Toxicology Program (NTP) (U.S. DHHS, 1991).
3. It is listed under Group 1 ("carcinogenic to humans") by the International Agency for Research on Cancer (IARC).
4. It is listed under IARC Group 2A ("probably carcinogenic to humans") or 2B ("possibly carcinogenic to humans"), or under the category "reasonably anticipated to be a carcinogen" by the NTP, *and* causes statistically significant tumor incidence in experimental animals in accordance with any of the following criteria: (a) after inhalation exposure of 6 to 7 h per day, 5 days per week, for a significant portion of a lifetime to dosages of less than 10 mg/m^3; (b) after repeated skin application of less than 300 mg/kg of body weight per week; or (c) after oral dosages of less than 50 mg/kg of body weight per day.

Table 3.4 lists some representative substances that meet the above criteria for classification as OSHA select carcinogens. These chemicals are classified as particularly hazardous substances and should be handled using the basic prudent practices given in Chapter 5, section 5.C, supplemented by the additional special practices outlined in section 5.D. Work with compounds that are *possible* human carcinogens may or may not require the additional precautions given in section 5.D. For these compounds, the LCSS should indicate whether or not the substance meets the additional criteria listed in category 4 and must therefore be treated as a select carcinogen. If an LCSS is not available, consultation with a safety professional such as a chemical hygiene officer may be necessary in order to determine whether a possible human carcinogen should be classified as a particularly hazardous substance.

Many chemical substances are encountered in the laboratory for which there is no animal test or human epidemiological data on carcinogenicity. In these cases, workers must evaluate the potential risk that the chemical in question is a carcinogenic substance. This determination can sometimes be made on the basis of knowledge of the specific classes of compounds and functional group types that have previously been correlated with carcinogenic activity. For example, chloromethyl methyl ether is a known human carcinogen and therefore is regarded as an OSHA select carcinogen requiring the handling procedures outlined in section 5.D. On the other hand, the carcinogenicity of ethyl chloromethyl ether and certain other alkyl chloromethyl ethers is not established, and these substances do not necessarily have to be treated as select carcinogens. However, because of the chemical similarity of

TABLE 3.4   Examples of Select Carcinogens

2-Acetylaminofluorene	Dimethyl sulfate
Acrylamide	Ethylene dibromide
Acrylonitrile	Ethylene oxide
Aflatoxins	Ethylenimine
4-Aminobiphenyl	Formaldehyde
Arsenic and certain arsenic compounds	Hexamethylphosphoramide
Asbestos	Hydrazine
Azathioprine	Melphalan
Barium chromate	4,4'-Methylene-bis[2-chloroaniline]
Benzene	Mustard gas (bis(2-chloroethyl)sulfide)
Benzidine	N,N-Bis(2-chloroethyl)-2-naphthylamine
Bis(chloromethyl)ether	(chlornaphazine)
1,4-Butanediol dimethylsulfonate (myleran)	α-Naphthylamine
Chlorambucil	β-Naphthylamine
Chloromethyl methyl ether	Nickel carbonyl
Chromium and certain chromium compounds	4-Nitrobiphenyl
Cyclophosphamide	N-Nitrosodimethylamine
1,2-Dibromo-3-chloropropane	β-Propiolactone
3,3'-Dichlorobenzidine (and its salts)	Thorium dioxide
Diethylstilbestrol	Treosulfan
4-Dimethylaminoazobenzene	Vinyl chloride

NOTE: Compounds on this list are classified as select carcinogens on the basis of OSHA Laboratory Standard criteria. See accompanying text for details.

these compounds to chloromethyl methyl ether, it is possible that these substances have comparable carcinogenicity, and it is therefore prudent to regard them as select carcinogens requiring the special handling procedures outlined in section 5.D.

Table 3.5 lists important general classes of chemicals for which some members (but not necessarily all) have been identified as being carcinogenic substances. Listed for each general class are representative compounds that are "reasonably anticipated to be carcinogens" based on animal tests, selected from lists of substances identified as carcinogens or potential carcinogens by OSHA, IARC, and the *Annual Report on Carcinogens* (U.S. DHHS, 1991) published by the National Toxicology Program.

The determination of whether a suspected carcinogenic chemical must be treated as a "particularly hazardous substance" in the context of a particular laboratory use will be affected by the scale and circumstances associated with the intended experiment. The laboratory worker must decide whether the amount and frequency of use, as well as other circumstances, are such that additional precautions beyond the basic prudent practices of section 5.C are required. For example, the large-scale or recurring use of such a chemical might suggest that the special precautions of section 5.D be followed to control exposure, whereas adequate protection from a single use of a small amount of such a substance may be obtained through the use of the basic procedures in section 5.C.

When evaluating the carcinogenic potential of chemicals, it should be noted that exposure to certain combinations of compounds (not necessarily simultaneously) can cause cancer even at exposure levels where neither of the individual compounds would have been carcinogenic. 1,8,9-Trihydroxyanthracene and certain phorbol esters are examples of "tumor promoters." Although not carcinogenic themselves, they can dramatically amplify the carcinogenicity of other compounds. It should also be understood that the response of an organism to a toxicant typically increases with the dose given, but the relationship is not always a linear one. Some carcinogenic alkylating agents exhibit a dose threshold above which the tendency to cause mutations increases markedly. At lower doses, natural protective systems prevent genetic damage, but when the capacity of these systems is overwhelmed, the organism becomes much more sensitive to the toxicant. However, there are differences between individuals in the levels of protection against genetic damage as well as in other defense systems. These differences are determined in part by genetic factors and in part by the aggregate exposure of the individual to all chemicals within and outside of the laboratory.

3.C.3.4   *Reproductive and Developmental Toxins*

Reproductive toxins are defined by the OSHA Laboratory Standard as substances that cause chromosomal damage (mutagens) and substances with lethal or

TABLE 3.5　Classes of Carcinogenic Substances

Alkylating agents
 α-Halo ethers
  Bis(chloromethyl) ether
  Methyl chloromethyl ether
 Sulfonates
  1,4-Butanediol dimethanesulfonate (myleran)
  Diethyl sulfate
  Dimethyl sulfate
  Ethyl methanesulfonate
  Methyl methanesulfonate
  Methyl trifluoromethanesulfonate
  1,3-Propanesultone
 Epoxides
  Ethylene oxide
  Diepoxybutane
  Epichlorohydrin
  Propylene oxide
  Styrene oxide
 Aziridines
  Ethylenimine
  2-methylaziridine
 Diazo, azo, and azoxy compounds
  4-Dimethylaminoazobenzene
 Electrophilic alkenes and alkynes
  Acrylonitrile
  Acrolein
  Ethyl acrylate

Acylating agents
 β-Propiolactone
 β-Butyrolactone
 Dimethylcarbamyl chloride

Organohalogen compounds
 1,2-Dibromo-3-chloropropane
 Mustard gas (bis(2-chloroethyl)sulfide)
 Vinyl chloride
 Carbon tetrachloride
 Chloroform
 3-Chloro-2-methylpropene
 1,2-Dibromoethane
 1,4-Dichlorobenzene
 1,2-Dichloroethane
 2,2-Dichloroethane
 1,3-Dichloropropene
 Hexachlorobenzene
 Methyl iodide
 Tetrachloroethylene
 Trichloroethylene
 2,4,6-Trichlorophenol

Hydrazines
 Hydrazine (and hydrazine salts)
 1,2-Diethylhydrazine
 1,1-Dimethylhydrazine
 1,2-Dimethylhydrazine

N-Nitroso compounds
 N-Nitrosodimethylamine
 N-Nitroso-N-alkylureas

Aromatic amines
 4-Aminobiphenyl
 Benzidine (4, 4'-diaminobiphenyl)
 α-Naphthylamine
 β-Naphthylamine
 Aniline
 o-Anisidine (2-methoxyaniline)
 2,4-Diaminotoluene
 o-Toluidine

Aromatic hydrocarbons
 Benzene
 Benz[a]anthracene
 Benzo[a]pyrene

Natural products (including antitumor drugs)
 Adriamycin
 Aflatoxins
 Bleomycin
 Cisplatin
 Progesterone
 Reserpine
 Safrole

Miscellaneous organic compounds
 Formaldehyde (gas)
 Acetaldehyde
 1,4-Dioxane
 Ethyl carbamate (urethane)
 Hexamethylphosphoramide
 2-Nitropropane
 Styrene
 Thiourea
 Thioacetamide

Miscellaneous inorganic compounds
 Arsenic and certain arsenic compounds
 Chromium and certain chromium compounds
 Thorium dioxide
 Beryllium and certain beryllium compounds
 Cadmium and certain cadmium compounds
 Lead and certain lead compounds
 Nickel and certain nickel compounds
 Selenium sulfide

teratogenic (malformation) effects on fetuses. Many reproductive toxins are chronic toxins that cause damage after repeated or long-duration exposures with effects that become evident only after long latency periods. Developmental toxins act during pregnancy and cause adverse effects on the fetus; these effects include embryo lethality (death of the fertilized egg, embryo, or fetus), teratogenic effects, and postnatal functional

TABLE 3.6 Examples of Reproductive Toxins

Arsenic and certain arsenic compounds	Ethylene oxide
Benzene	Lead compounds
Cadmium and certain cadmium compounds	Mercury compounds
	Toluene
Carbon disulfide	Vinyl chloride
Ethylene glycol monomethyl and ethyl ethers	Xylene

defects. Embryotoxins have the greatest impact during the first trimester of pregnancy. *Because a woman often does not know that she is pregnant during this period of high susceptibility, women of childbearing potential are advised to be especially cautious when working with chemicals, especially those rapidly absorbed through the skin (e.g., formamide).* Pregnant women and women intending to become pregnant should seek advice from knowledgeable sources before working with substances that are suspected to be reproductive toxins. As minimal precautions, the general procedures outlined in Chapter 5, section 5.D, should then be followed for work with such compounds.

Information on reproductive toxins can be obtained from LCSSs, MSDSs, and by consulting safety professionals in the environmental safety department, industrial hygiene office, or medical department of the worker's institution. Literature sources of information on reproductive and developmental toxins include the *Catalog of Teratogenic Agents* (Shepard, 1992), *Reproductively Active Chemicals: A Reference Guide* (Lewis, 1991), and "What Every Chemist Should Know About Teratogens" in the *Journal of Chemical Education* (Beyler and Meyers, 1982). Table 3.6 lists some common materials that are suspected to be reproductive toxins. In some cases it will be appropriate to handle these compounds as particularly hazardous substances using the special additional precautions outlined in section 5.D.

3.D FLAMMABLE, REACTIVE, AND EXPLOSIVE HAZARDS

In addition to the hazards due to the toxic effects of chemicals, hazards due to flammability, explosibility, and reactivity need to be considered in risk assessment. These hazards are described in detail in the following sections. Further information can be found in *Bretherick's Handbook of Reactive Chemical Hazards* (Bretherick, 1990), an extensive compendium that is the basis for the lists of incompatible chemicals included in various reference works. Bretherick describes computational protocols that consider thermodynamic and kinetic parameters of a system to arrive at quantitative measures such as the Reaction Hazard Index (RHI). So-called "reactive" hazards arise when the release of energy from a chemical reaction occurs in quantities or at rates too great for the energy to be absorbed by the immediate environment of the reacting system, and material damage results. In addition, the "Letters to the Editor" column of *Chemical & Engineering News* routinely reports incidents with explosive reaction mixtures or conditions.

3.D.1 Flammable Hazards

3.D.1.1 Flammable Substances

Flammable substances, those that readily catch fire and burn in air, may be solid, liquid, or gaseous. The most common fire hazard in the laboratory is a flammable liquid or the vapor produced from such a liquid. An additional hazard is that a compound can enflame so rapidly that it produces an explosion. Proper use of substances that can cause fires requires knowledge of their tendencies to vaporize, ignite, or burn under the variety of conditions of use in the laboratory.

For a fire to occur, three conditions must exist simultaneously: an oxidizing atmosphere, usually air; a concentration of flammable gas or vapor that is within the flammable limits of the substance; and a source of ignition. In most situations, oxygen or air is present. Prevention of the coexistence of flammable vapors and an ignition source is the optimal way to deal with the hazard. When the vapors of a flammable liquid cannot always be controlled, strict control of ignition sources is the principal approach to reduction of the risk of flammability. The rates at which different liquids produce flammable vapors depend on their vapor pressures, which increase with increasing temperature. The degree of fire hazard of a substance depends also on its ability to form combustible or explosive mixtures with air and on the ease of ignition of these mixtures. Also important are the relative density and solubility of a liquid with respect to water and of a gas with respect to air. These characteristics can be evaluated and compared in terms of the following specific properties.

3.D.1.2 Flammability Characteristics

3.D.1.2.1 Flash Point

The flash point is the lowest temperature at which a liquid has a sufficient vapor pressure to form an ignitable mixture with air near the surface of the liquid. Note that many common organic liquids have a flash point below room temperature: for example, acetone (−18 °C), benzene (−11.1 °C), diethyl ether (−45 °C),

QUICK GUIDE
TO RISK ASSESSMENT FOR HAZARDOUS CHEMICALS

The following outline provides a summary of the steps discussed in this chapter that laboratory workers should use to assess the risks of handling toxic chemicals. Note that if a Laboratory Chemical Safety Summary is not already available, then following the protocol outlined here should enable a worker to prepare his or her own LCSS.

1. Identify chemicals to be used and circumstances of use. Identify the chemicals involved in the proposed experiment and determine the amounts that will be used. Is the experiment to be done once, or will the chemicals be handled repeatedly? Will the experiment be conducted in an open laboratory, in an enclosed apparatus, or in a fume hood? Is it possible that new or unknown substances will be generated in the experiment? Are any of the workers involved in the experiment pregnant or likely to become pregnant? Do they have any known sensitivities to specific chemicals?

2. Consult sources of information. Consult an up-to-date LCSS for each chemical involved in the planned experiment. Examine an up-to-date MSDS if an LCSS is not available. In cases where substances with significant or unusual potential hazards are involved, it may also be advisable to consult more detailed references such as Mackison et al. (U.S. DHHS, 1981), Patnaik (1992), *Patty's* (Clayton and Clayton, 1993), and other sources discussed in section 3.B. Depending on the worker's level of experience and the degree of potential hazard associated with the proposed experiment, it may also be necessary to obtain the assistance of supervisors and safety professionals before proceeding with risk assessment.

3. Evaluate type of toxicity. Use the above sources of information to determine the type of toxicity associated with each chemical involved in the proposed experiment. Are any of the chemicals to be used acutely toxic or corrosive? Are any of the chemicals to be used irritants or sensitizers? Will any select carcinogens or possibly carcinogenic substances be encountered? For many substances, it will be necessary to consult the listings of carcinogens in this chapter (see Tables 3.4 and 3.5) to identify chemical similarities to known carcinogens. Are any chemicals involved in the proposed experiment suspected to be reproductive or developmental toxins or neurotoxins?

4. Consider possible routes of exposure. Determine the potential routes of exposure for each chemical. Are the chemicals gases, or are they volatile enough to present a significant risk of exposure through inhalation? If liquid, can the substances be absorbed through the skin? Is it possible that dusts or aerosols will be formed in the experiment? Does the experiment involve a significant risk of inadvertent ingestion or injection of chemicals?

5. Evaluate quantitative information on toxicity. Consult the information sources to determine the LD_{50} for each chemical via the relevant routes of exposure. Determine the acute toxicity hazard level for each substance, classifying each chemical as highly toxic, moderately toxic, slightly toxic, and so forth. For substances that pose inhalation hazards, take note of the threshold limit value time-weighted average (TLV-TWA), short-term exposure limit (STEL), and permissible exposure limit (PEL) values.

6. Select appropriate procedures to minimize exposure. Use the "basic prudent practices for handling chemicals," which are discussed in Chapter 5, section 5.C, for all work with chemicals in the laboratory. In addition, determine whether any of the chemicals to be handled in the planned experiment meet the definition of a particularly hazardous substance due to high acute toxicity, carcinogenicity, and/ or reproductive toxicity. If so, consider the total amount of the substance that will be used, the expected frequency of use, the chemical's routes of exposure, and the circumstances of its use in the proposed experiment. As discussed in this chapter, use this information to determine whether it is appropriate to apply the additional procedures for work with highly toxic substances and whether additional consultation with safety professionals is warranted (see Chapter 5, section 5.D).

7. Prepare for contingencies. Note the signs and symptoms of exposure to the chemicals to be used in the proposed experiment. Note appropriate measures to be taken in the event of exposure or accidental release of any of the chemicals.

TABLE 3.7 NFPA Fire Hazard Ratings, Flash Points, Boiling Points, Ignition Temperatures, and Flammable Limits of Some Common Laboratory Chemicals

	NFPA Rating[a]	Flash Point (°C)	Boiling Point (°C)	Ignition Temperature (°C)	Flammable Limits (percent by volume)	
					Lower	Upper
Acetaldehyde	4	−37.8	21.1	175	4.0	60
Acetic acid (glacial)	2	39	118	463	4.0	19.9
Acetone	3	−18	56.7	465	2.6	12.8
Acetonitrile	3	6	82	524	3	16
Carbon disulfide	3	−30.0	46.1	90	1.3	50.0
Cyclohexane	3	−20.0	81.7	245	1.3	8.0
Diethylamine	3	−23	57	312	1.8	10.1
Diethyl ether	4	−45.0	35.0	160	1.9	36.0
Dimethyl sulfoxide	1	95	189	215	2.6	42
Ethyl alcohol	3	12.8	78.3	365	3.3	19.0
Heptane	3	−3.9	98.3	204	1.05	6.7
Hexane	3	−21.7	68.9	225	1.1	7.5
Hydrogen	4	—	−252	500	4	75
Isopropyl alcohol	3	11.7	82.8	398	2.0	12.0
Methyl alcohol	3	11.1	64.9	385	6.7	36.0
Methyl ethyl ketone	3	−6.1	80.0	515	1.8	10.0
Pentane	4	−40.0	36.1	260	1.5	7.8
Styrene	3	32.2	146.1	490	1.1	6.1
Tetrahydrofuran	3	−14	66	321	2	11.8
Toluene	3	4.4	110.6	480	1.2	7.1
p-Xylene	3	27.2	138.3	530	1.1	7.0

[a]0, will not burn; 1, must be preheated to burn; 2, ignites when moderately heated; 3, ignites at normal temperature; 4, extremely flammable.
SOURCE: Adapted from NFPA (1991b), pp. 325M-11 to 94.

and methyl alcohol (11.1 °C). The degree of hazard associated with a flammable liquid also depends on other properties, such as its ignition point and boiling point. Commercially obtained chemicals are now clearly labeled as to flammability and flash point. Consider the example of acetone given in section 3.C.1.3.1. At ambient pressure and temperature, an acetone spill can produce a concentration as high as 23.7% acetone in air. Acetone is not particularly toxic. However, with a flash point of − 18 °C and upper and lower flammable limits of 2.6% and 12.8% acetone in air, respectively (see Table 3.7), it is clear that an acetone spill produces an extreme fire hazard. Thus the major hazard given for acetone in the LCSS is flammability.

3.D.1.2.2 Ignition Temperature

The ignition temperature (autoignition temperature) of a substance, whether solid, liquid, or gaseous, is the minimum temperature required to initiate or cause self-sustained combustion independent of the heat source. The lower the ignition temperature, the greater the potential for a fire started by typical laboratory equipment. A spark is not necessary for ignition when

the flammable vapor reaches its autoignition temperature. For instance, carbon disulfide has an ignition temperature of 90 °C, and it can be set off by a steam line or a glowing lightbulb. Diethyl ether has an ignition temperature of 160 °C and can be ignited by the surface of a hot plate.

3.D.1.2.3 Limits of Flammability

Each flammable gas and liquid (as a vapor) has two fairly definite limits of flammability defining the range of concentrations in mixtures with air that will propagate a flame and cause an explosion. At the low extreme, the mixture is oxygen rich but contains insufficient fuel. The lower flammable limit (lower explosive limit (LEL)) is the minimum concentration (percent by volume) of the fuel (vapor) in air at which a flame is propagated when an ignition source is present. The upper flammable limit (upper explosive limit (UEL)) is the maximum concentration (percent by volume) of the vapor in air above which a flame is not propagated. The flammable range (explosive range) consists of all concentrations between the LEL and the UEL. This range becomes wider with increasing temperature and

HEALTH HAZARD
4 - Deadly
3 - Extreme danger
2 - Hazardous
1 - Slightly Hazardous
0 - Normal Material

FIRE HAZARD
Flash Points:
4 - Below 73 °F
3 - Below 100 °F
2 - Above 100 °F , Not
    Exceeding 200 °F
1 - Above 200 °F
0 - Will not burn

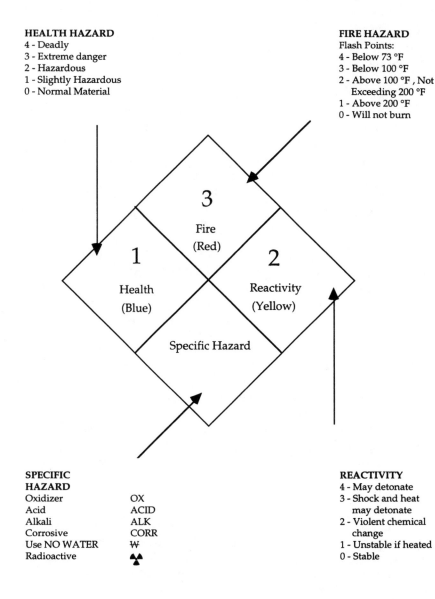

**SPECIFIC
HAZARD**
Oxidizer   OX
Acid   ACID
Alkali   ALK
Corrosive   CORR
Use NO WATER   W̶
Radioactive   ☢

REACTIVITY
4 - May detonate
3 - Shock and heat
    may detonate
2 - Violent chemical
    change
1 - Unstable if heated
0 - Stable

FIGURE 3.1 National Fire Protection Association system for classification of hazards. SOURCE: National Fire Protection Association (1990).

in oxygen-rich atmospheres and also changes depending on the presence of other components. The limitations of the flammability range, however, provide little margin of safety from the practical point of view because, when a solvent is spilled in the presence of an energy source, the LEL is reached very quickly and a fire or explosion will ensue before the UEL can be reached.

3.D.1.3 Classes of Flammability

Several systems are in use for classifying the flammability of materials. Some (e.g., Class I—flammable liquid, etc., see Chapter 4) apply to storage or transportation considerations. Another (Class A, B, C—paper, liquid, electrical fire) concerns the type of fire extinguisher to be used (see Chapter 6, section 6.F.2 on

emergency equipment). To assess risk quickly, the most direct indicator is the NFPA (National Fire Protection Association) system, which classifies flammables according to the severity of the fire hazard with numbers 0 to 4 in order of increasing hazard: 0, will not burn; 1, must be preheated to burn; 2, ignites when moderately heated; 3, ignites at normal temperature; 4, extremely flammable (Figure 3.1). Substances rated 3 or 4 under this system require particularly careful handling and storage in the laboratory. Some vendors include the NFPA hazard diamond on the labels of chemicals. The *Fire Protection Guide on Hazardous Materials* (NFPA, 1991) is a comprehensive listing of flammability data and ratings.

The NFPA fire hazard ratings, flash points, boiling points, ignition temperatures, and flammability limits of a number of common laboratory chemicals are given in Table 3.7 and in the LCSSs (see Appendix B). The data illustrate the range of flammability found for liquids commonly in use in laboratories. Dimethyl sulfoxide and glacial acetic acid (NFPA fire hazard ratings of 1 and 2, respectively) can be handled in the laboratory without great concern about their fire hazards. By contrast, both acetone (NFPA 3) and diethyl ether (NFPA 4) have flash points well below room temperature.

It should be noted, however, that tabulations of properties of flammable substances are based on standard test methods, which may have very different conditions from those encountered in practical laboratory use. Large safety factors should be applied. For example, the published flammability limits of vapors are for uniform mixtures with air. In a real situation, local concentrations that are much higher than the average may exist. Thus, it is good practice to set the maximum allowable concentration for safe working conditions at some fraction of the tabulated LEL; 20% is a commonly accepted value.

Among the most hazardous liquids are those that have flash points near or below 38 °C (100 °F) because these materials can be hazardous in the common laboratory environment. There is particular risk if their range of flammability is broad. It is important to note, as shown in Table 3.7, that some commonly used substances are potentially very hazardous, even under relatively cool conditions. Some flammable liquids will maintain their flammability even at concentrations of 10% by weight in water. Methanol and isopropyl alcohol have flash points below 38 °C (100 °F) at concentrations as low as 30% by weight in water. HPLC users generate acetonitrile/water mixtures that contain from 15 to 30% acetonitrile in water, a waste that is considered toxic and flammable and thus cannot be added to a sewer.

Because of its extreme flammability and tendency for peroxide formation, diethyl ether should be available for laboratory use only in metal containers. Carbon disulfide is almost as hazardous.

3.D.1.4 Causes of Ignition

3.D.1.4.1 Spontaneous Combustion

Spontaneous ignition (autoignition) or combustion takes place when a substance reaches its ignition temperature without the application of external heat. The possibility of spontaneous combustion should always be considered, especially when storing or disposing of materials. Examples of materials susceptible to spontaneous combustion include oily rags, dust accumulations, organic materials mixed with strong oxidizing agents (e.g., nitric acid, chlorates, permanganates, peroxides, and persulfates), alkali metals (e.g., sodium and potassium), finely divided pyrophoric metals, and phosphorus.

3.D.1.4.2 Ignition Sources

Potential ignition sources in the laboratory include the obvious torch and Bunsen burner, as well as a number of less obvious, electrically powered, sources ranging from refrigerators, stirring motors, and heat guns to microwave ovens (see section 6.C). Whenever possible, open flames should be replaced by electrical heating.

The vapors of most flammable liquids are heavier than air and capable of traveling considerable distances. This possibility should be recognized, and special note should be taken of ignition sources situated at a lower level than that at which the substance is being used. Flammable vapors from massive sources such as spills have been known to descend into stairwells and elevator shafts and ignite on a lower story. If the path of vapor within the flammable range is continuous, as along a floor or benchtop, the flame will propagate itself from the point of ignition back to its source. Metal lines and vessels discharging flammable substances should be bonded and grounded properly to discharge static electricity. There are many sources of static electricity, particularly in cold, dry atmospheres, and caution should be exercised.

3.D.1.4.3 Oxidants Other Than Oxygen

The most familiar fire involves a combustible material burning in air. However, the oxidant driving a fire or explosion need not be oxygen itself, depending on the nature of the reducing agent. All oxidants have the ability to accept electrons, and fuels are reducing agents or electron donors (see Young, 1991).

Examples of nonoxygen oxidants are shown in Table 3.8. When potassium ignites on being added to water, the metal is the reducing agent and water is the oxidant. If the hydrogen produced is ignited, it becomes the fuel for a conventional fire, with oxygen as the oxidant. In ammonium nitrate explosions, the ammonium cation is oxidized by the nitrate anion. These

TABLE 3.8 Examples of Oxidants

• Gases:	fluorine, chlorine, ozone, nitrous oxide, steam, oxygen
• Liquids:	hydrogen peroxide, nitric acid, perchloric acid, bromine, sulfuric acid, water
• Solids:	nitrites, nitrates, perchlorates, peroxides, chromates, dichromates, picrates, permanganates, hypochlorites, bromates, iodates, chlorites, chlorates

hazardous combinations are treated further in section 3.D.2.

(See Chapter 5, section 5.F, for a more detailed discussion on flammable substances.)

3.D.1.5 Special Hazards

Compressed or liquefied gases present hazards in the event of fire because the heat will cause the pressure to increase and the container may rupture (Braker and Mossman, 1980; Braker et al., 1988; Matheson Gas Products, 1983). Leakage or escape of flammable gases can produce an explosive atmosphere in the laboratory. Acetylene, hydrogen, ammonia, hydrogen sulfide, propane, and carbon monoxide are especially hazardous.

Even if not under pressure, a substance in the form of a liquefied gas is more concentrated than in the vapor phase and may evaporate extremely rapidly. Oxygen is an extreme hazard. Liquefied air is almost as dangerous because nitrogen boils away first, leaving an increasing concentration of oxygen. Liquid nitrogen standing for some time may have condensed enough oxygen to require careful handling. When a liquefied gas is used in a closed system, pressure may build up. Hence adequate venting is required. If the liquid is flammable (e.g., hydrogen and methane), explosive concentrations may develop without warning unless an odorant has been added. Flammability, toxicity, and pressure buildup may become more serious on exposure of gases to heat.

(Also see Chapter 5, section 5.G.2.5, for more information.)

3.D.2 Reactive Hazards

3.D.2.1 Water Reactives

Water reactive materials are those that react violently with water. Alkali metals (e.g., lithium, sodium, and potassium), many organometallic compounds, and some hydrides react with water to produce heat and flammable hydrogen gas, which can ignite or combine explosively with atmospheric oxygen. Some anhydrous metal halides (e.g., aluminum bromide), oxides (e.g., calcium oxide), and nonmetal oxides (e.g., sulfur trioxide) and halides (e.g., phosphorus pentachloride) react exothermically with water, and the reaction can be violent if there is insufficient coolant water to dissipate the heat produced.

(See Chapter 5, section 5.G, for further information.)

3.D.2.2 Pyrophorics

For pyrophoric materials, oxidation of the compound by oxygen or moisture in air proceeds so rapidly that ignition occurs. Many finely divided metals are pyrophoric, and their degree of reactivity depends on particle size, as well as factors such as the presence of moisture and the thermodynamics of metal oxide or metal nitride formation. Many other reducing agents, such as metal hydrides, alloys of reactive metals, low-valent metal salts, and iron sulfides, are also pyrophoric.

3.D.2.3 Incompatible Chemicals

Accidental contact of incompatible substances could result in a serious explosion or the formation of substances that are highly toxic or flammable or both. Many laboratory workers question the necessity of following storage compatibility guidelines. The reasons for such guidelines can be made obvious by reading descriptions of the condition of laboratories following California earthquakes in recent decades (see Pine, 1988, 1994). Those who do not live in seismically active zones should take these accounts to heart, as well. Other natural disasters and chemical explosions themselves can set off shock waves that empty chemical shelves and result in inadvertent mixing of chemicals.

Some compounds can pose either a reactive or a toxic hazard, depending on the conditions. Thus, hydrocyanic acid (HCN), when used as a pure liquid/gas in industrial applications, is incompatible with bases because it is stabilized against (violent) polymerization by the addition of acid inhibitor. HCN can also be formed when cyanide salt is mixed with an acid. In this case, the toxicity of hydrogen cyanide gas, rather than the instability of the liquid, is the characteristic of concern.

Some general guidelines can be applied to lessen the risks involved with these substances. Concentrated oxidizing agents are incompatible with concentrated reducing agents. Indeed, either may pose a reactive hazard even with chemicals that are not strongly oxidizing or reducing. For example, sodium or potassium, strong reducing agents frequently used to dry organic solvents, are extremely reactive toward halocarbon solvents (which are not strong oxidizing agents). Strong oxidizing agents are frequently used to clean glassware. Clearly, it is prudent to use such potent reagents only on the last traces of contaminating material. Tables 3.9 and 3.10 are guides to avoiding accidents involving incompatible substances. Chemicals or classes of chemicals in one column can be hazardous when mixed with those opposite them in the adjacent column. The magnitude of the risk obviously depends on quantities. In ordinary laboratory use, chemical incompatibilities will not usually pose much, if any, risk if the quantity of the substance is small (a solution in an NMR tube or a microscale synthesis). However, storage of commercially obtained chemicals (e.g., in

TABLE 3.9   Partial List of Incompatible Chemicals (Reactive Hazards)

Substances in the left hand column should be stored and handled so that they cannot accidentally contact corresponding substances in the right hand column under uncontrolled conditions.

Acetic acid	Chromic acid, nitric acid, peroxides, permanganates
Acetic anhydride	Hydroxyl-containing compounds such as ethylene glycol, perchloric acid
Acetone	Concentrated nitric and sulfuric acid mixtures, hydrogen peroxide
Acetylene	Chlorine, bromine, copper, silver, fluorine, mercury
Alkali and alkaline earth metals, such as sodium, potassium, lithium, magnesium, calcium, powdered aluminum	Carbon dioxide, carbon tetrachloride, other chlorinated hydrocarbons (also prohibit the use of water, foam, and dry chemical extinguishers on fires involving these metals—dry sand should be employed)
Ammonia (anhydrous)	Mercury, chlorine, calcium hypochlorite, iodine, bromine, hydrogen fluoride
Ammonium nitrate	Acids, metal powders, flammable liquids, chlorates, nitrites, sulfur, finely divided organics, combustibles
Aniline	Nitric acid, hydrogen peroxide
Bromine	Ammonia, acetylene, butadiene, butane, other petroleum gases, sodium carbide, turpentine, benzene, finely divided metals
Calcium oxide	Water
Carbon, activated	Calcium hypochlorite, other oxidants
Chlorates	Ammonium salts, acids, metal powders, sulfur, finely divided organics, combustibles
Chromic acid and chromium trioxide	Acetic acid, naphthalene, camphor, glycerol, turpentine, alcohol, other flammable liquids
Chlorine	Ammonia, acetylene, butadiene, butane, other petroleum gases, hydrogen, sodium carbide, turpentine, benzene, finely divided metals
Chlorine dioxide	Ammonia, methane, phosphine, hydrogen sulfide
Copper	Acetylene, hydrogen peroxide
Fluorine	Isolate from everything
Hydrazine	Hydrogen peroxide, nitric acid, any other oxidant
Hydrocarbons (benzene, butane, propane, gasoline, turpentine, etc.)	Fluorine, chlorine, bromine, chromic acid, peroxides
Hydrocyanic acid	Nitric acid, alkalis
Hydrofluoric acid (anhydrous) Hydrogen fluoride	Ammonia (aqueous or anhydrous)

(continued on facing page)

TABLE 3.9 Partial List of Incompatible Chemicals (Reactive Hazards) *(continued)*

Substances in the left hand column should be stored and handled so that they cannot accidentally contact corresponding substances in the right hand column under uncontrolled conditions.

Hydrogen peroxide	Copper, chromium, iron, most metals or their salts, any flammable liquid, combustible materials, aniline, nitromethane
Hydrogen sulfide	Fuming nitric acid, oxidizing gases
Iodine	Acetylene, ammonia (anhydrous or aqueous)
Mercury	Acetylene, fulminic acid,[a] ammonia
Nitric acid (concentrated)	Acetic acid, acetone, alcohol, aniline, chromic acid, hydrocyanic acid, hydrogen sulfide, flammable liquids, flammable gases, nitratable substances
Nitroparaffins	Inorganic bases, amines
Oxalic acid	Silver and mercury and their salts
Oxygen	Oils, grease, hydrogen, flammable liquids, solids, gases
Perchloric acid	Acetic anhydride, bismuth and its alloys, alcohol, paper, wood, grease, oils (all organics)
Peroxides, organic	Acids (organic or mineral), (also avoid friction, store cold)
Phosphorus (white)	Air, oxygen
Phosphorus pentoxide	Alcohols, strong bases, water
Potassium chlorate	Acids (see also chlorates)
Potassium perchlorate	Acids (see also perchloric acid)
Potassium permanganate	Glycerol, ethylene glycol, benzaldehyde, sulfuric acid
Silver and silver salts	Acetylene, oxalic acid, tartaric acid, fulminic acid,[a] ammonium compounds
Sodium	See alkali metals (above)
Sodium nitrite	Ammonium nitrate and other ammonium salts
Sodium peroxide	Any oxidizable substance, such as ethanol, methanol, glacial acetic acid, acetic anhydride, benzaldehyde, carbon disulfide, glycerol, ethylene glycol, ethyl acetate, methyl acetate, furfural
Sulfuric acid	Chlorates, perchlorates, permanganates

[a]Produced in nitric acid–ethanol mixtures.
SOURCE: Reproduced, by permission, from *Hazards in the Chemical Laboratory,* 4th edition, L. Bretherick, Ed. (1986).

500-g jars or 1-L bottles) should be carefully managed from the standpoint of chemical compatibility.

3.D.3 Explosive Hazards

3.D.3.1 Explosives

An explosive is any chemical compound or mechanical mixture that, when subjected to heat, impact, friction, detonation, or other suitable initiation, undergoes rapid chemical change, evolving large volumes of highly heated gases that exert pressure on the surrounding medium. The term applies to materials that either detonate or deflagrate. Heat, light, mechanical shock, and certain catalysts initiate explosive reactions. Hydrogen and chlorine react explosively in the presence of light. Acids, bases, and other substances catalyze the explosive polymerization of acrolein, and many metal ions can catalyze the violent decomposition of hydrogen peroxide. Shock-sensitive materials include acetylides, azides, nitrogen triiodide, organic nitrates, nitro compounds, perchlorate salts (especially those of heavy metals such as ruthenium and osmium), many organic peroxides, and compounds containing diazo, halamine, nitroso, and ozonide functional groups.

Table 3.11 lists a number of explosive compounds. Some are set off by the action of a metal spatula on the solid; some are so sensitive that they are set off by the action of their own crystal formation. Diazomethane (CH_2N_2) and organic azides, for example, may decompose explosively when exposed to a ground glass joint. The mechanisms of the explosions of nitroaromatic compounds have been reviewed by Brill and James (1993).

3.D.3.2 Peroxides

Organic peroxides are among the most hazardous substances handled in the chemical laboratory. They are generally low-power explosives that are sensitive to shock, sparks, or other accidental ignition. They are far more shock-sensitive than most primary explosives such as TNT.

Also potentially hazardous are compounds that undergo autooxidation to form organic hydroperoxides and/or peroxides when exposed to the oxygen in air (see Table 3.12). Especially dangerous are ether bottles that have evaporated to dryness. A peroxide present as a contaminant in a reagent or solvent can be very hazardous and change the course of a planned reaction. Autooxidation of organic materials (solvents and other liquids are most frequently of primary concern) proceeds by a free-radical chain mechanism. For the substrate R—H, the chain is initiated by ultraviolet

TABLE 3.10 Classes of Incompatible Chemicals

A incompatible with	B

Alkali and alkaline earth
  Carbides
  Hydrides
  Hydroxides
  Metals
  Oxides
  Peroxides
	Water
	Acids
	Halogenated organic compounds
	Halogenating agents
	Oxidizing agents[a]

Azides, inorganic	Acids Heavy metals and their salts Oxidizing agents[a]
Cyanides, inorganic	Acids Strong bases
Nitrates, inorganic	Acids Reducing agents[a]
Nitrites, inorganic	Acids Oxidizing agents[a]
Organic compounds Organic acyl halides	Oxidizing agents[a] Bases Organic hydroxy and amino compounds
Organic anhydrides	Bases Organic hydroxy and amino compounds
Organic halogen compounds	Group IA and IIA metals Aluminum
Organic nitro compounds	Strong bases

Oxidizing agents[a]
  Chlorates
  Chromates
  Chromium trioxide
  Dichromates
  Halogens
  Halogenating agents
  Hydrogen peroxide
  Nitric acid
  Nitrates
  Perchlorates
  Peroxides
  Permanganates
  Persulfates
	Reducing agents[a]
	Ammonia, anhydrous and
	  aqueous
	Carbon
	Metals
	Metal hydrides
	Nitrites
	Organic compounds
	Phosphorus
	Silicon
	Sulfur

Reducing agents[a]	Oxidizing agents[a] Arsenates Arsenites Phosphorus Selenites Selenates Tellurium salts and oxides
Sulfides, inorganic	Acids

[a] The examples of oxidizing and reducing agents are illustrative of common laboratory chemicals; they are not intended to be exhaustive.

TABLE 3.11 Functional Groups in Some Explosive Compounds

Structural Feature	Compound
—C≡C—	Acetylenic compound
—C≡C–M	Metal acetylide or carbide
—C≡C–X	Haloacetylide
$\diagdown$CN₂	Diazo compounds
$\diagup$C–N=O	Nitroso compounds
$\diagup$C–NO₂	Nitroalkanes, C-nitro and polynitroaryl compounds, polynitroalkyl compounds, trinitroethyl compounds
C–O–N=O	Acyl or alkyl nitrites
C–O–NO₂	Acyl or alkyl nitrates
C–O–O–C	Alkyl or acyl peroxides
$\diagup$C–O–O–H	Alkyl hydroperoxides
$\diagup$C–O–C(=O)–O–O–C$\diagup$	Dialkyl peroxycarbonates
CNO–M	Metal fulminates or *aci*-nitro salts, oximates
—N₃	Organic azides, acyl azides Metal azides, metal azide complexes
M(CO)ₙ	Transition metal–carbonyl compounds
—C≡N	Metal cyanides, organic nitriles, cyanogen halides

SOURCE: Adapted from Bretherick (1990), pp. S20–S22.

light, by the presence of a radical source, and by the peroxide itself. Oxygen adds to the R radical, producing the peroxy radical R—O—O. The chain is propagated when the peroxy radical abstracts a hydrogen atom from R—H. Excluding oxygen by storing potential peroxide-formers under an inert atmosphere (N_2 or argon) or under vacuum greatly increases their safe storage lifetime. In some cases, stabilizers or inhibitors (free-radical scavengers that terminate the chain reaction) have been added to the liquid to extend its storage lifetime. Because distillation of the stabilized liquid will remove the stabilizer, the distillate must be stored with care and monitored for peroxide formation.

Note that alkali metals and their amides may form peroxides on their surfaces. **Do not apply standard peroxide tests to such materials because they are both water and oxygen reactive!**

For purposes of managing the storage of chemicals that can form peroxides upon aging, the three classes given in Table 3.13 provide useful distinctions. As part of its Chemical Hygiene Plan (CHP), an institution should provide guidelines for handling these three classes. For example, if on-site incineration is available, disposal of chemicals in Class III after 3 months might be recommended. Various time limits for disposal of the different classes have been given.

3.D.3.3 Other Oxidizers

Oxidizing agents may react violently when they come into contact with reducing materials, and sometimes with ordinary combustibles. Such oxidizing agents include the halogens, oxyhalogens and organic peroxyhalogens, chromates, and persulfates as well as peroxides. Inorganic peroxides are generally stable. However, they may generate organic peroxides and hydroperoxides in contact with organic compounds, react violently with water (alkali metal peroxides), and form superoxides and ozonides (alkali metal peroxides). Perchloric acid is a powerful oxidizing agent with organic compounds and other reducing agents. Perchlorate salts can be explosive and should be treated as potentially hazardous compounds.

For many years, sulfuric acid–dichromate mixtures were used to clean glassware (a sulfuric acid–peroxydisulfate solution is now recommended because disposal of chromate is a problem). Confusion about cleaning baths has led to explosions on mixing potas-

TABLE 3.12 Types of Compounds Known to Autooxidize to Form Peroxides

- Aldehydes

- Ethers, especially cyclic ethers and those containing primary and secondary alkyl groups (*never* distill an ether before it has been shown to be free of peroxide)

- Compounds containing benzylic hydrogens

- Compounds containing allylic hydrogens (C=C—CH), including most alkenes; vinyl and vinylidene compounds

- Compounds containing a tertiary C—H group (e.g., decalin and 2,5-dimethylhexane)

TABLE 3.13   Classes of Chemicals That Can Form Peroxides Upon Aging

Class I: Unsaturated materials, especially those of low molecular weight, may polymerize violently and hazardously due to peroxide initiation.

Acrylic acid	Tetrafluoroethylene
Acrylonitrile	Vinyl acetate
Butadiene	Vinyl acetylene
Chlorobutadiene (chloroprene)	Vinyl chloride
Chlorotrifluoroethylene	Vinyl pyridine
Methyl methacrylate	Vinylidene chloride
Styrene	

Class II: The following chemicals are a peroxide hazard upon concentration (distillation/evaporation). A test for peroxide should be performed if concentration is intended or suspected.

Acetal	Dioxane (*p*-dioxane)
Cumene	Ethylene glycol dimethyl ether (glyme)
Cyclohexene	Furan
Cyclooctene	Methyl acetylene
Cyclopentene	Methyl cyclopentane
Diacetylene	Methyl-*i*-butyl ketone
Dicyclopentadiene	Tetrahydrofuran
Diethylene glycol dimethyl ether (diglyme)	Tetrahydronaphthalene
Diethyl ether	Vinyl ethers

Class III: Peroxides derived from the following compounds may explode without concentration.

Organic	Inorganic
Divinyl ether	Potassium metal
Divinyl acetylene	Potassium amide
Isopropyl ether	Sodium amide (sodamide)
Vinylidene chloride	

NOTE: Lists are illustrative but not exhaustive.

sium permanganate with sulfuric acid and nitric acid with alcohols.

3.D.3.4   Dusts

Suspensions of oxidizable particles (e.g., flour, coal dust, magnesium powder, zinc dust, carbon powder, and flowers of sulfur) in the air can constitute a powerful explosive mixture. These materials should be used with adequate ventilation and should not be exposed to ignition sources. Some solid materials, when finely divided, are spontaneously combustible if allowed to dry while exposed to air. These materials include zirconium, titanium, Raney nickel, finely divided lead (such as prepared by pyrolysis of lead tartrate), and catalysts such as activated carbon containing active metals and hydrogen.

3.D.3.5   Explosive Boiling

Not all explosions result from chemical reactions. A dangerous, physically caused explosion can occur if a hot liquid or a collection of very hot particles comes into sudden contact with a lower-boiling-point material. Sudden boiling eruptions occur when a nucleating agent (e.g., charcoal, "boiling chips") is added to a liquid heated above its boiling point. Even if the material does not explode directly, the sudden formation of a mass of explosive or flammable vapor can be very dangerous.

3.D.3.6   Other Considerations

The hazards of running a new reaction should be considered especially carefully if the chemical species involved contain functional groups associated with explosions (see Table 3.11) or are unstable near the reaction or work-up temperature, if the reaction is subject to an induction period, or if gases are by-products. Modern analytical techniques (see Chapter 5, section 5.G) can be used to determine reaction exothermicity under suitable conditions.

Even a small sample may be dangerous. Furthermore, the hazard is associated not with the total energy released, but rather with the remarkably high rate of a detonation reaction. A high-order explosion of even milligram quantities can drive small fragments of glass or other matter deep into the body. It is important to use minimum amounts of these hazardous materials with adequate shielding and personal protection. A compound is apt to be explosive if its heat of formation is more than about 100 calories per gram (cal/g) less than the sum of the heats of formation of its products. In making this calculation, a reasonable reaction should be used in order to yield the most exothermic products.

Scaling up reactions can introduce several hazards. The current use of microscale teaching methods in undergraduate laboratories unfortunately increases the likelihood that graduate students and others may be unprepared for a number of problems that can arise when a reaction is run on a larger scale. These include heat buildup and serious hazard of explosion from the use of incompatible materials. The rate of heat input and production must be weighed against that of heat removal. Bumping of the solution or a runaway reaction can result when heat builds up too rapidly. Exothermic reactions can "run away" if the heat evolved is not dissipated. When scaling up experiments, sufficient cooling and surface for heat exchange should be provided, and mixing and stirring rates should be considered. Detailed guidelines for circumstances that require a systematic hazard evaluation and thermal analysis are given in Chapter 5, section 5.G.

Another situation that can lead to problems is a reaction susceptible to an induction period; particular care must be given to the rate of reagent addition versus its rate of consumption. Finally, the hazards of exothermic reactions or unstable or reactive chemicals are exacerbated under extreme conditions, such as high temperature or high pressure used for hydrogenations, oxygenations, or work with supercritical fluids.

3.D.4 The Dirty Dozen

In laboratories carrying out moderate- to large-scale synthetic chemistry, it is generally recognized that certain substances tend to be responsible for more than their share of accidents (see also Chapter 5, section 5.G.6). In some laboratories these perennial "bad actors" are known as the "Dirty Dozen" (see Table 3.14). Although accident statistics for such laboratories show that most accidents lead to cut hands and back injuries (Kaufmann, 1990), enough workers have had incidents with these elements and compounds to make extreme caution advisable. Inappropriate mixing or handling of certain compounds can also produce hazardous toxic gases. Institutions might find it useful to prepare their own lists as part of their Chemical Hygiene Plans.

3.E PHYSICAL HAZARDS

3.E.1 Compressed Gases

Compressed gases can expose the worker to both mechanical and chemical hazards, depending on the gas. Hazards can result from the flammability, reactivity, or toxicity of the gas, from the possibility of asphyxiation, and from the gas compression itself, which could lead to a rupture of the tank or valve.

3.E.2 Nonflammable Cryogens

Nonflammable cryogens (chiefly liquid nitrogen) can cause tissue damage from extreme cold because of contact with either liquid or boil-off gases. In poorly ventilated areas, inhalation of gas due to boil-off or spills can result in asphyxiation. Another hazard is explosion from liquid oxygen condensation in vacuum traps or from ice plug formation or lack of functioning vent valves in storage Dewars. Because 1 volume of liquid nitrogen at atmospheric pressure vaporizes to 694 volumes of nitrogen gas at 20 °C, the warming of such a cryogenic liquid in a sealed container produces enormous pressure, which can rupture the vessel.

(See Chapter 5, section 5.G, for detailed discussion.)

3.E.3 High-Pressure Reactions

Experiments carried out at pressures above one atmosphere can lead to explosion from equipment failure. Hydrogenation reactions are frequently carried out at elevated pressures. A potential hazard is the formation of explosive O_2/H_2 mixtures and the reactivity/pyrophoricity of the catalyst (see section 3.D). High pressures can also be associated with the growing use of supercritical fluids (see McHugh and Krukonis, 1994; Bright and McNally, 1992).

3.E.4 Vacuum Work

Precautions to be taken when working with vacuum lines and other glassware used at subambient pressure are mainly concerned with the substantial danger of injury in the event of glass breakage. The degree of hazard does not depend significantly on the magnitude of the vacuum because the external pressure leading to implosion is always one atmosphere. Thus, evacuated systems using aspirators merit as much respect as high-

TABLE 3.14 The "Dirty Dozen"

1. Organic azides	Explosion hazards, especially with ground glass joints
2. Perchlorate salts of organic, organometallic, and inorganic complexes	Explosion hazards
3. Diethyl ether	Fires (see also entry 10 below)
4. Lithium aluminum hydride	Fires on quenching
5. Sodium, potassium	Fires on quenching
6. Potassium metal	Fires on quenching
7. Sodium–benzophenone ketyl still pots	Fires on quenching
8. Palladium on carbon	Fires on removal from the inert atmosphere, especially if wet with organic solvent or when contacting combustible materials such as filter paper
9. Heat	Exothermic reactions causing violent spills on scale-up due to inadequate provision for heat removal
10. Ethers with α-hydrogen atoms	Dangerous peroxide concentration during distillation; explosion hazards, especially with ground glass joints
11. Carbon monoxide	Toxicity and role in forming nickel tetracarbonyl from steel gas lines and autoclaves
12. Organic peroxides	Sensitivity to shock, sparks, and other forms of accidental detonation; sensitivity to heat, friction, impact, and light, as well as to strong oxidizing and reducing agents

vacuum systems. Injury due to flying glass is not the only hazard in vacuum work. Additional dangers can result from possible toxicity of the chemicals contained in the vacuum system, as well as from fire following breakage of a flask (e.g., of a solvent stored over sodium or potassium).

Because vacuum lines typically require cold traps (generally liquid nitrogen) between the pumps and the vacuum line, precautions regarding the use of cryogens should be observed also. Health hazards associated with vacuum gauges have recently been reviewed (Peacock, 1993). The hazards include the toxicity of mercury used in manometers and McLeod gauges, overpressure and underpressure situations arising with thermal conductivity gauges, electric shock with hot cathode ionization systems, and the radioactivity of the thorium dioxide used in some cathodes.

3.E.5 Ultraviolet, Visible, and Near-Infrared Radiation

Ultraviolet, visible, and near-infrared radiation from lamps and lasers in the laboratory can produce a number of hazards. Medium-pressure Hanovia 450 Hg lamps are commonly used for ultraviolet irradiation in photochemical experiments. Powerful arc lamps can cause eye damage and blindness within seconds. Some compounds, for example, chlorine dioxide, are explosively photosensitive.

When incorrectly used, the ultraviolet, visible, or near-infrared light from lasers poses a hazard to the eyes of the operators and other people present in the room and is also a potential fire hazard. Depending on the type of laser, the associated hazards can include mutagenic, carcinogenic, or otherwise toxic laser dyes and solvents or flammable solvents, ultraviolet or visible radiation from the pump lamps, and electric shock from power supplies for lamps.

Lasers are classified according to their relative hazards: Class I lasers, including laser printers, compact disc players, and *unfocused* laser diodes, are either completely enclosed or have such a low output of power that even a direct beam in the eye could not cause damage. Class II lasers, including supermarket scanners and visible laser bar code scanners, are visible light lasers with power of less than 1 milliwatt (mW). These can be a hazard if a person stares into the beam and resists the natural reaction to blink or turn away. Class IIIA lasers have powers between 1 and 5 mW and can present an eye hazard if a person stares into the beam and resists the natural reaction to blink or turn away, or views the beam with focusing optical instruments. Class IIIB lasers are visible, ultraviolet, and infrared lasers with powers in the 5 to 500 mW range and produce eye injuries instantly from both direct and specularly reflected beams. Class IV lasers are visible, ultraviolet, and infrared lasers with continuous powers in excess of 500 mW or pulse energies in excess of a threshold that depends on wavelength and pulse duration. Class IV lasers present all of the hazards of Class III lasers and may also produce eye or skin damage from diffuse scattered light. Anyone who is not the authorized operator of a laser system should never enter a posted laser-controlled laboratory if the laser is in use.

3.E.6 Radiofrequency and Microwave Hazards

Radiofrequency (RF) and microwaves occur within the range 10 kilohertz (kHz) to 300,000 megahertz (MHz) and are used in RF ovens and furnaces, induction heaters, and microwave ovens. Extreme overexposure to microwaves can result in the development of cataracts and/or sterility. Microwave ovens are increasingly being used in laboratories for organic synthesis and digestion of analytical samples. Use of metal in microwave ovens can result in arcing and, if a flammable solvent is present, in fire or explosion. Superheating of liquids can occur. Capping of vials and other containers used in the oven can result in explosion from pressure buildup within the vial. Inappropriately selected plastic containers may melt.

3.E.7 Electrical Hazards

The electrocution hazards of electrically powered instruments, tools, and other equipment can almost be eliminated by taking reasonable precautions, and the presence of electrically powered equipment in the laboratory need not pose a significant risk. Many electrically powered devices are used in homes and workplaces in the United States, often with little awareness of the safety features incorporated in their design and construction. But, in the laboratory, as well as elsewhere, it is critical that these features not be defeated by thoughtless or ignorant modification. The possibility of serious injury or death by electrocution is a very real one if careful attention is not paid to engineering, maintenance, and personal work practices. Equipment malfunctions can lead to electrical fires. Every worker should know the location of electrical shutoff switches and/or circuit breaker switches and should know how to turn off power to burning equipment by using these switches.

Some special concerns arise in laboratory settings. The insulation on wires can be eroded by corrosive chemicals, organic solvent vapors, or ozone (from ultraviolet lights, copying machines, and so forth). Eroded insulation on electrical equipment in wet locations such as cold rooms or cooling baths must be repaired immediately. In addition, sparks from electrical equipment can serve as an ignition source in the presence of flammable vapor. Operation of certain equipment (e.g., lasers, electrophoresis equipment) may involve high voltages and stored electrical energy. The large capacitors used in many flash lamps and other systems are capable of storing lethal amounts of electrical energy and should be regarded as "live" even if the power source has been disconnected.

Loss of electrical power can produce extremely hazardous situations. Flammable or toxic vapors may be released from freezers and refrigerators as chemicals stored there warm up; certain reactive materials may decompose energetically upon warming. Hoods may cease to function and to protect workers. Stirring (motor or magnetic) required for safe reagent mixing may cease. Return of power to an area containing flammable vapors may ignite them.

3.E.8 Magnetic Fields

Increasingly, instruments that generate large static magnetic fields (e.g., frequently, NMR spectrometers) are present in research laboratories. Such magnets typically have fields of 25,000 to 160,000 gauss (2.5 to 16 teslas), far above Earth's magnetic field, which is about 0.5 G. The magnitude of these large static magnetic fields falls off rapidly with distance, which is fortunate, because effects on magnetic media such as credit cards and computer disks are thus limited (see Chapter 6, Table 6.1). Strong attraction occurs when the magnetic field is above 50 to 100 G and increases by the seventh power as the separation is reduced. However, this highly nonlinear falloff of magnetic field with distance results in an insidious hazard. Objects made of ferromagnetic materials such as ordinary steel may be scarcely affected beyond a certain distance but at a slightly shorter distance may experience a significant attraction to the field. If the object is able to move still closer, the attractive force increases rapidly, and the object can become a projectile aimed at the magnet. Objects ranging from scissors, knives, wrenches, and other tools and keys to oxygen cylinders, buffing machines, and wheelchairs have been pulled from a considerable distance to the magnet itself.

Superconducting magnets use liquid nitrogen and liquid helium coolants. Thus, the hazards associated with cryogenic liquids (see section 3.E.2) are of concern, as well.

There is no epidemiological evidence that exposure to static magnetic fields results in adverse effects on human health (Persson and Stahlberg, 1989; Budinger, 1992). The health effects of electromagnetic fields remain unresolved (Hileman, 1993). The effects of electromagnetic fields on protein biosynthesis, similar to those seen in response to heat shock, and the response of cells to changes in electrical stimulation have been reported (Blank, 1983).

3.E.9   Cuts, Slips, Trips, and Falls

Among the most common injuries in laboratories are back injuries and injuries arising from broken glass and from slipping or tripping. Cuts can be minimized by the use of correct procedures (e.g., the procedure for inserting glass tubing into rubber stoppers and tubing, which is taught in introductory laboratories), through the appropriate use of protective equipment, and by careful attention to manipulation. Spills resulting from dropping chemicals not stored in protective rubber buckets or laboratory carts can be serious because the worker can fall or slip into the spilled chemical, thereby risking injury from both the fall and exposure to the chemical. Chemical spills resulting from tripping over bottles of chemicals stored on laboratory floors are part of a general pattern of bad housekeeping that can also lead to serious accidents. Wet floors around ice, dry ice, or liquid nitrogen dispensers can be slippery if the areas are not carpeted and if drops or small puddles are not wiped up as soon as they form. Attempts to retrieve 5-gallon bottles of distilled water, jars of bulk chemicals, and rarely used equipment stored on high shelves have often led to back injuries in laboratory environments. Careful planning of where to store difficult-to-handle equipment and containers (because of weight, shape, or overall size) can therefore be expected to reduce the incidence of back injuries.

3.F   BIOHAZARDS

Biohazards are a concern in laboratories in which microorganisms or material contaminated with them is handled. These hazards are usually present in clinical and infectious disease research laboratories, but may also be present in any laboratory in which bodily fluids or tissues of human or animal origin are handled. Occasionally, biohazards are present in testing and quality control laboratories, particularly those associated with water and sewage treatment plants and facilities involved in the production of biological products and disinfectants. Teaching laboratories may introduce low-risk infectious agents as part of a course of study in microbiology for advanced students.

A consensus code of practice for controlling biohazards, *Biosafety in Microbiological and Biomedical Laboratories*, was first produced by the Centers for Disease Control and Prevention and the National Institutes of Health in 1984; the third and most recent edition was published in 1993 (U.S. DHHS, 1993).

(Also see Chapter 5, section 5.E.)

3.G   HAZARDS FROM RADIOACTIVITY

The discussion in this section provides a brief primer on the hazards arising from radioactivity. A comprehensive treatment of radiation laboratory safety is given in Shapiro (1990).

Unstable atomic nuclei eventually achieve a more stable form by emission of some type of radiation. These nuclei or isotopes are termed radioactive. The energy emitted from a decaying nucleus may be alpha, beta, or gamma particles or electromagnetic radiation gamma rays or x-rays, as discussed below. Radiation that has enough energy to ionize atoms into ions and electrons is denoted ionizing radiation. Ionizing radiation can also be produced by machines such as particle accelerators and x-ray machines.

- Alpha particles are charged particles containing two protons and two neutrons and are emitted from certain heavy atoms such as uranium and thorium. An alpha particle can be stopped by a sheet of paper but is very damaging inside the body.
- Beta particles are electrons emitted with very high energy from many radioisotopes. Positively charged counterparts of beta particles are called positrons. Positronic and electron emissions from radioactive atoms can be shielded by thin metal foils or one-quarter inch of plastic. Tritium (^3H), phosphorus-32, and carbon-14

are beta emitters. Beta particles are usually stopped by the skin but can cause serious damage to skin and eyes.

• Gamma rays and x-rays, extremely energetic photons, have no mass or charge. Gamma rays are generally emitted from the nucleus during nuclear decay, and x-rays are emitted from the electron shells. Gamma rays are also produced by particle accelerators and nuclear reactors. Extremely dense materials such as lead or depleted uranium are required to shield against these very energetic, penetrating forms of radiation.

• Neutrons, uncharged particles, are emitted from the nucleus during decay. Shielding materials for neutrons include water, paraffin, boron, and concrete.

Radioactive decay rates are reported in curies (1 curie (Ci) = 3.7×10^{10} disintegrations per second) or in the International System of Units (SI) in becquerels (Bq) (1 becquerel = 1 disintegration per second). The decay rate provides a characterization of a given source, but provides no absolute guide as to the hazard of the material. The hazard depends on the nature of, as well as the rate of production of, the ionizing radiation. In characterizing human exposure to ionizing radiation, it is assumed that the damage is proportional to the energy absorbed. The radiation absorbed dose (rad) is defined in terms of energy absorbed per unit mass: 1 rad = 100 ergs/g (SI: 1 gray (Gy) = 1 joule/kg = 100 rads). For electromagnetic energy, the roentgen (R) produces 1.61×10^{12} ion pairs per gram of air (SI: 1 coulomb/kg = 3.876 R).

For evaluation of the risk of exposure to ionizing radiation in humans, the dose equivalent in rem (roentgen equivalent man) is defined as

$$rem = rads \times Q \times N$$

where the absorbed dose is given in rads, Q is the quality factor, and N is the tissue factor. Q is 1 for x-rays and gamma radiation of any energy, and for beta radiation. For alpha radiation, Q is 20. For neutrons, Q is 2 to 10, depending on their energy. In the United States, the applicable *Standards for Protection Against Radiation from Sealed Gamma Sources* (U.S. National Committee on Radiation Protection and Measurements, 1960), defines dose equivalents as follows: for x-ray, gamma ray, and electron radiations, $Q \times N = 1$ and so 1 rad = 1 rem; for neutrons or high-energy protons, $Q \times N = 10$ and 1 rem = 0.1 rad.

Damage may occur directly as a result of the radiation interacting with a part of the cell or indirectly by the formation of toxic substances within the cell. The extent of damage incurred depends on many factors, including the dose rate, the size of the dose, and the site of exposure. Effects may be short-term or long-term. The acute short-term effects associated with large doses and high dose rates, for example, 100,000 mrads (100 rads) in less than 1 week, may include nausea, diarrhea, fatigue, hair loss, sterility, and easy bruising. In appropriately managed workplaces, such exposures are impossible unless various barriers, alarms, and other safety systems are deliberately destroyed or bypassed. Above 600 rads, all exposures are probably fatal. Long-term effects, which develop years after the exposure, are primarily observed as cancer. Exposure of the fetus in utero to radiation is of concern, and the risk of damage to the fetus increases significantly when doses exceed 15,000 mrems. The U.S. Nuclear Regulatory Commission has set limits for whole-body occupational exposure at 500 mrems per quarter and 2,000 mrems/year and recommends that student exposures not exceed 500 mrems/year. Exposure limits are lower in facilities operated by the Department of Energy and other agencies. No completely safe limit of exposure is known.

As with all laboratory work, protection of the worker against the hazard consists of good facility design, operation, and monitoring, as well as good work practices on the part of the worker. The ALARA (as low as reasonably achievable) exposure principle is central to both levels of protection. The amount of radiation or radioactive material used should be minimized. Exposures should be minimized by shielding radiation sources and workers and visitors and by use of emergency alarm and evacuation procedures. Physical distance between personnel and radiation sources should be maximized, and whenever possible, robotic or other remote operations should be used to reduce exposure of personnel.

(Also see Chapter 5, section 5.E.)

4 Management of Chemicals

4.A INTRODUCTION

This chapter organizes the discussion of managing laboratory chemicals into five main topics: source reduction, acquisition, inventory and tracking, storage in stockrooms and laboratories, and recycling of chemicals and laboratory materials. As Chapter 1 makes clear, the concept of prudence in these areas requires knowledge of the hazards posed by laboratory chemicals and the formulation of reasonable measures to control and minimize the risks associated with their handling and disposal. It is not possible to eliminate risk altogether, but through informed risk assessment and careful risk management, laboratory safety can be greatly enhanced.

Laboratory workers, laboratory supervisors, and individuals who handle chemicals all will find essential information in this chapter. Each of these people has an important role to play in a chemical's life cycle at an institution, and each one of them should be aware that the wise management of that life cycle can not only minimize risks to humans and to the environment, but also decrease costs.

4.B SOURCE REDUCTION

Prudent management of chemicals in laboratories must begin long before the actual arrival of the chemicals. When experiments have been carefully planned, laboratory workers can be confident that they have chosen the procedures for working with chemicals that meet the following goals:

- to minimize quantities of chemicals to be used,
- to minimize disposal of hazardous materials, and
- to minimize risks.

Strategies for achieving the first three goals generally also are effective in achieving a fourth:

- to minimize exposure of laboratory workers and storeroom and receiving personnel to hazardous materials.

4.B.1 Importance of Minimizing Chemical Orders

In order to cut costs, manufacturing firms are increasingly asking for "just-in-time" delivery of raw materials. Laboratories might well borrow this strategy. A quantity of hazardous chemical not ordered is one to which workers are not exposed, for which appropriate storage need not be found, which need not be tracked in an inventory control system, and

which will not end up requiring costly disposal when it becomes a waste.

In acquiring a chemical, it is important to do a life cycle analysis. All costs associated with the progress of each chemical through its lifetime at an institution must be considered. The purchase cost is only the beginning; the handling costs, human as well as financial, and the disposal costs must be taken into account as well. Without close attention to this aspect of managing chemicals in laboratories, orders are not likely to be minimized and unused chemicals can become a significant fraction of the laboratory's hazardous waste.

Institutions also need to minimize the amount of chemical accepted as a gift or as part of a research contract. More than one laboratory has been burdened with the cost of disposing of a donated chemical that was not needed. A "free" material can become a significant liability. The American Chemical Society's booklet *Less Is Better: Laboratory Chemical Management for Waste Reduction* (ACS, 1993) gives several reasons for ordering chemicals in smaller containers, even if that means using several containers of a material for a single experiment:

- The risk of breakage is substantially reduced for small package sizes.
- The risk of accident and exposure to the hazardous material is less when handling smaller containers.
- Storeroom space needs are reduced when only a single size is inventoried.
- Containers are emptied faster, resulting in less chance for decomposition of reactive compounds.
- The large "economy size" often dictates a need for other equipment, such as smaller transfer containers, funnels, pumps, and labels. Added labor to subdivide the larger quantities into smaller containers, as well as additional personal protective equipment for the hazards involved, also may be needed.
- If unused hazardous material must be disposed

> Donated material can easily become a liability. A chemical engineering researcher accepted a 55-gallon drum of an experimental diisocyanate as part of a research contract. The ensuing research project used less than 1 gallon of the material, and the grantor would not take the material back for disposal. No commercial incinerator would handle the material in its bulk form. The remaining material had to be transferred to 1-liter containers and sent as Lab Packs for disposal, at a cost of $4,000 to $5,000.

of, the disposal cost per container is less for smaller containers.

Later in this chapter (section 4.D.2), the exchange or transfer of chemicals to other laboratory workers is discussed. The use of smaller containers increases the chance that chemicals to be transferred will still be in sealed containers, which increases the receiver's confidence that the chemicals are pure.

4.B.2 Strategies to Minimize Hazardous Waste Generation

Experimental design and execution are central in strategies to minimize the generation of hazardous waste, just as they were in the section above. The design should evaluate all potential sources of hazardous waste expected from the proposed experiment and incorporate strategies to minimize those sources. Examples of such strategies include

• carrying out chemical reactions and other laboratory procedures on a smaller scale;
• considering the use to which a reaction product will be put and then making only the amount needed for that use;
• appreciating the price that may be paid for making and storing an unneeded material;
• thinking about minimization of material used in each step of an experiment;
• improving yields;
• using less solvent to rinse equipment, for example, by carrying out several rinses with small volumes of solvent, rather than using only one or two rinses with larger volumes;
• using more sensitive analytical equipment;
• substituting nonhazardous, or less hazardous, chemicals where possible by considering alternate synthetic routes and alternate procedures for working up reaction mixtures;
• recycling and reusing materials where possible, and coordinating laboratory work with co-workers who may be using some of the same chemicals (section 4.D.2);
• isolating nonhazardous waste from hazardous waste; and
• including in the experiment plan the reaction work-up steps that deactivate hazardous materials or reduce toxicity (see Chapter 7—examples include oxidation of carcinogens in situ or treating excess potassium metal with *t*-butyl alcohol).

Clearly, some of these steps have become important only recently as a result of the changing requirements and economics of laboratory management. Three of these critical strategies are elaborated on below.

4.B.2.1 Microscale Work

In microscale chemistry the amounts of materials used are reduced to 25 to 100 milligrams (mg) for solids and 100 to 200 microliters (μL) for liquids, compared to the usual 10 to 50 g for solids or 100 to 500 milliliters (mL) for liquids. Carrying out synthetic and analytical work on a small scale requires that smaller amounts of materials be ordered. Working with smaller amounts of materials can promote more attention to detail, which improves the quality of the science being done. The smaller scale also means that there will be less to recycle or dispose of from reaction work-up. Smaller quantities of items such as used filter papers, used filter cakes and filtrate from washings of the cakes, residues from distillation, and solvents to be redistilled will be produced. The glassware used in smaller-scale procedures is also generally not as easily broken as that required by procedures on a larger scale. Broken glassware contaminated with hazardous materials is itself a waste that must be disposed of. Microscale work also reduces the likelihood and severity of accidents resulting in personal exposure to hazardous chemicals. Fire hazards are also likely to be reduced.

As an example of the benefits of microscale work, consider the typical Kjeldahl reaction, which uses mercury as a catalyst. The mercury waste produced by this procedure creates a difficult disposal problem. Converting to micro-Kjeldahl equipment and quantities reduces the waste by 90%, which could result in a reduction of several liters of waste per day in laboratories that routinely run Kjeldahl reactions.

If 30,000 educational institutions that currently generate more than 4,000 metric tons of hazardous waste per year in the United States were to convert to microscale chemistry, 3,960 metric tons of that waste would be eliminated, at a savings of hundreds of millions of dollars per year. Many industrial research and development laboratories could achieve comparable financial and environmental savings.

The committee recognizes that enormous quantities of hazardous waste can be minimized by converting to microscale chemistry with proportionate environmental and financial savings. Many tons of waste and millions of dollars would be saved by going to the microscale level. At the same time it must be recognized that multigram laboratory preparation is often required to provide sufficient material for further work. Precaution appropriate to the scale, as well as the inherent hazard, of a laboratory operation must be exercised.

4.B.2.2   Step-by-Step Planning for Minimization

Experiment planning in the new culture of laboratory safety should include minimization of the material used at each step of an experiment. Consider two simple examples: (1) Transferring a liquid reaction mixture or other solution from one flask to another container usually requires the use of a solvent to rinse out the flask. During this procedure, the worker should use the smallest amount of solvent possible that will enable a complete transfer. (2) Celite® is often used during filtrations to keep the pores of filter papers or filter frits from becoming clogged. When putting the Celite® in place, the worker should carefully determine the minimum amount needed to be effective.

4.B.2.3   Substitution of Materials

To enhance safety and minimize the environmental consequences of an experiment, careful thought should be given to the materials to be used and the scale of an experiment. Traditionally, chemists have chosen reagents and materials for experiments to meet scientific criteria without always giving careful consideration to waste minimization or environmental objectives. In synthetic procedures, overall yield and purity of the desired product were usually regarded as the most important factors. Material substitution emerged as an important consideration in manufacturing process design because of the large quantities of chemicals involved. The following questions should now be considered when choosing a material to be used as a reagent or solvent in an experimental procedure:

• Can this material be replaced by one that will expose the experimenter, and others who handle it, to a lower order of potential hazard?

• Can this material be replaced by one that will reduce or eliminate the generation of hazardous waste and the consequent cost of waste disposal?

The following examples illustrate applications of these principles to common laboratory procedures:

• A standard general chemistry experiment designed to study Beer's law involves the use of a considerable volume of a copper–ammonia complex. When this volume is multiplied by the number of students in a general chemistry class, a waste disposal problem is created, because a large quantity of copper should not be released directly into a sewage treatment system. The experiment has been modified to use an iron–salicylic acid complex instead, resulting in a waste product that can be disposed of via the sanitary sewer without causing environmental harm (although specific regulations must be consulted).

• Liquid scintillation counting of low-level radioactive samples using flammable solvent-based cocktails (e.g., based on xylene, toluene, or dioxane) requires precautions because of the flammability of the solvent and generates large volumes of waste, which must be disposed of by incineration. Substitution of nonflammable, water-miscible cocktails eliminates the fire hazard and generates aqueous waste, which can be disposed of via the sanitary sewer rather than by incineration in many localities. Implementation of this strategy at one major university resulted in a substantial reduction in the volume of flammable organic waste sent out for incineration. Acceptance of the new practice was achieved following demonstrations by key research groups that the new cocktails gave results comparable to those from the flammable solvent-based cocktails.

• Phosgene is a highly toxic gas used as a reagent in many organic transformations. Its use requires proper precautions to deal with the containment of the gas and the handling and disposal of cylinders. Commercially available substitutes such as diphosgene (trichloromethyl) chloroformate, a liquid, or triphosgene bis(trichloromethyl)carbonate, a low-melting solid, can often be substituted for phosgene by appropriate adjustment of experimental conditions or can be used to generate phosgene only on demand. Both chemicals are highly toxic themselves, but they offer a means to avoid the problems associated with handling a toxic gas.

• Many widely used reagents contain toxic heavy metals, such as chromium and mercury. Waste containing these materials cannot be incinerated and must be handled separately for disposal. Thus, substitution of other reagents for heavy metal reagents will almost always be beneficial with respect to hazard and waste minimization. Chromic acid cleaning solutions for glassware can be replaced by proprietary detergents used, if necessary, along with ultrasonic baths. Various chromium(VI) oxidants have been important in synthetic organic chemistry, but their use can often be avoided by the substitution of organic oxidants. The Swern oxidation of alcohols (oxalyl chloride/dimethyl sulfoxide) produces relatively innocuous by-products, which can be handled with other organic waste. Other oxidation reagents tailored to the specific needs of a given transformation are available.

• Fluorine and fluorinating reagents such as perchloryl fluoride are among the most demanding reagents to handle because of their high reactivity and toxicity. Accordingly, there has been considerable incentive to develop substitutes for these materials.

One example is F-TEDA-BF4, or 1-chloromethyl-4-fluoro-1,4-diazonia [2.2.2] bicycloctane bis(tetrafluoroborate). This reagent can be substituted for more hazardous reagents in many fluorination procedures.

• Organic solvents for liquid-liquid extraction or chromatography can often be replaced by other solvents with significant benefit. Benzene, once a widely used solvent, is now recognized as a human carcinogen and must be handled accordingly. Toluene can often serve as a satisfactory substitute. Diethyl ether is a flammable solvent whose handling must take into account its tendency to form explosive peroxides. Methyl *t*-butyl ether (MTBE) offers only slight advantages over diethyl ether with respect to flammability, but its greatly reduced tendency to form peroxides eliminates the need to monitor peroxide formation during handling and storage.

The technology for handling supercritical fluids has developed rapidly in recent years. Supercritical carbon dioxide can replace organic solvents for high-performance chromatography and is beginning to find use as a reaction solvent. While supercritical solvents require specialized equipment for handling, they offer the potential benefit of large reductions in organic solvent waste.

4.B.3 Strategies to Avoid Multihazardous Waste Generation

Because handling and disposal of multihazardous waste require special waste management, it is especially prudent to develop strategies to minimize its generation. Chapter 7, section 7.C.1.1, provides information on eliminating or minimizing the components of waste that are radioactive or biological hazards. The strategies discussed include substituting nonradioactive materials for radioactive materials, substituting radioisotopes having shorter decay times (e.g., using iodine-131, with a half-life of 8 days, instead of iodine-125, with a half-life of 60 days, in thyroid research), and carrying out procedures with smaller amounts of materials.

4.C ACQUISITION OF CHEMICALS

4.C.1 Ordering Chemicals

Before purchasing a chemical, several questions should be asked:

• Is the material already available from another laboratory within the institution or from a surplus-chemical stockroom? If so, waste is reduced, and the purchase price is saved. The tendency to require the use of new chemicals because of their purity should be scrutinized and that requirement should be carefully justified to ensure that materials already on hand are used whenever possible.

• What is the minimum quantity that will suffice for the current use? Chemical purchases should not be determined by the cheaper unit price basis of large quantities, but rather by the amount needed for the experiment. The cost of disposing of the excess is likely to exceed any potential savings gained in a bulk purchase (i.e., in the present economic climate, the cost of getting rid of a chemical may exceed its acquisition cost). If a quantity smaller than the minimum offered by a supplier is needed, the supplier should be contacted and repackaging requested. Compressed gas cylinders, including lecture bottles, should normally be purchased only from suppliers who will accept return of empty cylinders.

• What is the maximum size container allowed in the areas where the material will be used and stored? Fire codes and institutional policies regulate quantities of certain chemicals, most notably flammables and combustibles. For these materials, a maximum allowable quantity for laboratory storage has been established (see also sections 4.E.3, 4.E.4, and 4.E.5).

• Can the chemical be managed safely when it arrives? Does it require special storage, such as in a dry box or freezer? Do receiving personnel need to be notified of the order and given special instructions for receipt? Will the equipment necessary to use the chemical be ready when it arrives? An effort should be made to order chemicals for just-in-time delivery, by purchasing all necessary materials from the same supplier with a request for delivery all together at the best time for performing an experiment.

• Is the chemical unstable? Inherently unstable materials may have very short storage times and should be purchased just before use to avoid losing a reagent and creating an unnecessary waste of material and time. Some materials may require express or overnight delivery and will not tolerate being held in transit over a weekend or holiday.

• Can the waste be managed satisfactorily? A chemical that will produce a new category of waste may cause a great deal of trouble for the waste management program. An appropriate waste disposal mechanism should be identified before the chemical is ordered.

More detail on all of these questions should be reviewed as necessary to arrive at satisfactory answers. Only when these issues have been identified and resolved can ordering proceed.

Authority to place orders for chemicals may be centralized in one purchasing office or may be dispersed to varying degrees throughout the institution. The advent of highly computerized purchasing systems, and even on-line ordering, has made it attractive to allow ordering at the departmental or research group level. However, the ability to control ordering of certain types of materials through a central purchasing system (e.g., prohibiting flammables in containers over a certain size or ensuring appropriate licensing of radioactive material users) is almost completely lost as the purchasing function is decentralized. In these cases, other, creative ways of exercising control need to be found.

One of the advantages of computerization of ordering is the information that can be retrieved from the chemical supplier. Some institutions have included in their annual contracts with suppliers a requirement to report on a monthly, quarterly, or annual basis the quantity of each type of chemical purchased and the location to which it was delivered. This information can be helpful in preparing the various annual reports on chemical use that may be required by federal, state, or local agencies.

A purchase order for a chemical should include a request for a Material Safety Data Sheet (MSDS). However, many of the larger laboratory chemical suppliers have established a policy of sending each MSDS only once, when the chemical is first ordered. Subsequent orders of the same chemical may not be accompanied by the MSDS. Therefore, a central network of accessible MSDSs should be established if feasible.

4.C.2 Receiving Chemicals

Chemicals arrive at institutions in a variety of ways, including U.S. mail, commercial package delivery, express mail services, and direct delivery from chemical warehouses. It is important to confine deliveries of chemicals to areas that are equipped to handle them, usually a loading dock, receiving room, or laboratory. Proper equipment for receipt of chemicals includes chains for temporary holding of cylinders and carts designed to safely move various types of chemical containers. Shelves, tables, or caged areas should be designated for packages to avoid damage by receiving room vehicles. Chemical deliveries should not normally be made to departmental offices because, in general, they are unlikely to be equipped to receive these packages. However, if delivery to such an office is the only option, a separate, undisturbed location, such as a table or shelf, should be identified for chemical deliveries, and the person ordering the material should be notified immediately upon its arrival.

Receiving room, loading dock, and clerical personnel need to be trained adequately to recognize hazards that may be associated with chemicals coming into the facility. They need to know what is expected of them if a package is leaking or if there is a spill in the receiving facility, and they need to know whom to call for assistance when a problem develops. The Department of Transportation (DOT) requires training for anyone involved in the movement (including receiving) of hazardous materials (see Chapter 9, section 9.D.10).

Transportation of chemicals within the facility, whether by internal staff or outside delivery personnel, must be done safely. Single boxes of chemicals, in their original packaging, can be hand carried to their destination if they are not too heavy to manage easily. Groups of packages or heavy packages should be transported by a cart that is stable, has straps or sides to contain packages securely, and has wheels large enough to negotiate uneven surfaces easily. Suitable carriers should be used when transporting individual containers of liquids.

Cylinders of compressed gases should always be secured on specially designed carts and should never be dragged or rolled. The cap should always be securely in place. Whenever possible, chemicals and gas cylinders should be moved on "freight-only" elevators.

If outside delivery personnel do not handle materials according to the receiving facility's standards, immediate correction should be sought, or other carriers or suppliers should be used. Delivery criteria can be specified in the original purchase order.

When packages are opened in the laboratory, laboratory personnel should verify that the container is intact and is labeled, at a minimum, with an accurate name on a well-adhered label. For unstable materials, and preferably for all materials, the date of receipt should be placed on the label. Labels placed by the manufacturer should not be obliterated or removed. New chemicals should be entered into the laboratory's inventory promptly and placed in the appropriate storage area.

4.C.3 Responsibilities for Chemicals Being Shipped or Transported

The DOT regulates shipment of chemicals by a specific set of hazardous materials regulations (49 CFR 100-199). These regulations contain detailed instructions on how hazardous materials have to be identified, packaged, marked, labeled, documented, and placarded. Shipments not in compliance with the applicable regulations may not be offered or accepted for transportation. Since October 1, 1993, HM126F, a new, more stringent set of regulations on training for safe transportation of hazardous materials (49 CFR 172.700-704), has been in effect. It is essential that all individuals who are preparing hazardous materials for shipment communicate with their institution's transportation coordinators. Shipment of experimental materials is also discussed in Chapter 9, section 9.D.10.

The use of personal vehicles, company or institutional vehicles (including airplanes), and customer vehicles for transporting regulated materials, which may be hazardous, is a serious concern. *Most businesses and academic institutions forbid the use of privately owned personal vehicles, due to the serious insurance consequences if an accident occurs. Most individuals will find that their personal vehicle insurance does not cover them when they are transporting hazardous materials.* Anyone who needs to transport regulated materials personally between buildings within an institution should walk. (Secondary containment, such as a rubber bucket, should always be used for carrying bottled chemicals.)

4.D INVENTORY AND TRACKING OF CHEMICALS

4.D.1 General Considerations

Prudent management of chemicals in any laboratory is greatly facilitated by keeping an inventory of the chemicals stored. An inventory is a database that tabulates the chemicals in the laboratory, along with information essential for their proper management. Without an inventory of chemicals stored in a particular location, many important questions pertinent to prudent management of chemicals can be answered only by visually scanning container labels. A well-managed inventory system can promote economical use of chemicals by making it possible to determine immediately what chemicals are on hand. The scope of an inventory need not be limited to materials obtained from commercial sources, but can include chemicals synthesized in a laboratory that may be available for sharing. If the need for a chemical can be filled from a supply already on hand, the time and expense of procuring new material can be avoided. Information on chemicals that present particular storage or disposal problems can facilitate appropriate planning for their handling. While a detailed listing of hundreds of chemicals stored in a particular location may not be directly useful to emergency responders, it can be used to prepare a summary of the types of chemicals stored and the hazards that might be encountered. In larger organizations where chemicals are stored in multiple locations, the inventory system should include information on the storage location for each container of each chemical.

If procedures for the facile updating of information on storage locations are developed, the system becomes a tracking system. Such a system can promote the sharing of chemicals originally purchased by different research groups or laboratories. The more that laboratories in an organization agree to share chemicals, the greater the likelihood that items unneeded in one location will find a use elsewhere. Tracking systems are more complex to establish than simple inventories and require more effort to maintain, but their favorable impact on the economics and efficiency of chemical use in a large organization will often justify their use.

Each record in a chemical inventory database generally corresponds to a single container of a chemical rather than merely to the chemical itself. This approach allows for a more logical correspondence between the records in the database and the chemicals stored in the laboratory. The following data fields for each item are probably essential in any system:

- name as printed on the container,
- molecular formula, for further identification and to provide a simple means of searching,
- Chemical Abstract Service (CAS) registry number, for unambiguous identification of chemicals despite the use of different naming conventions,
- source, and
- size of container.

In addition, the following information may be useful:

- hazard classification, as a guide to safe storage, handling, and disposal,
- date of acquisition, to ensure that unstable chemicals are not stored beyond their useful life, and
- storage location, in laboratories where multiple locations exist.

In a chemical tracking system, the means by which the consumption of chemicals is tracked must be considered. The effort involved in maintaining data on the precise contents of each container must be weighed against the potential benefit such a system would provide. Many tracking systems ignore this information and record only the size of the container.

A simple inventory system can be established by recording the above information for each container on index cards, which are then kept in an accessible location in some logical order, such as by molecular formula. The ease of searching such a card file is limited by its size and the order in which it is sorted. This type of system has obvious advantages in terms of simplicity and low cost, but it suffers several limitations. Listings of chemicals must be prepared manually, and the integrity of the database depends on how well the card file is maintained.

For an inventory of more than a few hundred chemicals, a computer-based system offers many advantages. Many spreadsheet and database programs can be used to maintain an effective chemical inventory system. The integrity of the inventory system can be enhanced by the ease of making backup copies of the database. Searches for desired chemicals can be carried out in a number of ways, depending on the capability of the software. The ability to sort the database, for example, by hazard classification, acquisition date, or other parameters, and to prepare lists of the results of such a sort, can contribute to efficiency in a variety of chemical management tasks.

Section 4.C.1 above notes the prudence of establishing a central network of MSDSs. Including MSDSs and Laboratory Chemical Safety Summaries (LCSSs) (see Chapter 3 and Appendix B) in the inventory's database is highly desirable. The quality of MSDSs varies significantly from one manufacturer to another. LCSSs, which are targeted to the needs of the typical laboratory worker, are a useful supplement to the information provided by MSDSs.

Having a fully capable chemical tracking system depends on careful selection of more sophisticated database software. Such a package should permit access from multiple terminals or networked computers and, most importantly, have a foolproof, efficient method for rapidly recording the physical transfer of a chemical from one location to another. Barcode labeling of chemical containers as they are received provides a means of rapid, error-free entry of information for a chemical tracking system. If reagent chemical suppliers were to adopt a system in which chemical containers were labeled with bar codes providing essential information on their products, the maintenance of chemical tracking systems would be greatly facilitated. Proprietary software packages for tracking of chemicals are available. The investment in hardware, software, and personnel to set up and maintain a chemical inventory tracking system is considerable, but it can pay significant dividends in terms of economical and prudent management of chemicals.

As with any database, the utility of an inventory or chemical tracking system depends on the integrity of the information it contains. If an inventory system is used as a means of locating chemicals for use or sharing in the laboratory, even a moderate degree of inaccuracy will erode confidence in the system and discourage use. The need for high fidelity of data is greater for a tracking system, because laboratory workers will rely on it to save time locating chemicals using the system rather than physically searching. For these reasons, appropriate measures should be employed periodically to purge any inventory or tracking system of inaccurate data. A physical inventory of chemicals stored, verification of the data on each item, and reconciliation of differences can be

performed annually. This procedure can coincide with an effort to identify unneeded, outdated, or deteriorated chemicals and to arrange for their disposal. The following guidelines for culling inventory may be helpful:

• Consider disposing of materials anticipated not to be needed within a reasonable period, say, 2 years. Stable, relatively nonhazardous substances may have indefinite shelf lives; a decision to retain them in storage should take into account their economic value, scarcity, availability, and storage costs.

• Make sure that deteriorating containers, or containers in which evidence of a chemical change in the contents is apparent (e.g., appearance of peroxide crystals in a bottle of an ether), are inspected and handled by someone experienced in the possible hazards inherent in such situations.

• Dispose of or recycle chemicals before the expiration date on the container.

• Replace deteriorating labels before information is obscured or lost.

• Because many odoriferous substances will make their presence known despite all efforts to contain them, aggressively purge such items from storage and inventory.

• Aggressively cull the inventory of chemicals that require storage at reduced temperature in environmental rooms or refrigerators. Because these chemicals may include air- and moisture-sensitive materials, they are especially prone to problems that can be exacerbated by the effects of condensation.

• Dispose of, or remove to storage, all hazardous chemicals at the completion of the laboratory supervisor's tenure or transfer to another laboratory. The institution's cleanup policy for departing laboratory researchers and students should be enforced strictly to avoid accumulation of expensive orphaned unknowns.

• Develop and enforce procedures relating to transfer or disposal of chemicals and other materials when decommissioning laboratories because of renovation or relocation.

4.D.2 Exchange or Transfer of Chemicals

The exchange or transfer of chemicals between laboratories at an institution depends on the kind of inventory system and central stockroom facilities in place. Some institutions encourage laboratory workers to return materials to the central stockroom for redistribution to other workers. The containers may be sealed or open; a portion of the material may have been used. Materials from containers that have been opened are often of sufficient purity to be used "as is" in many procedures. If the purity is in doubt, the worker who returned the material should be consulted. The stockroom personnel can update the central inventory periodically to indicate which containers of which materials are available for this exchange or transfer. For an exchange program to be effective, all contributors to and users of the facility must reach a consensus on the standards to be followed concerning the labeling and purity of stored chemicals.

A word of caution needs to be offered in regard to surplus-chemical stockrooms. It is essential that such a facility be managed with the same degree of control that is afforded a new-chemical storage area. The surplus-chemical stockroom must not be operated as a depository for any chemical that probably will not be wanted in the laboratory within a reasonable period (e.g., 2 to 3 years); such materials should be disposed of properly. Rooms that are used as general depositories of unwanted chemicals are likely to become "mini-Superfund" sites because of lack of control. Academic institutions should consider recycling common organic solvents from one research laboratory to another, or from research laboratories to teaching laboratories. For example, chromatography effluents such as toluene could be collected from research laboratories, distilled, and checked for purity before reuse.

Such laboratory-to-laboratory exchange can be an effective alternative to a central surplus-chemical stockroom in organizations unwilling or unable to manage a central storeroom properly. In such a system, workers in the individual laboratory retain responsibility for the storage of unwanted chemicals but notify colleagues periodically of available materials. A chemical tracking system as described above can facilitate an exchange system greatly. If colleagues within the same laboratory are using the same hazardous material, particularly one that is susceptible to decomposition upon contact with air or water, they should try to coordinate the timing of their experiments.

4.D.3 Labeling Commercially Packaged Chemicals

Commercially packaged chemical containers received from 1986 onward generally meet current labeling requirements. The label usually includes the name of the chemical and any necessary handling and hazard information. Inadequate labels on older containers should be updated to meet current standards. To avoid ambiguity about chemical names, many labels carry the CAS registry number as an unambiguous identifier. This information should be added to any label that does not include it. On receipt of a chemical, the manufacturer's label should be supplemented by the date received and possibly the name and location of the individual respon-

sible for purchasing the chemical. If chemicals from commercial sources are repackaged into secondary containers, the new containers should be labeled with all essential information on the original container. *Warning: Do not remove or deface any existing labels on incoming containers of chemicals and other materials.*

4.D.4 Labeling Other Chemical Containers

The contents of all chemical containers, including, but not limited to, beakers, flasks, reaction vessels, and process equipment, should be properly identified. The overriding goal of prudent practice in the identification of laboratory chemicals is to avoid orphaned containers of unknown materials that may be expensive or dangerous to dispose of. The labels should be understandable to laboratory workers, members of well-trained emergency response teams, and others. Labels or tags should be resistant to fading from aging, chemical exposure, temperature, humidity, and sunlight.

Chemical identification and hazard warning labels on containers used for storing chemicals should include the following information:

- name, address, and telephone number of the chemical manufacturer, importer, or responsible party (including researcher),
- chemical identification and identity of hazard component(s), and
- appropriate hazard warnings.

Containers in immediate use, such as beakers and flasks, should, at a minimum, be labeled with the name of the chemical contents. Labeled materials transferred from primary (labeled) containers to secondary containers (e.g., safety cans and squeeze bottles) should include chemical identification and synonyms, precautions, and first aid information.

4.D.5 Labeling Experimental Materials

Labeling of all containers of experimental chemical materials is prudent. Because the properties of an experimental material are generally not completely known, its label cannot be expected to provide all necessary information to ensure safe handling.

The most important information on the label of an experimental material is the name of the researcher responsible, as well as any other information, such as a laboratory notebook reference, that can readily lead to what is known about the material. For items that are to be stored and retained within a laboratory where the properties of materials are likely to be well understood, only the sample identification and/or name may be needed. Samples that will be transferred outside the laboratory, or that may be handled by individuals not generally familiar with the type of material involved, should be labeled as completely as possible, including the name, address, and telephone number of the sender and recipient for samples in transit. In addition, samples that are sent to individuals at another institution must be accompanied by appropriate labeling and a Material Safety Data Sheet, according to OSHA's Hazard Communication Standard amendments and OSHA's Laboratory Standard "hazard identification" provision. When available, the following information should accompany experimental materials:

- Originator: give the name, location, and telephone number of the person to contact for safe handling information.
- Identification: include, at least, the laboratory notebook reference.
- Hazardous components: list primary components that are known to be hazardous.
- Potential hazards: indicate all the known or suspected potential hazards.
- Date: note the date that the material was placed in the container and labeled.
- Ship: indicate the name, location, and telephone number of the person to whom the material is being transferred.

4.D.6 Use of Inventory and Tracking Systems in Emergency Planning

The most important assistance to have in an emergency is access to a researcher who is knowledgeable about the chemical(s) involved. In addition, an organization's emergency preparedness plan should include components on what to do in the event of a hazardous materials release. The information in the inventory and tracking systems and the ability of individuals to access and make use of it are essential to proper functioning of the plan in an emergency. The care taken in labeling chemicals is also extremely important. See Chapter 5, section 5.C.11, for a detailed discussion of what to do in laboratory emergencies.

4.E STORAGE OF CHEMICALS IN STOCKROOMS AND LABORATORIES

The storage requirements and limitations for stockrooms and laboratories vary widely depending on

- level of expertise of the employees,
- level of safety features designed into the facility,
- location of the facility,

- nature of the chemical operations,
- accessibility of the stockroom,
- local and state regulations,
- insurance requirements, and
- building and fire codes.

Many local, state, and federal regulations have specific requirements that affect the handling and storage of chemicals in laboratories and stockrooms. For example, radioactive materials, controlled substances (drugs), consumable alcohol, explosives, needles, hazardous waste, and so forth have requirements ranging from locked storage cabinets and controlled access to specified waste containers and "regulated" areas. Stringent requirements may also be placed on an institution by its insurance carriers. Of particular applicability are fire and building codes, which may be either local or statewide and which have become considerably more rigorous in the past several years. All of these specific requirements must be identified when designing procedures for laboratory and stockroom storage. Other elements to consider include the safety features designed into the facility, the location of the facility, and the nature of the chemical operations.

4.E.1 General Considerations

In general, store materials and equipment in cabinets and on shelving provided for such storage.

- Avoid storing materials and equipment on top of cabinets. If you must place things there, however, maintain a clearance of *at least 18 inches from the sprinkler heads* to allow proper functioning of the sprinkler system.
- Do not store materials on top of high cabinets where they will be hard to see or reach.
- Avoid storing heavy materials up high.
- Keep exits, passageways, areas under tables or benches, and emergency equipment areas free of stored equipment and materials.

Storing chemicals in stockrooms and laboratories requires consideration of a number of health and safety factors. In addition to the inventory control and storage area considerations as discussed above, proper use of containers and equipment is crucial (see section 4.E.2).

In addition to the basic storage area guidelines above, these general guidelines should be followed when storing chemicals:

- Label all chemical containers appropriately.
- Place the user's name and the date received on all

TABLE 4.1 Related and Compatible Storage Groups

Inorganic Family	Nitric acid, other inorganic acids
Metals, hydrides	Sulfur, phosphorus, arsenic, phosphorus pentoxide
Halides, sulfates, sulfites, thiosulfates, phosphates, halogens	**Organic Family**
Amides, nitrates (except ammonium nitrate), nitrites, azides	Acids, anhydrides, peracids
Hydroxides, oxides, silicates, carbonates, carbon	Alcohols, glycols, amines, amides, imines, imides
Sulfides, selenides, phosphides, carbides, nitrides	Hydrocarbons, esters, aldehydes
	Ethers, ketones, ketenes, halogenated hydrocarbons, ethylene oxide
Chlorates, perchlorates, perchloric acid, chlorites, hypochlorites, peroxides, hydrogen peroxide	Epoxy compounds, isocyanates
Arsenates, cyanides, cyanates	Peroxides, hydroperoxides, azides
	Sulfides, polysulfides, sulfoxides, nitrites
Borates, chromates, manganates, permanganates	Phenols, cresols

NOTE: Store flammables in a storage cabinet for flammable liquids or in safety cans.

Separate chemicals into their organic and inorganic families and then related and compatible groups, as shown. Separation of chemical groups *can be by different shelves within the same cabinet.*

Do NOT store chemicals alphabetically as a general group. This may result in incompatibles appearing together on a shelf. Rather, store alphabetically within compatible groups.

This listing is only a suggested method of arranging chemical materials for storage and is not intended to be complete.

purchased materials in order to facilitate inventory control of the materials.

• Provide a definite storage place for each chemical and return the chemical to that location after each use.

• Avoid storing chemicals on bench tops, except for those chemicals being used currently.

• Avoid storing chemicals in laboratory hoods, except for those chemicals being used currently.

• Store volatile toxics and odoriferous chemicals in a ventilated cabinet. Check with the institution's environmental health and safety officer.

• Provide ventilated storage near laboratory hoods.

• If a chemical does not require a ventilated cabinet, store it inside a closable cabinet or on a shelf that has a lip to prevent containers from sliding off in the event of a fire, serious accident, or earthquake.

• Do not expose stored chemicals to heat or direct sunlight.

• Observe all precautions regarding the storage of incompatible chemicals.

• Separate chemicals into compatible groups and store alphabetically within compatible groups. See Table 4.1 for one suggested method for arranging chemicals in this way.

• Store flammable liquids in approved flammable liquid storage cabinets.

In seismically active regions, storage of chemicals requires additional consideration for the stability of shelving and containers. Shelving and other storage units should be secured. Shelving should contain a front-edge lip to prevent containers from falling. Ideally, containers of liquids should be placed on a metal or plastic tray that could hold the liquid if the container broke while on the shelf. All laboratories, not only those in seismically active regions, can benefit from these additional storage precautions.

4.E.2 Containers and Equipment

Specific guidelines regarding containers and equipment to use in storing chemicals are as follows:

• Use corrosion-resistant storage trays or secondary containers to retain materials if the primary container breaks or leaks.

• Provide vented cabinets beneath laboratory hoods for storing hazardous materials. (This encourages the use of the hoods for transferring such materials.)

• Use chemical storage refrigerators *only* for storing chemicals.

• Label these refrigerators with the following signage:

NO FOOD—CHEMICAL STORAGE ONLY

• Seal containers to minimize escape of corrosive, flammable, or toxic vapors.

• Label all materials in the refrigerator with contents, owner, date of acquisition or preparation, and nature of any potential hazard.

• Do not store flammable liquids in a refrigerator unless it is approved for such storage. Such refrigerators are designed not to spark inside the refrigerator. If refrigerated storage is needed inside a flammable-storage room, it is advisable to choose an explosion-proof refrigerator.

4.E.3 Storing Flammable and Combustible Liquids

The National Fire Protection Association (NFPA) Standard 45 (NFPA, 1991d) limits the quantity of flammable and combustible liquids per 100 square feet of laboratory space. (Local regulations should also be consulted.) The quantity depends on these safety factors:

• construction of the laboratory,
• fire protection systems built into the laboratory,
• storage of flammable liquids in flammable liquid storage cabinets or safety cans, and
• type of laboratory (i.e., instructional or research and development).

Many laboratories have a B (business) classification with sprinkler systems and have a flammable and com-

bustible liquid storage limitation, as shown in Table 4.2.

The container size for storing flammable and combustible liquids is limited both by NFPA Standards 30 and 45 and by OSHA. Limitations are based on the type of container and the flammability of the liquid, as shown in Table 4.3.

The following precautions should be taken when storing flammable liquids:

• When possible, store quantities of flammable liquids greater than 1 L (approximately 1 quart, or 32 ounces) in safety cans. Refer to Table 4.3.

• Store combustible liquids either in their original (or other NFPA- and DOT-approved) containers or in safety cans. Refer to Table 4.3.

4.E.4 Storing Gas Cylinders

The following precautions should be taken when storing compressed gas cylinders:

• Always label cylinders so you know their contents; do not depend on the manufacturer's color code.

• Securely strap or chain gas cylinders to a wall or bench top. In seismically active areas, it may be advisable to use more than one strap or chain.

• When cylinders are no longer in use, shut the valves, relieve the pressure in the gas regulators, remove the regulators, and cap the cylinders.

• Segregate gas cylinder storage from the storage of other chemicals.

• Keep incompatible classes of gases stored separately. Keep flammables from reactives, which include oxidizers and corrosives.

• Segregate empty cylinders from full cylinders.

• Keep in mind the physical state—compressed, cryogenic, and/or liquefied—of the gases.

• *Warning: Do not abandon cylinders in the dock storage*

TABLE 4.2   Storage Limits for Flammable and Combustible Liquids for Laboratories: B Classification with Sprinkler System

Class of Liquid	Flash Point (°C)	Amount (gallons per 100 square feet)
Class I Flammable	Below 38	4
Class II Combustible	38–60	4
Class IIIA Combustible	60–93	12
Class IIIB Combustible	Above 93	Unlimited

NOTE: Liquid (pumpable) flammable waste is included in the storage limitation. Non-pumpable waste is not included. Locations with an H (hazard) classification have much higher limits. Inside storage rooms for flammable liquids, the limits are from 5 to 10 gallons per square foot, depending on the size and construction of the room.

SOURCE: NFPA (1991c), Chapter 2-2, "Laboratory Unit Fire Hazard Classification."

TABLE 4.3 Container Size for Storage of Flammable and Combustible Liquids

| Container | Flammable Liquids[a] | | | | | | Combustible Liquids[b] | | | |
| | Class IA | | Class IB | | Class IC | | Class II | | Class IIIA | |
	Liters	Gallons	Liters	Gallons	Liters	Gallons	Liters	Gallons	Liters	Gallons
Glass[c]	0.5	0.12	1	0.25	4	1	4	1	4	1
Metal or approved plastic	4	1	20	5	20	5	20	5	20	5
Safety cans	7.5	2	20	5	20	5	20	5	20	5

NOTE: Label safety cans with contents and hazard warning information. Safety cans containing flammable or combustible liquid waste must have appropriate waste labels. Place 20-L (5-gallon) and smaller containers of flammable liquids that are not in safety cans into storage cabinets for flammable liquids. Do not vent these cabinets unless they also contain volatile toxics or odoriferous chemicals. Aerosol cans that contain 21% (by volume), or greater, alcohol or petroleum base liquids are considered Class IA flammables. When space allows, store combustible liquids in storage cabinets for flammable liquids. Otherwise, store combustible liquids in their original (or other Department of Transportation-approved) containers according to Table 4.2. Store 55-gallon drums of flammable and combustible liquids in special storage rooms for flammable liquids. Keep flammable and combustible liquids away from strong oxidizing agents, such as nitric or chromic acid, permanganates, chlorates, perchlorates, and peroxides. Keep flammable and combustible liquids away from an ignition source. Remember that most flammable vapors are heavier than air and can travel to ignition sources.

[a]Class IA includes those flammable liquids having flashpoints below 73 °F and having a boiling point below 100 °F, Class IB includes those having flashpoints below 73 °F and having a boiling point at or above 100 °F, and Class IC includes those having flashpoints at or above 73 °F and below 100 °F.

[b]Class II includes those combustible liquids having flashpoints at or above 100 °F and below 140 °F, Class IIIA includes those having flashpoints at or above 140 °F and below 200 °F, and Class IIIB includes those having flashpoints at or above 200 °F.

[c]Glass containers as large as 1 gallon can be used if needed and if the required purity would be adversely affected by storage in a metal or approved plastic container, or if the liquid would cause excessive corrosion or degradation of a metal or approved plastic container.

SOURCE: NFPA (1991c), Chapter 7-2.3, "Storage."

areas. Return them to the supplier when you are finished with them.

For commonly used laboratory gases, it is prudent to consider the installation of in-house gas systems. Such systems remove the need for transport and in-laboratory handling of compressed gas cylinders.

Chapter 5, section H, provides additional information on working with compressed gases in the laboratory.

4.E.5 Storing Highly Reactive Substances

The following guidelines should be followed when storing highly reactive substances:

• Consider the storage requirements of each highly reactive chemical prior to bringing it into the laboratory.

• Consult the MSDSs or other literature in making decisions about storage of highly reactive chemicals.

• Bring into the laboratory only the quantities of material you will need for your immediate purposes (less than a 3- to 6-month supply, the length depending on the nature and sensitivity of the materials).

• Label, date, and inventory all highly reactive materials as soon as received. Make sure the label states,

DANGER! HIGHLY REACTIVE MATERIAL!

• Do not open a container of highly reactive material that is past its expiration date. Call your institution's hazardous waste coordinator for special instructions.

• Do not open a liquid organic peroxide or peroxide former if crystals or a precipitate are present. Call your institution's hazardous waste coordinator for special instructions.

• Dispose of (or recycle) highly reactive material prior to expiration date.

• Segregate the following materials:
—oxidizing agents from reducing agents and combustibles,
—powerful reducing agents from readily reducible substrates,
—pyrophoric compounds from flammables, and
—perchloric acid from reducing agents.

• Store highly reactive liquids in trays large enough to hold the contents of the bottles.

• Store perchloric acid bottles in glass or ceramic trays.

• Store peroxidizable materials away from heat and light.

• Store materials that react vigorously with water away from possible contact with water.

• Store thermally unstable materials in a refrigerator. Use a refrigerator with these safety features:
—all spark-producing controls on the outside,
—a magnetic locked door, and
—an alarm to warn when the temperature is too high.

• Store liquid organic peroxides at the lowest possible temperature consistent with the solubility or freezing point. Liquid peroxides are particularly sensitive during phase changes.

• Inspect and test peroxide-forming chemicals periodically (these should be labeled with an acquisition or expiration date) and discard containers that have exceeded their safe storage lifetime.

• Store particularly sensitive materials or larger amounts of explosive materials in explosion relief boxes.

• Restrict access to the storage facility.

• Assign responsibility for the storage facility to one primary person and a backup person. Review this responsibility at least yearly.

4.E.6 Storing Toxic Substances

The following precautions should be taken when storing toxic substances:

• Store chemicals known to be highly toxic (including carcinogens) in ventilated storage in unbreakable, chemically resistant secondary containers.

• Keep quantities at a minimum working level.

• Label storage areas with appropriate warning signs, such as

CAUTION! REPRODUCTIVE TOXIN STORAGE

or

CAUTION! CANCER-SUSPECT AGENT STORAGE

and limit access to those areas.

• Maintain an inventory of all highly toxic chemicals. Some localities require that inventories be maintained of all hazardous chemicals in laboratories.

4.F RECYCLING OF CHEMICALS, CONTAINERS, AND PACKAGING

4.F.1 General Considerations

Recycling of chemicals can take many forms, from solvent distillation to cleaning of mercury to precipitation and purification of heavy metal salts. In each case a material that is not quite clean enough to be used as is must be brought to a higher level of purity or changed to a different physical state. Recycling processes can be very time- and energy-intensive and may not be economically justifiable. Before a decision on recycling is made, the cost of avoided waste disposal must be figured into the equation.

Another significant issue is whether recycling activities require a waste treatment permit under the Re-

source Conservation and Recovery Act (RCRA). This issue is discussed in Chapter 9. State and local regulations must also be considered.

A general comment applicable to all recycling is that a recyclable waste stream needs to be kept as clean as possible. If a laboratory produces a large quantity of waste xylene, small quantities of other organic solvents should be collected in a different container, because the distillation process will give a better product with fewer materials to separate. Steps should also be taken to avoid getting mercury into oils used in vacuum systems, oil baths, and other applications. Similarly, certain ions in a solution of waste metal salts may have a serious negative impact on the recrystallization process.

It is also important to identify users for a recycled product so that time and energy are not wasted on producing a product that must still be disposed of as a waste. Recycling some of the chemicals used in large undergraduate courses may be especially cost effective because the needs of the users are known well in advance.

Among the factors to be considered when ordering from a supplier of laboratory chemicals is whether the supplier will accept return of unopened chemicals, including highly reactive chemicals. Materials other than chemicals, such as containers or packaging materials and parts of laboratory instruments, can also be recycled. Examples include certain glass and plastic containers, drums and pails, plastic scrap and film scrap, cardboard, office paper, circuit boards, and metals such as steel and aluminum.

4.F.2 Solvent Recycling

The choice of a distillation unit for solvent recycling is controlled largely by the level of purity desired in the solvent, and so it is useful to know the intended use of the redistilled solvent before equipment is purchased. A simple flask, column, and condenser setup may be adequate for a solvent that will be used for crude separations or for initial glassware cleaning. For a much higher level of purity, a spinning band column will probably be required. Stills with automatic controls that shut down the system under conditions such as loss of cooling or overheating of the still pot enhance the safety of the distillation operation greatly. Overall, distillation is likely to be most effective when fairly large quantities (roughly 5 L) of relatively clean single-solvent waste can be accumulated before the distillation process is begun.

4.F.3 Mercury Recycling

The simplest method of cleaning mercury of entrained particulates or small quantities of water is to allow the mercury to run very slowly through a tiny hole in a conical filter paper. The filter should be covered to reduce the evaporation of mercury. This method is slow but does produce a reasonably clean product that is adequate for a number of uses. Commercial recycling of mercury usually involves multiple distillations, resulting in a high-purity product. Distillation within the laboratory should be discouraged because it is very difficult to avoid contaminating the surrounding area with spilled or vaporized mercury.

4.F.4 Reclamation of Heavy Metals

Inorganic qualitative analysis experiments typically include some toxic metal elements, such as cadmium, chromium, and lead. If the single-element "knowns" and "unknowns" can be collected separately, they can be readily precipitated for reuse in the next term's classes. Mixed samples can be precipitated and used as a starting material in a separations experiment. Although the amount of this type of waste may be quite small, it can require very expensive disposal if a commercial vendor must be used.

Many recycling processes will result in some amount of residue that will not be reusable and will probably have to be handled as a hazardous waste. See Chapter 7 for further information.

5 Working with Chemicals

5.A INTRODUCTION

Prudent execution of experiments requires not only sound judgment and an accurate assessment of the risks involved in laboratory work, but also the selection of appropriate work practices to reduce risk and protect the health and safety of the laboratory workers as well as the public and the environment. Chapter 3 provides specific guidelines to enable laboratory workers to evaluate the hazards and assess the risks associated with laboratory chemicals, equipment, and operations. Chapter 4 demonstrates how to control those risks when managing the inventory of chemicals in the laboratory. How the protocols outlined in Chapter 3 are put to use in the execution of a carefully planned experiment is the subject of Chapter 5.

Chapter 5 presents general guidelines for laboratory work with hazardous chemicals rather than specific standard operating procedures for individual substances. Hundreds of thousands of different chemicals are encountered in the research conducted in laboratories, and the specific health hazards associated with most of these compounds are generally not known. Also, laboratory work frequently generates new substances of unknown properties and unknown toxicity. Consequently, the only prudent course is for laboratory personnel to conduct their work under conditions that minimize the risks due to both known and unknown hazardous substances. The general work practices outlined in this chapter are designed to achieve this purpose.

Specifically, section 5.C describes basic prudent practices that should be employed in *all* laboratory work with chemicals. These guidelines are the standard operating procedures for all work conducted in laboratories where hazardous chemicals are stored or are in use.

In section 5.D, additional special procedures are presented for work with highly toxic substances. How to determine when these additional procedures are necessary is discussed in detail in Chapter 3, section 3.C. Section 5.E gives detailed special procedures for work with chemicals that pose risks due to biohazards and radioactivity; section 5.F, flammability; and section 5.G, reactivity and explosibility. Special considerations for work with compressed gases are the subject of section 5.H.

Chapter 6 provides precautionary methods for handling laboratory equipment commonly used in conjunction with hazardous chemicals. Chapters 3, 5, and 6 should all be consulted before working with hazardous chemicals.

Four fundamental principles underlie all of the work practices discussed in this chapter:

- **Plan ahead.** Determine the potential hazards associated with an experiment before beginning it.

- **Minimize exposure to chemicals.** Do not allow laboratory chemicals to come in contact with skin. Use laboratory hoods and other ventilation devices to prevent exposure to airborne substances whenever possible.

- **Do not underestimate risks.** Assume that any mixture of chemicals will be more toxic than its most toxic component. Treat all new compounds and substances of unknown toxicity as toxic substances.

- **Be prepared for accidents.** Before beginning an experiment, know what specific action to take in the event of the accidental release of any hazardous substance. Know the location of all safety equipment and the nearest fire alarm and telephone, and know what telephone numbers to call and whom to notify in the event of an emergency. Be prepared to provide basic emergency treatment. Keep your co-workers informed of your activities so that they can respond appropriately.

5.B PRUDENT PLANNING

The risk associated with an experiment should be determined before the laboratory work begins. The hypothetical question that should be posed before an experiment is, "What would happen if . . . ?" For the possible contingencies, preparations should be made to take the appropriate emergency actions. The worker should know the location of emergency equipment and how to use it. He or she should be familiar with emergency procedures and should know how to obtain help in an emergency. Any special safety precautions that may be required should be addressed before the experiment is begun. The consequences of loss of electrical power or water pressure should also be considered.

The physical and health hazards associated with chemicals should be determined before working with them. This determination may involve consulting literature references, Laboratory Chemical Safety Summaries (LCSSs), Material Safety Data Sheets (MSDSs), or other reference materials (see also Chapter 3, section 3.B) and may require discussions with the laboratory supervisor and consultants such as safety and industrial hygiene officers. Every step of the waste minimization and removal processes should be checked against federal, state, and local regulations. Production of mixed chemical-radioactive-biological waste (see Chapter 7, section 7.C.1.3) should not be considered without discussions with environmental health and safety experts.

Many of the general practices applicable to working

with hazardous chemicals are given elsewhere in this volume (as discussed in Chapter 2). The reader is referred to Chapter 4, section 4.C, for detailed instructions on the transport of chemicals; Chapter 4, section 4.E on storage; Chapter 6 for information on use and maintenance of equipment and glassware; and Chapter 7 for information on disposal of chemicals.

5.C GENERAL PROCEDURES FOR WORKING WITH HAZARDOUS CHEMICALS

5.C.1 Personal Behavior

Professional standards of personal behavior are required in any laboratory:

- Avoid distracting or startling other workers.
- Do not allow practical jokes and horseplay at any time.
- Use laboratory equipment only for its designated purpose.
- Do not allow visitors, including children and pets, in laboratories where hazardous substances are stored or are in use or hazardous activities are in progress.
- If children are permitted in laboratories, for example, as part of an educational or classroom activity, ensure that they are under the direct supervision of qualified adults.
- Make sure that teaching materials and publicity photographs show people wearing appropriate safety gear, in particular, eye protection.

5.C.2 Minimizing Exposure to Chemicals

Precautions should be taken to avoid exposure by the principal routes, that is, contact with skin and eyes, inhalation, and ingestion, which are discussed in detail in Chapter 3, section 3.C.

5.C.2.1 Avoiding Eye Injury

Eye protection should be required for all personnel and visitors in all locations where chemicals are stored or used. Eye protection is required whether or not one is actually performing a chemical operation. Visitor safety glasses should be made available at the entrances to all laboratories.

Researchers should assess the risks associated with an experiment and use the appropriate level of eye protection:

- Safety glasses with side shields provide the minimum protection acceptable for regular use. Safety glasses must meet the American National Standards Institute (ANSI) standard Z87.1-1989, Standard for Occupational and Educational Eye and Face Protection, which specifies a minimum lens thickness, certain impact resistance requirements, and so on.
- Safety splash goggles or face shields should be worn when carrying out operations in which there is any danger from splashing chemicals or flying particles. These thin shields do not provide protection from projectiles, however.
- Goggles are preferred over regular safety glasses to protect against hazards such as projectiles, as well as when working with glassware under reduced or elevated pressures (e.g., sealed tube reactions), when handling potentially explosive compounds (particularly during distillations), and when employing glassware in high-temperature operations.
- Because goggles offer little protection to the face and neck, full-face shields should be worn when conducting particularly hazardous laboratory operations. In addition, glassblowing and the use of laser or ultraviolet light sources require special glasses or goggles.

Ordinary prescription glasses do not provide adequate protection against injury. Prescription safety glasses and goggles can be obtained.

Contact lenses offer no protection against eye injury and cannot be substituted for safety glasses and goggles. It is best not to wear contact lenses when carrying out operations where chemical vapors are present or a chemical splash to the eyes or chemical dust is possible because contact lenses can increase the degree of harm and can interfere with first aid and eye-flushing procedures. If an individual must wear contact lenses for medical reasons, then safety glasses with side shields or tight-fitting safety goggles must be worn over the contact lenses.

5.C.2.2 Avoiding Ingestion of Hazardous Chemicals

Eating, drinking, smoking, gum chewing, applying cosmetics, and taking medicine in laboratories where hazardous chemicals are used should be strictly prohibited. Food, beverages, cups, and other drinking and eating utensils should not be stored in areas where hazardous chemicals are handled or stored. Glassware used for laboratory operations should never be used to prepare or consume food or beverages. Laboratory refrigerators, ice chests, cold rooms, ovens, and so forth should not be used for food storage or preparation. Laboratory water sources and deionized laboratory water should not be used for drinking water.

Laboratory chemicals should never be tasted. A pipet bulb or aspirator should be used to pipet chemicals or to start a siphon; pipetting should never be done by mouth. Hands should be washed with soap and water immediately after working with any laboratory chemicals, even if gloves have been worn.

5.C.2.3 Avoiding Inhalation of Hazardous Chemicals

Toxic chemicals or compounds of unknown toxicity should never be smelled. Procedures involving volatile toxic substances and operations involving solid or liquid toxic substances that may result in the generation of aerosols should be conducted in a laboratory hood. Dusts should be recognized as potentially contaminated and hazardous. Hoods should not be used for disposal of hazardous volatile materials by evaporation. Such materials should be treated as chemical waste and disposed of in appropriate containers in accord with institutional procedures.

The following general rules should be followed when using laboratory hoods:

• For work involving hazardous substances, use only hoods that have been evaluated for adequate face velocity and proper operation. Hood operation should be inspected regularly, and the inspection certified in a visible location.

• Keep reactions and hazardous chemicals at least 6 inches behind the plane of the hood sash.

• Never put your head inside an operating laboratory hood to check an experiment. The plane of the sash is the barrier between contaminated and uncontaminated air.

• On hoods where sashes open vertically, work with the hood sash in the *lowest possible position*. On hoods where sashes open horizontally, position one of the doors to act as a shield in the event of an accident in the hood. When the hood is not in use, keep the sash closed to maintain laboratory airflow.

• Keep hoods clean and clear; do not clutter with bottles or equipment. If there is a grill along the bottom slot or a baffle in the back of the hood, clean them regularly so they do not become clogged with papers and dirt. Allow only materials actively in use to remain in the hood. Following this rule will provide optimal containment and reduce the risk of extraneous chemicals being involved in any fire or explosion. Support any equipment that needs to remain in hoods on racks or feet to provide airflow under the equipment.

• Report suspected hood malfunctions promptly to the appropriate office, and make sure they are corrected. Post the name of the individual responsible for use of the hood in a visible location. Clean hoods before maintenance personnel work on them.

(See Chapter 8, section 8.C, for more information on hoods.)

5.C.2.4 Avoiding Injection of Hazardous Chemicals

Solutions of chemicals are often transferred in syringes, which for many uses are fitted with sharp needles. The risk of inadvertent injection is significant, and vigilance is required to avoid that accident. Needles must be properly disposed of in "sharps" containers. *Use special care when handling solutions of chemicals in hypodermic syringes.*

5.C.2.5 Minimizing Skin Contact

Wear gloves whenever handling hazardous chemicals, sharp-edged objects, very hot or very cold materials, toxic chemicals, and substances of unknown toxicity. The following general guidelines apply to the selection and use of protective gloves:

• Wear gloves of a material known to be resistant to permeation by the substances in use. Wearing the wrong type of glove can be more hazardous than wearing no gloves at all, because if a chemical seeps through, the glove can hold it in prolonged contact with the wearer's hand.

• Inspect gloves for small holes or tears before use.

• Wash gloves appropriately before removing them. (Note: some gloves, e.g., leather and polyvinyl alcohol, are water-permeable.)

• In order to prevent the unintentional spread of hazardous substances, remove gloves before handling objects such as doorknobs, telephones, pens, and computer keyboards.

• Replace gloves periodically, depending on the frequency of use and their permeation and degradation characteristics relative to the substances handled.

(For more information, see OSHA Personal Protective Equipment Standard (29 CFR 1910.132-138) regarding hand protection.)

5.C.2.6 Clothing and Protective Apparel

Long hair and loose clothing or jewelry must be confined when working in the laboratory. Unrestrained long hair, loose or torn clothing, and jewelry can dip into chemicals or become ensnared in equipment and moving machinery. Clothing and hair can catch fire. Sandals and open-toed shoes should never be worn in a laboratory in which hazardous chemicals are in use.

It is advisable to wear a laboratory coat when working with hazardous chemicals. This is particularly important if personal clothing leaves skin exposed. Apparel giving additional protection (e.g., nonpermeable laboratory aprons) is required for work with certain hazardous substances. Because many synthetic fabrics are flammable and can adhere to the skin, they can increase the severity of a burn. Therefore, cotton is the preferred fabric.

5.C.3 Housekeeping

There is a definite correlation between orderliness and level of safety in the laboratory. In addition, a disorderly laboratory can hinder or endanger emergency response personnel. The following housekeeping rules should be adhered to:

• Never obstruct access to exits and emergency equipment such as fire extinguishers and safety showers.
• Clean work areas (including floors) regularly. Properly label (see Chapter 3, section 3.B.4) and store (see Chapter 4, section 4.E) all chemicals. Accumulated dust, chromatography adsorbents, and other chemicals pose respiratory hazards.
• Secure all compressed gas cylinders to walls or benches.
• Do not store chemical containers on the floor.
• Do not use floors, stairways, and hallways as storage areas.

5.C.4 Transport of Chemicals

Chemicals being transported outside the laboratory or between stockrooms and laboratories should be in break-resistant secondary containers. Secondary containers commercially available are made of rubber, metal, or plastic, with carrying handle(s), and are large enough to hold the contents of the chemical containers in the event of breakage. When transporting cylinders of compressed gases, the cylinder should always be strapped in a cylinder cart and the valve protected with a cover cap. When cylinders must be transported between floors, passengers should not be in the elevator.

5.C.5 Storage of Chemicals

The accumulation of excess chemicals can be avoided by purchasing the minimum quantities necessary for a research project. All containers of chemicals should be labeled properly. Any special hazards should be indicated on the label. For certain classes of compounds (e.g., ethers as peroxide formers), the date

the container was opened should be written on the label. Peroxide formers should have the test history and date of discard written on the label as well. Only small quantities (less than 1 liter (L)) of flammable liquids should be kept at workbenches. Larger quantities should be stored in approved storage cabinets. Quantities greater than 1 L should be stored in metal or break-resistant containers. Large containers (more than 1 L) should be stored below eye level on low shelves. Hazardous chemicals and waste should never be stored on the floor.

Refrigerators used for storage of flammable chemicals must be explosion-proof, laboratory-safe units. Materials placed in refrigerators should be clearly labeled with water-resistant labels. Storage trays or secondary containers should be used to minimize the distribution of material in the event a container should leak or break. It is good practice to retain the shipping can for such secondary containers.

All chemicals should be stored with attention to incompatibilities so that if containers break in an accident, reactive materials do not mix and react violently. (See Chapter 4, section 4.E, and Chapter 7, section 7.C.1.2, for more information.)

5.C.6 Disposal of Chemicals

Virtually every laboratory experiment generates some waste, which may include such items as used disposable labware, filter media and similar materials, aqueous solutions, and hazardous chemicals. The overriding principle governing the handling of waste in prudent laboratory practice is that *no activity should begin unless a plan for the disposal of nonhazardous and hazardous waste has been formulated.* Application of this simple rule will ensure that the considerable regulatory requirements for waste handling are met and that unexpected difficulties, such as the generation of a form of waste (e.g., chemical-radioactive-biological) that the institution is not prepared to deal with, are avoided.

Each category of waste has certain appropriate disposal methods. In choosing among these methods, several general principles apply, but local considerations can strongly influence the application of these rules:

• Hazardous or flammable waste solvents should be collected in an appropriate container pending transfer to the institution's central facility or satellite site for chemical waste handling or pickup by an outside disposal agency.
• Waste solvents can usually be mixed for disposal, with due regard for the compatibility of the compo-

nents. Sometimes halogenated and nonhalogenated wastes must be segregated for separate handling.

• The container used for the collection of liquid waste must be appropriate for its use. Glass bottles are impervious to most chemicals but present a breakage hazard, and narrow necks can cause difficulty in emptying the bottles. The use of plastic (e.g., polyethylene jerrycans) or metal (galvanized or stainless steel) safety containers for the collection of liquid waste is strongly encouraged and, indeed, required for flammable liquids.

• Galvanized steel safety cans should not be used for halogenated waste solvents because they tend to corrode and leak. Flame arresters in safety cans can easily become plugged if there is sediment and may need to be cleaned occasionally.

• Waste containers should be clearly and securely labeled as to their contents and securely capped when not in immediate use.

• Aqueous waste should be collected separately from organic solvent waste. Some laboratories may be served by a wastewater treatment facility that allows the disposal of aqueous waste to the sanitary sewer if it falls within a narrow range of acceptable waste types. Thus, solutions of nonhazardous salts or water-miscible organic materials may be acceptable in some localities. Solutions containing flammable or hazardous waste, even if water-miscible, are almost never allowed, and water-immiscible substances must never be put down the drain. Aqueous waste for nonsewer disposal should be collected in a container selected for resistance to corrosion. Glass should not be used for aqueous waste if there is danger of freezing. Depending on the requirements of the disposal facility, adjustment of the pH of aqueous waste may be required. Such adjustment requires consideration of the possible consequences of the neutralization reaction that might take place: gas evolution, heat generation, or precipitation.

• Solid chemical waste, such as reaction by-products, or contaminated filter or chromatography media, should be placed in an appropriately labeled container to await disposal or pickup. Unwanted reagents should be segregated for disposal in their original containers, if possible. If original containers are used, labels should be intact and fully legible. Every effort should be made to use, share, or recycle unwanted reagents rather than commit them to disposal. (See Chapter 4, sections 4.D and 4.E, for a discussion of labeling alternatives.)

• Nonhazardous solid waste can be disposed of in laboratory trash or segregated for recycling. Institutional policy should be consulted for these classifications. (See Chapter 7 for further information regarding disposal, and check the appropriate LCSS to determine toxicity.)

5.C.7 Use and Maintenance of Equipment and Glassware

Good equipment maintenance is essential for safe and efficient operations. Laboratory equipment should be inspected and maintained regularly and serviced on schedules that are based on both the likelihood of and the hazards from failure. Maintenance plans should ensure that any lockout procedures cannot be violated.

Careful handling and storage procedures should be used to avoid damaging glassware. Chipped or cracked items should be discarded or repaired. Vacuum-jacketed glassware should be handled with extreme care to prevent implosions. Evacuated equipment such as Dewar flasks or vacuum desiccators should be taped or shielded. Only glassware designed for vacuum work should be used for that purpose.

Hand protection should be used when picking up broken glass. Small pieces should be swept up with a brush into a dustpan. Glassblowing operations should not be attempted unless proper annealing facilities are available. Adequate hand protection should be used when inserting glass tubing into rubber stoppers or corks or when placing rubber tubing on glass hose connections. Cuts from forcing glass tubing into stoppers or plastic tubing are the most common kind of laboratory accident and are often serious. Tubing should be fire polished or rounded and lubricated, and hands should be protected with toweling and held close together to limit movement of glass should it fracture. The use of plastic or metal connectors should be considered.

(Refer to Chapter 6 for more discussion.)

5.C.8 Handling Flammable Substances

Flammable substances present one of the most widespread hazards encountered in the laboratory. Because flammable materials are employed in so many common laboratory operations, basic prudent laboratory practice should always assume the presence of fire hazard unless a review of the materials and operations in the laboratory verifies the absence of significant hazard. For example, simple operations with aqueous solutions in a laboratory where no flammable organic liquids are present involve no appreciable fire hazard. In all other circumstances, the risk of fire should be recognized and kept to a minimum.

For a fire to start, an ignition source, fuel, and oxidizer must be present. Prudent laboratory practice in avoiding fire is based on avoiding the presence of one of these components. The flammability and explosive characteristics of the materials being used should be known. Solvent labels, LCSSs, or other sources of information can be consulted to learn the flash point, vapor pressure, and explosive limit in air of each chemical handled. While all flammable substances should be handled prudently, the extreme flammability of some materials requires additional precautions.

To ensure that laboratory workers respond appropriately, they should be briefed on the necessary steps to take in case of a fire. The laboratory should be set up in such a way that the locations of fire alarms, pull stations, fire extinguishers, safety showers, and other emergency equipment are marked and all laboratory personnel alerted to them (see section 5.C.11 below). Exit routes in case of fire should be reviewed. Fire extinguishers in the immediate vicinity of an experiment should be appropriate to the particular fire hazards. Proper extinguishers must be used because fires can be exacerbated by use of an inappropriate extinguisher. Telephone numbers to call in case of an accident should be readily available.

(Refer to Chapter 3, section 3.B, for further information.)

5.C.9 Working with Scaled-up Reactions

Scale-up of reactions from those producing a few milligrams or grams to those producing more than 100 g of a product may represent several orders of magnitude of added risk. The attitudes, procedures, and controls applicable to large-scale laboratory reactions are fundamentally the same as those for smaller-scale procedures. However, differences in heat transfer, stirring effects, times for dissolution, and effects of concentration and the fact that substantial amounts of materials are being used introduce the need for special vigilance for scaled-up work. Careful planning and consultation with experienced workers to prepare for any eventuality are essential for large-scale laboratory work.

Although it is not always possible to predict whether a scaled-up reaction has increased risk, hazards should be evaluated if the following conditions exist:

• The starting material and/or intermediates contain functional groups that have a history of being explosive—e.g., N–N, N–O, N–halogen, O–O, and O–halogen bonds—or that could explode to give a large increase in pressure.
• A reactant or product is unstable near the reaction or work-up temperature. A preliminary test consists of heating a small sample in a melting point tube.
• A reaction is delayed; that is, an induction period is required.
• Gaseous by-products are formed.

• A reaction is exothermic. What can be done to provide cooling if the reaction begins to run away?
• A reaction requires a long reflux period. What will happen if solvent is lost owing to poor condenser cooling?
• A reaction requires temperatures below 0 °C. What will happen if the reaction warms to room temperature?

In addition, thermal phenomena that produce significant effects on a larger scale may not have been detected in smaller-scale reactions and therefore could be less obvious than toxic and/or environmental hazards. Thermal analytical techniques should be used to determine whether any process modifications are necessary.

(See sections 5.D.1 and 5.G.1 and Chapter 4, section 4.B, for more information.)

5.C.10 Responsibility for Unattended Experiments and Working Alone

Generally, it is prudent to avoid working alone at the bench in a laboratory building. Individuals working in separate laboratories outside of working hours should make arrangements to check on each other periodically, or ask security guards to check on them. Experiments known to be hazardous should not be undertaken by a worker who is alone in a laboratory. Under unusually hazardous conditions, special rules may be necessary.

Laboratory operations involving hazardous substances are sometimes carried out continuously or overnight with no one present. It is the responsibility of the worker to design these experiments so as to prevent the release of hazardous substances in the event of interruptions in utility services such as electricity, cooling water, and inert gas. Laboratory lights should be left on, and signs should be posted identifying the nature of the experiment and the hazardous substances in use. If appropriate, arrangements should be made for other workers to periodically inspect the operation. Information should be posted indicating how to contact the responsible individual in the event of an emergency.

(See also Chapter 3, section 3.A.)

5.C.11 Responding to Accidents and Emergencies

5.C.11.1 General Preparation for Emergencies

All laboratory personnel should know what to do in case of an emergency. Laboratory work should not be undertaken without knowledge of the following points:

- How to report a fire, injury, chemical spill, or other emergency to summon emergency response;
- The location of emergency equipment such as safety showers and eyewashes;
- The location of fire extinguishers and spill control equipment; and
- The locations of all available exits for evacuation from the laboratory.

The above information should be available in descriptions of laboratory emergency procedures and in the institution's Chemical Hygiene Plan. Laboratory supervisors should ensure that all laboratory workers are familiar with all of this information.

Inappropriate action by individuals inadequately trained in emergency procedures can make the consequences of an emergency worse. Laboratory workers should be aware of their level of expertise with respect to use of fire extinguishers and emergency equipment, dealing with chemical spills, and dealing with injuries. They should not take actions outside the limits of their expertise but instead should rely on trained personnel.

Names and telephone numbers of responsible individuals should be posted on the laboratory door.

5.C.11.2 Handling the Accidental Release of Hazardous Substances

Experiments should always be designed so as to minimize the possibility of an accidental release of hazardous substances. Experiments should use the minimal amounts of hazardous compounds practical, and such materials should be transported properly, using break-resistant bottles or secondary containers. Personnel should be familiar with the properties (physical, chemical, and toxicological) of hazardous substances before working with them. A contingency plan to deal with the accidental release of each hazardous substance should be in place. The necessary safety equipment, protective apparel, and spill control materials should be readily available. In the event of a laboratory-scale spill, the following general guidelines for handling it should be followed in the indicated order:

1. Notify other laboratory personnel of the accident and, if necessary, evacuate the area (see section 5.C.11.3).
2. Tend to any injured or contaminated personnel and, if necessary, request help (see section 5.C.11.4).
3. Take steps to confine and limit the spill if this can be done without risk of injury or contamination (see section 5.C.11.5).

4. Clean up the spill using appropriate procedures (see section 5.C.11.6). Dispose of contaminated materials properly, according to the procedures described in Chapter 7, section 7.B.8.

(See Chapter 6, section 6.F.2, for more information on emergency procedures.)

5.C.11.3 Notification of Personnel in the Area

Other nearby workers should be alerted to the accident and the nature of the chemicals involved. In the event of the release of a highly toxic gas or volatile material, the laboratory should be evacuated and personnel posted at entrances to prevent other workers from inadvertently entering the contaminated area. In some cases (e.g., incidents involving the release of highly toxic substances and spills occurring in nonlaboratory areas), it may be appropriate to activate a fire alarm to alert personnel to evacuate the entire building. The proper authorities should be called on for emergency assistance.

5.C.11.4 Treatment of Injured and Contaminated Personnel

If an individual is injured or contaminated with a hazardous substance, tending to him or her generally takes priority over implementing the spill control measures outlined in section 5.A.11.5 below. It is important to obtain medical attention as soon as possible by calling the posted number.

For spills covering small areas of skin, follow these procedures:

1. Immediately flush with flowing water for no less than 15 minutes.
2. If there is no visible burn, wash with warm water and soap, removing any jewelry to facilitate clearing of any residual materials.
3. Check the Material Safety Data Sheet (MSDS) to see if any delayed effects should be expected.
4. Seek medical attention for even minor chemical burns.
5. Do not use creams, lotions, or salves.

Take the following steps for spills on clothes:

1. Do not attempt to wipe the clothes.
2. Quickly remove all contaminated clothing, shoes, and jewelry while using the safety shower.
3. Seconds count, so do not waste time because of modesty.
4. Take care not to spread the chemical on the skin or, especially, in the eyes.

5. Use caution when removing pullover shirts or sweaters to prevent contamination of the eyes; it may be better to cut the garments off.

6. Immediately flood the affected body area with warm water for at least 15 minutes. Resume if pain returns.

7. Get medical attention as soon as possible.

8. Discard contaminated clothes or have them laundered separately from other clothing.

For splashes into the eye, take these steps:

1. Immediately flush with tepid potable water from a gently flowing source for at least 15 minutes.

2. Hold the individual's eyelids away from the eyeball, and instruct him or her to move the eye up and down and sideways to wash thoroughly behind the eyelids.

3. Use an eyewash. If one is not available, place the injured person on his or her back and pour water gently into the eyes for at least 15 minutes.

4. Follow first aid by prompt treatment by a member of a medical staff or an ophthalmologist who is acquainted with chemical injuries.

5.C.11.5   Spill Containment

Every laboratory in which hazardous substances are used should have spill control kits tailored to deal with the potential risk associated with the materials being used in the laboratory. These kits are used to confine and limit the spill if such actions can be taken without risk of injury or contamination. A specific individual should be assigned to maintain the kit. Spill control kits should be located near laboratory exits for ready access. Typical spill control kits might include these items:

- Spill control pillows. These commercially available pillows generally can be used for absorbing solvents, acids, and caustic alkalis, but not hydrofluoric acid.
- Inert absorbents such as vermiculite, clay, sand, kitty litter, and Oil Dri®. Paper is not an inert material and should not be used to clean up oxidizing agents such as nitric acid.
- Neutralizing agents for acid spills such as sodium carbonate and sodium bicarbonate.
- Neutralizing agents for alkali spills such as sodium bisulfate and citric acid.
- Large plastic scoops and other equipment such as brooms, pails, bags, and dust pans.
- Appropriate personal protective equipment, warnings, barricade tapes, and protection against slips or falls on wet floor during and after cleanup.

5.C.11.6   Spill Cleanup

Specific procedures for cleaning up spills vary depending on the location of the accident, the amount and physical properties of the spilled material, the degree and type of toxicity, and the training of the personnel involved. Outlined below are some general guidelines for handling several common spills:

- *Materials of low flammability that are not volatile or that have low toxicity.* This category of hazardous substances includes inorganic acids (e.g., sulfuric and nitric acid) and caustic bases (e.g., sodium and potassium hydroxide). For cleanup, appropriate protective apparel, including gloves, goggles, and (if necessary) shoe coverings should be worn. Absorption of the spilled material with an inert absorbent and appropriate disposal are recommended. The spilled chemicals can be neutralized with materials such as sodium bisulfate (for alkalis) and sodium carbonate or bicarbonate (for acids), absorbed on Floor-Dri® or vermiculite, scooped up, and disposed of according to the procedures detailed in Chapter 7, section 7.B.8.
- *Flammable solvents.* Fast action is crucial when a flammable solvent of relatively low toxicity is spilled. This category includes petroleum ether, pentane, diethyl ether, dimethoxyethane, and tetrahydrofuran. Other workers in the laboratory should be alerted, all flames extinguished, and any spark-producing equipment turned off. In some cases the power to the laboratory should be shut off with the circuit breaker, but the ventilation system should be kept running. The spilled solvent should be soaked up with spill absorbent or spill pillows as quickly as possible. These should be sealed in containers and disposed of properly. Nonsparking tools should be used in cleanup.
- *Highly toxic substances.* The cleanup of highly toxic substances should not be attempted alone. Other personnel should be notified of the spill, and the appropriate safety or industrial hygiene office should be contacted to obtain assistance in evaluating the hazards involved. These professionals will know how to clean up the material and may perform the operation.

5.C.11.7   Handling Leaking Gas Cylinders

Leaking gas cylinders constitute hazards that may be so serious as to require an immediate call for outside help. Workers should not apply extreme tension to close a stuck valve. Personal protective equipment should be worn. The following guidelines cover leaks of various types of gases:

- *Flammable, inert, or oxidizing gases.* The cylinder

should be moved to an isolated area, away from combustible material if the gas is flammable or an oxidizing agent, and signs should be posted that describe the hazards and state warnings. Care should be taken when moving leaking cylinders of flammable gases so that accidental ignition does not occur. If feasible, leaking cylinders should always be moved into laboratory hoods until exhausted.

• *Corrosive gases.* Corrosive gases may increase the size of the leak as they are released, and some corrosives are also oxidants, flammable, and/or toxic. The cylinder should be moved to an isolated, well-ventilated area, and suitable means used to direct the gas into an appropriate chemical neutralizer. If there is apt to be a reaction with the neutralizer that could lead to a "suck-back" into the valve (e.g., aqueous acid into an ammonia tank), a trap should be placed in the line before starting neutralization. Signs should be posted that describe the hazards and state warnings.

• *Toxic gases.* The same procedure should be followed for toxic gases as for corrosive gases, but for the protection of personnel, a special warning should be given for the added hazard of exposure. The cylinder should be moved to an isolated, well-ventilated area, and suitable means used to direct the gas into an appropriate chemical neutralizer. Signs should be posted that describe the hazards and state warnings. Appropriate personal protective equipment should be worn. (See also section 5.D.6.)

5.C.11.8 Handling Spills of Elemental Mercury

Mercury spills can be avoided by using supplies and equipment that do not contain mercury. However, most mercury spills do not pose a high risk. The initial response to a spill of elemental mercury should be to isolate the spill area and begin the cleanup procedure. Those doing the cleanup should wear protective gloves. The cleanup should begin with collecting the droplets. The large droplets can be consolidated by using a scraper or a piece of cardboard, and the pool of mercury removed with a pump or other appropriate equipment. A standard vacuum cleaner should never be used to pick up mercury. If a house vacuum system is used, it can be protected from the mercury by a charcoal filter in a trap. For cleaning up small mercury droplets, a special vacuum pump may be used, or the mercury may be picked up on wet toweling, which consolidates the small droplets to larger pieces, or picked up with a piece of adhesive tape. Commercial mercury spill cleanup sponges and spill control kits are available. The common practice of using sulfur should be discontinued because the practice is ineffective and the resulting waste creates a disposal problem. The mercury should be placed in a thick-wall high-density polyethylene bottle and transferred to a central depository for reclamation. After a mercury spill the exposed work surfaces and floors should be decontaminated by using an appropriate decontamination kit.

5.C.11.9 Responding to Fires

Fires are one of the most common types of laboratory accidents. Accordingly, all personnel should be familiar with general guidelines (as stated below) to prevent and minimize injury and damage from fires. Hands-on experience with common types of extinguishers and proper choice of extinguisher should be part of basic laboratory training.

(See also Chapter 6, section 6.F.2.)

The following should be noted:

• Preparation is essential! Make sure all laboratory personnel know the locations of all fire extinguishers in the laboratory, what types of fires they can be used for, and how to operate them correctly. Also ensure that they know the location of the nearest fire alarm pull station, safety showers, and emergency blankets.

• Even though a small fire that has just started can sometimes be extinguished with a laboratory fire extinguisher, attempt to extinguish such fires only if you are confident that you can do it successfully and quickly, and from a position in which you are always between the fire and an exit to avoid being trapped. Do not underestimate the danger from a fire, and remember that toxic gases and smoke may present additional hazards. Notify trained professionals.

• Fires in small vessels can usually be put out by covering the vessel loosely. Never pick up a flask or container of burning material.

• Extinguish small fires involving reactive metals and organometallic compounds (e.g., magnesium, sodium, potassium, and metal hydrides) with Met-L-X® or Met-L-Kyl® extinguishers or by covering with dry sand. Because these fires are very difficult to extinguish, sound the fire alarms before you attempt to extinguish the fire.

• In the event of a more serious fire, evacuate the laboratory and activate the nearest fire alarm. Upon their arrival, tell the fire department and emergency response team what hazardous substances are in the laboratory.

• If a person's clothing catches fire, have him or her immediately drop to the floor and roll. Dousing with water from the safety shower can be effective. Use fire blankets only as a last resort because they tend to hold in heat and to increase the severity of burns. Remove contaminated clothing quickly, douse the person with water, and place clean, wet, cold cloth on burned areas.

Wrap the injured person in a blanket to avoid shock, and get medical attention promptly.

5.D WORKING WITH SUBSTANCES OF HIGH TOXICITY

Individuals who are working with highly toxic chemicals, as identified in Chapter 3, section 3.C, should be thoroughly familiar with the general guidelines for the safe handling of chemicals in laboratories (see section 5.C). They should also have acquired through training and experience the knowledge, skill, and discipline to carry out safe laboratory practices consistently. But these guidelines alone are not sufficient when handling substances that are known to be highly toxic and chemicals that, when combined in an experimental reaction, may generate highly toxic substances or produce new substances with the potential for high toxicity. Additional precautions are needed to set up *multiple lines of defense to minimize the risks* posed by these substances. As discussed in section 5.B, preparations for handling highly toxic substances must include sound and thorough planning of the experiment, understanding the intrinsic hazards of the substances and the risks of exposure inherent in the planned processes, selecting additional precautions that may be necessary to minimize or eliminate these risks, and reviewing all emergency procedures to ensure appropriate response to unexpected spills and accidents. Each experiment must be evaluated individually because assessment of the level of risk for work with any substance depends on how the substance will be used. Therefore, it would not be prudent for the planner to rely solely on a list of "highly toxic" chemicals to determine the level of the risk; under certain conditions, even chemicals not on these lists may become highly toxic.

In general, the guidelines in section 5.C reflect the minimum standards for handling hazardous substances. They should become standard practice when highly toxic substances are handled in the laboratory. For example, it is always preferable to avoid working alone in laboratories. However, when highly toxic materials are being handled, it is essential that more than one person be present and that all people working in the area be familiar with the hazards of the experiments being conducted and with the appropriate emergency response procedures. Personal protective equipment to safeguard the hands, forearms, and face from exposure to chemicals, while desirable in most circumstances, is essential in handling highly toxic materials. Good housekeeping creates an intrinsically safer workplace and should be maintained scrupulously in areas where highly toxic substances are handled. Source reduction is always a prudent practice, but in the case

of highly toxic chemicals it may mean the difference between working with toxicologically dangerous amounts of materials and working with quantities that can be handled safely with routine practice. Similarly, emergency response planning and training become very important when working with highly toxic compounds. Additional hazards from these materials (e.g., flammability and high vapor pressures) can complicate the situation, making operational safety all the more important.

5.D.1 Planning

Careful planning needs to precede any experiment involving a highly toxic substance whenever the substance is to be used for the first time or whenever an experienced user carries out a new protocol that increases substantially the risk of exposure. Planning should include consultations with colleagues who have experience in handling the substance safely and in protocols of use. Experts in the institution's environmental health and safety program are a valuable source of information on the hazardous properties of chemicals and safe practice. They also need to be consulted for guidance regarding those chemicals that are regulated by federal, state, and local agencies or by institutional policy. Training and documentation requirements may have to be incorporated into the experiment plan.

Effective planning is always guided by two principles: substitution of highly toxic substances with less toxic alternatives whenever appropriate and use of the smallest amount of material that is practicable for the conduct of the experiment. Other important factors to be considered in determining the need for additional safeguards are the likelihood of exposure inherent in the proposed experimental process, the toxicological and physical properties of the chemical substances being used, the concentrations and amounts involved, the duration of exposure, and known toxicological effects. It is also important to plan for careful management of the substances throughout their life cycle—from acquisition and storage through destruction or safe disposal.

5.D.2 Experiment Protocols Involving Highly Toxic Chemicals

Experiment plans that involve the use of highly toxic substances or high-risk protocols should be considered carefully, and experienced personnel or an appropriate source should be consulted about the risk. An experiment plan that describes the additional safeguards that will be used for all phases of the experiment from acquisition of the chemical to its final safe disposal should be in place before the experiment begins. The amounts of materials used and the names of the people involved in the laboratory work should be included in the written summary and recorded in the laboratory notebook.

The planning process may determine that area monitoring and/or medical surveillance is necessary for ensuring the safety of the experimenters. Such a determination is likely to be made only when there is reason to believe that exposure levels for the substances planned to be used in an experiment could exceed OSHA-established regulatory action levels or similar guidelines established by other authoritative organizations. It would be prudent to review the amounts of material to be used, the toxicological properties of the substances, the opportunity for and duration of exposure, and plans for waste disposal for any experiment plans involving highly hazardous chemicals.

5.D.3 Designated Areas

Most experimental procedures involving highly toxic chemicals, including their transfer from storage containers to reaction vessels, should be confined to a designated work area in the laboratory. This area, which could be a hood or glove box, a portion of a laboratory, or the entire laboratory module, should be recognized by everyone in the laboratory or institution as a place where special precautions, laboratory skill, and safety discipline are required. Conspicuous signs should clearly indicate which areas are designated. It is not necessary to restrict the use of a designated area to the handling of highly toxic chemicals as long as laboratory personnel are aware of the nature of the substances being used and of the precautions that are necessary, and have been trained appropriately for emergency response. It may also be prudent to post relevant Laboratory Chemical Safety Summaries (LCSSs) outside the laboratory door.

The laboratory supervisor should determine which procedures need to be confined to designated areas. The general guidelines (section 5.C) for handling hazardous chemicals in laboratories may be sufficient for procedures involving low concentrations and small amounts of highly toxic chemicals, depending on the experiment, the reagents, and their toxicological and physical properties.

5.D.4 Access Control

Only persons who are directly involved in the laboratory work and who have been advised of the special precautions that may apply should have access to laboratories where highly toxic chemicals are handled. Administrative procedures or even physical barriers may be required to prevent unauthorized personnel from entering these laboratories.

The use of locks and barricades may be appropriate to limit access to unattended areas where large amounts of highly toxic materials are being handled routinely or stored. However, it is important that locks not prevent emergency exits from the laboratory or hinder entrance for emergency response. Locks are generally more appropriate for securing storage areas and unattended laboratories than for preventing access to laboratories in which toxic chemicals are being actively used.

Some long experiments involving highly toxic compounds may require unattended operations. In such cases, securing the laboratory from access by untrained personnel is essential. These operations should also include fail-safe backup options such as shutoff devices in case a reaction overheats or pressure builds up. Additionally, equipment should include interlocks that shut down experiments by turning off devices such as heating baths or reagent pumps, or that close solenoid valves if cooling water stops flowing through an apparatus or if airflow through a fume hood becomes restricted or stops. An interlock should be constructed carefully in such a way that if a problem develops, it places the experiment in a safer mode and will not reset even if the hazardous condition is reversed. Protective devices should include alarms that indicate their activation. Security guards and untrained personnel should never be asked or allowed to check on the status of unattended experiments involving highly toxic materials. Warning signs on locked doors should list the trained laboratory workers who can be contacted in case an alarm sounds within the laboratory.

5.D.5 Special Precautions for Minimizing Exposure to Highly Toxic Chemicals

The practices listed below help build the necessary multiple lines of defense to enable laboratory work with highly toxic chemicals to be conducted safely:

1. Procedures involving highly toxic chemicals that can generate dust, vapors, or aerosols must be conducted in a hood, glove box, or other suitable containment device. Hoods should be checked for acceptable operation prior to conducting experiments with toxic chemicals. If experiments are to be ongoing over a significant period of time, the hood should be rechecked at least quarterly for integrity of flow. Hoods in continuous or long-term use with toxic materials

should be equipped with flow-sensing devices that can show at a glance or by an audible signal whether they are performing adequately. When toxic chemicals are used in a glove box, it should be operated under negative pressure, and the gloves should be checked for integrity and appropriate composition before use. Any effluent from these reactions should be reactively or chemically scrubbed and/or cleaned with HEPA (high-efficiency particulate air) filters prior to discharge into the hood atmosphere. Hoods should not be used as waste disposal devices, particularly when toxic substances are involved. In order to offer maximum protection, hoods should be operated with sashes closed whenever possible, and experiments involving toxic materials should be shielded further. Monitoring equipment might include both active and passive devices to sample laboratory working environments. (See Chapter 8, section 8.C, for detailed discussion on hoods and environmental control.)

2. When working with toxic liquids or solids, it is critical that gloves be worn to protect the hands and forearms. These gloves must be carefully selected to ensure that they are impervious to the chemicals being used and are of appropriate thickness to allow reasonable dexterity while also ensuring adequate barrier protection. Double gloves can provide a multiple line of defense and are likely to be appropriate for many situations with highly toxic chemicals. When risks from toxicity are only one facet of working with a given chemical or experimental apparatus, it is important to find a glove or combination of gloves that addresses all of the hazards present.

When using gloves, it is important to exercise proper hygiene. Reusable gloves should be washed and inspected before and after each use. Gloves that might be contaminated with toxic materials should not be removed from the immediate area (usually a hood) in which the chemicals are located. They should never be worn when handling common items such as doorknobs, elevator buttons, handles, or switches on common equipment. Other types of personal protective equipment, such as aprons of reduced permeability and disposable laboratory coats, can offer additional safeguards when working with large quantities of toxic materials.

3. Face and eye protection is also essential in preventing ingestion, inhalation, and skin absorption of toxic chemicals in the case of unexpected events. Safety glasses with side shields are a minimum standard for all laboratory work. When using toxic substances that could generate vapors, aerosols, or dusts, additional levels of protection, including full-face shields and respirators, are appropriate, depending on the degree of hazard represented. Transparent explosion shields in hoods offer additional protection from splashes. Medi-

cal supervision or surveillance may be warranted when using some toxic substances, particularly when large quantities of chemicals are involved or experiments are conducted with smaller quantities over an extended period of time. Medical certification may also be required if respirators are worn.

4. Equipment used for the handling of highly toxic chemicals should be suitably isolated from the general laboratory environment. Laboratory vacuum pumps used with these substances should be protected by high-efficiency scrubbers or HEPA filters and vented into an exhaust hood. Motor-driven vacuum pumps are recommended because they are easy to decontaminate (decontamination should be conducted in a designated hood).

5. Good laboratory hygiene should never be compromised in laboratories where highly toxic chemicals are handled. After using toxic materials the laboratory worker should wash his or her face, hands, neck, and arms. Equipment (including personal protective equipment such as gloves) that might be contaminated must never be removed from the environment reserved for handling toxic materials without complete decontamination. When possible, laboratory equipment and glassware should be chosen with an eye toward the ease of cleaning and decontamination. Mixtures that contain toxic chemicals or substances of unknown toxicity must never be smelled or tasted.

6. Transportation of very toxic chemicals from one location to another should be planned carefully, and handling of these materials outside the specially designated laboratory area should be minimized. When these materials are transported, the full complement of personal protective equipment appropriate to the chemicals in question should be worn, and the samples should be carried in unbreakable secondary containers.

5.D.6 Preventing Accidents and Spills with Substances of High Toxicity

Emergency response procedures must cover highly toxic substances because such procedures provide the last line of defense in working with these chemicals. Spill control and appropriate emergency response kits should be nearby, and laboratory workers should be trained in their proper use. To avoid their being contaminated or made inaccessible in an emergency, these kits should not be located within the immediate area where highly toxic substances are handled. Spill control absorbents, impermeable ground covers (to prevent the spread of contamination while conducting emergency response), warning signs, emergency barriers, first aid supplies, and antidotes should be in these kits. The contents of the kits should be validated before starting experiments. Safety showers, eyewashes, and fire extinguishers should be readily available nearby. Self-contained impermeable suits, a self-contained breathing apparatus, and cartridge respirators may also be appropriate for spill response preparedness, depending on the physical properties and toxicity of the materials being used (see section 5.C.2.3).

Experiments conducted with highly toxic chemicals should be carried out in work areas designed to contain accidental releases (see also section 5.D.3). Hood trays and other types of secondary containers should be used to contain inadvertent spills, and careful technique must be observed to minimize the potential for spills and releases.

All toxicity and emergency response information about the highly toxic chemicals being used should be readily available both before and during experimentation and should be located outside the immediate work area to ensure accessibility in emergencies. All laboratory workers who could potentially be exposed must be properly trained to participate in first aid or emergency response operations. In some cases the frequency with which highly toxic chemicals are used or the quantities involved might make formal emergency response drills warranted. Such "dry runs" may involve medical personnel as well as emergency cleanup crews.

(See also sections 5.C.11.5 and 5.C.11.6.)

5.D.7 Storage and Waste Disposal

Highly toxic chemicals should be stored in unbreakable secondary containers. If the materials are volatile or could react with moisture or air to form volatile toxic compounds, these secondary containers should be placed in a ventilated environment under negative pressure. All containers of highly toxic chemicals should be labeled clearly with chemical composition, known hazards, and warnings for handling. Chemicals that can combine to make highly toxic materials (e.g., acids and inorganic cyanides, which can generate hydrogen cyanide) should not be stored together in the same secondary container. A list of highly toxic compounds, their locations, and contingency plans for dealing with spills should be displayed prominently at any storage facility. Access to areas where highly toxic compounds are stored should be restricted to workers who are familiar with the risks they pose and who have been trained to handle these chemicals. Highly toxic chemicals that have a limited shelf life need to be tracked and monitored for deterioration in the storage facility. Those that require refrigeration should be stored in a ventilated refrigeration facility.

Procedures for disposal of highly toxic materials should be established before experiments begin, preferably before the chemicals are ordered. The procedures should address methods for decontamination of all laboratory equipment that contacts (or could contact) highly toxic chemicals. Waste should be accumulated in clearly labeled, impervious containers that are stored in unbreakable secondary containers. Volatile or reactive waste should always be covered to minimize release to the hood environment in which it is being handled.

It is the responsibility of the experimenter and the laboratory supervisor to ensure that waste is disposed of in a manner that renders it innocuous. This may involve pretreatment of the waste either before or during accumulation. In other circumstances, prudence might dictate that highly toxic compounds never be moved from an enclosed environment and might suggest in-laboratory destruction as the safest and most effective way of dealing with the waste. Regulatory requirements may have an impact on this decision (see Chapter 9). If waste cannot be rendered harmless in the laboratory, then accumulation in closed, impervious containers within secondary containment systems is prudent. The choice of methods for final disposal must ensure that these chemicals are completely destroyed or rendered harmless in some manner.

5.D.8 Multihazardous Materials

Some highly toxic materials present additional hazards because of their flammability (see Chapter 3, sections 3.D.1 and 3.D.4; see also section 5.F), volatility (see sections 5.E and 5.G.6), explosibility (see Chapter 3, section 3.D.3; see also section 5.G.4), or reactivity (see Chapter 3, section 3.D.2; see also section 5.G.2). These materials warrant special attention to ensure that risks are minimized and that plans to deal effectively with all potential hazards and emergency response are implemented. (Tables 3.9 and 3.14 give information regarding incompatible chemicals and substances requiring extreme caution.)

5.E WORKING WITH BIOHAZARDOUS AND RADIOACTIVE MATERIALS

5.E.1 Biohazardous Materials

For even the most experienced laboratory worker, a careful review of the publication *Biosafety in Microbiological and Biomedical Laboratories* (U.S. DHHS, 1993) should be a prerequisite for beginning any laboratory activity involving a microorganism. It defines four levels of control that are appropriate for safe laboratory work with microorganisms that present occupational risks ranging from no risk of disease for normal healthy individuals to high individual risk of life-threatening disease, and it recommends guidelines for handling specific agents. The four levels of control, referred to as biosafety levels 1 through 4, describe microbiological practices, safety equipment, and features of laboratory facilities for the corresponding level of risk associated with handling a particular agent. The selection of a biosafety level is influenced by several characteristics of the infectious agent, the most important of which are the severity of the disease, the documented mode of transmission of the infectious agent, the availability of protective immunization or effective therapy, and the relative risk of exposure created by manipulations used in handling the agent.

Biosafety level 1 is the basic level of protection appropriate only for agents that are not known to cause disease in normal, healthy humans. Biosafety level 2 is appropriate for handling a broad spectrum of moderate-risk agents that cause human disease by ingestion or through percutaneous or mucous membrane exposure. Hepatitis B virus, human immunodeficiency virus (HIV), and salmonellae and toxoplasma spp. are representative of agents assigned to this biosafety level. Extreme precaution with needles or sharp instruments is emphasized at this level. A higher level of control may be indicated when some of these agents, especially HIV, are grown and concentrated.

Biosafety level 3 is appropriate for agents with a potential for respiratory transmission and for agents that may cause serious and potentially lethal infections. Emphasis is placed on the control of aerosols by containing all manipulations. At this level, the facility is designed to control access to the laboratory and includes a specialized ventilation system, such as a biological safety cabinet, that minimizes the release of infectious aerosols from the laboratory. The bacterium *Mycobacterium tuberculosis* is an example of an agent for which this higher level of control is appropriate. Exotic agents that pose a high individual risk of life-threatening disease by the aerosol route and for which no treatment is available are restricted to high containment laboratories that meet biosafety level 4 standards. Worker protection in these laboratories is provided by the use of physically sealed glove boxes or fully enclosed barrier suits that supply breathing air.

Several authoritative reference works are available that provide excellent guidance for the safe handling of infectious microorganisms in the laboratory, one of which is *Biosafety in the Laboratory—Prudent Practices for the Handling and Disposal of Infectious Materials* (NRC, 1989). Standard microbiological practices described in these references are consistent with the prudent practices used for the safe handling of chemicals.

Practices that are most helpful for preventing laboratory-acquired infections are as follows:

- Wear protective gloves and a laboratory coat or gown.
- Wash hands after infectious material is handled, after gloves are removed, and before leaving the laboratory.
- Perform procedures carefully to reduce the possibility of creating splashes or aerosols.
- Contain in biological safety cabinets operations that generate aerosols.
- Use mechanical pipetting devices.
- Promptly decontaminate work surfaces after spills of infectious materials and when procedures are completed.
- Never eat, drink, smoke, handle contact lenses, apply cosmetics, or take or apply medicine in the laboratory.
- Wear eye protection.
- Take special care when using "sharps," that is, syringes, needles, Pasteur pipets, capillary tubes, scalpels, and other sharp instruments.
- Keep laboratory doors closed when experiments are in progress.
- Use secondary leak-proof containers to move or transfer cultures.
- Decontaminate infectious waste before disposal.

5.E.2 Radioactive Materials

Prudent practices for working with radioactive materials are similar to those needed to reduce the risk of exposure to toxic chemicals (section 5.C has similiar information) and to biohazards:

- Know the characteristics of the radioisotopes that are being used, including half-life, types and energies of emitted radiations, the potential for exposure, how to detect contamination, and the annual limit on intake.
- Protect against exposure to airborne and ingestible radioactive materials.
- Never eat, drink, smoke, handle contact lenses, apply cosmetics, or take or apply medicine in the laboratory, and keep food, drinks, cosmetics, and tobacco products out of the laboratory entirely so that they cannot become contaminated.
- Do not pipet by mouth.
- Provide for safe disposal of waste radionuclides and their solutions.
- Use protective equipment to minimize exposures.
- Use equipment that can be manipulated remotely, as well as shielding, glove boxes, and personal protective equipment, including gloves, clothing, and respirators, as appropriate.
- Plan experiments so as to minimize exposure by reducing the time of exposure, using shielding against exposure, increasing your distance from the radiation,

and paying attention to monitoring and decontamination.

- Keep an accurate inventory of radioisotopes.
- Record all receipts, transfers, and disposals of radioisotopes.
- Record surveys.
- Check workers and the work area each day that radioisotopes are used.
- Minimize radioactive waste.
- Plan procedures to use the smallest amount of radioisotope possible.
- Check waste materials for contamination before discarding.
- Place only materials with known or suspected radioactive contamination in appropriate radioactive waste containers.
- Do not generate multihazardous waste (combinations of radioactive, biological, and chemical waste) without first consulting with the designated radiation and chemical safety officers.

(See Chapter 7 for more information on waste and disposal.)

5.F WORKING WITH FLAMMABLE CHEMICALS

All laboratory personnel should know the properties of chemicals they are handling as well as have a basic understanding of how these properties might be affected by the variety of conditions found in the laboratory. As stated in section 5.B, Laboratory Chemical Safety Summaries (LCSSs) or other sources of information should be consulted for further information such as vapor pressure, flash point, and explosive limit in air. The use of flammable substances is common, and their properties are also discussed in Chapter 3, section 3.D.

General prudent practices include minimizing the amounts used, storing chemicals properly, keeping appropriate fire extinguishing equipment readily available, physically separating flammable materials from other operations and sources of ignition, properly grounding static sources of ignition, and using the least hazardous alternative available.

Ignition sources should be eliminated from any area where flammable substances are handled. Open flames, such as Bunsen burners, matches, and smoking tobacco, are obvious ignition sources. Gas burners should not be used as a source of heat in any laboratory where flammable substances are used. Less obvious ignition sources include gas-fired space heating or water-heating equipment and electrical equipment, such as stirring devices, motors, relays, and switches, which can all produce sparks that will ignite flammable

vapors. Because the location of this equipment is often fixed, operations with flammable substances may have to be carried out elsewhere.

Even low-level sources of ignition, such as hot plates, steam lines, or other hot surfaces, can provide a sufficiently energetic ignition source for the most flammable substances in general laboratory use, such as diethyl ether and carbon disulfide (see Chapter 3, section 3.D.1.3). Flammable substances that require low-temperature storage should be stored only in refrigerators designed for that purpose. Ordinary refrigerators are a hazard because of the presence of potential ignition sources, such as switches, relays, and, possibly, sparking fan motors, and should never be used for storing chemicals. When transferring flammable liquids in metal containers, sparks from accumulated static charge must be avoided by grounding.

Fire hazards posed by water-reactive substances such as alkali metals and metal hydrides, pyrophoric substances such as metal alkyls, strong oxidizers such as perchloric acid, and flammable gases such as acetylene require procedures beyond the standard prudent practices for handling chemicals described here (see sections 5.C and 5.D) and should be researched in LCSSs or other references before work begins. In addition, emergency response to incidents involving these substances must take their special hazards into account.

5.F.1 Flammable Materials

The basic precautions for safe handling of flammable materials include the following:

- Handle flammable substances only in areas free of ignition sources. Besides open flames, ignition sources include electrical equipment (especially motors), static electricity, and, for some materials (e.g., carbon disulfide), even hot surfaces. Check the work area for flames or ignition sources before using a flammable substance. Before igniting a flame, check for the presence of a flammable substance.
- Never heat flammable substances with an open flame. Preferred heat sources include steam baths, water baths, oil and wax baths, salt and sand baths, heating mantles, and hot air or nitrogen baths.
- Ventilation by diluting the vapors until they are no longer flammable is one of the most effective ways to prevent the formation of flammable gaseous mixtures. Use appropriate and safe exhaust whenever appreciable quantities of flammable substances are transferred from one container to another, allowed to stand in open containers, heated in open containers, or handled in any other way. In using dilution techniques, make certain that equipment (e.g., fans) used to pro-

vide dilution is explosion proof and that sparking items are located outside the air stream.

- Keep containers of flammable substances tightly closed at all times when not in use.
- Use only refrigeration equipment certified for storage of flammable materials.
- Use the smallest quantities of flammable substances compatible with the need, and, especially when the flammable liquid must be stored in glass, purchase the smallest useful size bottle.

5.F.2 Flammable Liquids

Flammable liquids burn only when their vapor is mixed with air in the appropriate concentration. Therefore, such liquids should always be handled so as to minimize the creation of flammable vapor concentrations. Dilution of flammable vapors by ventilation is an important means of avoiding flammable concentrations. Containers of liquids should be kept closed except during transfer of contents. Transfers should be carried out only in fume hoods or in other areas where ventilation is sufficient to avoid a buildup of flammable vapor concentrations. Spillage or breakage of vessels or containers of flammable liquids or sudden eruptions from nucleation of heated liquid can result in a sudden release of vapor, which will produce an unexpected quantity of flammable vapor.

Metal lines and vessels discharging flammable liquids should be grounded properly and also grounded to discharge static electricity. For instance, when transferring flammable liquids in metal equipment, avoid static-generated sparks by grounding and the use of ground straps. Development of static electricity is related closely to the level of humidity and may become a problem on very cold, dry winter days. When nonmetallic containers (especially plastic) are used, the contact should be made directly to the liquid with the grounding device rather than to the container. In the rare circumstance that static electricity cannot be avoided, all processes should be carried out as slowly as possible to give the accumulated charge time to disperse, or should be handled in an inert atmosphere.

Note that vapors of many flammable liquids are heavier than air and capable of traveling considerable distances along the floor. This possibility should be recognized, and special note should be taken of ignition sources at a lower level than that at which the substance is being used. Close attention should be given to nearby potential sources of ignition.

5.F.3 Flammable Gases

Leakage or escape of flammable gases can produce an explosive atmosphere in the laboratory. Acetylene, hydrogen, ammonia, hydrogen sulfide, propane, and carbon monoxide are especially hazardous. Acetylene, methane, and hydrogen have very wide flammability limits, which adds greatly to their potential fire and explosion hazard. Installation of flash arresters on hydrogen cylinders is recommended. Prior to introduction of a flammable gas into a reaction vessel, the equipment should be purged by evacuation or with an inert gas. The flush cycle should be repeated three times to reduce residual oxygen to about 1%.

(See section 5.H for specific precautions on the use of compressed gases.)

5.F.4 Catalyst Ignition of Flammable Materials

Palladium or platinum on carbon, platinum oxide, Raney nickel, and other hydrogenation catalysts should be filtered carefully from hydrogenation reaction mixtures. The recovered catalyst is usually saturated with hydrogen, is highly reactive, and, thus, inflames spontaneously on exposure to air. Especially for large-scale reactions, the filter cake should not be allowed to become dry. The funnel containing the still-moist catalyst filter cake should be put into a water bath immediately after completion of the filtration. Use of a purge gas (nitrogen or argon) is strongly recommended for hydrogenation procedures so that the catalyst can then be filtered and handled under an inert atmosphere.

5.G WORKING WITH HIGHLY REACTIVE OR EXPLOSIVE CHEMICALS

An explosion results when a material undergoes rapid reaction that results in a violent release of energy. Such reactions can occur spontaneously or be initiated and can produce pressures, gases, and fumes that are hazardous. Highly reactive and explosive materials used in the laboratory require appropriate procedures. In this section, techniques for identifying and handling potentially explosive materials are discussed.

5.G.1 Overview

Light, mechanical shock, heat, and certain catalysts can be initiators of explosive reactions. Hydrogen and chlorine react explosively in the presence of light. Examples of shock-sensitive materials include acetylides, azides, organic nitrates, nitro compounds, perchlorates, and many peroxides. Acids, bases, and other substances can catalyze the explosive polymerizations. The catalytic effect of metallic contamination can lead to explosive situations. Many metal ions can catalyze the violent decomposition of hydrogen peroxide.

Many highly reactive chemicals can polymerize vigorously, decompose, condense, and/or become self-reactive. The improper handling of these materials may result in a runaway reaction that could become violent. Careful planning is essential to avoid serious accidents. When highly reactive materials are in use, emergency equipment should be at hand. The apparatus should be assembled in such a way that if the reaction begins to run away, immediate removal of any heat source, cooling of the reaction vessel, cessation of reagent addition, and closing of laboratory hood sashes are possible. Evacuation of personnel until the reaction is under control is advisable. A heavy, transparent plastic explosion shield should be in place to provide extra protection in addition to the hood window.

Highly reactive chemicals can lead to reactions with rates that increase rapidly as the temperature increases. If the heat evolved is not dissipated, the reaction rate can increase until an explosion results. Such an event must be prevented, particularly when scaling up experiments. Sufficient cooling and surface for heat exchange should be provided to allow control of the reaction. It is also important that the concentrations of the solutions used not be excessive, especially when a reaction is being attempted or scaled up for the first time. Use of too highly concentrated reagents has led to runaway conditions and to explosions. Particular care must also be given to the rate of reagent addition versus its rate of consumption, especially if the reaction is subject to an induction period.

Large-scale reactions with organometallic reagents and reactions that produce flammables as products and/or are carried out in flammable solvents require special attention. Active metals, such as sodium, magnesium, lithium, and potassium, are a serious fire and explosion risk because of their reactivity with water, alcohols, and other compounds containing acidic OH. These materials require special storage, handling, and disposal procedures. Where active metals are present, Class D fire extinguishers that use special extinguishing materials such as a plasticized graphite-based powder or a sodium chloride-based powder (Met-L-X®) are required.

Some chemicals decompose when heated. Slow decomposition may not be noticeable on a small scale, but on a large scale with inadequate heat transfer, or if the evolved heat and gases are confined, an explosive situation can develop. The heat-initiated decomposition of some substances, such as certain peroxides, is almost instantaneous. In particular, reactions that are subject to an induction period can be dangerous because there is no initial indication of a risk, but after the induction a violent process can result.

Oxidizing agents may react violently when they come in contact with reducing materials, trace metals, and sometimes ordinary combustibles. These compounds include the halogens, oxyhalogens, and peroxyhalogens, permanganates, nitrates, chromates, and persulfates, as well as peroxides (see also section 5.G.3). Inorganic peroxides are generally considered to be stable. However, they may generate organic peroxides and hydroperoxides in contact with organic compounds, react violently with water (alkali metal peroxides), or form superoxides and ozonides (alkali metal peroxides). Perchloric acid and nitric acid are powerful oxidizing agents with organic compounds and other reducing agents. Perchlorate salts can be explosive and should be treated as potentially hazardous compounds. "Dusts"—suspensions of oxidizable particles (e.g., magnesium powder, zinc dust, carbon powder, or flowers of sulfur) in the air—constitute a powerful explosive mixture.

Scale-up of reactions can create difficulties in dissipation of heat that are not evident on a smaller scale. Evaluation of observed or suspected exothermicity can be achieved by differential thermal analysis (DTA) to identify exothermicity in open reaction systems; differential scanning calorimetry (DSC), using a specially designed sealable metal crucible, to identify exothermicity in closed reaction systems; or syringe injection calorimetry (SIC) and reactive systems screening tool (RSST) calorimetry to determine heats of reaction on a microscale and small scale. (For an expanded discussion of identifying process hazards using thermal analytical techniques, see Tuma (1991).) When it becomes apparent that an exotherm exists at a low temperature and/or a large exotherm occurs that might present a hazard, large-scale calorimetry determination of exothermic onset temperatures and drop weight testing are advisable. In situations where formal operational hazard evaluation or reliable data from any other source suggest a hazard, review or modification of the scale-up conditions by an experienced group is recommended to avoid the possibility that an individual might overlook a hazard or the most appropriate procedural changes.

Any given sample of a highly reactive material may be dangerous. Furthermore, the risk is associated not with the total energy released, but rather with the remarkably high rate of a detonation reaction. A high-order explosion of even milligram quantities can drive small fragments of glass or other matter deep into the eye. It is important to use minimum amounts of hazardous materials with adequate shielding and personal protection.

Not all explosions result from chemical reactions. A dangerous, physically caused explosion can occur if a hot liquid is brought into sudden contact with a lower-boiling-point one. The instantaneous vaporization of the lower-boiling-point substance can be hazardous to personnel and destructive to equipment. The presence or inadvertent addition of water to the hot fluid of a heating bath is an example of such a hazard. Explosions can also occur when warming a cryogenic material in a closed container or overpressurizing glassware with nitrogen (N_2) or argon when the regulator is incorrectly set. Violent physical explosions have also occurred when a collection of very hot particles is suddenly dumped into water. For this reason, dry sand should be used to catch particles during laboratory thermite reaction demonstrations.

5.G.2 Reactive or Explosive Compounds

Occasionally, it is necessary to handle materials that are known to be explosive or that may contain explosive impurities such as peroxides. Since explosive chemicals might be detonated by mechanical shock, elevated temperature, or chemical action with forces that release large volumes of gases, heat, and often toxic vapors, they must be treated with special care.

The proper handling of highly energetic substances without injury demands attention to the most minute detail. The unusual nature of work involving such substances requires special safety measures and handling techniques that must be understood thoroughly and followed by all persons involved. The practices listed in this section are a guide for use in any laboratory operation that might involve explosive materials.

Work with explosive (or potentially explosive) materials generally requires the use of special protective apparel (e.g., face shields, gloves, and laboratory coats) and protective devices such as explosion shields, barriers, or even enclosed barricades or an isolated room with a blowout roof or window (see Chapter 6, sections 6.F.1 and 6.F.2). Before work with a potentially explosive material is begun, the experiment should be discussed with a supervisor or an experienced co-worker, and/or the relevant literature consulted (see Chapter 3, sections 3.B.2, 3.B.5, and 3.B.6). A risk assessment should be carried out.

Various state and federal regulations cover the transportation, storage, and use of explosives. These regulations should be consulted before explosives (and related dangerous materials) are used or generated in the laboratory. Explosive materials should be brought into the laboratory only as required and then in the smallest quantities adequate for the experiment (see Chapter 4, section 4.B). Insofar as possible, direct handling should be minimized. Explosives should be segregated from other materials that could create a serious risk to life or property should an accident occur.

5.G.2.1 Personal Protective Apparel

When explosive materials are handled, the following items of personal protective apparel are needed:

• Safety glasses that have solid side shields or goggles should be worn by all personnel, including visitors, in the laboratory.
• Full-length shields that fully protect the face and throat should be worn whenever the worker is in a hazardous or exposed position. Special care is required when operating or manipulating synthesis systems that may contain explosives (e.g., diazomethane), when bench shields are moved aside, and when handling or transporting such systems. In view of the special hazard to life that results from severing the jugular vein, extra shielding around the throat is recommended.
• Heavy leather gloves should be worn if it is necessary to reach behind a shielded area while a hazardous experiment is in progress or when handling reactive compounds or gaseous reactants. Proper planning of experiments should minimize the need for such activities.
• Laboratory coats should be worn at all times in explosives laboratories. The coat should be made of flame-resistant material and should be quickly removable. A coat can help reduce minor injuries from flying glass as well as the possibility of injury from an explosive flash.

5.G.2.2 Protective Devices

Barriers such as shields, barricades, and guards should be used to protect personnel and equipment from injury or damage from a possible explosion or fire. The barrier should completely surround the hazardous area. On benches and hoods, a 0.25-inch-thick acrylic sliding shield, which needs to be screwed together in addition to being glued, can effectively protect a worker from glass fragments resulting from a laboratory-scale detonation. The shield should be in place whenever hazardous reactions are in progress or whenever hazardous materials are being stored temporarily. However, such shielding is not effective against metal shrapnel. The laboratory hood sash provides a safety shield only against chemical splashes or sprays, fires, and minor explosions. If more than one hazardous reaction is carried out, the reactions should be shielded from each other and separated as far as possible.

Dry boxes should be fitted with safety glass windows overlaid with 0.25-inch-thick acrylic when potentially explosive materials capable of detonation in an inert atmosphere are to be handled. This protection is adequate against most internal 5-g detonations. Protective gloves should be worn over the rubber dry box gloves to provide additional protection. Other safety devices that allow remote manipulation should be used with the gloves. Detonation of explosives from static sparks can be a considerable problem in dry boxes, so adequate grounding is essential, and an antistatic gun is recommended.

Armored hoods or barricades made with thick (1.0 inch) polyvinylbutyral resin shielding and heavy metal walls give complete protection against detonations not in excess of the acceptable 20-g limit. These hoods are designed to contain a 100-g explosion, but an arbitrary 20-g limit is usually set because of the noise level in the event of a detonation. Such hoods should be equipped with mechanical hands that enable the operator to manipulate equipment and handle adduct containers remotely. A sign, such as

CAUTION: NO ONE MAY ENTER AN ARMORED HOOD FOR ANY REASON DURING THE COURSE OF A HAZARDOUS OPERATION

should be posted.

Miscellaneous protective devices such as both long- and short-handled tongs for holding or manipulating hazardous items at a safe distance and remote control equipment (e.g., mechanical arms, stopcock turners, labjack turners, remote cable controllers, and closed-circuit television monitors) should be available as required to prevent exposure of any part of the body to injury.

5.G.2.3 Evaluating Potentially Reactive Materials

Potentially reactive materials must be evaluated for their possible explosive characteristics by consulting the literature and considering their molecular structures. The presence of functional groups or compounds listed in sections 5.C.9 or 5.G.6 indicates a possible explosion hazard. New compounds can be screened for explosiveness by cautious heating and hammering of very small samples. Highly reactive chemicals should be segregated from materials that might interact with them to create a risk of explosion. Highly reactive chemicals should not be used past their expiration date.

5.G.2.4 Determining Reaction Quantities

When a possibly hazardous reaction is attempted, small quantities of reactants should be used. When handling highly reactive chemicals, it is advisable to use the smallest quantities needed for the experiment. In conventional explosives laboratories, no more than 0.1 g of product should be prepared in a single run.

During the actual reaction period, no more than 0.5 g of reactants should be present in the reaction vessel. This means that the diluent, the substrate, and the energetic reactant must all be considered when determining the total explosive power of the reaction mixture. Special formal risk assessments should be established to examine operational and safety problems involved in scaling up a reaction in which an explosive substance is used or could be generated.

5.G.2.5 Conducting Reaction Operations

The most common heating devices are heating tapes and mantles and sand, water, steam, wax, silicone oil, and air (or nitrogen) baths. These should be used in such a way that if an explosion were to occur the heating medium would be contained. Heating baths should consist of nonflammable materials. All controls for heating and stirring equipment should be operable from outside the shielded area. (See Chapter 6, section 6.C.5, for further information.)

Vacuum pumps should carry tags indicating the date of the most recent oil change. Oil should be changed once a month, or sooner if it is known that the oil has been exposed to reactive gases. All pumps should either be vented into a hood or trapped. Vent lines may be Tygon, rubber, or copper. If Tygon or rubber lines are used, they should be supported so that they do not sag, causing a trap for condensed liquids. (See Chapter 6, section 6.C.2, for details.)

When potentially explosive materials are being handled, the area should be posted with a sign such as

WARNING: VACATE THE AREA AT THE FIRST INDICATION OF [the indicator for the specific case]. **STAY OUT. CALL** [responsible person] **AT** [phone number].

When condensing explosive gases, the temperature of the bath and the effect on the reactant gas of the condensing material selected must be determined experimentally (see Chapter 6, section 6.D). Very small quantities should be used because detonations may occur. A taped and shielded Dewar flask should always be used when condensing reactants. Maximum quantity limits should be observed. A dry ice solvent bath is not recommended for reactive gases; liquid nitrogen is recommended. (See also Chapter 3, section 3.D.3.1.)

5.G.3 Organic Peroxides

Organic peroxides are a special class of compounds whose unusually low stability makes them among the most hazardous substances commonly handled in laboratories, especially as initiators for free-radical reactions. Although they are low-power explosives, they are hazardous because of their extreme sensitivity to shock, sparks, and other forms of accidental detonation. Many peroxides that are handled routinely in laboratories are far more sensitive to shock than most primary explosives (e.g., TNT), although many have been stabilized by the addition of compounds that inhibit reaction. Nevertheless, even low rates of decomposition may automatically accelerate and cause a violent explosion, especially in bulk quantities of peroxides (e.g., benzoyl peroxide). These compounds are sensitive to heat, friction, impact, and light, as well as to strong oxidizing and reducing agents. All organic peroxides are highly flammable, and fires involving bulk quantities of peroxides should be approached with extreme caution.

Precautions for handling peroxides include the following:

• Limit the quantity of peroxide to the minimum amount required. Do not return unused peroxides to the container.

• Clean up all spills immediately. Solutions of peroxides can be absorbed on vermiculite or other absorbing material and disposed of harmlessly according to institutional procedures.

• The sensitivity of most peroxides to shock and heat can be reduced by dilution with inert solvents, such as aliphatic hydrocarbons. However, do not use aromatics (such as toluene), which are known to induce the decomposition of diacyl peroxides.

• Do not use solutions of peroxides in volatile solvents under conditions in which the solvent might be vaporized because this will increase the peroxide concentration in the solution.

• Do not use metal spatulas to handle peroxides because contamination by metals can lead to explosive decomposition. Magnetic stirring bars can unintentionally introduce iron, which can initiate an explosive reaction of peroxides. Ceramic, Teflon, or wooden spatulas and stirring blades may be used if it is known that the material is not shock-sensitive.

• Do not permit smoking, open flames, and other sources of heat near peroxides. It is important to label areas that contain peroxides so that this hazard is evident.

• Avoid friction, grinding, and all forms of impact near peroxides, especially solid peroxides. Glass containers that have screw-cap lids or glass stoppers should not be used. Polyethylene bottles that have screw-cap lids may be used.

• To minimize the rate of decomposition, store per-

oxides at the lowest possible temperature consistent with their solubility or freezing point. Do not store liquid peroxides or solutions at or lower than the temperature at which the peroxide freezes or precipitates because peroxides in these forms are extremely sensitive to shock and heat.

• If a container of peroxide-forming material is past its expiration date, and there is a risk that peroxides may be present, open it with caution and dispose of it according to institutional procedures (see section 5.G.3.2).

• Test for the presence of peroxides if there is a reasonable likelihood of their presence and the expiration date has not passed (see section 5.G.3.1).

(Also refer to Chapter 7, section 7.D.2.5.)

5.G.3.1 Peroxide Detection Tests

The following tests can detect most (but not all) peroxy compounds, including all hydroperoxides:

• Add 1 to 3 milliliters (mL) of the liquid to be tested to an equal volume of acetic acid, add a few drops of 5% aqueous potassium iodide solution, and shake. The appearance of a yellow to brown color indicates the presence of peroxides. Alternatively, addition of 1 mL of a freshly prepared 10% solution of potassium iodide to 10 mL of an organic liquid in a 25-mL glass cylinder should produce a yellow color if peroxides are present.

• Add 0.5 mL of the liquid to be tested to a mixture of 1 mL of 10% aqueous potassium iodide solution and 0.5 mL of dilute hydrochloric acid to which has been added a few drops of starch solution just prior to the test. The appearance of a blue or blue-black color within a minute indicates the presence of peroxides.

• Peroxide test strips, which turn to an indicative color in the presence of peroxides, are available commercially. Note that these strips must be air dried until the solvent evaporates and then exposed to moisture for proper operation.

None of these tests should be applied to materials (such as metallic potassium) that may be contaminated with inorganic peroxides.

5.G.3.2 Disposal of Peroxides

Pure peroxides should never be disposed of directly but must be diluted before disposal. Small quantities (25 g or less) of peroxides are generally disposed of by dilution with water to a concentration of 2% or less, after which the solution is transferred to a polyethylene bottle containing an aqueous solution of a reducing agent, such as ferrous sulfate or sodium bisulfite. The material can then be handled as a waste chemical; however, it must not be mixed with other chemicals for disposal. Spilled peroxides should be absorbed on vermiculite or other absorbent as quickly as possible. The vermiculite-peroxide mixture can be burned directly or may be stirred with a suitable solvent to form a slurry that can be treated according to institutional procedures. Organic peroxides should never be flushed down the drain.

Large quantities (more than 25 g) of peroxides require special handling. Each case should be considered separately, and handling, storage, and disposal procedures should be determined by the physical and chemical properties of the particular peroxide (see also Hamstead, 1964).

Peroxides can be formed during storage of some materials in air, and a peroxide present as a contaminant in a reagent or solvent (e.g., 1,4-dioxane) can be very hazardous and change the course of a planned reaction. Especially dangerous are ether bottles that have evaporated to dryness. Excluding oxygen by storing potential peroxide-formers under an inert atmosphere (N_2 or argon) or under vacuum greatly increases their safe storage lifetime. In many instances, it is possible to purchase the chemical stored under nitrogen in septum-capped bottles. In some cases, stabilizers or inhibitors (free-radical scavengers that terminate the chain reaction) are added to the liquid to extend its storage lifetime. Because distillation of the stabilized liquid removes the stabilizer, the distillate must be stored with care and monitored for peroxide formation. Furthermore, HPLC-grade solvents generally contain no stabilizer, and the same considerations apply to their handling.

(Also see Chapter 3, section 3.D.3.2; for disposal information, see Chapter 7, section 7.D.2.5.)

5.G.4 Explosive Gases and Liquefied Gases

A substance is more concentrated in the form of a liquefied gas than in the vapor phase and may evaporate extremely rapidly. Contact with liquid oxygen, in particular, may introduce extreme risk. Liquefied air is almost as dangerous as liquid oxygen because the nitrogen boils away, and as it does, it leaves an increasing concentration of oxygen. Other cryogenic liquids, such as nitrogen and helium, if they have been open to air, may have absorbed and condensed enough atmospheric oxygen to be very hazardous. When a liquefied gas is used in a closed system, pressure may build up, so that adequate venting is required. Relief devices are required to prevent this dangerous buildup

of pressure. If the liquid is flammable (e.g., hydrogen), explosive concentrations in air may develop. Because flammability, toxicity, and pressure buildup may become serious when gases are exposed to heat, gases should be stored only in specifically designed and designated areas (see Chapter 8, section 8.E).

5.G.5 Hydrogenation Reactions

Hydrogenation reactions are often carried out under pressure with a reactive catalyst and so require special attention. Along with observation of the precautions for the handling of gas cylinders and flammable gases, additional attention must be given to carrying out hydrogenation reactions at pressures above 1 atm. The following precautions are applicable:

- Make sure that the autoclave, pressure bottle, or other apparatus is appropriate for the experiment. Most preparative hydrogenations of substances such as alkenes can be carried out safely in a commercial hydrogenation apparatus using a heterogeneous catalyst (e.g., Pt and Pd) under moderate (<80 psi H_2) pressure.

- Review the operating procedures for the apparatus, and inspect the container before each experiment. Glass reaction vessels are subject to scratches or chips that render them unsuitable for use under pressure. Never fill the vessel to capacity with the solution; filling it about half full (or less) is much safer.

- One of the most important precautions to be taken with any reaction involving hydrogen is to remove as much oxygen from the solution as possible before adding hydrogen. Failure to do this could result in an explosive oxygen-hydrogen (O_2/H_2) mixture. Normally, the oxygen in the vessel is removed by pressurizing the vessel with inert gas (N_2 or argon), followed by venting the gas. If available, vacuum can be applied to the solution. Repeat this procedure of filling with inert gas and venting several times before the hydrogen or other high-pressure gas is introduced.

- Do not approach the rated safe pressure limit of the bottle or autoclave, with due regard to increased pressure upon heating. A limit of 75% of the rating in a high-pressure autoclave is advisable, but if this limit is exceeded accidentally, replace the rupture disk upon completion of the experiment.

- Monitor the pressure of the high-pressure device periodically as the heating proceeds to avoid too high a pressure in case of unintentional overheating.

- Purge the system of hydrogen by repeated "rinsing" with inert gas at the end of the experiment to avoid producing hydrogen-oxygen mixtures in the presence of the catalyst during work-up. Handle catalyst that has been used in the reaction with special care because it can be a source of spontaneous ignition upon contact with air.

(Also see section 5.C.)

5.G.6   Reactive or Explosive Materials Requiring Special Attention

The following list is not intended to be all-inclusive. Further guidance on reactive and explosive materials should be sought from pertinent sections of this book (see Chapter 3, sections 3.D.2 and 3.D.3) and other sources of information (note sources included in Chapter 3, section 3.B).

Acetylenic compounds can be explosive in mixtures of 2.5 to 80% with air. At pressures of 2 or more atmospheres, acetylene (C_2H_2) subjected to an electrical discharge or high temperature decomposes with explosive violence. Dry acetylides detonate on receiving the slightest shock. **Acetylene must be handled in acetone solution and never stored alone in a cylinder.**

Aluminum chloride ($AlCl_3$) should be considered a potentially dangerous material. If moisture is present, there may be sufficient decomposition to form hydrogen chloride (HCl) and build up considerable pressure. If a bottle is to be opened after long storage, it should first be completely enclosed in a heavy towel.

Ammonia (NH_3) reacts with iodine to give nitrogen triiodide, which detonates on touch. Ammonia reacts with hypochlorites to give chlorine. Mixtures of NH_3 and organic halides sometimes react violently when heated under pressure. Ammonia is combustible. Inhalation of concentrated fumes can be fatal.

Azides, both organic and inorganic, and some azo compounds can be heat- and shock-sensitive. Azides such as sodium azide can displace halide from chlorinated hydrocarbons such as dichloromethane to form highly explosive organic polyazides; this substitution reaction is facilitated in solvents such as dimethyl sulfoxide (DMSO).

Carbon disulfide (CS_2) is both very toxic and very flammable; mixed with air, its vapors can be ignited by a steam bath or pipe, a hot plate, or a lightbulb.

Chlorine (Cl_2) is toxic and may react violently with hydrogen (H_2) or with hydrocarbons when exposed to sunlight.

Chromium trioxide–pyridine complex ($CrO_3 \cdot C_5H_5N$) may explode if the CrO_3 concentration is too high. The complex should be prepared by addition of CrO_3 to excess C_5H_5N.

Diazomethane (CH_2N_2) and related diazo compounds should be treated with extreme caution. They are very toxic, and the pure gases and liquids explode readily even from contact with sharp edges of glass. Solutions in ether are safer from this standpoint. An ether solution of diazomethane is rendered harmless by dropwise addition of acetic acid.

Diethyl, diisopropyl, and other ethers, including tetrahydrofuran and 1,4-dioxane and particularly the branched-chain type of ethers, sometimes explode during heating or refluxing because the presence of peroxides has developed from air oxidation. Ferrous salts or sodium bisulfite can be used to decompose these peroxides, and passage over basic active alumina can remove most of the peroxidic material. In general, however, old samples of ethers should be disposed of after testing, following procedures for disposal of peroxides (see Chapter 7, section 7.D.2.5).

Dimethyl sulfoxide (DMSO), ($CH_3)_2SO$, decomposes violently on contact with a wide variety of active halogen compounds, such as acyl chlorides. Explosions from contact with active metal hydrides have been reported. Dimethyl sulfoxide does penetrate and carry dissolved substances through the skin membrane.

Dry benzoyl peroxide ($C_6H_5CO_2)_2$ is easily ignited and sensitive to shock. It decomposes spontaneously at temperatures above 50 °C. It is reported to be desensitized by addition of 20% water.

Dry ice should not be kept in a container that is not designed to withstand pressure. Containers of other substances stored over dry ice for extended periods generally absorb carbon dioxide (CO_2) unless they have been sealed with care. When such containers are removed from storage and allowed to come rapidly to room temperature, the CO_2 may develop sufficient pressure to burst the container with explosive violence. On removal of such containers from storage, the stopper should be loosened or the container itself should be wrapped in towels and kept behind a shield. Dry ice can produce serious burns, as is also true for all types of dry-ice-cooled cooling baths.

Drying agents, such as Ascarite® (sodium hydroxide-coated silica), should not be mixed with phosphorus pentoxide (P_2O_5) because the mixture may explode if it is warmed with a trace of water. Because the cobalt salts used as moisture indicators in some drying agents may be extracted by some organic solvents, the use of these drying agents should be restricted to drying gases.

Dusts that are suspensions of oxidizable particles (e.g., magnesium powder, zinc dust, carbon powder, and flowers of sulfur) in the air can constitute powerful explosive mixtures. These materials should be used with adequate ventilation and should not be exposed to ignition sources. When finely divided, some solids, including zirconium, titanium, Raney nickel, lead (such as prepared by pyrolysis of lead tartrate), and catalysts (such as activated carbon containing active metals and hydrogen), can combust spontaneously if allowed to dry while exposed to air and should be handled wet.

Ethylene oxide (C_2H_4O) has been known to explode when heated in a closed vessel. Experiments using ethylene oxide under pressure should be carried out behind suitable barricades.

Halogenated compounds, such as chloroform ($CHCl_3$), carbon tetrachloride (CCl_4), and other halogenated solvents, should not be dried with sodium, potassium, or other active metal; violent explosions usually result. Many halogenated compounds are toxic. Oxidized halogen compounds—chlorates, chlorites, bromates, and iodates—and the corresponding peroxy compounds may be explosive at high temperatures.

Hydrogen peroxide (H_2O_2) stronger than 3% can be dangerous; in contact with the skin, it can cause severe burns. Thirty percent H_2O_2 may decompose violently if contaminated with iron, copper, chromium, or other metals or their salts. Stirring bars may inadvertently bring metal into a reaction and should be used with caution.

Liquid nitrogen-cooled traps open to the atmosphere condense liquid air rapidly. Then, when the coolant is removed, an explosive pressure buildup occurs, usually with enough force to shatter glass equipment if the system has been closed. Hence, only sealed or evacuated equipment should be so cooled.

Lithium aluminum hydride ($LiAlH_4$) should not be used to dry methyl ethers or tetrahydrofuran; fires from reaction with damp ethers are often observed. The reaction of $LiAlH_4$ with carbon dioxide has reportedly generated explosive products. Carbon dioxide or bicarbonate extinguishers should not be used for $LiAlH_4$ fires; instead such fires should be smothered with sand or some other inert substance.

Nitrates, nitro and nitroso compounds may be explosive, especially if more than one nitro group is present. Alcohols and polyols can form highly explosive nitrate esters (e.g., nitroglycerine) from reaction with nitric acid.

Organometallics are hazardous because some organometallic compounds burn vigorously on contact with air or moisture. For example, solutions of *t*-butyl lithium can cause ignition of some organic solvents on exposure to air. The pertinent information should be obtained for a specific compound.

Oxygen tanks should be handled with care because serious explosions have resulted from contact between oil and high-pressure oxygen. Oil or grease should not be used on connections to an O_2 cylinder or gas line carrying O_2.

Ozone (O_3) is a highly reactive and toxic gas. It is formed by the action of ultraviolet light on oxygen (air), and, therefore, certain ultraviolet sources may require venting to the exhaust hood. Ozonides can be explosive.

Palladium **(Pd)** or *platinum* **(Pt)** on carbon, platinum oxide, Raney nickel, and other catalysts present the danger of explosion if additional catalyst is added to a flask in which an air-flammable vapor mixture and/or hydrogen is present. The use of flammable filter paper should be avoided.

Parr bombs used for hydrogenations should be handled with care behind a shield, and the operator should wear goggles and a face shield.

Perchlorates should be avoided insofar as possible. Perchlorate salts of organic, organometallic, and inorganic cations are potentially explosive and have been set off either by heating or by shock.

Perchlorates should not be used as drying agents if there is a possibility of contact with organic compounds or of proximity to a dehydrating acid strong enough to concentrate the perchloric acid ($HClO_4$) (e.g., in a drying train that has a bubble counter containing sulfuric acid). Safer drying agents should be used.

Seventy percent $HClO_4$ can be boiled safely at approximately 200 °C, but contact of the boiling undiluted acid or the hot vapor with organic matter, or even easily oxidized inorganic matter, will lead to serious explosions. Oxidizable substances must never be allowed to contact $HClO_4$. This includes wooden benchtops or hood enclosures, which may become highly flammable after absorbing $HClO_4$ liquid or vapors. Beaker tongs, rather than rubber gloves, should be used when handling fuming $HClO_4$. Perchloric acid evaporations should be carried out in a hood that has a good draft. The hood and ventilator ducts should be washed with water frequently (weekly; but see also section 8.C.7.5) to avoid danger of spontaneous combustion or explosion if this acid is in common use. Special perchloric acid hoods are available from many manufacturers. Disassembly of such hoods must be preceded by washing of the ventilation system to remove deposited perchlorates.

Permanganates are explosive when treated with sulfuric acid. If both compounds are used in an absorption train, an empty trap should be placed between them and monitored for entrapment.

Peroxides (inorganic) should be handled carefully. When mixed with combustible materials, barium, sodium, and potassium peroxides form explosives that ignite easily.

Phosphorus (P) (red and white) forms explosive mixtures with oxidizing agents. White phosphorus should be stored under water because it ignites spontaneously in air. The reaction of phosphorus with aqueous hydroxides gives phosphine, which may either ignite spontaneously or explode in air.

Phosphorus trichloride (PCl₃) reacts with water to form phosphorous acid with HCl evolution; the phosphorous acid decomposes on heating to form phosphine, which may either ignite spontaneously or explode. Care should be taken in opening containers of PCl₃, and samples that have been exposed to moisture should not be heated without adequate shielding to protect the operator.

Potassium (K) is much more reactive than sodium; it ignites quickly on exposure to humid air and, therefore, should be handled under the surface of a hydrocarbon solvent such as mineral oil or toluene (see *Sodium*). Potassium can form explosive peroxides on contact with air. If this happens, the act of cutting a surface crust off the metal can cause a severe explosion.

Residues from vacuum distillations have been known to explode when the still was vented suddenly to the air before the residue was cool. Such explosions can be avoided by venting the still pot with nitrogen, by cooling it before venting, or by restoring the pressure slowly. Sudden venting may produce a shockwave that can detonate sensitive materials.

Sodium (Na) should be stored in a closed container under kerosene, toluene, or mineral oil. Scraps of sodium or potassium should be destroyed by reaction with *n*-butyl alcohol. Contact with water should be avoided because sodium reacts violently with water to form hydrogen (H₂) with evolution of sufficient heat to cause ignition. Carbon dioxide, bicarbonate, and carbon tetrachloride fire extinguishers should not be used on alkali metal fires. Metals like sodium become more reactive as the surface area of the particles increases. Prudence dictates using the largest particle size consistent with the task at hand. For example, use of sodium ''balls'' or cubes is preferable to use of sodium ''sand'' for drying solvents.

Sodium amide (NaNH₂) can undergo oxidation on exposure to air to give sodium nitrite in a mixture that is unstable and may explode.

Sulfuric acid (H₂SO₄) should be avoided, if possible, as a drying agent in desiccators. If it must be used, glass beads should be placed in it to help prevent splashing when the desiccator is moved. To dilute H₂SO₄, the acid should be added slowly to cold water. Addition of water to the denser H₂SO₄ can cause localized surface boiling and spattering on the operator.

Trichloroethylene (Cl₂CCHCl) reacts under a variety of conditions with potassium or sodium hydroxide to form dichloroacetylene, which ignites spontaneously in air and detonates readily even at dry ice temperatures. The compound itself is highly toxic, and suitable precautions should be taken when it is used.

5.G.7 Chemical Hazards of Incompatible Chemicals

When transporting, storing, using, or disposing of any substance (see Chapter 4, and section 5.C), utmost care must be exercised to ensure that it cannot accidentally come into contact with an incompatible substance (see Chapter 3, section 3.D). Such contact could result in a serious explosion or the formation of substances that are highly toxic or flammable or both. Oxidizing agents and reducing agents should be separated from one another so that no contact is possible in the event of an accident. These reagents can also pose a risk upon contact with the atmosphere. Storage should be appropriate for the chemical under consideration. Glass systems that are to be evacuated should be taped to prevent danger of flying glass on implosion.

5.H   WORKING WITH COMPRESSED GASES

5.H.1   Chemical Hazards of Compressed Gases

Compressed gases expose the worker to both chemical and physical hazards. Such hazards and the equipment required for the safe use of compressed gases are discussed in Chapter 6, section 6.D.

Safe storage, monitoring for leaks, and proper labeling are essential for the prudent use of compressed gases. If the gas is flammable, flash points lower than room temperature compounded by rapid diffusion throughout the laboratory present the danger of fire or explosion. Additional hazards can arise from the reactivity and toxicity of the gas, and asphyxiation can be caused by high concentrations of even inert gases such as nitrogen. An additional risk of simple asphyxiants is head injury resulting from falls following rapid loss of oxygen from the brain. Death can also occur after asphyxiation if oxygen levels remain too low to sustain life. Finally, the large amount of potential energy resulting from the compression of the gas makes a highly compressed gas cylinder a potential rocket or fragmentation bomb. On-site chemical generation of a gas should be considered as an alternative to use of a compressed gas if relatively small amounts are needed. Monitoring compressed gas inventories and disposal or return of gases not in current or likely future use are advisable to avoid the development of hazardous situations.

5.H.2   Specific Chemical Hazards of Select Gases

Workers are advised to consult the Laboratory Chemical Safety Summary (LCSS) and the Material Safety Data Sheet

(MSDS) for specific gases. Certain hazardous substances that may be supplied as compressed gases are listed below:

Boron halides are powerful Lewis acids and hydrolyze to strong protonic acids. **Boron trichloride** (BCl_3) reacts with water to give HCl, and its fumes are corrosive, toxic, and irritating to the eyes and mucous membranes.

Chlorine trifluoride (ClF_3) in liquid form is corrosive and very toxic. It is a potential source of explosion and causes deep, penetrating burns on contact with the body. The effect may be delayed and progressive, as in the case of burns caused by hydrogen fluoride.

Chlorine trifluoride reacts vigorously with water and most oxidizable substances at room temperature, frequently with immediate ignition. It reacts with most metals and metal oxides at elevated temperatures. In addition, it reacts with silicon-containing compounds and thus can support the continued combustion of glass, asbestos, and other such materials. Chlorine trifluoride forms explosive mixtures with water vapor, ammonia, hydrogen, and most organic vapors. The substance resembles elemental fluorine in many of its chemical properties and handling procedures, which include precautionary steps to prevent accidents.

Hydrogen selenide (H_2Se) is a colorless gas with an offensive odor. It is a dangerous fire and explosion risk and reacts violently with oxidizing materials. It may flow to ignition sources. Hydrogen selenide is an irritant to eyes, mucous membranes, and pulmonary system. Acute exposures can cause symptoms such as pulmonary edema, severe bronchitis, and bronchial pneumonia. Symptoms also include gastrointestinal distress, dizziness, increased fatigue, and a metallic taste in the mouth.

Methyl chloride (CH_3Cl) has a slight, not unpleasant odor that is not irritating and may pass unnoticed unless a warning agent has been added. Exposure to excessive concentrations of CH_3Cl is indicated by symptoms similar to those of alcohol intoxication, that is, drowsiness, mental confusion, nausea, and possibly vomiting.

Methyl chloride may, under certain conditions, react with aluminum or magnesium to form materials that ignite or fume spontaneously with air, and contact with these metals should be avoided.

Phosphine (PH_3) is a spontaneously flammable, explosive, poisonous, colorless gas with the foul odor of decaying fish. The liquid can cause frostbite. Phosphine is a dangerous fire hazard and ignites in the presence of air and oxidizers. It reacts with water, acids, and halogens. If heated, it will form hydrogen phosphides, which are explosive and toxic. There may be a delay between exposure and the appearance of symptoms.

Silane (SiH_4) is a pyrophoric, colorless gas that ignites spontaneously in air. It is incompatible with water, bases, oxidizers, and halogens. The gas has a choking, repulsive odor.

Silyl halides are toxic, colorless gases with a pungent odor that are corrosive irritants to the skin, eyes, and mucous membranes. When silyl halides are heated, toxic fumes can be emitted.

6 Working with Laboratory Equipment

6.A INTRODUCTION

Proper use of laboratory equipment is required to work safely with hazardous chemicals. Maintenance and regular inspection of laboratory equipment are an essential part of this activity. Many of the accidents that occur in the laboratory can be attributed to improper use or maintenance of laboratory equipment. This chapter discusses prudent practices for handling the apparatus often used in laboratories.

The most common equipment hazards in laboratories come from electrically powered devices, and these are followed by hazards with devices for work with compressed gases and high/low pressures and temperatures. Other physical hazards include electromagnetic radiation hazards from such equipment as lasers and radio-frequency generating devices. Seemingly ordinary hazards such as floods from water-cooled equipment, accidents with rotating equipment and machines or tools for cutting and drilling, noise extremes, slips, trips, and falls, lifting, and poor ergonomics probably account for the greatest frequency of laboratory accidents and injuries.

6.B WORKING WITH WATER-COOLED EQUIPMENT

The use of cooling water in laboratory condensers and other equipment is common laboratory practice, but can create a flooding hazard. The most common source of the problem is disconnection of the tubing supplying the water to the condenser. Hoses can pop off under irregular flows when building water pressure fluctuates or can break when the hose material has deteriorated from long-term use. Floods also result when exit hoses jump out of the sink from a strong flow pulse or sink drains are blocked by an accumulation of extraneous material. Proper use of hose clamps and maintenance of the entire cooling system or alternate use of a portable cooling bath with suction feed can resolve such problems. Plastic locking disconnects can make it easy to disconnect water lines without having to unclamp and reclamp secured lines. Some quick disconnects also incorporate check valves, which when disconnected do not allow flow into or out of either half of the connection. This feature allows for disconnecting and reconnecting with minimal spillage of water.

6.C WORKING WITH ELECTRICALLY POWERED LABORATORY EQUIPMENT

Electrically powered laboratory equipment is used routinely for laboratory operations requiring heating, cooling, agitation or mixing, and pumping. Electrically powered equipment found in the laboratory includes fluid and vacuum pumps, lasers, power supplies, both electrophoresis and electrochemical apparatus, x-ray equipment, stirrers, hot plates, heating mantles, and, more recently, microwave ovens and ultrasonicators. Attention must be paid to both the mechanical and the electrical hazards inherent in these devices. High voltage and high power requirements are increasingly prevalent; therefore prudent practices for handling these devices are increasingly necessary.

Electric shock is the major electrical hazard. A relatively low current of 10 milliamperes (mA) poses some danger, and 80 to 100 mA can be fatal. In addition, if improperly used, electrical equipment can serve as an ignition source for flammable or explosive vapors. Most of the risks involved can be minimized by regular, proper maintenance and a clear understanding of the correct use of the device.

6.C.1 General Principles

Particular caution must be exercised during installation, modification, and repair, as well as during use of the equipment. In order to ensure safe operation, all electrical equipment must be installed and maintained in accordance with the provisions of the National Electrical Code (NEC) of the National Fire Protection Association (NFPA, 1991a). Laboratory workers should also consult state and local codes and regulations, which may contain special provisions and be more stringent than the NEC and NFPA rules. All repair and calibration work on electrical equipment must be carried out by properly trained and qualified personnel. Before modification, installation, or even minor repairs of electrical equipment are carried out, the devices must be deenergized and all capacitors discharged safely. Furthermore, this deenergized and/or discharged condition must be verified before proceeding (note that OSHA Control of Hazardous Energy Standard (29 CFR 1910.147; Lock out/Tag out) applies).

It is imperative that each person participating in any experiment involving the use of electrical equipment be aware of all applicable equipment safety issues and be briefed on any potential problems. Workers can significantly reduce hazards and dangerous behavior by following some basic principles and techniques: checking and rechecking outlet receptacles (section 6.C.1.1), making certain that wiring complies with national standards and recommendations (section 6.C.1.2), and reviewing general precautions (section 6.C.1.3) and personal safety techniques (section 6.C.1.4).

6.C.1.1 Outlet Receptacles

All 110-volt (V) outlet receptacles in laboratories should be of the standard design that accepts a three-

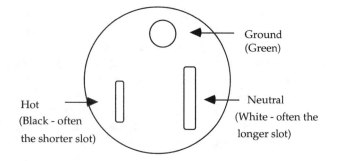

FIGURE 6.1 Standard design for a three-wire grounded outlet.

prong plug and provides a ground connection. Two-prong receptacles should be replaced as soon as feasible, and a separate ground wire should be added so that each receptacle is wired as shown in Figure 6.1. The ground wire should be on top so that anything falling onto the plug will not fall onto either the hot or the neutral line.

It is also possible to fit a receptacle with a ground-fault circuit interrupter (GFCI), which disconnects the current if a ground fault is detected. GFCI devices are required by local electrical codes for outdoor receptacles and for selected laboratory receptacles located less than 6 feet (1.83 meters) from sinks if maintenance of a good ground connection is essential for safe operation. These devices differ in operation and purpose from fuses and circuit breakers, which are designed primarily to protect equipment and prevent electrical fires due to short circuits or other abnormally high current draw situations. Certain types of GFCIs can cause equipment shutdowns at unexpected and inappropriate times; hence, their selection and use need careful planning.

Receptacles that provide electric power for operations in hoods should be located outside the hood. This location prevents the production of electrical sparks inside the hood when a device is plugged in or disconnected, and it also allows a laboratory worker to disconnect electrical devices from outside the hood in case of an accident. Cords should not dangle outside the hood in such a way that they can accidentally be pulled out of their receptacles or tripped over. Simple, inexpensive plastic retaining strips and ties can be used to route cords safely. For fume hoods with airfoils, the electrical cords should be routed under the bottom airfoil so that the sash can be closed completely. Most airfoils can be easily removed and replaced with a screwdriver.

6.C.1.2   Wiring

Laboratory equipment plugged into a 110-V (or higher) receptacle should be fitted with a standard three-conductor line cord that provides an independent ground connection to the chassis of the apparatus (see Figure 6.2). All electrical equipment should be grounded unless it is "double-insulated." This type of equipment has a two-conductor line cord that meets national codes and standards. The use of two-pronged "cheaters" to connect equipment with three-prong grounded plugs to old-fashioned two-wire outlets should be prohibited.

The use of extension cords should be limited to temporary (less than one day) setups, if they are permitted at all. A standard three-conductor extension cord of sufficient rating for the connected equipment with an independent ground connection should be used. Electrical cables should be installed properly, even if only for temporary use, and should be kept out of aisles and other traffic areas. Overhead racks and floor channel covers should be installed if wires must pass over or under walking areas. Signal and power cables should not be intermingled in cable trays or panels. Special care is needed when installing and placing water lines (used, for example, to cool such equipment as flash lamps for lasers) so that they do not leak or produce condensation, which can dampen power cables nearby.

Equipment plugged into an electrical receptacle should include a fuse or other overload protection device to disconnect the circuit if the apparatus fails or is overloaded. This overload protection is particularly useful for equipment likely to be left on and unattended for a long time, such as variable autotransformers (e.g., Variacs and powerstats), vacuum pumps, drying ovens, stirring motors, and electronic instruments. Equipment that does not contain its own built-in overload protection should be modified to provide such protection or replaced with equipment that provides it. Overload protection does not protect the worker from electrocution, but it does reduce the risk of fire.

6.C.1.3   General Precautions for Working with Electrical Equipment

Laboratory personnel should be certain that all electrical equipment is maintained well, properly located, and safely used. In order to do this, the following precautions should be reviewed and the necessary adjustments made prior to working in the laboratory:

• Insulate all electrical equipment properly. Visually inspect all electrical cords monthly, especially in any laboratory where flooding can occur. Keep in mind that rubber-covered cords can be eroded by organic solvents and by ozone (produced by ultraviolet lamps).
• Replace all frayed or damaged cords before any

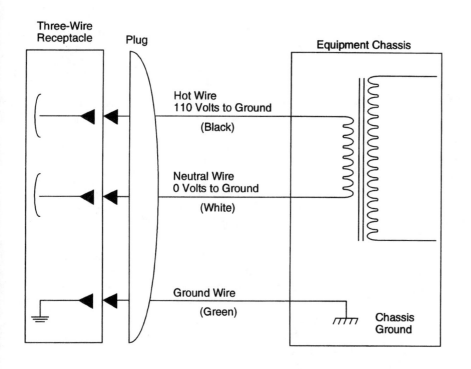

FIGURE 6.2 Standard wiring convention for 110-V electric power to equipment.

further use of the equipment is permitted. Replacement should be conducted by qualified personnel.

• Ensure the complete electrical isolation of electrical equipment and power supplies. Enclose all power supplies in a manner that makes accidental contact with power circuits impossible. In every experimental setup, including temporary ones, employ suitable barriers or enclosures to protect against accidental contact with electrical circuits.

• Equip motor-driven electrical equipment used in a laboratory where volatile flammable materials may be present (e.g., a hydrogenation room) with either nonsparking induction motors that meet Class 1, Division 2, Group C-D electrical standards (U.S. DOC, 1993) or air motors instead of series-wound motors that use carbon brushes, such as those generally used in vacuum pumps, mechanical shakers, stirring motors, magnetic stirrers, and rotary evaporators.

• Do not use variable autotransformers to control the speed of an induction motor because such operation will cause the motor to overheat and perhaps start a fire.

• Because series-wound motors cannot be modified to make them spark-free, do not use kitchen appliances (refrigerators, mixers, blenders, and so on) with such motors in laboratories where flammable materials may be present.

• When bringing ordinary electrical equipment such as vacuum cleaners and portable electric drills having series-wound motors into the laboratory for special purposes, take specific precautions to ensure that no flammable vapors are present before such equipment is used (see Chapter 5, section G).

• Locate electrical equipment so as to minimize the possibility of spills onto the equipment or flammable vapors carried into it. If water or any chemical is spilled on electrical equipment, shut off the power immediately at a main switch or circuit breaker and unplug the apparatus.

• Minimize the condensation that may enter electrical equipment if it is placed in a cold room or a large refrigerator. Cold rooms pose a particular risk in this respect because the atmosphere is frequently at a high relative humidity, and the potential for water condensation is significant.

• If electrical equipment must be placed in such areas, mount the equipment on a wall or vertical panel. This precaution will reduce, though not eliminate, the condensation problem.

• Condensation can also cause electrical equipment to overheat, smoke, or catch fire. In such a case, shut off the power to the equipment immediately at a main switch or circuit breaker and unplug the apparatus.

• To minimize the possibility of electrical shock, carefully ground the equipment using a suitable flooring material, and install ground-fault circuit interrupters (GFCIs).

• Always unplug equipment before undertaking any adjustments, modifications, or repairs (with the exception of certain instrument adjustments as indicated in section 6.C.7). When it is necessary to handle equipment that is plugged in, be certain hands are dry and, if feasible, wear nonconductive gloves and shoes with insulated soles.

• Ensure that all workers know the location and operation of power shutoffs (i.e., main switches and circuit breaker boxes) for areas in which they work. Do not use equipment again until it has been cleaned and properly inspected.

ACETONE SPILLED UNDER AN ELECTRONIC BALANCE

Acetone spilled out of a reaction vessel during the addition of dry ice. It seeped underneath a nearby electronic balance and ignited. The balance was severely damaged, but the fire was extinguished before the reaction vessel broke.

All laboratories should have access to a qualified technician who can make routine repairs to existing equipment and modifications to new or existing equipment so that it will meet acceptable standards for electrical safety. The National Fire Protection Association's *National Electrical Code Handbook* (NFPA, 1993) provides guidelines.

6.C.1.4 Personal Safety Techniques for Use with Electrical Equipment

Each individual working with electrical equipment should be informed of basic precautionary steps that should be taken to ensure personal safety:

• Avoid contact with energized electrical circuits. Electrical equipment should be serviced only by qualified individuals.

• Before qualified individuals service electrical equipment in any way, disconnect the power source to avoid the danger of electric shock. Ensure that any capacitors are, in fact, discharged.

• Before reconnecting electrical equipment to its power source after servicing, check the equipment with a suitable tester, such as a multimeter, to ensure that it is properly grounded.

• Do not reenergize a circuit breaker until there is assurance that the short circuit that activated it has been corrected.

• Install ground-fault circuit interrupters (GCFIs) as required by code to protect users from electric shock, particularly if an electrical device is hand-held during a laboratory operation.

• If a person is in contact with a live electrical con-

ductor, first disconnect the power source and then remove the person from the contact and administer first aid.

6.C.1.5 Additional Safety Techniques for Equipment Using High Current or High Voltage

Unless laboratory personnel are specially trained to install or repair high-current or high-voltage equipment, such tasks should be reserved for trained electrical workers. The following reminders are included for qualified personnel.

• Always assume that a voltage potential exists within a device while servicing it, even if it is deenergized and disconnected from its power source. For example, a device may contain capacitors, which retain a potentially harmful electrical charge.

• If it is not awkward or otherwise unsafe to do so, try to work with only one hand while keeping the other hand at your side or in a pocket, away from all conducting materials. This precaution reduces the likelihood of accidents that result in current passing through the chest cavity.

• Avoid becoming grounded by staying at least 6 inches away from walls, water, and all metal materials including pipes.

• Use voltmeters and test equipment with ratings and leads sufficient to measure the highest potential voltage to be found inside the equipment being serviced.

6.C.2 Vacuum Pumps

Distillations or concentration operations that involve significant quantities of volatile substances should normally be performed with the use of a facility vacuum system, a water aspirator, or a steam aspirator—each system protected by a suitable trapping device—rather than a mechanical vacuum pump. However, the distillation of less-volatile substances, removal of final traces of solvents, and some other operations that require pressures lower than those obtainable with a water aspirator are normally performed with a mechanical vacuum pump. The suction line from the system to the vacuum pump should be fitted with a cold trap to collect volatile substances from the system and to minimize the amount of material that enters the vacuum pump and dissolves in the pump oil. A cold trap should also be used with a water aspirator to minimize contamination of discharged water. The possibility that mercury will be swept into the pump as a result of a sudden loss of vacuum can be minimized by placing a trap in the line to the pump. Vacuum pump oil contaminated with mercury must be treated as hazardous waste. (See Chapter 5, sections 5.C.11.8 and 5.D.)

The output of each pump should be vented to an air exhaust system. This procedure is essential when the pump is being used to evacuate a system containing a volatile toxic or corrosive substance. Failure to observe this precaution would result in pumping any of the substance that is not trapped into the laboratory atmosphere. It is also recommended to scrub or absorb the gases exiting the pump. Even with these precautions, however, volatile toxic or corrosive substances may accumulate in the pump oil and, thus, be discharged into the laboratory atmosphere during future pump use. This hazard can be avoided by draining and replacing the pump oil when it becomes contaminated. The contaminated pump oil should be disposed of by following standard RCRA procedures for the safe disposal of toxic or corrosive substances. General-purpose laboratory vacuum pumps should have a record of use in order to prevent cross-contamination or reactive chemical incompatibility problems.

Belt-driven mechanical pumps with exposed belts must have protective guards. Such guards are particularly important for pumps installed on portable carts or tops of benches where laboratory workers might accidentally entangle clothing or fingers in the moving belt, but they are not necessary for enclosed pumps.

6.C.3 Refrigerators and Freezers

The potential hazards posed by laboratory refrigerators involve vapors from the contents, the possible presence of incompatible chemicals, and spillage. As general precautions, laboratory refrigerators should be placed against fire-resistant walls, should have heavy-duty cords, and preferably should be protected by their own circuit breaker. The contents of a laboratory refrigerator should be enclosed in unbreakable secondary containers. Because there is almost never a satisfactory arrangement for continuously venting the interior atmosphere of a refrigerator, any vapors escaping from vessels placed in one will accumulate in the refrigerated space and will gradually be absorbed into the surrounding insulation. Thus, the atmosphere in a refrigerator could contain an explosive mixture of air and the vapor of a flammable substance or a dangerously high concentration of the vapor of a toxic substance or both. The potential for exposure to toxic substances can be aggravated when a worker places his or her head inside a refrigerator while searching for a particular sample. The placement of potentially explosive (see Chapter 5, sections 5.C and 5.G) or highly toxic substances (see Chapter 5, sections 5.D and 5.E)

in a laboratory refrigerator is strongly discouraged. If this precaution must be violated, then a clear, prominent warning sign should be placed on the outside of the refrigerator door. Storage of these types of materials in a refrigerator should be kept to a minimum and monitored regularly. As noted in Chapter 5, section 5.C, laboratory refrigerators and freezers should never be used to store food or beverages for human consumption.

AMPOULE EXPLOSION IN A REFRIGERATOR

The door to a refrigerator used for storage of chemicals in a laboratory was left open for 10 minutes while a researcher searched through chemicals. Suddenly, an ampoule stored in the door exploded, spraying the contents in all directions, including toward the researcher. Fortunately, only one other container was ruptured, and the researcher received only a cut on his face from flying glass. A review of the incident concluded that the ampoule had been sealed at a relatively low temperature. When the ampoule warmed up in the open door, pressure built up inside it, causing it to rupture.

There should be no potential sources of electrical sparks on the inside of a laboratory refrigerator where volatile or flammable chemicals are stored. Only refrigerators that have been Underwriters-approved for flammable storage by the manufacturer should be used for this purpose. If this is not possible, all new or existing manual defrost refrigerators should be modified by

• removing the interior light and switch mounted on the door frame, if present, and
• moving the contacts of the thermostat controlling the fan and temperature outside the refrigerated compartment.

Although a prominent sign warning against the storage of flammable substances can be permanently attached to the door of an unmodified refrigerator, this alternative is less desirable than modifying the equipment by removing any spark sources from the refrigerated compartment. "Frost-free" refrigerators are not suitable for laboratory use, owing to the problems associated with attempts to modify them. Many of these refrigerators have a drain tube or hole that carries water (and any flammable material present) to an area adjacent to the compressor and, thus, present a spark hazard. The electric heaters used to defrost the freezing coils are also a potential spark hazard (see Chapter 5, section 5.G.1). To ensure its effective functioning, a freezer should be defrosted manually when ice builds up.

Uncapped containers of chemicals should never be placed in a refrigerator. Caps should provide a vapor-tight seal to prevent a spill if the container is tipped over. Aluminum foil, corks, corks wrapped with aluminum foil, and glass stoppers usually do not meet these criteria, and, therefore, their use should be discouraged. The most satisfactory temporary seals are normally screw-caps lined with either a conical polyethylene insert or a Teflon insert. The best containers for samples that are to be stored for longer periods of time are sealed, nitrogen-filled glass ampoules. At a minimum, catch pans should be used for secondary containment.

Careful labeling of samples placed in refrigerators and freezers with both the contents and the owner's name is essential. Water-soluble ink should not be used, and labels should be waterproof or covered with transparent tape. Storing samples with due consideration of chemical compatibility is important in these often small, crowded spaces.

6.C.4   Stirring and Mixing Devices

The stirring and mixing devices commonly found in laboratories include stirring motors, magnetic stirrers, shakers, small pumps for fluids, and rotary evaporators for solvent removal. These devices are typically used in laboratory operations that are performed in a hood, and it is important that they be operated in a way that precludes the generation of electrical sparks. Furthermore, it is important that, in the event of an emergency, such devices can be turned on or off from a location outside the hood. Heating baths associated with these devices (e.g., baths for rotary evaporators) should also be spark-free and controllable from outside the hood. (See sections 6.C.1 and 6.C.5; also see Chapter 5, section 5.C.7.)

Only spark-free induction motors should be used in power stirring and mixing devices or any other rotating equipment used for laboratory operations. Although the motors in most of the currently marketed stirring and mixing devices meet this criterion, their on-off switches and rheostat-type speed controls can produce an electrical spark any time they are adjusted, because they have exposed contacts. Many of the magnetic stirrers and rotary evaporators currently on the market have this disadvantage. An effective solution is to remove any switches located on the device and insert a switch in the cord near the plug end; because the electrical receptacle for the plug should be outside the hood, this modification ensures that the switch will also be outside the hood. The speed of an induction motor operating under a load should not be controlled by a variable autotransformer.

Because stirring and mixing devices, especially stirring motors and magnetic stirrers, are often operated for fairly long periods without constant attention, the consequences of stirrer failure, electrical overload, or blockage of the motion of the stirring impeller should be considered. It is good practice to attach a stirring impeller to the shaft of the stirring motor by using lightweight rubber tubing. If the motion of the impeller becomes impeded, the rubber can twist away from the motor shaft. If this occurs, the motor will not stall. However, this practice does not always prevent binding the impeller. Hence, it is also desirable to fit unattended stirring motors with a suitable fuse or thermal-protection device. (Also see section 6.C.1.)

6.C.5   Heating Devices

Perhaps the most common types of electrical equipment found in a laboratory are the devices used to supply the heat needed to effect a reaction or a separation. These include ovens, hot plates, heating mantles and tapes, oil baths, salt baths, sand baths, air baths, hot-tube furnaces, hot-air guns, and microwave ovens. The use of steam-heated devices rather than electrically heated devices is generally preferred whenever temperatures of 100 °C or less are required. Because they do not present shock or spark risks, they can be left unattended with assurance that their temperature will never exceed 100 °C.

A number of general precautions need to be taken when working with heating devices in the laboratory. First, new or existing variable autotransformers should be wired (or rewired) as illustrated in Figure 6.3. The actual heating element in any laboratory heating device should be enclosed in a glass, ceramic, or insulated metal case in such a fashion as to prevent a laboratory worker or any metallic conductor from accidentally touching the wire carrying the electric current. This type of construction minimizes the risk of electric shock and of accidentally producing an electrical spark near a flammable liquid or vapor (see Chapter 5, section 5.G.1). It also diminishes the possibility that a flammable liquid or vapor will come into contact with any wire whose temperature may exceed its ignition temperature. If any heating device becomes so worn or damaged that its heating element is exposed, the device should be either discarded or repaired to correct the damage before it is used again. Because many household appliances (e.g., hot plates and space heaters) do not meet this criterion, they should not be used in a

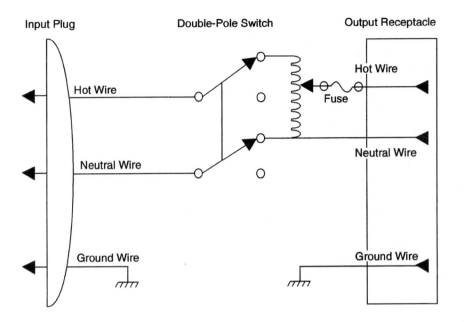

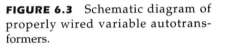

FIGURE 6.3 Schematic diagram of properly wired variable autotransformers.

laboratory. Resistance devices used to heat oil baths should not contain bare wires.

Laboratory heating devices should be used with a variable autotransformer to control the input voltage by supplying some fraction of the total line voltage, typically 110 V, to the heating element of the device. If a variable autotransformer is not wired in this manner, the switch on it may or may not disconnect both wires of the output from the 110 V line when it is switched to the off position. Also, if this wiring scheme has not been followed, and especially if the grounded three-prong plug is not used, even when the potential difference between the two output lines is only 10 V, each output line may be at a relatively high voltage (e.g., 110 V and 100 V) with respect to an electrical ground. *Because these potential hazards exist, whenever a worker uses a variable autotransformer whose wiring scheme is not known, it is prudent to assume that either of the output lines carries a potential of 110 V and is capable of delivering a lethal electric shock.*

The external cases of all variable autotransformers have perforations for cooling by ventilation, and some sparking may occur whenever the voltage adjustment knob is turned. Therefore, these devices should be located where water and other chemicals cannot be spilled onto them and where their movable contacts will not be exposed to flammable liquids or vapors. Variable autotransformers should be mounted on walls or vertical panels and outside of hoods; they should not simply be placed on laboratory benchtops.

Because the electrical input lines, including lines from variable transformers, to almost all laboratory heating devices have a potential of 110 V with respect

to any electrical ground, these lines should always be viewed both as potential shock hazards and as potential spark hazards. Thus, any connection from these lines to a heating device should be both mechanically and electrically secure and completely covered with insulating material. Alligator clips should not be used to connect a line cord from a variable autotransformer to a heating device, especially to an oil bath or an air bath, because such connections pose a shock hazard. They also may slip off, creating an electrical spark and, perhaps, contacting other metal parts to create an additional hazard. All connections should be made by using, preferably, a plug and receptacle combination, or wires with insulated terminals firmly secured to insulated binding posts.

Whenever an electrical heating device is used, it is essential to use either a temperature controller or a temperature-sensing device that will turn off the electric power if the temperature of the heating device exceeds some preset limit. Similar control devices are available that will turn off the electric power if the flow of cooling water through a condenser is stopped owing to the loss of water pressure or loosening of the water supply hose to a condenser. Fail-safe devices, which can be either purchased or fabricated, can prevent the more serious problems of fires or explosions that may arise if the temperature of a reaction increases significantly because of a change in line voltage, the accidental loss of reaction solvent, or loss of cooling. Fail-safe devices should be used for stills employed to purify reaction solvents, because such stills are often left unattended for significant periods of time.

(See section 6.C.1 for additional information.)

6.C.5.1   Ovens

Electrically heated ovens are commonly used in the laboratory to remove water or other solvents from chemical samples and to dry laboratory glassware. *Never use laboratory ovens for human food preparation.*

Laboratory ovens should be constructed such that their heating elements and their temperature controls are physically separated from their interior atmospheres. Small household ovens and similar heating devices usually do not meet these requirements and, consequently, should not be used in laboratories. With the exception of vacuum drying ovens, laboratory ovens rarely have a provision for preventing the discharge of the substances volatilized in them into the laboratory atmosphere. Thus, it should be assumed that these substances will escape into the laboratory atmosphere and may also be present in concentrations sufficient to form explosive mixtures with the air inside the oven (see Chapter 5, section 5.G). This hazard can be reduced by connecting the oven vent directly to an exhaust system.

Ovens should not be used to dry any chemical sample that has even moderate volatility and might pose a hazard because of acute or chronic toxicity unless special precautions have been taken to ensure continuous venting of the atmosphere inside the oven. Thus, most organic compounds should not be dried in a conventional unvented laboratory oven.

To avoid explosion, glassware that has been rinsed with an organic solvent should not be dried in an oven until it has been rinsed again with distilled water. Potentially explosive mixtures can be formed from volatile substances and the air inside an oven.

Bimetallic strip thermometers are preferred for monitoring oven temperatures. Mercury thermometers should not be mounted through holes in the tops of ovens so that the bulb hangs into the oven. Should a mercury thermometer be broken in an oven of any type, the oven should be closed and turned off immediately, and it should remain closed until cool. All mercury should be removed from the cold oven with the use of appropriate cleaning equipment and procedures (see Chapter 5, section 5.C.11.8) in order to avoid mercury exposure. After removal of all visible mercury, the heated oven should be monitored in a fume hood until the mercury vapor concentration drops below the threshold limit value (TLV).

6.C.5.2   Hot Plates

Laboratory hot plates are normally used when solutions are to be heated to 100 °C or above and the inherently safer steam baths cannot be used as the source of heat. As previously noted, only hot plates that have completely enclosed heating elements should be used in laboratories. Although almost all laboratory hot plates now sold meet this criterion, many older ones pose an electrical spark hazard arising from either the on-off switch located on the hot plate, the bimetallic thermostat used to regulate the temperature, or both. Normally, these two spark sources are both located in the lower part of the hot plate in a region where any heavier-than-air and possibly flammable vapors evolved from a boiling liquid on the hot plate would tend to accumulate. In principle, these spark hazards can be alleviated by enclosing all mechanical contacts in a sealed container or by using solid-state circuitry for switching and temperature control. However, in practice, such modifications are difficult to incorporate into many of the hot plates now in use. Laboratory workers should be warned of the spark hazard associated with these hot plates. Any newly purchased hot plates should be set up in a way that avoids electrical sparks. In addition to the spark hazard, old and corroded bimetallic thermostats in these devices can eventually fuse shut and deliver full, continuous current to a hot plate. This risk can be avoided by wiring a fusible coupling into the line inside the hot plate. If the device does overheat, then the coupling will melt and interrupt the current (see Section 6.C.1).

On many brands of combined stirrer/hot plates, the controls for the stirrer and temperature control look alike. Care must be taken to distinguish their functions. A fire or explosion may occur if the temperature rather than the stirrer speed is increased inadvertently.

6.C.5.3   Heating Mantles

Heating mantles are commonly used for heating round-bottomed flasks, reaction kettles, and related reaction vessels. These mantles enclose a heating element in a series of layers of fiberglass cloth. As long as the fiberglass coating is not worn or broken, and as long as no water or other chemicals are spilled into the mantle (see section 6.C.1), heating mantles pose no shock hazard. They are normally fitted with a male plug that fits into a female receptacle on an output line from a variable autotransformer. This plug combination provides a mechanically and electrically secure connection.

Heating mantles should always be used with a variable autotransformer to control the input voltage. They must never be plugged directly into a 110-V line. Workers should be careful not to exceed the input voltage recommended by the mantle manufacturer. Higher voltages will cause it to overheat, melting the fiberglass insulation and exposing the bare heating element.

Some heating mantles are constructed by encasing the fiberglass mantle in an outer metal case that provides physical protection against damage to the fiberglass. If such metal-enclosed mantles are used, it is good practice to ground the outer metal case either by using a grounded, three-conductor cord from the variable autotransformer or by securely affixing one end of a heavy, braided conductor to the mantle case and the other end to a known electrical ground. This practice protects the worker against an electric shock if the heating element inside the mantle shorts against the metal case.

6.C.5.4 Oil, Salt, and Sand Baths

In the use of oil, salt, and sand baths, care must be taken to avoid spilling water and other volatile substances into the baths. Such an accident can splatter hot material over a wide area and cause serious injuries.

Electrically heated oil baths are often used to heat small or irregularly shaped vessels or when a stable heat source that can be maintained at a constant temperature is desired. For temperatures below 200 °C, a saturated paraffin oil is often used; a silicone oil should be used for temperatures up to 300 °C. Care must be taken with hot oil baths not to generate smoke or have the oil burst into flames from overheating. An oil bath should always be monitored by using a thermometer or other thermal sensing device to ensure that its temperature does not exceed the flash point of the oil being used. For the same reason, oil baths left unattended should be fitted with thermal sensing devices that will turn off the electric power if the bath overheats. These baths should be heated by an enclosed heating element, such as a knife heater, a tubular immersion heater such as a Calrod®, or its equivalent. The input connection for this heating element should be a male plug that will fit a female receptacle from a variable autotransformer (e.g., Variac) output line. Alternatively, a temperature controller can be used to control the temperature of the bath precisely. Temperature controllers are now available that can provide a variety of heating and cooling options.

Oil baths must be well mixed to ensure that there are no "hot spots" around the elements that take the surrounding oil to unacceptable temperatures. This problem can be minimized by placing the thermoregulator fairly close to the heater. Heated oil should be contained in either a metal pan or a heavy-walled porcelain dish; a Pyrex dish or beaker can break and spill hot oil if struck accidentally with a hard object. The oil bath should be mounted carefully on a stable horizontal support such as a laboratory jack that can be

---

THERMITE REACTION EXPLOSION

An explosion injuring 27 people occurred when a thermite reaction was being demonstrated as part of a magic show at an engineering open house. The demonstration, which generated molten iron in a 2,500 to 3,000 °C reaction, was being carried out in a clay flowerpot above a beaker of water and sand to show the heat produced by the reaction when molten iron particles fall into water. Suddenly, the demonstration exploded, sending hot metal and water toward the audience.

The most likely cause of the accident was thought to be a physical vapor explosion, which can occur when a very hot liquid comes into contact with a second liquid. In this case, the water may have turned to steam so rapidly that an explosion resulted. The injuries consisted of minor burns.

---

raised or lowered easily without danger of the bath tipping over. It is also important that equipment always be clamped high enough above a hot plate or oil bath that if the reaction begins to overheat, the heater can be lowered immediately and replaced with a cooling bath without having to readjust the clamps holding the equipment setup. A bath should never be supported on an iron ring because of the greater likelihood of accidentally tipping the bath over. Secondary containment should be provided in the event of a spill of hot oil. Proper protective gloves should be worn when handling a hot bath.

Molten salt baths, like hot oil baths, offer the advantages of good heat transfer, commonly have a higher operating range (e.g., 200 to 425 °C), and may have a high thermal stability (e.g., 540 °C). The reaction container used in a molten salt bath must be able to withstand a very rapid heat-up to a temperature above the melting point of the salt. Care must be taken to keep salt baths dry, because they are hygroscopic, a property that can cause hazardous popping and splattering if the absorbed water vaporizes during heat-up.

6.C.5.5 Hot Air Baths and Tube Furnaces

Hot air baths can be useful heating devices. Nitrogen is preferred for reactions in which flammable materials are used. Electrically heated air baths are frequently used to heat small or irregularly shaped vessels. Because of their inherently low heat capacity, such baths normally must be heated considerably above the desired temperature (100 °C or more) of the vessel being heated. These baths should be constructed so that the heating element is completely enclosed and the connection to the air bath from the variable auto-transformer is both mechanically and electrically secure. These baths can be constructed from metal, ceramic, or, less desirably, glass vessels. If a glass vessel is used, it should be wrapped thoroughly with a heat-resistant tape so that if the vessel breaks accidentally, the glass will be contained and the bare heating element will not be exposed. Fluidized sand baths are usually preferred over air baths.

Tube furnaces are often used for high-temperature reactions under reduced pressure. The proper choice of glassware or metal tubes and joints is required, and the procedures should conform to safe practice with electrical equipment and evacuated apparatus.

(See also section 6.C.1 and Chapter 5, section 5.G.2.5.)

6.C.5.6   Heat Guns

Laboratory heat guns are constructed with a motor-driven fan that blows air over an electrically heated filament. They are frequently used to dry glassware or to heat the upper parts of a distillation apparatus during distillation of high-boiling materials. The heating element in a heat gun typically becomes red-hot during use and, necessarily, cannot be enclosed. Also, the on-off switches and fan motors are not usually spark-free. For these reasons, heat guns almost always pose a serious spark hazard (see Chapter 5, section 5.G.1). They should never be used near open containers of flammable liquids, in environments where appreciable concentrations of flammable vapors may be present, or in hoods used to remove flammable vapors. Household hair dryers may be substituted for laboratory heat guns only if they have three-conductor line cords or are double-insulated. Any hand-held heating device of this type that will be used in a laboratory should have ground-fault circuit interrupter (GFCI) protection to ensure against electric shock.

6.C.5.7   Microwave Ovens

To avoid exposure to microwaves, ovens should never be operated with doors open. Wires and other objects should not be placed between the sealing surface and the door on the oven's front face. The sealing surfaces must be kept absolutely clean. To avoid electrical hazards, the oven must be grounded. If use of an extension cord is necessary, only a three-wire cord with a rating equal to or greater than that for the oven should be used. To reduce the risk of fire in the oven, samples must not be overheated. The oven must be closely watched when combustible materials are in it.

Metal containers or metal-containing objects (e.g., stir bars) should not be used in the microwave, because they can cause arcing.

Generally, sealed containers should not be heated in the oven, because of the danger of explosion. If sealed containers must be used, their materials must be selected carefully and the containers properly designed. Commercially available microwave acid digestion bombs, for example, incorporate a Teflon sample cup, a self-sealing Teflon O-ring, and a compressible pressure-relief valve. The manufacturer's loading limits must not be exceeded. For such applications, the microwave oven should be properly vented using an exhaust system. Placing a large item such as an oven inside a fume hood is not recommended.

Heating a container with a loosened cap or lid poses a significant risk. Microwave ovens can heat material (e.g., solidified agar) so quickly that, even though container lids may be loosened to accommodate expansion, the lid can seat upward against the threads and containers can explode. Screw-caps must be removed from containers being microwaved. If the sterility of the contents must be preserved, screw-caps may be replaced with cotton or foam plugs.

6.C.6   Ultrasonicators, Centrifuges, and Other Electrical Equipment

6.C.6.1   Ultrasonicators

The use of high-intensity ultrasound in the chemical laboratory has grown enormously during the past decade and has a diverse set of applications. Human exposure to ultrasound with frequencies of between 16 and 100 kilohertz (kHz) can be divided into three distinct categories: airborne conduction, direct contact through a liquid coupling medium, and direct contact with a vibrating solid.

Ultrasound through airborne conduction does not appear to pose a significant health hazard to humans. However, exposure to the associated high volumes of audible sound can produce a variety of effects, including fatigue, headaches, nausea, and tinnitus. When ultrasonic equipment is operated in the laboratory, the apparatus must be enclosed in a 2-cm-thick wooden box or in a box lined with acoustically absorbing foam or tiles to substantially reduce acoustic emissions (most of which are inaudible).

Direct contact of the body with liquids or solids subjected to high-intensity ultrasound of the sort used to promote chemical reactions should be avoided. (In contrast, ultrasound used for medical diagnostic imaging is relatively benign.) Under sonochemical conditions, cavitation is created in liquids, and it can induce high-energy chemistry in liquids and tissues. Cell death from membrane disruption can occur even at relatively low acoustic intensities. Exposure to ultrasonically vibrating solids, such as an acoustic horn, can lead to rapid frictional heating and potentially severe burns.

6.C.6.2 Centrifuges

Centrifuges should be properly installed and must be operated only by trained personnel. It is important that the load be balanced each time the centrifuge is used and the lid be closed while the rotor is in motion. The disconnect switch must be working properly to shut off the equipment when the top is opened, and the manufacturer's instructions for safe operating speeds must be followed.

For flammable and/or hazardous materials, the centrifuge should be under negative pressure to a suitable exhaust system.

6.C.6.3 Electrical Instruments

Most modern electronic instruments have a cord that contains a separate ground wire for the chassis and are supplied with a suitable fuse or other overload protection. Any existing instrument that lacks these features should be modified to incorporate them. As is true for any electrical equipment, special precautions should be taken to avoid the possibility that water or other chemicals could be spilled into these instruments.

Under most circumstances, any repairs to, adjustments to, or alterations of such instruments should be made only by a qualified individual. Laboratory workers should not undertake such adjustments unless they have received certification as well as specific training for the particular instrument to be serviced. If laboratory workers do undertake repairs, the cord should always be unplugged before any disassembly begins. However, certain adjustments can be made only when the instrument is connected to a power source. Appropriate protective measures and due diligence are required when working on energized devices. Extra precautions are particularly important for instruments that incorporate high-voltage circuitry.

Many electrical instruments, such as lasers and x-ray, E-beam, radioactive, photochemical, and electrophoresis equipment, emit potentially harmful radiation, and, therefore, special precautions must be followed when they are used. This equipment should be used and serviced only by trained personnel. (See section 6.C.1 and Chapter 5, section 5.E.)

6.C.7 Electromagnetic Radiation Hazards

Equipment found in laboratories that can produce hazardous amounts of electromagnetic radiation includes ultraviolet lamps, arc lamps, heat lamps, lasers, microwave and radio-frequency sources, and x-rays and electron beams.

6.C.7.1 Visible, Ultraviolet, and Infrared Laser Light Sources

Overexposure to direct or reflected ultraviolet light, arc lamps, and infrared sources should be minimized by sealing or enclosing sources whenever possible. Appropriately rated safety glasses, goggles, or face shields should be worn for eye protection. Long-sleeved clothing and gloves should be worn to protect arms and hands.

Control measures for the safe use of lasers have been established by the American National Standards Institute (ANSI) and presented in *Safe Use of Lasers* (ANSI Z136.1-1993), which describes the different types of laser hazards and the appropriate measures to control each type. Class IIIB and IV lasers should be operated only in posted laser-controlled areas. No one but the authorized operator of a laser system should ever enter a posted laser-controlled laboratory when the laser is in use.

6.C.7.2 Radio-frequency and Microwave Sources

Section 6.C.5.7 provides guidelines for the safe use of microwave ovens in the laboratory. Other devices in the laboratory can also emit harmful microwave or radio-frequency emissions. People working with these types of devices should be trained in their proper operation as well as measures to prevent exposure to harmful emissions. Shields and protective covers should be in proper position when the equipment is operating. Warning signs to protect people wearing heart pacemakers should be posted on or near these devices.

A laboratory worker removed the shield from a high-powered microwave-generating device. Fortunately, the emissions triggered the fire alarm system, prompting an evacuation of the laboratory. Had the exposure continued, the person could have suffered severe injury.

6.C.7.3 X-rays, E-beams, and Sealed Sources

X-rays and electron beams (E-beams) are used in a variety of laboratory equipment, mostly for analytical operations. The equipment is government-regulated. In most cases, registration and licensing are required. Personnel operating or working in the vicinity of these types of equipment should have appropriate training to minimize the risk of their being exposed to harmful ionizing radiation.

6.C.7.4 Miscellaneous Physical Hazards Presented by Electrically Powered Equipment

6.C.7.4.1 Magnetic Fields

If an object moves into the attractive field of a strong magnet, it can become a projectile when it is pulled rapidly toward the magnet. Therefore, objects ranging from keys, scissors, knives, wrenches, and other tools to oxygen cylinders, buffing machines, and wheelchairs and other ferromagnetic objects must be excluded from the immediate vicinity of the magnet, for the sake of both safety and data quality, in the case of NMR.

Even relatively small peripheral magnetic fields can adversely affect credit cards, computer disks, and other magnetic objects, as summarized in Table 6.1. It is prudent to post warnings at the 5-gauss (G) line and to limit access to areas with more than 10 to 20 G to knowledgeable staff. People wearing heart pacemakers and other electronic or electromagnetic prosthetic devices should be kept away from strong electromagnetic sources.

Superconducting magnets use liquid nitrogen and liquid helium coolants. Thus, the precautions associated with the use of cryogenic liquids must be observed as well. (Also see section 6.E.2.)

6.C.7.4.2 Rotating Equipment and Moving Parts

Injuries can result from bodily contact with rotating or moving objects, including mechanical equipment, parts, and devices. The risk of injury can be reduced through improved engineering, good housekeeping, and safe work practice and personal behavior. Laboratory workers must know how to shut down equipment in the event of an emergency; must enclose or shield hazardous parts, such as belts, chains, gears, and pulleys, with appropriate guards; and must not wear loose clothing, jewelry, or unrestrained long hair around machinery.

6.C.7.4.3 Cutting and Puncturing Tools

Hand injuries are probably the most frequently encountered injuries in laboratories. Many of these injuries can be prevented by keeping all cutting and puncturing devices fully protected, avoiding the use of razor blades as cutting tools, and using utility knives that have a spring-loaded guard that covers the blade. Razor blades, needles, and other sharp objects or instruments should be disposed of carefully rather than simply thrown into the trash bin unprotected.

Glass cuts can be minimized by use of correct procedures (for example, that for inserting glass tubing into rubber stoppers and tubing, which is taught in introductory laboratories), through appropriate use of protective equipment, and by careful attention to manipulation.

6.C.7.4.4 Noise Extremes

Any laboratory operation that produces significant noise (85 decibels or greater) needs a hearing conservation program to protect employees from excessive exposure, that is, exposure to significant noise for an 8-hour average duration. An audiologist or industrial hygienist should be consulted to determine the need for such a program and to provide assistance in developing one.

6.C.7.4.5 Slips, Trips, and Falls

The risks of slips, trips, falls, and collisions between persons and objects can be reduced by cleaning up

TABLE 6.1 Summary of Magnetic Field Effects

	Level at Which Effects Occur (gauss)
Effects on electron microscopes	1
Disturbance of color computer displays	1–3
Disturbance of monochrome computer displays	3–5
Erasure of credit card and bank card coding	10
Effects on watches and micromechanical devices	10
Lowest known field effect on pacemakers	17
Saturation of transformers and amplifiers	50
Erasure of floppy disks	350

SOURCE: Adapted from *Site Planning Guide for Superconducting NMR Systems,* Bruker Instruments (1992).

liquid or solid spills immediately, keeping doors and drawers closed and passageways clear of obstructions, providing step stools, ladders, and lifts to reach high areas, and walking along corridors and on stairways at a normal pace. Wet floors around ice, dry ice, or liquid nitrogen dispensers should be carpeted and paper towel dispensers made available for wiping up drops or small puddles as soon as they form.

6.C.7.4.6 Ergonomics and Lifting

Both standing and sitting in a static posture and making repeated motions have been shown to cause a wide variety of musculoskeletal complaints. Problems due to poor ergonomics include eyestrain, stiff and sore back, leg discomfort, and hand and arm injuries. Each situation needs to be evaluated individually. However, personnel who spend significant time working on video display terminals should use furniture appropriate for these tasks, proper posture, and perhaps special eyeglasses. Also, people who use the same tools and/or hand motions for extended periods of time should take breaks at appropriate intervals to help prevent injuries.

Lifting injuries are one of the more common types of injuries for laboratory workers. The weight of the item to be lifted is a factor, but it is only one of several. The shape and size of an object as well as the lifting posture and the frequency of lifting are also key factors in determining the risks of lifting. The National Institute for Occupational Safety and Health (NIOSH) has developed a guide that should be consulted to help determine lifting safety (U.S. DHHS, 1994). Personnel who are at risk for lifting injuries should receive periodic training.

6.D WORKING WITH COMPRESSED GASES

6.D.1 Compressed Gas Cylinders

Precautions are necessary for handling the various types of compressed gases, the cylinders that contain them, the regulators used to control their flow, the piping used to confine them during flow, and the vessels in which they are ultimately used. *Regular inventories of cylinders and checks of their integrity with prompt disposal of those no longer in use are important.* (See Chapter 4, section 4.E.4, for information on storing gas cylinders, and Chapter 5, section 5.H, for discussion of the chemical hazards of gases.)

A compressed gas is defined as a material in a container with an absolute pressure greater than 276 kilopascals (kPa), or 40 psi (pounds per square inch) at 21 °C or an absolute pressure greater than 717 kPa (104 psi) at 54 °C, or both, or any liquid flammable material having a Reid vapor pressure greater than 276 kPa (40

psi) at 38 °C. The Department of Transportation (DOT) has established codes that specify the materials to be used for the construction and the capacities, test procedures, and service pressures of the cylinders in which compressed gases are transported. However, regardless of the pressure rating of the cylinder, the physical state of the material within it determines the pressure of the gas. For example, liquefied gases such as propane and ammonia will exert their own vapor pressure as long as any liquid remains in the cylinder and the critical temperature is not exceeded.

Prudent procedures for the use of compressed gas cylinders in the laboratory include attention to appropriate purchase, especially selecting the smallest cylinder compatible with the need, as well as proper transportation and storage, identification of contents, handling and use, and marking and return of the empty cylinder. The practice of purchasing unreturnable lecture bottles should be discouraged if that leads to accumulation of partially filled cylinders and disposal problems. Returnable cylinders should be purchased and returned as prescribed by the manufacturer.

6.D.1.1 Identification of Contents

The contents of any compressed gas cylinder should be identified clearly so as to be easily, quickly, and completely determined by any laboratory worker. Such identification should be stenciled or stamped on the cylinder itself, or *a durable label should be provided that cannot be removed from the cylinder*. No compressed gas cylinder should be accepted for use that does not identify its contents legibly by name. Color coding is not a reliable means of identification; cylinder colors vary from supplier to supplier, and labels on caps have no value because many caps are interchangeable. Care in the maintenance of cylinder labels is important because unidentified compressed gas cylinders may pose a high risk and present very high disposal costs. It is good practice to provide compressed gas cylinders with tags on which the names of users and dates of use can be entered. If the labeling on a cylinder becomes unclear or an attached tag is defaced and the contents cannot be identified, the cylinder should be marked "contents unknown" and the manufacturer contacted regarding appropriate procedures.

All gas lines leading from a compressed gas supply should be labeled clearly to identify the gas, the laboratory served, and relevant emergency telephone numbers. The labels, in addition to being dated, should be color-coded to distinguish hazardous gases—that is, flammable, toxic, or corrosive substances coded with a yellow background and black letters—from inert gases, which are coded with a green background and black letters. Signs should be posted conspicuously in areas in which flammable compressed gases are stored, identifying the substances and appropriate precautions, for example:

**HYDROGEN—FLAMMABLE GAS
NO SMOKING—NO OPEN FLAMES**

6.D.1.2   Handling and Use

Gas cylinders must be handled carefully to prevent accidents or damage to the cylinder. The valve protection cap should be left in place until the cylinder is secured and ready for use. Gas cylinders should not be dragged, rolled, slid, or allowed to strike each other forcefully. Cylinders should always be transported on wheeled cylinder carts with retaining straps or chains. The plastic mesh sleeves sometimes installed on cylinders by vendors are intended only to protect the paint on the cylinder and do not serve as a safety device.

Compressed gas cylinders should be secured firmly at all times. A clamp and belt or chain, securing the cylinder between "waist" and "shoulder" to a wall, are generally suitable for this purpose. In areas of seismic activity, gas cylinders should be secured both toward the top and toward the bottom. Cylinders should be individually secured; using a single restraint strap or chain around a number of cylinders is often not effective. Pressure-relief devices protecting equipment that is attached to cylinders of flammable, toxic, or otherwise hazardous gases should be vented to a safe place. (See Section 6.D.2.2.1 for details.)

Standard cylinder-valve outlet connections have been devised by the Compressed Gas Association (CGA) to prevent the mixing of incompatible gases due to an interchange of connections. The outlet threads used vary in diameter: some are internal and some are external; some are right-handed and some are left-handed. In general, right-handed threads are used for nonfuel and water-pumped gases, and left-handed threads are used for fuel and oil-pumped gases. Information on the standard equipment assemblies for use with specific compressed gases is available from the supplier. To minimize undesirable connections that may result in a hazard, only CGA standard combinations of valves and fittings should be used in compressed gas installations; the assembly of miscellaneous parts (even of standard approved types) should be avoided. Use of an "adapter" or cross-threading of a valve fitting should not be attempted. The threads on cylinder valves, regulators, and other fittings should be examined to ensure that they correspond to one another and are undamaged.

Cylinders should be placed so that the rotary cylinder valve handle at the top is accessible at all times. Cylinder valves should be opened slowly, and only when a proper regulator is firmly in place and the attachment has been shown to be leak-proof by an appropriate test (see Chapter 5, section 5.H). The cylinder valve should be closed as soon as the necessary amount of gas has been released. Valves should be either completely open or completely closed. Flow restrictors should be installed on gas cylinders to minimize the chance of excessive flows. The cylinder valve should never be left open when the equipment is not in use. This precaution is necessary not only for safety when the cylinder is under pressure, but also to prevent the corrosion and contamination that would result from diffusion of air and moisture into the cylinder after it has been emptied.

Most cylinders are equipped with hand-wheel valves. Those that are not should have a spindle key on the valve spindle or stem while the cylinder is in service. Only wrenches or other tools provided by the cylinder supplier should be used to remove a cylinder cap or to open a valve. In no case should a screwdriver be used to pry off a stuck cap or should pliers be used to open a cylinder valve. Some valve fittings require washers or gaskets, and the materials of construction should be checked before the regulator is fitted.

If the valve on a cylinder containing an irritating or toxic gas is being opened outside, the worker should stand upwind of the cylinder with the valve pointed downwind, away from him or herself, and should warn those working nearby in case of a possible leak. If the work is being done inside, the cylinder should be opened only in a fume hood or specially designed cylinder cabinet. A differential pressure switch with an audible alarm should be installed in any hood dedicated for use with toxic gases. In the event of hood failure, the pressure switch should activate an audible alarm warning the user of hood failure.

6.D.1.2.1   Preventing and Controlling Leaks

Cylinders, connections, and hoses should be checked regularly for leaks. To check for leaks, a flammable gas leak detector (for flammable gases only) or soapy water, or a 50% glycerin-water solution, is used to look for bubbles. At or below freezing temperatures, the glycerin solution should be used instead of soapy water. When the gas to be used in the procedure is a flammable oxidizing or highly toxic gas, the system should be checked first for leaks with an inert gas (helium or nitrogen) before introducing the hazardous gas.

The general procedures discussed in Chapter 5, section 5.C, can be used for relatively minor leaks, when

SPONTANEOUS IGNITION OF HYDROGEN

Late one evening in a chemical engineering facility, a student employee was working near a six-cylinder hydrogen gas manifold when she heard a cracking sound above her, followed by a whistling sound. She stepped away from the cylinder area to where she could see above the roof and noticed a flame on the roof above the cylinders. She immediately reported this to her supervisor, who went up on the roof and found that a rupture disk on the hydrogen manifold had ruptured. Had the disk ruptured in the daylight, the flame might not have been visible. The most likely cause of the flame was the spontaneous ignition of the hydrogen as it entered the air at high pressure. The hydrogen manifold was shut down, the rupture disk was replaced, and research was resumed.

the indicated action can be taken without exposing personnel to highly toxic substances. The leaking cylinder can be moved through populated portions of the building, if necessary, by placing a plastic bag, rubber shroud, or similar device over the top and taping it (preferably with duct tape) to the cylinder to confine the leaking gas. If there is any risk of exposure, the environmental health and safety office should be called and the area evacuated before the tank is moved.

If a leak at the cylinder valve handle cannot be remedied by tightening a valve gland or a packing nut, emergency action should be taken and the supplier should be notified. Laboratory workers should never attempt to repair a leak at the junction of the cylinder valve and the cylinder or at the safety device; rather, they should consult with the supplier for instructions.

When the nature of the leaking gas or the size of the leak constitutes a more serious hazard, a self-contained breathing apparatus and protective apparel may be required, and personnel may need to be evacuated (see Chapter 5, section 5.C.2). Cylinders leaking toxic gases always require protective equipment and evacuation of personnel. Cylinder coffins are also available to encapsulate leaking cylinders. (See Chapter 5, section 5.G, for more information.)

6.D.1.2.2 Pressure Regulators

Pressure regulators are strongly recommended to reduce a high-pressure supplied gas to a desirable lower pressure and to maintain a satisfactory delivery pressure and flow level for the required operating con-

ditions. They can be obtained to fit many operating conditions over a range of supply and delivery pressures, flow capacities, and construction materials. All regulators are of a diaphragm type and are spring-loaded or gas-loaded, depending on pressure requirements. They can be single-stage or two-stage. Under no circumstances should oil or grease be used on regulator valves or cylinder valves because these substances may be reactive with some gases (e.g., oxygen).

Each regulator is supplied with a specific CGA standard inlet connection to fit the outlet connection on the cylinder valve for the particular gas. Regulators should be checked before use to be sure they are free of foreign objects and correct for the particular gas.

Regulators for use with noncorrosive gases are usually made of brass. Special regulators made of corrosion-resistant materials can be obtained for use with such gases as ammonia, boron trifluoride, chlorine, hydrogen chloride, hydrogen sulfide, and sulfur dioxide. Because of freeze-up and corrosion problems, regulators used with carbon dioxide gas must have special internal design features and be made of special materials. Regulators used with oxidizing agents must be cleaned specially to avoid the possibility of an explosion on contact of the gas with any reducing agent or oil left from the cleaning process.

All pressure regulators should be equipped with spring-loaded pressure-relief valves (see section 6.D.2.2.1 for further information on pressure-relief devices) to protect the low-pressure side. When used on cylinders of flammable, toxic, or otherwise hazardous gases, the relief valve should be vented to a hood or other safe location. The use of internal-bleed-type regulators should be avoided. Regulators should be removed from corrosive gases immediately after use and flushed with dry air or nitrogen. Mercury bubblers should not be used.

6.D.1.2.3 Flammable Gases

It is important to keep all sources of ignition away from cylinders of flammable gases and to ensure that these cylinders will not leak. A solution of soapy water should be used to detect leaks except during freezing weather, when a 50% glycerin-water solution or its equivalent should be used. Connections to piping, regulators, and other appliances should always be kept tight to prevent leakage, and the tubing or hoses used should be kept in good condition. Regulators, hoses, and other appliances used with cylinders of flammable gases should not be interchanged with similar equipment intended for use with other gases. Cylinders should be grounded properly to prevent static electricity buildup, especially in very cold or dry environments. All cylinders containing flammable gases should be stored in a well-ventilated place. Reserve stocks of such cylinders should never be stored in the vicinity of cylinders containing oxygen, fluorine, chlorine, or other oxidizing gases. Reaction vessels should be equipped with pressure-relief devices.

6.D.1.2.4 *Empty Cylinders*

A cylinder should never be emptied to a pressure lower than 172 kPa (25 psi) because the residual contents may become contaminated with air if the valve is left open. Empty cylinders should never be refilled by the user. Rather, the regulator should be removed, and the valve cap should be replaced. The cylinder should be clearly marked as empty (MT) and returned to a storage area for pickup by the supplier. Empty and full cylinders should not be stored in the same place.

Cylinder discharge lines should be equipped with approved check valves to prevent inadvertent contamination of cylinders that are connected to a closed system where the possibility of flow reversal exists. Backflow is particularly troublesome in the case of gases used as reactants in a closed system. A cylinder in such a system should be shut off and removed from the system while the pressure remaining in the cylinder is still greater than the pressure in the closed system. If there is a possibility that a cylinder has become contaminated, it should be so labeled and returned to the supplier.

6.D.2 Other Equipment Used with Compressed Gases

6.D.2.1 *Records, Inspection, and Testing*

High-pressure operations should be carried out only with equipment specifically built for this use and only by those trained especially to use this equipment. Reactions should never be carried out in, nor heat applied to, an apparatus that is a closed system unless it has been designed and tested to withstand pressure. To ensure that the equipment has been properly designed, each pressure vessel should have stamped on it, or on an attached plate, its maximum allowable working pressure, the allowable temperature at this pressure, and the material of construction. Similarly, the relief pressure and setting data should be stamped on a metal tag attached to installed pressure-relief devices, and the setting mechanisms should be sealed. Relief devices used on pressure regulators do not require these seals or numbers.

All pressure equipment should be tested or inspected periodically. The frequency of tests and/or inspections varies, depending on the type of equip-

ment, how often it is used, and the nature of its usage. Corrosive or otherwise hazardous service requires more frequent tests and inspections. Inspection data should be stamped on or attached to the equipment.

Testing the entire assembled apparatus with soap solution and air or nitrogen pressure to the maximum allowable working pressure of the weakest section of the assembled apparatus can usually detect leaks at threaded joints, packings, and valves.

Before any pressure equipment is altered, repaired, stored, or shipped, it should be vented, and all toxic, flammable, or other hazardous material removed completely so it can be handled safely. Especially hazardous materials may require special cleaning techniques, which should be solicited from the distributor.

(See section 6.E.1 for further information.)

6.D.2.2 *Assembly and Operation*

During the assembly of pressure equipment and piping, only appropriate components should be used, and care should be taken to avoid strains and concealed fractures resulting from the use of improper tools or excessive force. Tubing in place in a pressure apparatus should not be used to support any significant weight.

Threads that do not fit exactly should not be forced (refer to section 6.D.1.2.1). Thread connections must match; tapered pipe threads cannot be joined with parallel machine threads. Teflon tape or a suitable thread lubricant should be used when assembling the apparatus (see section 6.D.2.2.6). However, oil or lubricant must never be used on any equipment that will be used with oxygen. Parts having damaged or partly stripped threads should be rejected (also see section 6.D.2.2.5).

In assembling copper tubing installations, sharp bends should be avoided and considerable flexibility should be allowed. Copper tubing hardens and cracks on repeated bending. Many metals can become brittle in hydrogen (H_2) or corrosive gas service. Nickel alloys can generate $Ni(CO)_4$ in some carbon monoxide atmospheres. All tubing should be inspected frequently and replaced when necessary.

Stuffing boxes and gland joints are a likely source of trouble in pressure installations. Particular attention should be given to the proper installation and maintenance of these parts, including the proper choice of lubricant and packing material.

Experiments carried out in closed systems and involving highly reactive materials, such as those subject to rapid polymerization (e.g., dienes or unsaturated aldehydes, ketones, or alcohols) should be preceded by small-scale tests using the exact reaction materials to determine the possibility of an unexpectedly rapid reaction or unforeseen side reactions. All reactions under pressure should be shielded and should be carried out as remotely as possible, for example, with valve extensions and behind a heavy shield or with closed-circuit TV monitoring if needed.

Autoclaves and other pressure-reaction vessels should not be filled more than half full to ensure that space remains for expansion of the liquid when it is heated. Leak corrections or adjustments to the apparatus should not be made while it is pressurized; rather, the system should be depressurized before mechanical adjustments are made.

Immediately after an experiment in which low-pressure equipment connected to a source of high pressure is pressurized, the low-pressure equipment should either be disconnected entirely or left independently vented to the atmosphere. Either action will prevent the accidental buildup of excessive pressure in the low-pressure equipment due to leakage from the high-pressure side.

Vessels or equipment made partly or entirely of silver, copper, or alloys containing more than 50% copper should not be used in contact with acetylene or ammonia. Those made of metals susceptible to amalgamation (e.g., copper, brass, zinc, tin, silver, lead, and gold) should not come into contact with mercury. This includes equipment that has soldered and brazed joints.

Prominent warning signs should be placed in any area where a pressure reaction is in progress so that people entering the area will be aware of the potential risk.

6.D.2.2.1 Pressure-Relief Devices

All pressure or vacuum systems and all vessels that may be subjected to pressure or vacuum should be protected by properly installed and tested pressure-relief devices. Experiments involving highly reactive materials that might explode may also require the use of special pressure-relief devices and may need to be operated at a fraction of the permissible working pressure of the system.

Examples of pressure-relief devices include the rupture-disk type used with closed-system vessels and the spring-loaded safety valves used with vessels for transferring liquefied gases. The following precautions are advisable in the use of pressure-relief devices:

• The maximum setting of a pressure-relief device is the rated working pressure established for the vessel or for the weakest member of the pressure system at the operating temperature. The operating pressure should be less than the allowable working pressure of the system. In the case of a system protected by a spring-loaded relief device, the maximum operating pressure should be from 5 to 25% lower than the rated working pressure, depending on the type of safety valve and the importance of leak-free operation. In the case of a system protected by a rupture-disk device, the maximum operating pressure should be about two-thirds of the rated working pressure; the exact figure is governed by the fatigue life of the disk used, the temperature, and load pulsations.

• Pressure-relief devices that may discharge toxic, corrosive, flammable, or otherwise hazardous or noxious materials should be vented in a safe and environmentally acceptable manner such as scrubbing and/or diluting with nonflammable streams.

• Shutoff valves must not be installed between pressure-relief devices and the equipment they are to protect.

• Only qualified persons should perform maintenance work on pressure-relief devices.

• Pressure-relief devices should be inspected and replaced periodically.

6.D.2.2.2 Pressure Gauges

The proper choice and use of a pressure gauge involve several factors, including the flammability, compressibility, corrosivity, toxicity, temperature, and pressure range of the fluid with which it is to be used. Generally, a gauge with a range that is double the working pressure of the system should be selected.

A pressure gauge is normally a weak point in any pressure system because its measuring element must operate in the elastic zone of the metal involved. The resulting limited factor of safety makes careful gauge selection and use mandatory and often dictates the use of accessory protective equipment. The primary element of the most commonly used gauges is a Bourdon tube, which is usually made of brass or bronze and has soft-soldered connections. More expensive gauges can be obtained that have Bourdon tubes made of steel, stainless steel, or other special metals and welded or silver-soldered connections. Accuracies vary from $\pm 2\%$ for less-expensive pressure gauges to $\pm 0.1\%$ for higher-quality gauges. A diaphragm gauge should be used with corrosive gases or liquids or with viscous fluids that would destroy a steel or bronze Bourdon tube.

Consideration should be given to alternative methods of pressure measurement that may provide greater safety than the direct use of pressure gauges. Such methods include the use of seals or other isolating devices in pressure tap lines, indirect observation devices, and remote measurement by strain-gauge transducers with digital readouts.

6.D.2.2.3 *Glass Equipment*

The use of glassware for work at high pressure should be avoided whenever possible. Glass is a brittle material subject to unexpected failures due to factors such as mechanical impact and assembly and tightening stresses. Poor annealing after glassblowing can leave severe strains. Glass equipment, such as rotameters and liquid-level gauges, incorporated in metallic pressure systems should be installed with shutoff valves at both ends to control the discharge of liquid or gaseous materials in the event of breakage. Mass flowmeters are available that can replace rotameters in desired applications.

6.D.2.2.4 *Plastic Equipment*

Except as noted below, the use of plastic equipment for pressure or vacuum work should be avoided unless no suitable substitute is available. These materials can fail under pressure or thermal stress.

Tygon and similar plastic tubing have quite limited applications in pressure work. These materials can be used for hydrocarbons and most aqueous solutions at room temperature and moderate pressure. Reinforced plastic tubing that can withstand higher pressures is also available. However, loose tubing under pressure can cause physical damage by its own whipping action. Details of permissible operating conditions must be obtained from the manufacturer. Because of their very large coefficients of thermal expansion, some polymers have a tendency to expand a great deal on heating and to contract on cooling. This behavior can create a hazard in equipment subjected to very low temperatures or to alternating low and high temperatures.

6.D.2.2.5 *Piping, Tubing, and Fittings*

The proper selection and assembly of components in a pressure system are critical safety factors. Considerations should include the materials used in manufacturing the components, compatibility with the materials to be under pressure, the tools used for assembly, and the reliability of the finished connections. No oil or lubricant of any kind should be used in a tubing system with oxygen because the combination produces an explosion hazard.

All-brass and stainless steel fittings should be used with copper or brass and steel or stainless steel tubings, respectively. It is important that fittings of this type be installed correctly. Different brands of tube fittings should not be mixed in the same apparatus assembly because construction parts are often not interchangeable.

6.D.2.2.6 *Teflon Tape Applications*

Teflon tape should be used on tapered pipe thread where the seal is formed in the thread area. Tapered pipe thread is commonly found in applications where fittings are not routinely taken apart (e.g., general building piping applications).

Teflon tape should not be used on straight thread where the seal is formed through gaskets or by other metal-to-metal contacts that are forced together when the fitting is tightened (e.g., CGA gas cylinder fittings or compression fittings). Metal-to-metal seals are machined to tolerances that seal without the need of Teflon tape or other gasketing materials. If used where not needed, as on CGA fittings, Teflon tape only spreads and weakens the threaded connections and can plug up lines that it enters accidentally.

6.E WORKING WITH HIGH/LOW PRESSURES AND TEMPERATURES

Work with hazardous chemicals at high/low pressures and/or high/low temperatures requires planning and special precautions. For many experiments, extremes of both pressure and temperature, such as reactions at elevated temperatures and pressures and work with cryogenic liquids and high vacuum, must be managed simultaneously. Procedures at high/low pressures should be carried out with protection against explosion or implosion by appropriate equipment selection and the use of safety shields. Appropriate temperature control and interlocks should be provided so that heating or cooling baths cannot exceed the desired limits even if the equipment fails. Care must be taken to select and use glass apparatus that can safely withstand thermal expansion or contraction at the designated pressure and temperature extremes.

6.E.1 Pressure Vessels

High-pressure operations should be performed only in special chambers equipped for this purpose. Laboratory workers should ensure that equipment for operations using pressure vessels is appropriately selected, properly labeled and installed, and protected by pressure-relief and necessary control devices. Vessels must be strong enough to withstand the stresses encountered at the intended operating pressures and temperatures. The vessel material must not corrode when it is in contact with the material(s) it contains. The material should not react with the process being studied, and the vessel must be of the proper size and configuration needed for the process. Reactions should never be carried out in, nor heat applied to, an apparatus that is a closed system unless it has been designed and tested to withstand pressure.

Pressure-containing systems designed for use at elevated temperatures should have a positive temperature controller. Manual control using a simple variable autotransformer, such as a Variac, is not good practice. The use of both a back-up temperature controller capable of recording temperatures and shutting down an unattended system is strongly recommended.

(See section 6.D.2 above.)

6.E.1.1   Records, Inspection, and Testing

In some localities, adherence to national codes such as the ASME (American Society of Mechanical Engineers) Boiler and Pressure Vessel Code (ASME, 1992) is mandatory. Selection of containers, tubing, fittings, and other process equipment, along with the operational techniques and procedures, must adhere to the constraints necessary for high-pressure service. The proper selection and assembly of components in a pressure system are critical safety factors. Compatibility of materials, tools used for assembly, and the reliability of connections are all key considerations.

Each pressure vessel in a laboratory should have a stamped number or fixed label plate that uniquely identifies it. Information such as the maximum allowable working pressure, allowable temperature at this pressure, material of construction, and burst diagram should be readily available. Information on the vessel's history should include temperature extremes it has experienced, any modifications and repairs made to the original vessel, and all inspections or test actions it has undergone. Similarly, the relieving pressure and setting data should be stamped on a metal tag attached to installed pressure-relief devices, and the setting mechanisms should be sealed. Relief devices used on pressure regulators do not require these seals or numbers.

All pressure equipment should be tested or inspected periodically. The interval between tests or inspections is determined by the severity of the usage the equipment has received. Corrosive or otherwise hazardous service requires more frequent tests and inspections. Inspection data should be stamped on or attached to the equipment. Pressure vessels may be subjected to nondestructive inspections such as visual inspection, penetrant inspection, acoustic emissions recording, and radiography. However, hydrostatic proof tests are necessary for final acceptance. These tests should be as infrequent as possible. They should be performed before the vessel is placed in initial service, every 10 years thereafter, after a significant repair or modification, and if the vessel experiences overpressure or overtemperature.

Testing the entire apparatus with soap solution and air or nitrogen pressure to the maximum allowable working pressure of the weakest section of the assembled apparatus can usually detect leaks at threaded joints, packings, and valves.

Final assemblies should be pressure-tested and leak-tested to ensure their integrity. Laboratory workers are strongly advised to consult an expert on high-pressure work as they design, build, and operate a high-pressure process. Finally, extreme care should be exercised when disassembling pressure equipment for repair, modification, or decommissioning. Protective equipment should be worn just in case a line or vessel that is opened contains material under pressure.

6.E.1.2   Pressure Reactions in Glass Equipment

For any reaction run on a large scale (more than 10 g total weight of reactants) or at a maximum pressure in excess of 690 kPa (100 psi), only procedures involving a suitable high-pressure autoclave or shaker vessel should be used. Whenever possible, metal reactors with glass liners should be used instead of sealed glass tubes. Fisher-Porter-type tubes with a pressure gauge and release device are preferred for pressure reactions in glass equipment. However, it is sometimes convenient to run very small-scale reactions at low pressures in a small sealed glass tube or in a thick-walled pressure bottle of the type used for catalytic hydrogenation. For any such reaction, the worker should be fully prepared for the significant possibility that the sealed vessel will burst. Removal of any gas should be prepared for by appropriate ventilation. Every precaution should be taken to prevent injury from flying glass or from corrosive or toxic reactants by using suitable shielding. Centrifuge bottles should be sealed with rubber stoppers clamped in place, wrapped with friction tape and shielded with a metal screen or wrapped with friction tape and surrounded by multiple layers of loose cloth toweling, and clamped behind a good safety shield. Some bottles are typically equipped with a head containing inlet and exhaust gas valves, a pressure gauge, and a pressure-relief valve. When corrosive materials are being used, a Teflon pressure-relief valve should be used. The preferred source of heat for such vessels is steam, because an explosion in the vicinity of an electrical heater could start a fire and an explosion in a liquid heating bath would scatter hot liquid around the area. Any reaction of this type should be carried out in a hood and labeled with signs that indicate the contents of the reaction vessel and the explosion risk.

Glass tubes with high-pressure sealers should be no more than three-quarters full. Appropriate precautions using the proper shielding must be taken for condensing materials and sealing tubes. Vacuum work can be carried out on a Schlenck line as long as proper technique is used. The sealed glass tubes can be placed either inside pieces of brass or iron pipe capped at one end with a pipe cap or in an autoclave containing some of the reaction solvent (to equalize the pressure inside and outside the glass tube). The tubes can be heated with steam or in a specially constructed, electrically heated "sealed-tube" furnace that is controlled thermostatically and located such that the force of an explosion would be directed into a safe area. When the required heating has been completed, the sealed tube or bottle should be allowed to cool to room temperature. Sealed bottles and tubes of flammable materials should be wrapped with cloth toweling, placed behind a safety shield, and then cooled slowly, first in an ice bath and then in dry ice. After cooling, the clamps and rubber stoppers can be removed from the bottles prior to opening. Personal protective equipment and apparel, including shields, masks, coats, and gloves, should be used during tube-opening operations. It should be noted that NMR tubes are often thin-walled and should only be used for pressure reactions in a special high-pressure probe or in capillary devices.

Newly fabricated or repaired glass equipment for pressure or vacuum work should be examined for flaws and strains under polarized light. Corks, rubber stoppers, and rubber or plastic tubing should never be relied on as relief devices for protection of glassware against excess pressure; a liquid seal, Bunsen tube, or equivalent positive relief device should be used. When glass pipe is used, only proper metal fittings should be used.

6.E.2 Liquefied Gases and Cryogenic Liquids

Cryogenic liquids are materials with boiling points of less than $-73\,°C$ ($-100\,°F$). Liquid nitrogen, helium, and argon, and slush mixtures of dry ice with isopropanol are the materials most commonly used in cold traps to condense volatile vapors from a system. In addition, oxygen, hydrogen, and helium are often used in the liquid state.

The primary hazards of cryogenic liquids are fire or explosion, pressure buildup (either slowly or due to rapid conversion of the liquid to the gaseous state), embrittlement of structural materials, frostbite, and asphyxiation. The extreme cold of cryogenic liquids requires special care in their use. The vapor that boils off from a liquid can cause the same problems as the liquid itself.

The fire or explosion hazard is obvious when gases such as oxygen, hydrogen, methane, and acetylene are used. Air enriched with oxygen can greatly increase the flammability of ordinary combustible materials and may even cause some noncombustible materials to burn readily (see Chapter 5, sections 5.G.4 and 5.G.5). Oxygen-saturated wood and asphalt have been known to literally explode when subjected to shock. Because oxygen has a higher boiling point ($-183\,°C$) than nitrogen ($-195\,°C$), helium ($-269\,°C$), or hydrogen ($-252.7\,°C$), it can be condensed out of the atmosphere during the use of these lower-boiling-point cryogenic liquids. With the use of liquid hydrogen particularly, conditions may develop for an explosion. (See Chapter 5, sections 5.F.3 and 5.G.2, for further discussion.)

It is advisable to furnish all cylinders and equipment containing flammable or toxic liquefied gases (not vendor-owned) with a spring-loaded pressure-relief device (not a rupture disk) because of the magnitude of the potential risk that can result from activation of a non-resetting relief device. Commercial cylinders of liquefied gases are normally supplied only with a fusible-plug type of relief device, as permitted by DOT regulations. Pressurized containers that contain cryogenic material should be protected with multiple pressure-relief devices.

Cryogenic liquids must be stored, shipped, and handled in containers that are designed for the pressures and temperatures to which they may be subjected. Materials that are pliable under normal conditions can become brittle at low temperatures. Dewar flasks, which are used for relatively small amounts of material, should have a dust cap over the outlet to prevent atmospheric moisture from condensing and plugging the neck of the tube. Special cylinders insulated and vacuum-jacketed with pressure-relief valves and rupture devices to protect the cylinder from pressure buildup are available in capacities of 100 to 200 liters (L).

A special risk to personnel is skin or eye contact with the cryogenic liquid. Because these liquids are prone to splash in use owing to the large volume expansion ratio when the liquid warms up, eye protection, preferably a face shield, should be worn when handling liquefied gases and other cryogenic fluids. The transfer of liquefied gases from one container to another should not be attempted for the first time without the direct supervision and instruction of someone experienced in this operation. Transfers should be done very slowly to minimize boiling and splashing.

Unprotected parts of the body should not be in contact with uninsulated vessels or pipes that contain cryogenic liquids because extremely cold material may bond firmly to the skin and tear flesh if separation or withdrawal is attempted. Even very brief skin contact with a cryogenic liquid can cause tissue damage similar to that of frostbite or thermal burns, and prolonged contact may result in blood clots that have potentially very serious consequences. Gloves must be impervious to the fluid being handled and loose enough to be tossed off easily. A potholder may be a desirable alternative. Objects that are in contact with cryogenic liquids should also be handled with tongs or potholders. The work area should be well ventilated. Virtually all liquid gases present the threat of poisoning, explosion, or, at a minimum, asphyxiation in a confined space. Major harmful consequences of the use of cryogenic inert gases, including asphyxiation, are due to boiling off of the liquid and pressure buildup, which can lead to violent rupture of the container or piping.

In general, liquid hydrogen should not be transferred in an air atmosphere because oxygen from the air can condense in the liquid hydrogen, presenting a possible explosion risk. All precautions should be taken to keep liquid oxygen from organic materials; spills on oxidizable surfaces can be hazardous. Though nitrogen is inert, its liquefied form can be hazardous because of its cryogenic properties and because displacement of air oxygen in the vicinity can lead to asphyxiation followed by death with little warning. Rooms that contain appreciable quantities of liquid nitrogen (N_2) should be fitted with oxygen meters and alarms. Liquid nitrogen should not be stored in a closed room because the oxygen content of the room can drop to unsafe levels.

Cylinders and other pressure vessels used for the storage and handling of liquefied gases should not be filled to more than 80% of capacity, to protect against possible thermal expansion of the contents and bursting of the vessel by hydrostatic pressure. If the possibility exists that the temperature of the cylinder may increase to above 30°C, a lower percentage (e.g., 60%) of capacity should be the limit.

6.E.2.1 Cold Traps and Cold Baths

Cold traps should be chosen that are large enough and cold enough to collect the condensable vapors in a vacuum system. Cold traps should be checked frequently to make sure they do not become plugged with frozen material. Cold traps in a reduced-pressure system should be taped or placed in a metal can filled with vermiculite. After completion of an operation in which a cold trap has been used, the system should

be vented in a safe and environmentally acceptable way. Otherwise, pressure could build up, creating a possible explosion and sucking pump oil into the system. Cold traps under continuous use, such as those used to protect inert atmosphere dry boxes, should be cooled electrically and monitored by low-temperature probes.

Appropriate gloves and a face shield should be used to avoid contact with the skin when using cold baths. Dry gloves should be used when handling dry ice. Lowering of the head into a dry ice chest is to be avoided because carbon dioxide is heavier than air and asphyxiation can result. The preferred liquids for dry ice cooling baths are isopropyl alcohol or glycols, and the dry ice should be added slowly to the liquid portion of the cooling bath to avoid foaming. The common practice of using acetone-dry ice as a coolant should be avoided. Dry ice and liquefied gases used in refrigerant baths should always be open to the atmosphere. They should never be used in closed systems, where they may develop uncontrolled and dangerously high pressures.

Extreme caution should be exercised in using liquid nitrogen as a coolant for a cold trap. If such a system is opened while the cooling bath is still in contact with the trap, *oxygen may condense from the atmosphere*. The oxygen could then combine with any organic material in the trap to create a highly explosive mixture. Thus, a system that is connected to a liquid nitrogen trap should not be opened to the atmosphere until the trap has been removed. Also, if the system is closed after even a brief exposure to the atmosphere, some oxygen (or argon) may have already condensed. Then, when the liquid nitrogen bath is removed or when it evaporates, the condensed gases will vaporize, producing a pressure buildup and the potential for explosion. The same explosion hazard can be created if liquid nitrogen is used to cool a flammable mixture that is exposed to air.

6.E.2.2 Selection of Low-Temperature Equipment

Equipment used at low temperatures should be selected carefully. Temperature can dramatically change characteristics of materials. For example, even the impact strength of ordinary carbon steel is greatly reduced at low temperatures, and failure can occur at points of weakness, such as notches or abrupt changes in the material of construction, in cold equipment. When combinations of materials are required, it is important that the temperature dependence of their volumes be considered so that leaks, ruptures, and glass fractures can be avoided. For example, O-rings that provide a good seal at room temperature may lose resilience and fail to function on chilled equipment.

IMPROPER GLASSWARE IN A CRYOGENIC FLUID

A thin-walled Pyrex NMR sample tube containing absorbed hydrocarbons on platinum on an alumina support, which had been sealed under vacuum and annealed, was placed in a dry ice and chloroform mixture in a Dewar flask in a hood with horizontal sliding sashes. The tube exploded after approximately one minute in the bath, apparently due to thermal shock. Although the Dewar was not damaged, the researcher suffered severely irritated eyes and had to be transported to the trauma center. The researcher had been wearing glasses and a laboratory coat as personal protection, and the hood sash had been slid to the side. Face shields, goggles, gloves, and acrylic shielding were available in the laboratory but had not been used.

The 18% chromium/8% nickel stainless steels retain their impact resistance down to approximately −240 °C, the exact value depending heavily on special design considerations. The impact resistance of aluminum, copper, nickel, and many other nonferrous metals and alloys increases with decreasing temperatures. Special alloy steels should be used for liquids or gases containing hydrogen at temperatures greater than 200 °C or at pressures greater than 34.5 MPa (500 psi) because of the danger of weakening carbon steel equipment by hydrogen embrittlement.

6.E.2.3 Cryogenic Lines and Supercritical Fluids

Liquid cryogen transfer lines should be designed so that liquid cannot be trapped in any nonvented part of the system. Experiments in supercritical fluids include high pressure and should be carried out with appropriate protective systems.

6.E.3 Vacuum Work and Apparatus

Vacuum work can result in an implosion and the possible hazards of flying glass, spattering chemicals, and fire. All vacuum operations must be set up and operated with careful consideration of the potential risks.

Although a vacuum distillation apparatus may appear to provide some of its own protection in the form of heating mantles and column insulation, this is not sufficient because an implosion could scatter hot, flammable liquid. An explosion shield and a face mask should be used to protect the worker, and the procedure should be carried out in a hood.

Equipment at reduced pressure is especially prone to rapid pressure changes, which can create large pressure differences within the apparatus. Such conditions can push liquids into unwanted locations, sometimes with undesirable consequences.

Water, solvents, and corrosive gases should not be allowed to be drawn into a building vacuum system. When the potential for such a problem exists, a water aspirator with a solvent collection device and a trap with a check valve installed between the water aspirator and the apparatus, to prevent water from being drawn back into the apparatus, should be used as the vacuum source.

Mechanical vacuum pumps should be protected by cold traps, and their exhausts should be vented to an exhaust hood or to the outside of the building. If solvents or corrosive substances are inadvertently drawn into the pump, the oil should be changed before any further use. (Oil contaminated with solvents, mercury, corrosive substances, and so on, must be handled as hazardous waste.) It may be desirable to maintain a log of pump usage as a guide to length of use and potential contaminants in the pump oil. The belts and pulleys on vacuum pumps should be covered with guards.

(See section 6.C.2 for a discussion of vacuum pumps.)

6.E.3.1 Glass Vessels

Although glass vessels are frequently used in low-vacuum operations, evacuated glass vessels may collapse violently, either spontaneously from strain or from an accidental blow. Therefore, pressure and vacuum operations in glass vessels should be conducted behind adequate shielding. It is advisable to check for flaws such as star cracks, scratches, and etching marks each time a vacuum apparatus is used. Only round-bottomed or thick-walled (e.g., Pyrex) evacuated reaction vessels specifically designed for operations at reduced pressure should be used. Repaired glassware is subject to thermal shock and should be avoided. Thin-walled, Erlenmeyer, or round-bottomed flasks larger than 1 L should never be evacuated.

6.E.3.2 Dewar Flasks

Dewar flasks are under high vacuum and can collapse as a result of thermal shock or a very slight mechanical shock. They should be shielded, either by a layer of fiber-reinforced friction tape or by enclosure in a wooden or metal container, to reduce the risk of flying glass in case of collapse. Metal Dewar flasks

---

IMPLODING DEWAR

A researcher was about to prepare an ice trap in a Dewar to cool a stationary stainless steel receiver on a chemical reactor system. The researcher had positioned the Dewar on a laboratory jack stand and had raised the Dewar into position. The Dewar imploded, propelling glass shards toward the researcher, who fortunately was wearing prescription safety glasses and received only minor facial cuts. The researcher should have been wearing a full-length face shield and should have had a cover on the Dewar.

---

should be used whenever there is a possibility of breakage.

Styrofoam buckets with lids can be a safer form of short-term storage and conveyance of cryogenic liquids than glass vacuum Dewars. Although they do not insulate as well as Dewar flasks, they eliminate the danger of implosion.

6.E.3.3   Desiccators

If a glass vacuum desiccator is used, it should be made of Pyrex or similar glass, completely enclosed in a shield or wrapped with friction tape in a grid pattern that leaves the contents visible and at the same time guards against flying glass should the vessel implode. Plastic (e.g., polycarbonate) desiccators reduce the risk of implosion and may be preferable, but should also be shielded while evacuated. Solid desiccants are preferred. *An evacuated desiccator should never be carried or moved.* Care should be taken in opening the valve to avoid a shock wave into the desiccator.

6.E.3.4   Rotary Evaporators

Glass components of the rotary evaporator should be made of Pyrex or similar glass, completely enclosed in a shield to guard against flying glass should the components implode. Increase in rotation speed and application of vacuum to the flask whose solvent is to be evaporated should be gradual.

6.E.3.5   Assembly of Vacuum Apparatus

Vacuum apparatus should be assembled so as to avoid strain. Joints must be assembled so as to allow various sections of the apparatus to be moved if necessary without transmitting strain to the necks of the

flasks. Heavy apparatus should be supported from below as well as by the neck. The assembler should avoid putting pressure on a vacuum line. Failure to keep the pressure below 1 atmosphere could lead to the stopcocks popping out at high velocity or to an explosion of the glass apparatus. Such increased pressure could result from warming of the contents of the trap due to failure to maintain low temperatures.

Vacuum apparatus should be placed well back onto the bench or into the hood where they will not be inadvertently hit. If the back of the vacuum setup faces the open laboratory, it should be protected with panels of suitably heavy transparent plastic to prevent injury to nearby workers from flying glass in case of explosion.

6.F   USING PERSONAL PROTECTIVE, SAFETY, AND EMERGENCY EQUIPMENT

As outlined in previous chapters, it is essential for each laboratory worker to be proactive to ensure the laboratory is a safe working environment. This attitude begins with wearing appropriate apparel and using proper eye, face, hand, and foot protection when working with hazardous chemicals. It is the responsibility of the institution to provide appropriate safety and emergency equipment for laboratory workers and for emergency personnel. (See also section 5.C.)

6.F.1   Personal Protective Equipment and Apparel

6.F.1.1   Personal Clothing

Clothing that leaves large areas of skin exposed is inappropriate in laboratories where hazardous chemicals are in use. The worker's personal clothing should be fully covering. Appropriate laboratory coats should be worn, buttoned, with the sleeves rolled down. Laboratory coats should be fire-resistant. Those fabricated of polyester are not appropriate for glassblowing or work with flammable materials. Cotton coats are inexpensive and do not burn readily. Laboratory coats or laboratory aprons made of special materials are available for high-risk activities. Laboratory coats that have been used in the laboratory should be left there to minimize the possibility of spreading chemicals to public assembly, eating, or office areas, and they should be cleaned regularly. (For more information, see the OSHA Personal Protective Equipment Standard (29 CFR 1910.132) and the OSHA Laboratory Standard (29 CFR 1910.1450).)

Unrestrained long hair and loose clothing such as neckties, baggy pants, and coats are inappropriate in a laboratory where hazardous chemicals are in use. Such items can catch fire, be dipped in chemicals, and get caught in equipment. Similarly, rings, bracelets, watches, or other jewelry that could be damaged, trap chemicals close to the skin, come in contact with electrical sources, or get caught in machinery should not be worn. Leather clothing or accessories should not be worn in situations where chemicals could be absorbed in the leather and held close to the skin.

Protective apparel should always be worn if there is a possibility that personal clothing could become contaminated with chemically hazardous material. Washable or disposable clothing worn for laboratory work with especially hazardous chemicals includes special laboratory coats and aprons, jumpsuits, special boots, shoe covers, and gauntlets, as well as splash suits. Protection from heat, moisture, cold, and/or radiation may be required in special situations. Among the factors to be considered in choosing protective apparel, in addition to the specific application, are resistance to physical hazards, flexibility and ease of movement, chemical and thermal resistance, and ease of cleaning or disposal. Although cotton is a good material for laboratory coats, it reacts rapidly with acids. Plastic or rubber aprons can provide good protection from corrosive liquids but can be inappropriate in the event of a fire. Plastic aprons can also accumulate static electricity, and so they should not be used around flammable solvents, explosives sensitive to electrostatic discharge, or materials that can be ignited by static discharge. Disposable garments provide only limited protection from vapor or gas penetration. Disposable garments that have been used when handling carcinogenic or other highly hazardous material should be removed without exposing any individual to toxic materials and disposed of as hazardous waste. (See Chapter 5, sections 5.C.2.5 and 5.C.2.6.)

6.F.1.2 Foot Protection

Street shoes may not be appropriate in the laboratory, where both chemical and mechanical hazards may exist. Substantial shoes should be worn in areas where hazardous chemicals are in use or mechanical work is being done. Clogs, perforated shoes, sandals, and cloth shoes do not provide protection against spilled chemicals. In many cases, safety shoes are advisable. Shoe covers may be required for work with especially hazardous materials. Shoes with conductive soles are useful to prevent buildup of static charge, and insulated soles can protect against electrical shock.

6.F.1.3 Eye and Face Protection

Safety glasses with side shields that conform to ANSI standard Z87.1-1989 should be required for work with hazardous chemicals. Ordinary prescription glasses with hardened lenses do not serve as safety glasses. Contact lenses can sometimes be worn safely if appropriate eye and face protection is also worn **(see, however, section 5.C.2.1)**. Although safety glasses can provide satisfactory protection from injury from flying particles, they do not fit tightly against the face and offer little protection against splashes or sprays of chemicals. It is appropriate for a laboratory to provide impact goggles that include splash protection (splash goggles), full-face shields that also protect the throat, and specialized eye protection (i.e., protection against ultraviolet light or laser light). Splash goggles, which have splash-proof sides to fully protect the eyes, should be worn if there is a splash hazard in any operation involving hazardous chemicals. Impact protection goggles should be worn if there is a danger of flying particles, and full-face shields with safety glasses and side shields are needed for complete face and throat protection. When there is a possibility of liquid splashes, both a face shield and splash goggles should be worn; this is especially important for work with highly corrosive liquids. Full-face shields with throat protection and safety glasses with side shields should be used when handling explosive or highly hazardous chemicals. If work in the laboratory could involve exposure to lasers, ultraviolet light, infrared light, or intense visible light, specialized eye protection should be worn. It also is appropriate for a laboratory to provide visitor safety glasses and a sign indicating that eye protection is required in laboratories where hazardous chemicals are in use.

6.F.1.4 Hand Protection

Gloves appropriate to the hazard should be used at all times. It is important that the hands and any skin that is likely to be exposed to hazardous chemicals receive special attention. Proper protective gloves should be worn when handling hazardous chemicals, toxic materials, materials of unknown toxicity, corrosive materials, rough or sharp-edged objects, and very hot or very cold objects. Before the gloves are used, it is important that they be inspected for discoloration, punctures, or tears. A defective or improper glove can itself be a serious hazard in handling hazardous chemicals. If chemicals do penetrate glove material, they could then be held in prolonged contact with the hand and cause more serious damage than in the absence of a proper glove.

The degradation and permeation characteristics of the glove material selected must be appropriate for protection from the hazardous chemicals being handled. Glove selection guides (available from most manufacturers) should be consulted, with careful consideration given to the permeability of any material, particularly when working with organic solvents, which may be able to permeate or dissolve the glove materials. The thin latex "surgical" vinyl and nitryl gloves that are popular in many laboratories because of their composition and thin construction may not be appropriate for use with highly toxic chemicals or solvents. For example, because latex is readily permeated by carbon disulfide, a hand covered by a latex glove immersed in carbon disulfide would receive constant wetting by this toxic chemical, which would by then be absorbed through the skin. Gloves should be replaced immediately if they are contaminated or torn. The use of double gloves may be appropriate in situations involving chemicals of high or multiple hazards. Leather gloves are appropriate for handling broken glassware and inserting tubing into stoppers, where protection from chemicals is not needed. Insulated gloves should be used when working with very hot or very cold materials. With cryogenic fluids the gloves must be impervious to fluid, but loose enough to be tossed off easily. Absorbent gloves could freeze on the hand and intensify any exposure to liquefied gases. Turning up the cuffs on gloves can prevent liquids from running down the arms when hands are raised.

Gloves should be decontaminated or washed appropriately before they are taken off and should be left in the work area and not be allowed to touch any uncontaminated objects in the laboratory or any other area. Gloves should be replaced periodically, depending on the frequency of use. Regular inspection of their serviceability is important. If they cannot be cleaned, contaminated gloves should be disposed of according to institutional procedures.

Barrier creams and lotions can provide some skin protection but should never be a substitute for gloves, protective clothing, or other protective equipment. These creams should be used only to supplement the protection offered by personal equipment.

6.F.2 Safety and Emergency Equipment

Safety equipment, including spill control kits, safety shields, fire safety equipment, respirators, safety showers and eyewash fountains, and emergency equipment should be available in well-marked, highly visible locations in all chemical laboratories. Fire alarm pull stations and telephones with emergency telephone numbers clearly indicated must be readily accessible. In addition to the standard items, there may also be a need for other safety devices. It is the responsibility of the laboratory supervisor to ensure proper training and provide supplementary equipment as needed.

6.F.2.1 Spill Control Kits and Cleanup

In most cases, researchers are responsible for cleaning up their own spills. If a spill exceeds their ability or challenges their safety, they should leave the spill site and call the emergency telephone number for help. Emergency response spill cleanup personnel should be given all available information about the spill.

A spill control kit should be on hand. A typical cleanup kit may be a container on wheels that can be moved to the location of the spill and may include such items as instructions; absorbent pads; a spill absorbent mixture for liquid spills; a polyethylene scoop for dispensing spill absorbent; mixing it with the spill, and picking up the mixture; thick polyethylene bags for deposit of the mixture; and tags and ties for labeling the bags. Any kit should be used in conjunction with the personal protective equipment needed for the chemical that is to be cleaned up. Before beginning an operation that could produce a spill, the worker should locate the specialized spill control kits for that operation.

(Also see Chapter 5, section 5.C.11.5.)

6.F.2.2 Safety Shields

Safety shields should be used for protection against possible explosions or splash hazards. Laboratory equipment should be shielded on all sides so that there is no line-of-sight exposure of personnel. The front sashes of conventional laboratory exhaust hoods can provide shielding. However, a portable shield should also be used when manipulations are performed, particularly with hoods that have vertical-rising doors rather than horizontal-sliding sashes.

Portable shields can be used to protect against hazards of limited severity, such as small splashes, heat, and fires. A portable shield, however, provides no protection at the sides or back of the equipment, and many such shields not sufficiently weighted for forward protection may topple toward the worker when there is a blast. A fixed shield that completely surrounds the experimental apparatus can afford protection against minor blast damage.

Polymethyl methacrylate, polycarbonate, polyvinyl chloride, and laminated safety plate glass are all satisfactory transparent shielding materials. Where combustion is possible, the shielding material should be nonflammable or slow burning; if it can withstand the working blast pressure, laminated safety plate glass may be the best material for such circumstances. When cost, transparency, high tensile strength, resistance to bending loads, impact strength, shatter resistance, and burning rate are considered, polymethyl methacrylate offers an excellent overall combination of shielding characteristics.

Polycarbonate is much stronger and self-extinguishing after ignition but is readily attacked by organic solvents.

6.F.2.3   Fire Safety Equipment

6.F.2.3.1   Fire Extinguishers

All chemical laboratories should have carbon dioxide and dry chemical fire extinguishers. Other types of extinguishers should be available if required for the work being done. The four types of extinguishers most commonly used are classified by the type of fire for which they are suitable, as listed below. It should be noted that multipurpose class A, B, and C extinguishers are available.

- Water extinguishers are effective against burning paper and trash (class A fires). These should not be used for extinguishing electrical, liquid, or metal fires.
- Carbon dioxide extinguishers are effective against burning liquids, such as hydrocarbons or paint, and electrical fires (class B and C fires). They are recommended for fires involving computer equipment, delicate instruments, and optical systems because they do not damage such equipment. They are less effective against paper and trash fires and *must not be used* against metal hydride or metal fires. Care must be taken in using these extinguishers, because the force of the compressed gas can spread burning combustibles such as papers and can tip over containers of flammable liquids.
- Dry powder extinguishers, which contain ammonium phosphate or sodium bicarbonate, are effective against burning liquids and electrical fires (class B and C fires). They are less effective against paper and trash or metal fires. They are not recommended for fires involving delicate instruments or optical systems because of the cleanup problem. Computer equipment may need to be replaced if exposed to sufficient amounts of the dry powders. These extinguishers are generally used where large quantities of solvent may be present.
- Met-L-X® extinguishers and others that have special granular formulations are effective against burning metal (class D fires). Included in this category are fires involving magnesium, lithium, sodium, and potassium; alloys of reactive metals; and metal hydrides, metal alkyls, and other organometallics. These extinguishers are less effective against paper and trash, liquid, or electrical fires.

Every extinguisher should carry a label indicating what class or classes of fires it is effective against and the date last inspected. There are a number of other, more specialized types of extinguishers available for unusual fire hazard situations. Each laboratory worker should be responsible for knowing the location, operation, and limitations of the fire extinguishers in the work area. It is the responsibility of the laboratory supervisor to ensure that all workers are shown the locations of fire extinguishers and are trained in their use. After use, an extinguisher should be recharged or replaced by designated personnel.

6.F.2.3.2   Heat and Smoke Detectors

Heat sensors and/or smoke detectors may be part of the building safety equipment. If designed into the fire alarm system, they may automatically sound an alarm and call the fire department, they may trigger an automatic extinguishing system, or they may only serve as a local alarm. Because laboratory operations may generate heat or vapors, the type and location of the detectors must be carefully evaluated in order to avoid frequent false alarms.

6.F.2.3.3   Fire Hoses

Fire hoses are intended for use by trained firefighters against fires too large to be handled by extinguishers and are included as safety equipment in some structures. Water has a cooling action and is effective against fires involving paper, wood, rags, trash, and such (class A fires). Water should not be used directly on fires that involve live electrical equipment (class C fires) or chemicals such as alkali metals, metal hydrides, and metal alkyls that react vigorously with it (class D fires).

Streams of water should not be used against fires that involve oils or other water-insoluble flammable liquids (class B fires). Water will not readily extinguish such fires. Rather, it can cause the fire to spread or float to adjacent areas. These possibilities are minimized by the use of a water fog. Water fogs are used extensively by the petroleum industry because of their fire-controlling and extinguishing properties. A fog can be used safely and effectively against fires that involve oil products, as well as those involving wood, rags, rubbish, and such.

Because of the potential risks involved in using water around chemicals, laboratory workers should refrain from using fire hoses except in extreme emergencies. Otherwise, such use should be reserved for trained firefighters. Clothing fires can be extinguished by immediately dropping to the floor and rolling; however, if a safety shower is immediately available, it should be used (as noted in section 6.F.2.5).

6.F.2.3.4 *Automatic Fire-Extinguishing Systems*

In areas where fire potential and the risk of injury or damage are high, automatic fire-extinguishing systems are often used. These may be of the water sprinkler, foam, carbon dioxide, halon, or dry chemical type. If an automatic fire-extinguishing system is in place, laboratory workers should be informed of its presence and advised of any safety precautions required in connection with its use (e.g., evacuation before a carbon dioxide total-flood system is activated, to avoid asphyxiation).

6.F.2.4 Respiratory Protective Equipment

The primary method for the protection of laboratory personnel from airborne contaminants should be to minimize the amount of such materials entering the laboratory air. When effective engineering controls are not possible, suitable respiratory protection should be used after proper training. Respiratory protection may be needed in carrying out an experimental procedure, in dispensing or handling hazardous chemicals, in responding to a chemical spill or release in cleanup decontamination, or in hazardous waste handling.

Under Occupational Safety and Health Administration (OSHA) regulations, only equipment listed and approved by the Mine Safety and Health Administration (MSHA) and the National Institute for Occupational Safety and Health (NIOSH) may be used for respiratory protection. Also under the regulations, each site on which respiratory protective equipment is used must implement a respirator program (including training and medical certification) in compliance with OSHA's Respiratory Protection Standard (29 CFR 1910.134); see also ANSI standard Z88.2-1992, Practices for Respiratory Protection.

6.F.2.4.1 *Types of Respirators*

Several types of nonemergency respirators are available for protection in atmospheres that are not immediately dangerous to life or health but could be detrimental after prolonged or repeated exposure. Other types of respirators are available for emergency or rescue work in hazardous atmospheres from which the wearer needs protection. In either case, additional protection may be required if the airborne contaminant is of a type that could be absorbed through or irritate the skin. For example, the possibility of eye or skin irritation may require the use of a full-body suit and a full-face mask rather than a half-face mask. For some chemicals the dose from skin absorption can exceed the dose from inhalation.

The choice of the appropriate respirator to use in a given situation depends on the type of contaminant and its estimated or measured concentration, known exposure limits, and hazardous properties. The degree of protection afforded by the respirator varies with the type. Four main types of respirators are currently available:

• Chemical cartridge respirators can be used only for protection against particular individual (or classes of) vapors or gases as specified by the respirator manufacturer and cannot be used at concentrations of contaminants above that specified on the cartridge. Also, these respirators cannot be used if the oxygen content of the air is less than 19.5%, in atmospheres immediately dangerous to life, or for rescue or emergency work. These respirators function by trapping vapors and gases in a cartridge or canister that contains a sorbent material, with activated charcoal being the most common adsorbent. Because it is possible for significant breakthrough to occur at a fraction of the canister capacity, knowledge of the potential workplace exposure and length of time the respirator will be worn is important. It may be desirable to replace the cartridge after each use to ensure the maximum available exposure time for each new use. Difficulty in breathing or the detection of odors indicates plugged or exhausted filters or cartridges or concentrations of contaminants higher than the absorbing capacity of the cartridge, and the user should immediately leave the area of contamination. Chemical cartridge respirators must be checked and cleaned on a regular basis. New and used cartridges must not be stored near chemicals because they are constantly filtering the air. Cartridges should be stored in sealed containers to prevent chemical contamination.

Respirators must fit snugly on the face to be effective. Failure to achieve a good face-to-facepiece seal (for example, because of glasses or facial hair) can permit contaminated air to bypass the filter and create a dangerous situation for the user. Respirators requiring a face-to-facepiece seal should not be used by those with facial hair, for whom powered air-purifying or supplied-air respirators are at times appropriate. Tests for a proper fit must be conducted prior to selection of a respirator and verified before the user enters the area of contamination.

Organic vapor cartridges cannot be used for vapors that are not readily detectable by their odor or other irritating effects or for vapors that will generate substantial heat upon reaction with the sorbent materials in the cartridge.

• Dust, fume, and mist respirators can be used only for protection against particular, or certain classes of, dusts, fumes, and mists as specified by the manufacturer. The useful life of the filter depends on the concentration of contaminant encountered. Such particulate-removing respirators usually trap the particles in a filter composed of fibers; they are not 100% efficient in removing particles. Respirators of this type are generally disposable. Examples are surgical masks and 3M® toxic-dust and nuisance-dust masks. Some masks are NIOSH-approved for more specific purposes such as protection against simple or benign dust and fibrogenic dusts and asbestos.

Particulate-removing respirators afford no protection against gases or vapors and may give the user a false sense of security. They are also subject to the limitations of fit.

• Supplied-air respirators supply fresh air to the facepiece of the respirator at a pressure high enough to cause a slight buildup relative to atmospheric pressure. As a result, the supplied air flows outward from the mask, and contaminated air from the work environment cannot readily enter the mask. This characteristic renders face-to-facepiece fit less important than with other types of respirators. Fit testing is, however, required before selection and use.

Supplied-air respirators are effective protection against a wide range of air contaminants (gases, vapors, and particulates) and can be used where oxygen-deficient atmospheres are present. Where concentrations of air contaminants could be immediately dangerous to life, such respirators can be used provided (1) the protection factor of the respirator is not exceeded and (2) the provisions of OSHA's Respiratory Standard (which indicates the need for a safety harness and an escape system in case of compressor failure) are not violated.

The air supply of this type of respirator must be kept free of contaminants (e.g., by use of oil filters and carbon monoxide absorbers). Most laboratory air is not suitable for use with these units. These units usually require the user to drag lengths of hose connected to the air supply, and they have a limited range.

• The self-contained breathing apparatus (SCBA) is the only type of respiratory protective equipment suitable for emergency or rescue work. Untrained personnel should not attempt to use them.

6.F.2.4.2 *Procedures and Training*

Each area where respirators are used should have written information available that shows the limitations, fitting methods, and inspection and cleaning procedures for each type of respirator available. Personnel who may have occasion to use respirators in their work must be thoroughly trained, before initial use and annually thereafter, in the fit testing, use, limitations, and care of such equipment. Training should include demonstrations and practice in wearing, adjusting, and properly fitting the equipment. OSHA regulations require that a worker be medically certified before beginning work in an area where a respirator must be worn (OSHA Respiratory Standard, 29 CFR 1910.134(b)(10)).

6.F.2.4.3 *Inspections*

Respirators for routine use should be inspected before each use by the user and periodically by the laboratory supervisor. Self-contained breathing apparatus should be inspected at least once a month and cleaned after each use.

6.F.2.5 Safety Showers and Eyewash Fountains

6.F.2.5.1 *Safety Showers*

Safety showers should be available in areas where chemicals are handled. They should be used for immediate first aid treatment of chemical splashes and for extinguishing clothing fires. Every laboratory worker should know where the safety showers are located in the work area and should learn how to use them. Safety showers should be tested routinely to ensure that the valve is operable and to remove any debris in the system.

The shower should be capable of drenching the subject immediately and should be large enough to accommodate more than one person if necessary. It should have a quick-opening valve requiring manual closing; a downward-pull delta bar is satisfactory if long enough, but chain pulls are not advisable because they can hit the user and be difficult to grasp in an emergency. It is preferable to have drains under safety showers to reduce the risks associated with the water.

6.F.2.5.2 *Eyewash Fountains*

Eyewash fountains should be required in research or instructional laboratories if substances used there present an eye hazard or if unknown hazards may be encountered. An eyewash fountain should provide a soft stream or spray of aerated water for an extended period (15 minutes). These fountains should be located close to the safety showers so that, if necessary, the eyes can be washed while the body is showered.

6.F.2.6 Storage and Inspection of Emergency Equipment

It is often useful to establish a central location for storage of emergency equipment. Such a location should contain the following:

- self-contained breathing apparatus,
- blankets for covering the injured,
- stretchers (although it is generally best not to move a seriously injured person and to wait for qualified medical help to provide this service),
- first aid equipment (for unusual situations such as exposure to hydrofluoric acid or cyanide, where immediate first aid is required), and
- chemical spill cleanup kits and spill control equipment (e.g., spill pillows, booms, shoe covers, and a 55-gallon drum in which to collect sorbed material). (Also consult Chapter 5, sections 5.C.11.5 and 5.C.11.6.)

Safety equipment should be inspected regularly (e.g., every 3 to 6 months) to ensure that it will function properly when needed. It is the responsibility of the laboratory supervisor or safety coordinator to establish a routine inspection system and to verify that inspection records are being kept.

Inspections of emergency equipment should be performed as follows:

- Fire extinguishers should be inspected for broken seals, damage, and low gauge pressure (depending on type of extinguisher). Proper mounting of the extinguisher and its ready accessibility should also be checked. Some types of extinguishers must be weighed annually, and periodic hydrostatic testing may be required.
- Self-contained breathing apparatus should be checked at least once a month and after each use to determine whether proper air pressure is being maintained. The examiner should look for signs of deterioration or wear of rubber parts, harness, and hardware and make certain that the apparatus is clean and free of visible contamination.
- Safety showers and eyewash fountains should be examined visually and their mechanical function should be tested. They should be purged as necessary to remove particulate matter from the water line.

6.G EMERGENCY PROCEDURES

The following emergency procedures are recommended in the event of a fire, explosion, spill, or medical or other laboratory accident. These procedures are intended to limit injuries and minimize damage if an accident should occur. Telephone numbers to call in emergencies should be posted clearly at all telephones in hazard areas.

1. Have someone call for emergency help. State clearly where the accident has occurred and its nature.
2. Ascertain the safety of the situation. Do not enter or reenter an unsafe area.
3. Render assistance to the people involved and remove them from exposure to further injury.
4. Warn personnel in adjacent areas of any potential risks to their safety.
5. Render immediate first aid; appropriate measures include washing under a safety shower, administration of CPR by trained personnel if heartbeat and/or breathing have stopped, and special first aid measures.
6. Extinguish small fires by using a portable extinguisher. Turn off nearby equipment and remove combustible materials from the area. For larger fires, contact the appropriate fire department promptly.
7. Provide emergency personnel with as much information as possible about the nature of the hazard.

In case of medical emergency, laboratory personnel should remain calm and do only what is necessary to protect life.

1. Summon medical help immediately.
2. Do not move an injured person unless he or she is in danger of further harm.
3. Keep the injured person warm. If feasible, designate one person to remain with the injured person. The injured person should be within sight, sound, or physical contact of that person at all times.
4. If clothing is on fire and a safety shower is immediately available, douse the person with water; otherwise, move the person to the floor and roll him or her around to smother the flames.
5. If harmful chemicals have been spilled on the body, remove them, usually by flooding the exposed area with sufficient running water from the safety shower, and immediately remove any contaminated clothing.
6. If a chemical has splashed into the eye, immediately wash the eyeball and the inner surface of the eyelid with plenty of water for 15 minutes. An eyewash fountain should be used if available. Forcibly hold the eye open to wash thoroughly behind the eyelids.
7. If possible, determine the identity of the chemical and inform the emergency medical personnel attending the injured person.

7 Disposal of Waste

7.A INTRODUCTION

Within the broad theme of pollution prevention, earlier chapters of this book consider various management strategies to reduce the formation of waste during laboratory operations. These include reducing the scale of laboratory operations, cataloging and reusing excess materials, and recycling chemicals that can be recovered safely. Clearly, the best approach to laboratory waste is to not generate it. However, this ideal situation is seldom attained in the laboratory. Therefore, this chapter considers methods for dealing with the waste that is generated during laboratory operations and for accomplishing its ultimate disposal.

The earlier chapters are directed primarily at enhancing the safety of laboratory workers and visitors and focus on the laboratory environment. However, discussing prudent practices for disposal of waste requires a broader perspective. When waste is eventually removed from the laboratory, it affects individuals other than those who acquired or generated it, and, ultimately, society as a whole. Waste is disposed of by three routes: (1) into the atmosphere, either through evaporation or through the volatile effluent from incineration; (2) into rivers and oceans via the sewer system and wastewater treatment facilities; and (3) into landfills. Occasionally, waste has to be held indefinitely at the laboratory site or elsewhere until acceptable modes of disposal are developed. The laboratory worker who generates waste has an obligation to consider the ultimate fate of the materials resulting from his or her work. The high cost of disposal of many materials, the potential hazards to people outside the laboratory, and the impact on the environment are all important factors to be considered.

Because of the potential adverse impact on the public through pollution of the air, water, or land, society invariably regulates waste disposal. Disposal of household waste is usually regulated by municipalities, while hazardous waste disposal is regulated at the federal level and often also by states and municipalities. The focus in this chapter is on the disposal of waste that may present chemical hazards, as well as those multihazardous wastes that contain some combination of chemical, radioactive, and biological hazards. Many of the disposal solutions outlined in this chapter have been designed to take advantage of the fact that there is a normal stream of nonhazardous waste generated in the laboratory and other parts of the institution. In some instances, waste that is classified as hazardous can be modified to permit disposal as nonhazardous waste, which is usually a less expensive and less cumbersome undertaking. The scientist who generates hazardous waste must make decisions consistent with the institutional framework for handling such materials.

Generally, waste is defined as surplus, unneeded, or unwanted material. It is usually the laboratory worker or supervisor who decides whether to declare a given laboratory material a waste. However, specific regulatory definitions must be taken into account as well. Even the question of when an unwanted or excess material becomes a waste involves some regulatory considerations. Whereas some institutions have created glossaries of terms to label waste materials as co-products or surplus reagents, regulations state that a material may be declared a waste if it is "abandoned" or is considered "inherently wastelike." Spilled materials, for example, often fall into these latter categories. *Therefore, it is not necessarily up to the generator to decide whether or not a material is a waste.*

Once material becomes a waste by a generator's decision or by regulatory definition, the first responsibility for its proper disposal rests with the laboratory worker. These experimentalists are in the best position to know the characteristics of the materials they have used or synthesized. It is their responsibility to evaluate the hazards and assess the risks associated with the waste and to choose an appropriate strategy to handle, minimize, or dispose of it. As discussed earlier in this volume (see Chapter 3, section 3.B), there are numerous sources of information available to the laboratory worker to guide in the decision making, including those required under various Occupational Safety and Health Administration (OSHA) regulations.

7.B CHEMICALLY HAZARDOUS WASTE

7.B.1 Characterization of Waste

Because proper disposal requires information about the properties of the waste, it is recommended that all chemicals used or generated be identified clearly. In general, they must be retained in clearly marked containers, and if they have been generated within the laboratory, their source must be defined clearly on the container and ideally in some type of readily available notebook record. In academic laboratories where student turnover is frequent, it is particularly important that the materials used or generated be identified. This practice can be as important for small quantities as it is for large quantities of material.

It is usually quite simple to establish the hazardous characteristics of clearly identified waste. Unidentified materials present a problem, however, because treatment disposal facilities are prohibited from accepting materials whose hazards are not known. In those cases when the identity of the material is not known, it is possible to carry out simple tests to determine the hazard class into which the material should be categorized. Because the generator may be able to apply some general information, it is usually advisable to carry out the hazard categorization process before the materials are removed from the laboratory. Having the analysis done at the laboratory is also usually cheaper than having it performed by the treatment disposal facility or an outside contractor.

Generally, it is not necessary to determine the molecular structure of the unknown material precisely. Hazard classification information usually satisfies the regulatory requirements and those of the treatment disposal facility. However, it is important to establish that the disposal facility will accept the analytical data that are ultimately provided.

The first concern in identification of an unknown waste is safety. The laboratory worker who carries out the procedures should be familiar with the characteristics of the waste and any necessary precautions. Because the hazards of the materials being tested are unknown, it is imperative that proper personal protection and safety devices such as fume hoods and shields be employed. Older samples can be particularly dangerous because they may have changed in composition, for example, through the formation of peroxides. (See Chapter 3, section 3.D.3.2, and Chapter 5, section 5.G.3, for more information on peroxides.)

The following information is commonly required by treatment disposal facilities before they will consider handling unknown materials:

- physical description,
- water reactivity,
- water solubility,
- pH and possibly also neutralization information,
- ignitability (flammability),
- presence of oxidizer,
- presence of sulfides or cyanides,
- presence of halogens,
- presence of radioactive materials,
- presence of biohazardous materials, and
- presence of toxic constituents.

The following test procedures should be readily accomplished by a trained laboratory worker. The overall sequence for testing is depicted in Figure 7.1 for liquid and solid materials.

- *Physical description*. The physical description should include the state of the material (solid, liquid), the color, and the consistency (for solids) or viscosity (for liquids). For liquid materials, describe the clarity of the solution (transparent, translucent, or opaque). If an unknown material is a bi- or tri-layered liquid, describe each layer separately, giving an approximate percentage of the total for each layer.

After taking appropriate safety precautions for handling the unknown, including the use of personal protection devices, remove a small sample for use in the following tests.

- *Water reactivity*. Carefully add a small quantity of the unknown to a few milliliters of water. Observe any changes, including heat evolution, gas evolution, and flame generation.
- *Water solubility*. Observe the solubility of the unknown in water. If it is an insoluble liquid, note whether it is less or more dense than water (i.e., does it float or sink?). Most nonhalogenated organic liquids are less dense than water.
- *pH*. Test the material with multirange pH paper. If the sample is water-soluble, test the pH of a 10% aqueous solution. It may also be desirable or even required to carry out a neutralization titration.
- *Ignitability (flammability)*. Place a small sample of the material (<5 milliliters (mL)) in an aluminum test tray. Apply an ignition source, typically a propane torch, to the test sample for one-half second. If the material supports its own combustion, it is a flammable liquid with a flash point of less than 60 °C. If the sample does not ignite, apply the ignition source again for one second. If the material burns, it is combustible. Combustible materials have a flash point between 60 and 93 °C.
- *Presence of oxidizer*. Wet commercially available starch-iodide paper with 1 N hydrochloric acid, and then place a small portion of the unknown on the wetted paper. A change in color of the paper to dark purple is a positive test for an oxidizer. The test can also be carried out by adding 0.1 to 0.2 grams (g) of sodium or potassium iodide to 1 mL of an acidic 10% solution of the unknown. Development of a yellow-brown color indicates an oxidizer. To test for hydroperoxides in water-insoluble organic solvents, dip the test paper into the solvent, and then let it evaporate. Add a drop of water to the same section of the paper. Development of a dark color indicates the presence of hydroperoxides.
- *Presence of sulfide*. The test for inorganic sulfides is carried out only when the pH of an aqueous solution of the unknown is greater than 10. Add a few drops of concentrated hydrochloric acid to a sample of the unknown while holding a piece of commercial lead acetate paper, wetted with distilled water, over the sample. Development of a brown-black color on the paper indicates generation of hydrogen sulfide. Because of the toxicity of the hydrogen sulfide formed during this test, only a small sample should be tested, and appropriate ventilation should be used.
- *Presence of cyanide*. The test for inorganic cyanides is carried out only when the pH of an aqueous solution of the unknown is greater than 10. Prior to testing for

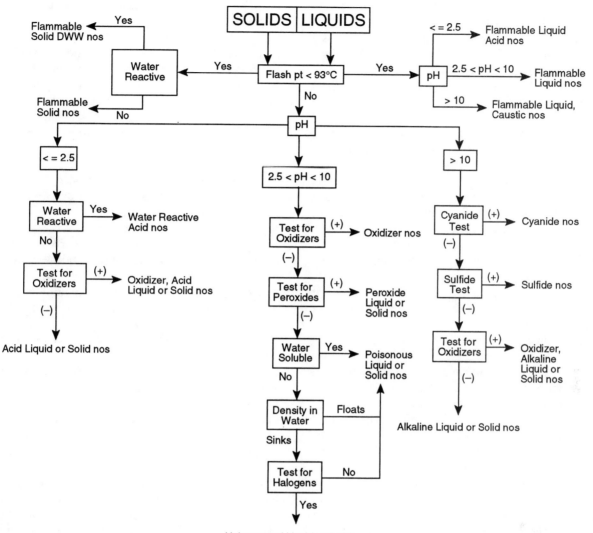

FIGURE 7.1 Flow chart for categorizing unknown chemicals. This decision tree shows the sequence of tests that may need to be performed to determine the appropriate hazard category of an unknown chemical. DWW, dangerous when wet; nos, not otherwise specified.

cyanides, the following stock solutions should be prepared: 10% aqueous sodium hydroxide (solution A), 10% aqueous ferrous sulfate (solution B), and 5% ferric chloride (solution C). Mix 2 mL of the sample with 1 mL of distilled water and 1 mL each of solutions A, B, and C. Add enough concentrated sulfuric acid to make the solution acidic. Development of a blue color (Prussian blue, from ferric ferrocyanide) indicates cyanide. Because of the toxicity of the hydrogen cyanide formed during this test, only a small sample should be tested, and appropriate ventilation should be used.

• *Presence of halogen.* Heat a piece of copper wire until red in a flame. Cool the wire in distilled or deionized water, and then dip the wire into the unknown.

Again heat the wire in the flame. The presence of halogen is indicated by a green color around the wire in the flame.

7.B.2 Regulated Chemically Hazardous Waste

An important question for planning within the laboratory is whether or not a waste is *regulated* as hazardous, because regulated hazardous waste must be handled and disposed of in rather specific ways. This determination has very important regulatory implications, which can lead to significant differences in disposal cost. The Environmental Protection Agency

(EPA) defines chemically hazardous waste under the Resource Conservation and Recovery Act (RCRA), and the U.S. Nuclear Regulatory Commission (U.S. NRC) defines radioactivity hazards. Biological hazards are generally not defined within federal regulations.

Although we must consider and pay close attention to the regulatory definitions and procedures that govern the handling and disposal of waste, primary importance must be given to the safe and prudent handling of this material. It is important to remember that the danger associated with a specific hazardous waste depends not only on the composition of the waste, but also on its quantity. In fact, regulations recognize quantity in many of their definitions for hazard compliance as well as in the definitions of waste generators. The concept of *de minimis* quantities, that is, very small amounts of material, though not defined clearly, is also a consideration in determining hazardous waste risk. Enlightened risk management dictates that the amount of material be one factor in the decisions on handling and disposal of waste.

Waste that is regulated as hazardous because of its chemical properties is defined by EPA in two ways: (1) waste that has certain hazardous characteristics and (2) waste that is on certain lists of chemicals. The first category is based on properties of materials that should be familiar to every laboratory worker. The second category is based on lists, established by EPA and certain states, of certain chemicals common to industry. These lists generally include materials that are widely used and recognized as hazardous. Chemicals are placed on these RCRA lists primarily on the basis of their toxicity. (To determine if waste is hazardous or not, see Chapter 9, section 9.D.2.)

Regardless of the regulatory definitions of hazard, understanding chemical characteristics that pose potential hazards should be a fundamental part of the education and training of any laboratory worker. These characteristics may be derived from knowledge of the properties and/or precursors of the waste. The characteristics may also be established by specific tests cited in the regulations.

(Regulatory issues, specifically RCRA, are discussed further in Chapter 9, section 9.D.)

7.B.2.1 Definition of Characteristic Waste

The properties that pose potential hazards are as follows:

1. *Ignitability.* Ignitable materials are defined as having one or more of the following characteristics:
   (a) Liquids that have a flash point of less than

60 °C or some other characteristic that has the potential to cause fire.
   (b) Materials other than liquids that are capable, under standard temperature and pressure, of causing fire by friction, adsorption of moisture, or spontaneous chemical changes and, when ignited, burn so vigorously and persistently as to create a hazard.
   (c) Flammable compressed gases, including those that form flammable mixtures.
   (d) Oxidizers that stimulate combustion of organic materials.

Ignitable materials include most common organic solvents, gases such as hydrogen and hydrocarbons, and certain nitrate salts.

2. *Corrosivity.* Corrosive liquids have a pH of 2 or less or 12.5 or greater or lead to corrosion of certain grades of steel. Most common laboratory acids and bases are corrosive. *Solid* corrosives, such as sodium hydroxide pellets and powders, are not legally considered by RCRA to be corrosive. However, laboratory workers must recognize that such materials can be extremely dangerous to skin and eyes and must be handled accordingly.

3. *Reactivity.* The reactivity classification includes substances that are unstable, react violently with water, are capable of detonation if exposed to some initiating source, or produce toxic gases. Alkali metals, peroxides and compounds that have peroxidized, and cyanide or sulfide compounds are classed as reactive.

4. *Toxicity.* Toxicity is established through the Toxicity Characteristic Leaching Procedure (TCLP), which measures the tendency of certain toxic materials to be leached (extracted) from the waste material under circumstances assumed to reproduce conditions of a landfill. The TCLP list includes a relatively small number of industrially important toxic chemicals and is based on the concentration above which a waste is considered hazardous. Failure to pass the TCLP results in classification of a material as a toxic waste.

7.B.2.2 Definition of Listed Waste

Although EPA has developed several lists of hazardous waste, three regulatory lists are of most interest to laboratory workers:

- the F list: waste from nonspecific sources (e.g., spent solvents and process or reaction waste),
- the U list: hazardous waste (e.g., toxic laboratory chemicals), and
- the P list: acutely hazardous waste (e.g., highly

toxic laboratory chemicals, that is, chemicals having an LD_{50} of less than 50 mg/kg (oral; rat)).

These lists may be updated periodically by EPA.

7.B.2.3 Determining the Status of a Waste

The EPA regulations place on the waste generator the burden of determining whether the waste is regulated as hazardous and in what hazard classification it falls. Testing is not necessarily required, and in most cases the laboratory worker should be able to provide sufficient information about the waste to allow the hazard classification to be assigned. If the waste is not a common chemical with known characteristics, enough information about it must be supplied to satisfy the regulatory requirements and to ensure that it can be handled and disposed of safely. Often, information on only the components present in amounts greater than 1% is required, but confirmation is needed from the treatment/disposal facility. The information needed to characterize a waste also depends on the method of ultimate disposal. (See the discussion of disposal methods in sections 7.B.6 to 7.B.8 below.)

7.B.3 Collection and Storage of Waste

7.B.3.1 At the Location of Generation

The first step in the disposal sequence usually involves the accumulation and temporary storage of waste in or near the laboratory (satellite accumulation). This step directly involves the laboratory workers who are familiar with the waste and its generation and is a most important part of ensuring that the disposal process proceeds safely and efficiently. It is often the time at which a decision can be made to recycle or reuse surplus materials rather than sending them for disposal. All of the costs and benefits of either decision should be evaluated here.

Again, safety considerations must be of primary concern. Waste should be stored in clearly labeled containers in a designated location that does not interfere with normal laboratory operations. Ventilated storage may be appropriate.

Federal regulations allow the indefinite accumulation of up to 55 gallons of hazardous waste or 1 quart of acutely hazardous waste at or near the point of generation. However, prudence dictates that the quantities accumulated should be consistent with good safety practices. Furthermore, satellite accumulation time must be consistent with the stability of the material. It is generally recommended that waste not be held for more than 1 year. Within 3 days of the time that the amount of waste exceeds the 55-gallon (or 1 quart) limit, it must be managed under the storage and accumulation time limits required at a central accumulation area. (See Chapter 9, section 9.D.4, for more information.)

Often, different kinds of waste can be accumulated within a common container. Such commingled waste must be chemically compatible to ensure that heat generation, gas evolution, or another reaction does not occur. (See the discussion of commingling in section 7.B.3.2 below.)

Packaging and labeling are a key part of this initial in-laboratory operation. Waste must be collected in dependable containers that are compatible with their contents. Glass containers have traditionally been the most resistant to chemical action, but they can break easily. Metal containers are sturdier than glass, but often are corroded by their contents. Various chemically resistant plastic containers are becoming preferable substitutes for containers of glass or metal. Safety cans, metal or plastic, should be considered for holding flammable solvents. It is advisable to use secondary containers, such as trays, in case of spills or leakage from the primary containers. Containers are required to remain closed except when their contents are being transferred. Containers of incompatible materials should be separated physically or otherwise stored in a protective manner.

Every container must be labeled with the material's identity and its hazard (e.g., flammable, corrosive). Although the identity need not be a complete listing of all chemical constituents, it should enable knowledgeable laboratory workers to evaluate the hazard. However, when compatible wastes are collected in a common container, it is advisable to keep a list of the components to aid in later disposal decisions. Labeling must be clear and permanent. Although federal regulations do not require posting the date when satellite accumulation begins, some states do require this. The institution may suggest that this information be recorded as part of its chemicals management plan.

7.B.3.2 At a Central Accumulation Area

The central accumulation area is an important component in the organization's chemicals management plan. In addition to being the primary location where waste management occurs, it may also be the location where excess chemicals are held for possible redistribution. Along with the laboratory, the central accumulation area is often where hazard reduction of waste takes place through allowable on-site treatment processes.

The central accumulation area is often the appropriate place to accomplish considerable cost savings by commingling (i.e., mixing) waste materials. This is the process where compatible wastes from various sources are combined prior to disposal. Commingling is particularly suitable for waste solvents because disposal of liquid in a 55-gallon drum is generally much less expensive than disposal of the same volume of liquid in small containers. Because mixing waste requires transfer of waste between containers, it is imperative that the identity of all materials be known and their compatibility be understood. Safety in carrying out the procedures, including the use of personal protective devices as well as engineering controls such as fume hoods, must be of high priority.

In some cases, the disposal method and ultimate fate of the waste may require that different wastes not be accumulated together. For example, if commingled waste contains significant amounts of halogenated solvents (usually above 1%), disposing of the mixture can be markedly more costly. In such cases, segregation of halogenated and nonhalogenated solvents is economically favorable.

Based on federal regulations, storage at a central accumulation area is normally limited to 90 days, although more time is allowed for small-quantity generators or other special situations (180 or 270 days). The count begins when the waste is brought to the central accumulation area from the laboratory or satellite accumulation area. *It is important to know that a special permit is required for long-term storage, that is, storage beyond the limit of 90 days (or 180 or 270 days, depending on the particular situation). Obtaining such a permit is usually too expensive and too time-consuming for most laboratory operations.* (See RCRA and Chapter 9, section 9.D.4, for more information.)

Waste materials stored within a central accumulation area should be held in appropriate and clearly labeled containers, separated according to chemical compatibility as noted in the previous section. The label must include the accumulation start date and the words "Hazardous Waste." Fire suppression systems, ventilation, and dikes to avoid sewer contamination in case of a spill should be considered when such a facility is planned. Training of employees in correct handling of the materials as well as contingency planning for emergencies is expected to be a part of the central accumulation area operations.

Transportation of waste between laboratories (satellite accumulation areas) and the central accumulation area also requires specific attention to safety. Materials transported must be held within appropriate and clearly labeled containers. There must be provision for spill control in case of an accident during transportation and handling. For larger institutions, it is advisable to have some kind of internal tracking system to follow the movement of waste. If public roads are used during the transportation process, additional Department of Transportation (DOT) regulations may apply.

Final preparations for off-site disposal usually occur at the central accumulation area. Decisions on disposal options are best made here, as the larger quantities of waste are gathered. Identification of unknown materials not carried out within the laboratory must be completed at this point because unidentified waste cannot be shipped to a disposal site.

Laboratory waste typically leaves the generator's facility commingled in drums as compatible wastes or within a Lab Pack. Lab Packs are containers, often 55-gallon drums, in which small containers of waste are packed with an absorbent material. Lab Packs had been used as the principal method for disposing of laboratory waste within a landfill. However, recent landfill disposal restrictions severely limit landfill disposal of hazardous materials. Thus, the Lab Pack has become principally a shipping container. Typically, the Lab Pack is taken to a disposal facility, where it is either incinerated or unpacked and the contents redistributed for safe, efficient, and legal treatment and disposal.

7.B.4 Records

Records are needed both to meet regulatory requirements and to help monitor the success of the hazardous waste management program. Because the central accumulation area is usually the last place where waste is dealt with before it leaves the facility, it is often the most suitable place for ensuring that all appropriate and required records have been generated.

For regulatory purposes, the facility needs to keep records for on-site activities that include

- the quantities and identification of waste generated and shipped,
- documentation of analyses of unknown materials if required,
- manifests for waste shipping as well as verification of disposal, and
- any other information required to ensure compliance and safety from long-term liability.

Records of costs, internal tracking, and so forth, can provide information on the success of the hazardous waste management program.

7.B.5 Hazard Reduction

Hazard reduction is part of the broad theme of pollution prevention that is addressed in previous chapters of this book. From a chemical point of view, it is feasible to reduce the volume or the hazardous characteristics of many chemicals by reactions within the laboratory. In fact, it is becoming common practice to include such reactions as the final steps in an experimental sequence. Such procedures, as part of an academic or industrial experiment, usually involve small amounts of materials, which can be handled easily and safely by the laboratory worker. Chemical deactivation as part of the experimental procedure can have considerable economic advantage by eliminating the necessity to treat small amounts of surplus materials as hazardous waste. Furthermore, the handling and deactivation of potential waste by the laboratory worker benefit from the expertise and knowledge about the materials of the person who has generated them.

The question of what is considered treatment under RCRA regulations has posed a dilemma for laboratory workers. RCRA regulations define treatment as "any method . . . designed to change the physical, chemical, or biological character or composition of any hazardous waste so as to neutralize such waste, or so as to recover energy or material resources from the waste, or so as to render the waste non-hazardous or less hazardous . . . " (U.S. Congress, 1978). Under RCRA, treatment, with very limited exceptions, must be permitted by EPA.

The regulatory procedures and costs to be a "permitted" treatment facility are beyond the resources and mission of most academic and industrial laboratories. Yet it is prudent to carry out small-scale "treatment" as a part of laboratory procedures. This fact has been recognized by state agencies and some regional EPA offices through "permit-by-rule," that is, by allowing categorical or blanket permitting of certain small-scale treatment methods. For example, elementary acid-base neutralization is usually allowed, as is treatment that is the last step of a chemical procedure. Most EPA regions also allow treatment in the waste collection container. It is important to note that treatment restrictions apply only to wastes that are addressed by EPA regulations.

A bill has been promoted in Congress to allow small-scale treatment by laboratory personnel. However, specific legislation has not been enacted at this time. The fact that regional EPA offices have interpreted such small-scale reactions differently further complicates decisions at the laboratory level. Because illegal treatment can lead to fines of up to $25,000 per day, it is most important that, before carrying out any processes that could be considered treatment, the responsible laboratory worker or the institution's environmental health and safety office check with the local, state, or regional EPA to clarify its interpretation of the rules.

(Section 7.D below provides methods for small-scale treatment of common chemicals.)

7.B.6 Disposal Options

Decisions on the ultimate disposal method are an important part of the on-site planning for handling of waste. The method of collection has an impact on, for example, how waste will be stored so as to most efficiently accomplish its transfer to the treatment, storage, and disposal facility (TSDF). Waste generators often use several disposal options because each has its own advantages for specific wastes. Disposal in the sanitary sewer, though appropriate in some cases, is becoming an unacceptable option in many communities. At the same time the options for landfill disposal are also disappearing rapidly. Incineration is becoming the most common disposal method. However, the long-term outlook for this method may be limited by increasing environmental concerns as well as the difficulty in obtaining permits for commercial incineration facilities. Waste minimization is the management strategy of the future. (See Chapter 4, section 4.B, for step-by-step instructions on source reduction and Chapter 7, section 7.C, for general information on minimizing hazardous waste.)

7.B.6.1 Incineration

Incineration is becoming the disposal method of choice for several reasons. It promises to give the generator the best assurance of long-term safety from liability. It also leads to a minimum amount of residues that must be disposed of in landfills. However, at this time, incineration is still one of the more expensive disposal options. It is becoming increasingly difficult to obtain a permit to establish a commercial incinerator because of local opposition (the "not in my backyard" syndrome) and environmental concerns centering on questions regarding the effectiveness of the incineration process.

Nevertheless, most disposal companies are moving toward incineration disposal, particularly for the kinds of hazardous waste generated by laboratories. Their typical variety of different wastes, usually in small quantities, makes incineration a favorable option. Laboratory waste can often be incinerated in its shipping Lab Packs without any further handling. Commingled flammable solvents are commonly blended with the incinerator fuel and thus destroyed as they provide energy for the burning process.

Earlier editions of this book were optimistic that small laboratory incinerator systems would be developed for efficient destruction of waste at the point of generation. That has not happened. Several factors, including the cost of development and concern about how regulatory agencies would view this kind of "treatment," surely have contributed to the lack of progress. A changing regulatory environment could provide a favorable basis for development of such thermal treatment systems.

7.B.6.2   Disposal in the Normal Trash

Laboratory workers may be surprised to learn the number of wastes they generate that can be disposed of in the normal trash. However, because the disposal of trash from households and businesses is normally controlled by the local municipality, the local agency should be approached to establish what is allowed.

When disposing of chemicals in the normal trash, certain precautions should be observed. Because custodians, who usually empty the trash containers, are not usually familiar with laboratory operations, no objects that could cause harm to them should be disposed of in those containers. Such objects include containers of chemicals, unless they are overpacked to avoid breakage, and powders, unless they are in closed containers. Free-flowing liquids are usually prohibited. Sharp metal and broken glassware, even though they may be considered nonhazardous trash, should be collected in specially marked containers, never in the normal trash baskets.

7.B.6.3   Disposal in the Sanitary Sewer

Disposal in the sewer system (down the drain) had been a common method of waste disposal until recent years. However, environmental concerns, the viability of publicly owned treatment works (POTW), and a changing disposal culture have changed that custom markedly. In fact, many industrial and academic laboratory facilities have completely eliminated sewer disposal. Again, like trash disposal, most sewer disposal is controlled locally, and it is therefore advisable to consult with the POTW to determine what is allowed. Yet, it is often reasonable to consider disposal of some chemical waste materials in the sanitary sewer. These include substances that are water-soluble, that do not violate the federal prohibitions on disposal of waste materials that interfere with POTW operations or pose a hazard, and that are allowed by the local sewer facility.

Chemicals that may be permissible for sewer disposal include aqueous solutions that readily biodegrade and low-toxicity solutions of inorganic substances. Water-miscible flammable liquids are frequently prohibited

from disposal in the sewer system. Water-immiscible chemicals should never go down the drain.

Disposal of regulated hazardous waste into the sanitary sewer is allowed only in limited situations. The total wastewater must be a mixture of domestic sewage along with the waste whose amount and concentration meet the regulations and limits of the POTW. If approved of by the local district, it may be allowable to dispose of dilute solutions of metals and other hazardous chemicals into the sanitary sewer.

Under the Clean Water Act, some exemption from regulation as a hazardous waste for wastewater containing laboratory-generated listed waste is allowed. In 1993, this exemption was expanded to include corrosive and ignitable wastes. For the exemption to apply, these laboratory wastes must be 1% or less of the annual total wastewater quantity reaching the facility's headworks or have an annualized average concentration of no more than 1 part per million (ppm) of the wastewater generated by the facility.

Waste should be disposed of in drains that flow to a POTW, never into a storm drain and seldom into a septic system. Waste should be flushed with at least a 100-fold excess of water, and the facility's wastewater effluent should be checked periodically to ensure that concentration limits are not being exceeded.

7.B.6.4   Release to the Atmosphere

The release of vapors to the atmosphere, via, for example, open evaporation or fume hood effluent, is not an acceptable disposal method. Apparatus for operations expected to release vapors should be equipped with appropriate trapping devices. Although the disposition of laboratories under the Clean Air Act is not established at this time, it is reasonable to expect that releases to the atmosphere will be controlled.

Fume hoods, the most common source of laboratory releases to the atmosphere, are designed as safety devices to transport vapors away from the laboratory in case of an emergency, not as a routine means for volatile waste disposal. Units containing absorbent filters have been introduced into some laboratories, but have limited absorbing capacity. Redirection of fume hood vapors to a common trapping device can completely eliminate discharge into the atmosphere. (See Chapter 8, sections 8.C.11 and 8.C.12, for more detail.)

7.B.7   Disposal of Nonhazardous and Nonregulated Waste

Many laboratories do not distinguish between waste that is hazardous and waste that neither poses a hazard nor is regulated as hazardous. If these different types of waste are combined, then the total must be treated as hazardous waste, and the price for disposal of the non-hazardous portion increases markedly. When safe and allowed by regulation, disposal of nonhazardous waste via the normal trash or sewer can substantially reduce disposal costs. This is the kind of waste segregation that makes economic as well as environmental sense.

It is wise to check the rules and requirements of the local solid waste management authority and develop a list of materials that can be disposed of safely and legally in the normal trash. This includes waste that is not regulated because it does not exhibit any of the hazardous characteristics (ignitability, corrosivity, reactivity, or toxicity) as defined by EPA and is not listed as hazardous. The common wastes usually not regulated as hazardous include certain salts (e.g., potassium chloride and sodium carbonate), many natural products (e.g., sugars and amino acids), and inert materials used in a laboratory (e.g., noncontaminated chromatography resins and gels).

7.B.8 Disposal of Spills

Most chemical spills can and should be cleaned up by laboratory workers themselves. In general, these are spills of known composition that do not involve injury, do not represent a fire or personal hazard, and are less than 1 gallon (or less for very toxic materials). Regulations allow laboratory workers to clean up such spills, although it is advisable that they have training to handle spills and adequate equipment to carry out the cleanup safely. Outside help, properly trained, should be requested if there is any doubt about the ability of the laboratory personnel to clean up the spill safely. But once help is requested from outside the immediate spill area, specific personnel training requirements and other regulatory control may apply.

General guidelines for cleaning up spills are as follows:

1. Assess the potential hazard presented by the spill to personnel within the work area as well as within other parts of the facility and the outside environment.
2. Remove possible sources of ignition if the spilled material is flammable:
   - Turn off hot plates, stirring motors, and flames.
   - Shut down equipment in the area that could increase danger.
3. Secure the area so that no one will walk through the spill or interfere with the cleanup efforts.
4. Choose appropriate personal protection devices:

   - Always wear protective gloves and goggles or a face shield.
   - If there is a chance of body contact with the spill, wear an apron or coveralls.
   - Wear rubber or plastic (not leather) boots if there is a chance of stepping into the spill.
   - Wear a respirator if there is danger of inhalation of toxic vapors, though only when proper training has preceded its use.
   - Note that protective devices must be chosen carefully to be appropriate for the anticipated hazard. Often training is appropriate or required (e.g., with respirators) prior to their use.
5. Locate a spill control kit or other appropriate absorbent and cleanup supplies.
6. Confine or contain the spill:
   - Do not let any of the spilled material enter the sewer system, for example, through a floor drain.
   - Cover the spill with an absorbent material; paper towels may be appropriate for small, unreactive materials.
   - Sweep up or in other ways collect the absorbed materials and place them in a container that can be securely closed.
   - If the spilled material is an acid or a base, use a neutralizing material; sodium bicarbonate is commonly used for acids, and sodium bisulfate for bases. Spill control kits are commercially available for the cleanup of many kinds of chemical spills. (Chapter 6, section 6.F.2.1, has further information on spill control kits and spill absorbents.)
7. Dispose of the absorbed spill appropriately as hazardous or nonhazardous waste.

(See Chapter 5, sections 5.C.11.5 and 5.C.11.6, for more detail on spill cleanup.)

7.B.9 Monitoring of Off-site Waste Disposal

The ultimate destination of waste is usually a treatment, storage, and disposal facility (TSDF). Here waste is held, treated (typically via chemical action or incineration), or actually disposed of. Although the waste has left the generator's facility, *the generator retains the final responsibility for the long-term fate of the waste*. It is imperative that the generator have complete trust and confidence in the TSDF, as well as in the transporter who carries the waste to the TSDF. In some cases the destination of waste is a recycler or reclaimer. The procedures for preparing and transporting the waste to such a facility are similar to those described above. (See section 7.B.3.)

7.B.9.1 Preparation for Off-site Disposal

How the waste has been handled at the generator site, usually at a central accumulation area, can have a significant impact on the cost of the off-site disposal operation. This usually depends on the method of preparation of the waste for disposal. Collecting containers of waste in a Lab Pack is usually much more expensive for ultimate disposal than is commingling of compatible wastes, partly because a 55-gallon Lab Pack only holds about 16 gallons of waste. However, factors other than economics may control those decisions. When small amounts of different wastes are generated, a situation typical of many academic laboratories, commingling compatible wastes may not be practical. The Lab Pack approach is usually chosen. On the other hand, waste solvents can often be combined advantageously, even in small operations.

On-site storage time limits also must be considered. For those generators that are governed by the 90-day limit, yet are relatively small operations, it is difficult to accumulate sufficient materials to make commingling a favorable option.

Safety can often be the determining factor. Using Lab Packs is quite simple. As described above, small containers of compatible waste materials are placed in a larger container, usually a 55-gallon drum, along with appropriate packing materials, as they are collected. When a drum is filled, it is sealed and ready for shipping. An inventory list of the contents of a Lab Pack is required for shipping and is usually requested by the TSDF.

In contrast, commingling requires opening of containers and transferring their contents from the smaller laboratory containers to a larger drum. Here the potential risk for workers is much greater. Furthermore, the containers should be rinsed before they are considered nonhazardous, and the rinsate must be treated as a hazardous waste. Drums of commingled waste usually require only a general hazard classification identification, although some TSDFs require an analysis or listing of contents.

7.B.9.2 Choice of Transporter and Disposal Facility

Because the long-term liability for the waste remains with the generator, it is imperative that the generator be thoroughly familiar with the experience and record of the transporter and TSDF. Economic factors alone should not govern choices, for the long-term consequences can be significant. The generator must obtain assurance, in terms of documentation, permits, records, insurance and liability coverage, and regulatory compliance history, that the chosen service provider is reliable.

There is often an advantage, particularly for smaller facilities, to contracting for all of the hazardous waste disposal operations. These include the packing and appropriate labeling of waste for off-site transportation and disposal, preparation of the shipping manifest, and arranging for the transporter and disposal facility. Again, *the liability remains with the generator*, and so the choice of such a contractor is critical.

In some states, Minnesota and Montana, for example, arrangements have been developed with local regulators to allow a large laboratory waste generator to handle the waste from very small laboratories such as those at small colleges and public schools. This plan results in informed assistance and cost savings for the smaller units. In Wisconsin, a statewide commercial contract that can be accessed by all state educational systems has been arranged. There is usually significant advantage to working with local and state agencies to develop acceptable plans for disposal methods that are environmentally and economically favorable for both large and small generators.

7.C MULTIHAZARDOUS WASTE

Multihazardous waste is waste that contains any combination of chemical, radioactive, or biological hazards. The combinations of these hazards are illustrated in Figure 7.2. Although many of the principles discussed for chemically hazardous waste earlier in this

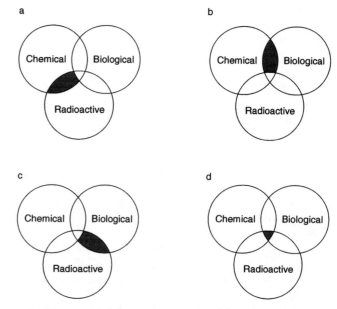

FIGURE 7.2 Multihazardous waste. (a) Chemical-radioactive waste, or "mixed waste," (b) chemical-biological waste, (c) radioactive-biological waste, and (d) chemical-radioactive-biological waste.

chapter (section 7.B) apply here also, multihazardous waste requires special management considerations because the treatment method for one of the hazards may be inappropriate for the treatment of another. For example, if a waste that contains a volatile organic solvent and infectious agents is autoclaved, it may release hazardous levels of solvent into the atmosphere. (For more on multihazardous waste, refer to Chapter 9, section 9.D.2.)

Management of multihazardous waste is complicated further by local or state requirements that may be inconsistent with the relative risk of each hazard and with sound waste management practices. Chemically hazardous waste that contains short-half-life radionuclides may, for example, be best managed by holding the waste in storage for decay, which may require up to 2 years. However, the EPA and state rules usually limit storage of chemically hazardous waste to 90 days.

Commercial treatment or disposal facilities for multihazardous waste from laboratories are scarce. Many of these waste types are unique to laboratories and are generated in such small volumes that there is little incentive for the development of a commercial market for their management.

While multihazardous waste is currently difficult and expensive to manage, it is generated in medical, biochemical, and other types of critically important research, as well as in clinical and environmental laboratories. As interdisciplinary techniques, technologies, and studies become more widely used, multihazardous waste will be more widely generated. Legally acceptable protocols for dealing with multihazardous waste need to be developed. (See also Chapter 4, section 4.B.3.)

Radioactive hazardous waste generated by laboratories is usually limited to low-level radioactive waste from the use of by-product material and naturally occurring or accelerator-produced radioactive material (NARM). By-product material, as defined by the U.S. Nuclear Regulatory Commission (U.S. NRC), is reactor-produced radioactive material and includes most purchased radiolabeled chemicals; NARM includes uranium and thorium salts. The use and disposal of by-product material are regulated by the U.S. NRC and usually require a license. NARM waste is not regulated by the U.S. NRC but may be regulated to some extent by some states. Common waste management methods for low-level radioactive waste from laboratories include storage for decay and indefinite on-site storage, burial at a low-level radioactive waste site, incineration, and sanitary sewer disposal.

Waste is considered biohazardous or infectious if it contains agents of sufficient virulence and quantity that exposure of a susceptible host could cause trans-

mission of an infectious disease. Unlike chemical and radioactive waste, infectious or medical waste is currently not subject to federal regulations that govern its treatment, storage, or disposal. OSHA regulates the collection and containment of certain laboratory waste that contains human blood or body fluids in order to prevent exposure of personnel to bloodborne pathogens. Although OSHA does not regulate waste treatment or disposal, its standard is often the impetus for managing infectious waste in laboratories. Putrescible waste, such as tissue and carcasses of laboratory animals, is also classified as biological waste, although putrescible laboratory waste is usually not biohazardous or infectious. Hypodermic needles, lancets, scalpel blades, and other medical laboratory sharps are considered biohazardous because of their potential for being contaminated with pathogens and the likelihood of accidental skin puncture. Biological waste may also include whole animals or plants made transgenic via recombinant DNA technology or into which recombinant DNA has been introduced. (See Chapter 5, section 5.E, on biohazards and radioactivity.)

Common management methods for biological waste include disinfection, autoclaving, and, for liquids, disposal in the sanitary sewer. Putrescible waste is usually disposed of by incineration, which destroys the unpleasant nature of these materials. Needles and sharps require destruction, typically by incineration or grinding. In general, if all hazards cannot be removed in one step, the goal is to reduce a multihazardous waste to a waste that presents a single hazard. This single-hazard waste can then be managed by standard methods for that category.

Most management principles apply to all types of waste. These universal management methods include waste minimization, training of laboratory personnel and waste handlers, reviewing proposed procedures, keeping dissimilar waste materials separate, identification of waste materials, and labeling of waste containers. However, multihazardous waste, because of its combination of hazards and regulatory controls, requires more complex attention, as detailed in the following guidelines:

- Assess the risk posed by the waste's inherent hazards. Laboratories often have flexibility to define those waste characteristics whose hazards are so low (*de minimis*) as to not present a significant risk. For example, the U.S. NRC or state authority may allow a licensee to propose limits below which laboratory waste can be designated as noncontaminated and disposed of as nonradioactive waste. (Some licensees have agreed with their regulators to consider some wastes nonradioactive if effluent concentrations are less than what the U.S. NRC specifies for an unrestricted area.) As a result, some chemical-radioactive waste with a U.S. NRC-approved level of radioactivity can be managed as chemical waste. Likewise, laboratory personnel can specify which biological wastes require no special handling and should not be mixed with infectious or biohazardous waste. Laboratory personnel may also determine if their waste meets the definition of a chemically hazardous waste. If the waste component is not regulated within this category, then it may be possible to manage it as only a radioactive or biological waste.

• Minimize the waste's hazards. Waste minimization methods specific to chemical, radioactive, or biological waste can be applied to multihazardous waste to mitigate or eliminate one hazard, which will then allow it to be managed as a single-hazard waste. For example, the substitution of nonignitable liquid scintillation fluid (LSF) for toluene-based LSF reduces a chemical-radioactive waste to a radioactive waste.

• Determine options for managing the waste's hazards. Waste management options include laboratory methods, management at institutional waste facilities, and treatment and disposal at commercial sites. Options can vary considerably between laboratories owing to institutional capabilities and state and local laws. It may be appropriate to manage the waste in order of risk priority, from high to low risk. Options must be compatible with all hazards, and combinations of waste management methods may be limited by their order of application. Reject any combination or sequence of methods that may create an unreasonable risk to waste handlers or the environment, or that might increase the overall risk. If an option has a clear advantage in efficiency and safety, it should have highest priority. For example, if safe facilities are available on site, hold short-half-life radioactive waste for decay before managing it as a chemical or biological waste. However, remember that in most cases a waste that has chemically hazardous characteristics may not be held beyond 90 days.

• When possible, select a single management option. Some waste management methods are appropriate for more than one waste hazard. Low-level radioactive animal tissue (a radioactive-biological waste) can often be incinerated on-site, which may be a satisfactory disposal option for both the radioactive and the biological characteristics of the waste. Some multihazardous waste can be disposed of safely in the sanitary sewer when allowed by the local publicly owned treatment works (POTW).

(See also Chapter 4 on chemical and waste management.)

7.C.1 Chemical-Radioactive (Mixed) Waste

"Mixed waste" is the regulatory term for multihazardous waste that contains chemical and radioactive hazards (see Figure 7.2a). Mixed waste is defined by EPA as "wastes that contain a chemically hazardous waste component regulated under Subtitle C of the Resource Conservation and Recovery Act of 1978 (RCRA) and a radioactive component consisting of source, special nuclear, or byproduct material regulated under the Atomic Energy Act of 1946 (AEA)" (U.S. EPA, 1986). The U.S. NRC defines mixed waste slightly differently. According to the U.S. NRC, mixed waste is a waste "that contains a chemically hazardous waste as defined in RCRA, and source, special nuclear, by-product material, low-level radioactive waste" (as defined in the Low-level Radioactive Waste Policy Act of 1980 (42 USC 2021(b) to 2021(j)), or "some types of naturally-occurring or accelerator-produced radioactive material (NARM), such as uranium and thorium." Regardless of the precise definition, well-informed laboratory workers should be prepared to deal with mixed waste in a prudent manner.

Examples of laboratory mixed waste include the following:

• Used flammable (e.g., toluene) liquid scintillation cocktails.

• Phenol-chloroform mixtures from extraction of nucleic acids from radiolabeled cell components.

• Aqueous solutions containing more than 6 ppm chloroform (which exceeds the limit set by the Toxicity Characteristic Leaching Procedure (TCLP) test) and radioactive material (typically found in solutions generated by the neutralization of radioactive trichloroacetic acid solutions).

• Certain gel electrophoresis waste (e.g., methanol or acetic acid containing radionuclides).

• Lead contaminated with radioactivity.

Mixed waste produced at university, hospital, and medical research laboratories is typically a mixture of a low-level radioactive waste and chemically hazardous waste. Mixed waste from nuclear and energy research laboratories can include both low- and high-level (e.g., spent nuclear fuels) radioactive materials combined with chemically hazardous waste. Disposal options for mixed waste are usually very expensive. For many types of mixed waste, there are no management options other than indefinite storage on site, or at an approved facility, in the hope that treatment or disposal options will be created in the near future.

An example of the mixed waste problem and the importance of keeping waste separate is illustrated by a researcher who accidentally combined a waste labeled as carbon-14 with 1 gallon of sulfuric acid-sodium dichromate solution. No disposal facility in the United States would accept the radioactive chromic acid waste. One option, the simple laboratory reduction of the dichromate, recovery of the chromium-containing precipitate, and neutralization of the acid, which would render the liquid waste only a radioactive hazard, may not be allowed without a permit in some states and EPA regions because it is considered to be treatment.

In large part, many of these regulatory dilemmas are unrelated to the real risks of laboratory mixed waste. Chief barriers to the safe and timely management of mixed waste include the following:

• EPA regulations that inhibit laboratory and on-site minimization, storage, and treatment of mixed waste.

• EPA regulations that discourage off-site minimization, treatment, storage, or disposal of mixed waste.

• The U.S. NRC's reluctance at this time to establish a national policy that defines the *de minimis* level of contamination for all types of laboratory radioactive waste, below which the risk to health and the environment is not significant.

• Community opposition to incinerators that could, with minimal risk, efficiently reduce hazard.

• The low volume, unusual character, and great variety of this laboratory waste stream, which, together with the above barriers, discourage the development of commercial markets for mixed waste.

(For more complete information on regulations, see Chapter 9.)

7.C.1.1 Minimization of Mixed Waste

Rigorous application of waste minimization principles can often solve the problems of managing mixed waste. Minimization of mixed waste can be achieved by modifying laboratory processes, improving operations, or using substitute materials. Such efforts are most successful when scientists and environmental health and safety staff work together to evaluate laboratory processes. Examples include the following:

• Use of 2.5-mL scintillation vials ("minivials") rather than 10-mL vials. Adapters are available for scintillation counters with 10-mL vial racks.

• Counting of phosphorus-32 (^{32}P) without scintillation fluid by the Cerenkov method on the tritium (^{3}H) setting of a liquid scintillation counter (approximately 40% efficiency); iodine-125 (^{125}I) can be counted without scintillation fluid in a gamma counter.

• Use of microscale chemistry techniques.

• Elimination of the methanol/acetic acid (chemical) and radioactive mixed hazards in gel electrophoresis work by skipping the gel fixing step if it is not required.

• Prevention of lead contamination by radioactivity by lining lead containers with disposable plastic or by using alternative shielding materials.

• Reducing the volume of dry waste by compaction of items such as contaminated gloves, absorbent pads, and glassware.

Some simple operational improvements can also help minimize mixed waste. Surpluses can be minimized by limiting the acquisition of chemicals and radioactive materials to immediate needs. Contaminated equipment can sometimes be reused within restricted areas or decontaminated. Establishing procedures for noncontaminated materials can enable generators to keep normal trash separate from contaminated waste.

When possible, a substitute can be used for either the chemical or the radioactive source of the mixed waste. With radioactivity, the experimenter should use the minimum activity necessary and select the radionuclide with the most appropriate decay characteristics. Examples include the following:

• Use of nonignitable scintillation fluid (e.g., phenylxylylethane, linear alkylbenzenes, and diisopropylnaphthalene) instead of flammable scintillation fluid (e.g., toluene, xylene, and pseudocumene). Liquid scintillation fluid that is sold as being "biodegradable" or "sewer disposable" is more appropriately labeled as "nonignitable" because biodegradability in the sanitary sewer can vary considerably with the local treatment facility.

• Use of nonradioactive substitutes such as scintillation proximity assays for ^{32}P or sulfur-35 (^{35}S) sequencing studies or ^{3}H cation assays, and enhanced chemiluminescence (ECL) as a substitute for ^{32}P and ^{35}S DNA probe labeling and southern blot analysis.

• Substitution of enriched stable isotopes for radionuclides in some cases. Mass spectrometry techniques, such as ICP-MS, are beginning to rival the sensitivity of some counting methods. Examples include use of oxygen-18 (^{18}O) and deuterium (^{2}H) with mass spectrometry detection as substitutes for ^{19}O and ^{3}H.

• Substitution of shorter-half-life radionuclides such as ^{32}P ($t_{1/2} = 14$ days) for ^{33}P ($t_{1/2} = 25$ days) in orthophosphate studies, or ^{33}P or ^{32}P for ^{35}S ($t_{1/2} = 87$ days) in nucleotides and deoxynucleotides. In many uses, ^{131}I ($t_{1/2} = 8$ days) can be substituted for ^{125}I ($t_{1/2} = 60$ days). Additional exposure precautions may be required.

7.C.1.2   Safe Storage of Mixed Waste

Waste containing short-half-life radionuclides should be stored for decay prior to subsequent waste management. The U.S. NRC refers to storage for decay as "decay-in-storage." Because on-site storage of low-level waste is very efficient and minimizes handling and transportation risks, laboratories and institutions should provide the space and safe storage facilities for decay-in-storage. In most cases, safe storage requires a designated room or facility that has been modified to contain the waste, protect workers, minimize the risk of a fire or spill, and minimize radiation levels outside the area. Proper ventilation and effluent trapping are critical needs for such a facility. Storage of mixed waste in the laboratory is not recommended because the required level of protection is difficult to achieve in a working area. Storage of such waste is not recommended when the waste may become more difficult to handle with age, such as with certain biological, putrescible, or reactive materials.

The specific U.S. NRC requirements for decay-in-storage of radioactive waste are usually detailed in the institution's license. Decay-in-storage is usually limited to half-lives of less than 65 days (although half-lives of up to 120 days are routinely approved by the U.S. NRC), but includes many of the radionuclides used in biomedical research. When the short-half-life radionuclides have decayed to background levels (the length of time depending on the initial radioactivity level but typically defined as a storage period of at least 10 half-lives), the chemical-radioactive waste can be managed as a chemical waste. After the decay period, U.S. NRC licenses usually require that the mixed waste be surveyed for external radiation prior to releasing it to the chemical waste stream.

EPA requirements for decay-in-storage of mixed waste have varied over time and by state and EPA region. Storage of mixed waste for more than 90 days, the period of time usually allowed for chemically hazardous waste, may require the approval of the state or regional EPA hazardous chemical waste authority. In permitted storage facilities, storage may be limited to 1 year for some types of mixed waste. Workers should contact their institution's environmental health and safety staff or local hazardous waste agency to determine their regulatory status and requirements for storing mixed waste for decay.

7.C.1.3   Hazard Reduction of Mixed Waste

Chemical hazards can be reduced by carrying out various common chemical reactions with the waste in the laboratory (see also section 7.B.5 and 7.D). How-ever, "treatment" of chemically hazardous waste has regulatory implications that must be considered. Many of the same considerations apply to treatment of mixed waste.

Nevertheless, there are still justifiable and legal reasons to carry out such operations in the laboratory when hazards can be reduced safely. Neutralization, oxidation, reduction, and various other chemical conversions as well as physical methods of separation and concentration can be applied prudently to many laboratory-scale mixed wastes. However, the dual character of the hazard, chemical and radioactive, requires that additional precautions be exercised. Treatment for the chemical hazard must not create a radioactivity risk for personnel or the environment. For example, vapors or aerosols from a reaction, distillation, or evaporation must not lead to escape of unsafe levels of radioactive materials into the atmosphere. Fume hoods appropriate for such operations should be designed to trap any radioactive effluent. When mixed waste is made chemically safe for disposal into the sanitary sewer, the laboratory must ensure that the radioactivity hazard is below the standards set by the publicly owned treatment works (POTW). Several examples for reducing the hazard of mixed waste are described below:

- The worker can reduce the chemical hazard to a safe level and then handle the material as only a radioactive hazardous waste. Many low-level radiation materials can then be allowed to decay to a safe level, following which simple disposal is allowable.
- Some trichloroacetic acid (TCA) solutions contain chloroform in excess of 6 ppm. Such a solution is considered a hazardous chemical waste because it fails the TCLP test. If the neutralized solution is not acceptable to the sewage treatment plant because of the presence of chloroform, it may be possible to remove that component from the solution by filtration through activated charcoal. The resulting radioactive filtrate can usually be disposed of in the sanitary sewer, and the contaminated charcoal can usually be disposed of as a chemical waste.
- Some radioactive methanol-acetic acid solutions from gel electrophoresis can be recycled via distillation and the methanol reused. The solution is neutralized prior to distillation to protect the distillation equipment from corrosion and to reduce the level of methyl acetate formed during the process.
- The volume of waste phenol, chloroform, methanol, and water containing radionuclides can be reduced by separating the nonaqueous portion using a separatory funnel. After separation, the organic phase can be distilled to produce chloroform waste,

which may contain levels of radioactivity below license limits for radioactive waste. The still bottom and aqueous phase must be handled as a mixed waste.

- High-performance liquid chromatography (HPLC), used to purify radiolabeled proteins and lipids, can generate a waste radioactive solution of acetonitrile, water, methanol, acetic acid, and often a small amount of dimethylformamide. When the solution is distilled by rotary flash evaporation, the distillate of acetonitrile, methanol, and water is nonradioactive and can be handled as a hazardous chemical waste. The radioactive still bottom, containing 1 to 5% methanol and acetic acid, can usually be neutralized, diluted, and disposed of in the sanitary sewer.

- Aqueous solutions containing uranyl or thorium compounds can be evaporated to dryness and the residues disposed of as radioactive waste. Because of their toxicity, solidification may be necessary prior to burial at a low-level radioactive waste site.

- Activated carbon, Molecular Sieves®, synthetic resins, and ion-exchange resins have been used with varying success in the separation of chemical and radioactive waste constituents. Activated carbon has been used to remove low concentrations of chloroform (less than 150 ppm) from aqueous mixed waste solutions. However, activated carbon is not suitable for high concentrations of phenol-chloroform or acetonitrile-water mixed waste. Amberlite® XAD resin, a series of Amberlite® polymeric absorbent resins used in chromatography, has been shown to be effective in removing the organic constituents from aqueous phenol, chloroform, and methanol solutions, leaving an aqueous solution that can be managed as a radioactive waste. Chemical constituents can be separated from mixed waste by using supercritical fluid extraction (e.g., carbon dioxide), which is now available commercially.

- Surface contamination from radioactively contaminated lead can be removed by dipping the contaminated lead into a solution of 1 M hydrochloric acid. After rinsing the lead with water, it usually can be documented as nonradioactive. The acidic wash and rinse solutions contain radionuclides and lead and must be handled accordingly. However, decontaminating the lead results in a smaller mass of mixed waste and allows the decontaminated lead to be reused or recycled. Commercial rinse products are also available for this purpose.

- Incineration is advantageous as a treatment for many types of chemical-radioactive waste, especially those that contain toxic or flammable organic chemicals. Incineration can destroy oxidizable organic chemicals in the waste. To comply with radionuclide release limits, U.S. NRC licensees need to control emissions and may need to restrict the incinerator's waste feed. Radioactive ash is typically managed as a radioactive waste. It is important to keep toxic metals (e.g., lead, mercury) out of the incinerable waste so that the ash is not chemically hazardous according to the TCLP test. On-site incineration minimizes handling and transportation risks; however, incineration of chemical waste is regulated by EPA and requires a permit, which is beyond the resources of most laboratory waste generators.

- Procedures for the solidification and stabilization of inorganic compounds from mixed waste (using concrete or epoxy resin) to meet federal land ban restrictions have been outlined (40 CFR 268). This method may also abate the waste's chemical hazard and render a chemical-radioactive waste a radioactive waste. For example, waste lead citrate and uranyl acetate mixtures from electron microscopy can be solidified with portland cement, which may be accepted for burial at a low-level radioactive waste site.

7.C.1.4 Commercial Disposal Services for Mixed Waste

Because of the great variety of laboratory mixed waste, it is often difficult to find a facility that can manage both the radioactive and the chemical hazards of the waste. In general, existing commercial disposal facilities are in business to manage mixed waste from the nuclear power industry, not waste from laboratories. Several commercial disposal facilities that accept mixed waste from off-site generators do exist in the United States. These sites have the capacity to manage liquid scintillation fluid, halogenated organics, and other organic waste. Treatment capacity exists for stabilization, neutralization, decontamination/macroencapsulation of lead, and reduction of chromium waste.

In spite of this capacity, many types of laboratory mixed waste have no commercial repository. No commercial mixed waste disposal facilities exist for waste contaminated with most toxic metals (such as mercury) or for lead-contaminated oils. Commercial disposal capacity likewise does not exist for high concentrations of halogen-containing organics and other TCLP waste, such as waste that contains chloroform.

(See also sections 7.B.5 and 7.D.)

7.C.2 Chemical-Biological Waste

Laboratory waste that is both chemically hazardous and exhibits a biological characteristic (depicted in Figure 7.2b) merits special disposal procedures. Animal and medical waste incinerators are usually not licensed to incinerate regulated chemical waste. Autoclaving or disinfection of chemical-biological waste usually does not destroy its chemically hazardous constituents, except for denaturing proteins and nucleic acids.

Conflicting laws are seldom a barrier to the safe management of chemical-biological waste. Although some states regulate the disposal of laboratory infectious waste, the federal government does not. Most laboratories using infectious agents abide by Centers for Disease Control and Prevention/National Institutes of Health (CDC/NIH) guidelines, which recommend on-site decontamination for agents in biosafety levels 3 and 4 (as defined in Chapter 5, section 5.E.1) but do not inhibit chemical waste storage or treatment (U.S. DHHS, 1993).

Although EPA is proposing to regulate transgenic plants that express insecticidal proteins, most types of chemical-biological waste are not regulated by EPA as a hazardous (RCRA) chemical waste. Waste (or spent) formalin that has been used in a process such as tissue preservation is not a discarded commercial chemical product and therefore is not regulated federally. However, its handling should be consistent with personal and environmental safety and within the limits set by local regulation. Animal tissue is regulated as chemical waste only in the unlikely circumstance that it contains a toxic chemical and the waste fails the TCLP test, or if the animal had been exposed to polychlorinated biphenyls (PCBs) in concentrations greater than 50 ppm.

Disposal is most difficult for the very small amount of chemical-biological waste that is EPA-regulated as chemically hazardous or contains a chemical, such as lead, that is inappropriate for an animal or medical waste incinerator. Disposal of tissue specimens preserved in ethanol or another flammable solvent is also difficult. In most cases, storage of this waste is limited to 90 days and must be managed at an EPA-permitted chemical waste facility. However, few chemical waste facilities are prepared to handle waste that is putrescible, infectious, or biohazardous.

7.C.2.1 Disposal of Chemically Contaminated Animal Tissue

Animal carcasses and tissues that contain a toxic chemical may be the most prominent chemical-biological laboratory waste. Such waste includes biological specimens preserved in formalin and rodents that have been fed lead, mercury, or PCBs in toxicity studies. If storage of such putrescible waste is necessary, refrigeration is usually advisable. Freezing potentially infectious animal tissue at the point of generation can add a margin of safety during waste handling, transport,

and prolonged storage. Infectious waste should be stored separately in a secure area.

Incineration is the most appropriate disposal method for this putrescible waste, and it can also destroy the infectious agents that such waste may contain. As discussed above (see beginning of section 7.C), federal law allows the incineration of most chemical-biological waste in an animal or medical waste incinerator. Most modern, efficiently run animal and medical waste incinerators can adequately destroy the small quantities of toxic organic chemicals present in chemically contaminated animal tissue. Large research institutions are likely to have an on-site animal incinerator. Medical waste incineration is also available through commercial waste haulers.

Incineration does not destroy lead and other inorganic chemicals, and they will be emitted or concentrated in the ash. In addition, some organic chemicals form products of incomplete combustion (PICs), which may be more toxic than the chemical contaminant. Incineration of PCBs and some other chlorinated aromatics, for example, can form extremely toxic polychlorinated dibenzo[*p*]dioxins and furans. Commercial disposal may be preferred for such waste.

(If animal or commercial incineration is unavailable, methods in section 7.C.3.3 below may be adaptable to chemical-biological waste.)

7.C.2.2 Sewer Disposal of Chemical-Biological Liquids

Laboratories that manipulate infectious agents, blood, or body fluids may generate waste that is contaminated with these materials and toxic chemicals. In most cases, blood and body fluids that contain toxic chemicals can be disposed of safely in a sanitary sewer, which is designed to accept biological waste. Approval for such disposal should be requested from the local wastewater treatment works. Chemical concentrations in such waste are typically low enough to be accepted by a local treatment works. OSHA recommends that a separate sink be used exclusively for disposal of human blood, body fluids, and infectious waste. It may be prudent to treat blood and body fluids with bleach (usually a 1:10 aqueous dilution of household bleach) prior to disposal in the sanitary sewer. The worker should take care to prevent personal exposure while waste is being discharged into the sewer.

7.C.2.3 Disinfection and Autoclaving of Contaminated Labware

Contaminated labware may include cultures, stocks, petri plates, and other disposable laboratory items (e.g., gloves, pipettes, and tips). In many cases, the small quantities of infectious waste on such labware can be disinfected safely with bleach or other chemical disinfectant (e.g., by soaking overnight). The disinfected waste can then be treated as a chemical waste. The worker must check with the state or regional EPA office to determine if a treatment permit is required for chemical disinfection of chemical-biological waste.

Autoclaves can be used to steam-sterilize infectious waste but should be tested routinely for efficacy. Autoclaving does not require an EPA permit. Care must be taken because autoclaving of chemical-biological waste at 120 to 130 °C may result in the volatilization or release of the chemical constituent. Additional waste containment may be needed to minimize chemical releases, but it can interfere with steam penetration into the waste load and sterilization. Before autoclaving untested waste streams containing volatile chemicals, a small load should be processed while monitoring the air emissions.

Autoclaving waste containing flammable liquids may result in a fire or explosion. (It should also be noted that steam sterilization of waste that contains bleach may harm an autoclave.) To autoclave voluminous chemical-biological waste streams, it may be appropriate to dedicate an autoclave room with ample ventilation and to restrict access.

The autoclaved chemical-biological waste can be managed as a chemical waste. After autoclaving, the biohazard markings on the container should be defaced or the material overpacked in a second container to indicate that the waste has been sterilized.

Similar precautions should be observed when using microwaves for decontamination. Although still under development, ultraviolet peroxidation may have the capacity to both sterilize and destroy certain chemicals in the waste. Chemical treatment permits may be required.

7.C.2.4 Disposal of Chemically Contaminated Medical Waste and Sharps

Laboratories that work with human blood must adhere to OSHA's Standard for Occupational Exposure to Bloodborne Pathogens (29 CFR 1910.1030), which requires waste containment, marking, and labeling. The OSHA standard also regulates waste disposal from laboratories that manipulate human immunodeficiency virus (HIV) or hepatitis B virus (HBV). In general, such waste that is chemically contaminated can be incinerated with other medical waste or can be autoclaved and managed as a chemical waste.

Waste hypodermic needles and other "sharps" (e.g, scalpels and razor blades) need to be contained in a puncture-resistant waste collection container. Sharps should be destroyed by incineration or by grinding as part of the disinfection treatment. Incineration of chemical or drug-contaminated needles in a medical waste incinerator is appropriate if the waste is not an EPA-regulated chemical waste and if the chemical's toxicity or contamination is low. Needles and other sharps that are contaminated with toxic chemicals and infectious agents or blood can be autoclaved or disinfected on site (see the precautions above), and then managed as a chemical waste. The waste container's biohazard symbol and markings should be defaced after autoclaving or disinfection to indicate that the waste has been sterilized. Noninfectious needles and sharps with high chemical toxicity or contamination are accepted by chemical incinerators.

Some biomedical research generates materials contaminated with blood and antineoplastic drugs. Incineration of these materials as medical waste is appropriate if the level of chemical contamination is low, which is typical. In some cases, chemical disinfection and treatment can be combined to destroy both infectious agents and the antineoplastic drug. It should be noted that unemptied source containers of some antineoplastic drugs are EPA-listed hazardous waste and must be managed as a regulated chemical waste.

7.C.2.5 Minimization Methods for Chemical-Biological Waste

Waste minimization methods used for chemical waste can be used to reduce or eliminate the chemical hazard of chemical-biological waste. Some laboratories that generate biohazardous waste have replaced disposable items with reusable supplies, which are disinfected between uses.

For biological waste, waste minimization can be accomplished best through careful source separation of biological waste from other waste streams. When state guidelines for defining infectious waste do not exist, it is important for laboratories to define carefully those biological wastes that can be disposed of safely as noninfectious within the framework of the CDC/NIH guidelines (U.S. DHHS, 1993). Training workers to identify and separate biological waste will prevent its inadvertent mixing with other waste streams and normal trash.

7.C.3 Radioactive-Biological Waste

The management of radioactive-biological laboratory waste (shown in Figure 7.2c) can be difficult because of limited on- and off-site disposal options.

Basic principles for the management of radioactive-biological waste include the following:

- Risk associated with the waste should be assessed. It may be prudent to disinfect highly biohazardous agents first to minimize handling risks and prevent growth of the waste's microbiological load. Appropriate containment, handling, and storage precautions should be taken prior to treatment.
- Radioactive-biological waste containing short-half-life radionuclides can be held for decay. After decay-in-storage, most U.S. NRC licenses allow the waste to be managed as biological waste. If the waste supports the growth of an infectious agent that it contains, storage should be in a freezer to prevent the waste's infectious risk from increasing.
- Refrigerated storage facilities or other preservation methods are necessary for putrescible waste.

7.C.3.1   On-site Incineration of Low-level Radioactive Waste

Laboratories that have an on-site radioactive waste incinerator have a great advantage in their ability to manage radioactive-biological waste. On-site incineration of radioactive-biological waste is practical and can be done with minimal impact to health or the environment. For waste that is putrescible or may be infectious, on-site incineration is ideal.

The institution's U.S. NRC license will identify which radionuclides can be incinerated and will set emission limits for those materials based on activity. It is usually beneficial to separate waste by radionuclide volatility and half-life. Incinerator emissions of waste containing short-half-life radionuclides can be minimized by refrigerated storage for decay prior to incineration. If the license allows, ash from the incineration of nonvolatile radionuclides that have short half-lives may be held for decay and disposed of as normal trash. Other ash must be disposed of at a radioactive waste site.

7.C.3.2   Off-site Management of Low-level Radioactive Waste

Many laboratories do not have an on-site incinerator for radioactive-biological waste. Communities tend to oppose waste incinerators, and on-site incineration is prohibitively costly for some radioactive-biological waste generators. Even institutions that have incinerators must usually rely on off-site disposal for some of their radioactive waste. For radioactive putrescible waste, off-site disposal requires special packaging, storage, and transport considerations.

Reliable access to off-site disposal will depend on the establishment of regional sites, which have been slow to develop under the Low-level Radioactive Waste Policy Act of 1980. Moreover, when established, regional low-level radioactive waste sites may not immediately accommodate laboratory radioactive-biological waste. As discussed earlier in this chapter, choice of off-site disposal must involve careful consideration of the safety record of the facility to ensure that the generator's long-term responsibility is liability-free.

7.C.3.3   Disposal of Radioactive Animal Carcasses and Tissue

Waste radioactive animal carcasses and tissue generated from biomedical research typically pose no significant infectious hazard, but they are putrescible. U.S. NRC regulations allow animal carcasses and tissue with less than 1.85 kilobecquerels per gram (kBq/g) of ^{3}H or ^{14}C to be disposed of without regard to radioactivity. Thus animal carcasses and tissue below this limit need not be managed as a radioactive-biological waste but only as a biological waste. Animal tissue with higher levels of activity or other radionuclides must be managed as a radioactive waste. As with all putrescible waste, waste should be either refrigerated, frozen, or otherwise preserved during accumulation, transport, and storage.

While on-site incineration is the preferred method of managing radioactive animal carcasses and tissue, several alternatives exist. Alkaline digestion of animal carcasses containing ^{3}H, ^{14}C, and formaldehyde, followed by neutralization, results in an aqueous radioactive stream that can usually be disposed of in the sanitary sewer. The process uses 1 M sodium hydroxide at 300 °C and pressures up to 150 psi. Commercial units are available for this process. Radioactive animal carcasses may be accepted at a low-level radioactive waste site when packed in lime.

Some institutions grind radioactive animal tissue for disposal in the sanitary sewer, although the U.S. NRC requires that all sewer-disposable waste be dispersible. Preventing contamination and exposure of waste handlers to dust or particles is an important safety measure in this operation.

Autoclaving of infectious animal carcasses is difficult because of the waste's high heat capacity and poor heat conductivity, and often unproductive because treated waste remains putrescible.

7.C.3.4   Disposal of Radioactive-Biological Contaminated Labware

Radioactive-biological contaminated labware (e.g., gloves and disposable laboratory articles) is generated from biomedical research using radioactive materials with infectious agents, blood, and body fluids. On-site incineration, autoclaving, and off-site disposal are the management options for this waste. Chemical decontamination (e.g., soaking in bleach) may be appropriate if it can be done without risking personal exposure, increasing waste volumes, or creating a waste that is difficult to handle (e.g., wet waste). After disinfection, radioactive-biological waste can be managed as radioactive waste.

Infectious waste and needle boxes that contain radionuclides can be autoclaved safely if the following precautions are satisfied:

• Monitor the air emissions of a test load to determine if the release of radioactive material is in compliance with U.S. NRC and license limits.
• Wipe-test the autoclave interior for surface contamination regularly.
• For ongoing treatment of this waste, dedicate an autoclave or autoclave room for this purpose. The room should have ample ventilation.
• Restrict access during autoclaving.
• Test the autoclave efficacy regularly.

Radioactive needles contaminated with infectious agents or blood should be autoclaved as described above, and then incinerated on site or shipped to a low-level radioactive waste site. To prevent injuries, it is important that hypodermic needles and other sharps be kept in waste containers that are puncture-resistant, leak-proof, and closable from the point of discard through ultimate disposal. To prevent airborne radioactive materials, destruction of needles by grinding or a similar means is not recommended.

7.C.3.5 Sewer Disposal of Radioactive-Biological Liquids

Radioactive blood, body fluids, and other sewer-compatible liquids may be disposed of in the sanitary sewer if quantities are within U.S. NRC license and treatment works limits. Precautions must be taken to prevent exposure of waste handlers. OSHA recommends that disposal of human blood and body fluids be done in a dedicated sink.

7.C.4 Chemical-Radioactive-Biological Waste

Chemical-radioactive-biological laboratory waste (depicted in Figure 7.2.d) is the most difficult multihazardous waste to manage. The strategies for managing the various other types of multihazardous waste described above are generally applicable to chemical-radioactive-biological waste. For example, toxicological research sometimes generates animal tissue that contains a radioactively labeled toxic chemical. However, the chemical toxicity of such waste is commonly inconsequential, both legally and in relation to the waste's other characteristics. It could be appropriate to dispose of such animal tissue as a radioactive-biological waste, without regard to its low toxic chemical content.

Reduction or elimination of one of the waste hazards through waste management methods is often an efficient first step. Decay-in-storage is a simple, low-cost way to reduce the radioactivity hazard of a waste with short-lived radionuclides. After decay, most U.S. NRC licenses allow the waste to be managed as a chemical-biological waste. Similarly, autoclaves are readily available to most laboratories for destruction of infectious agents. As described above, autoclaving multihazardous waste requires certain precautions, but renders a chemical-radioactive-biological waste a chemical-radioactive waste. Autoclaving or disinfection makes sense when any of the waste's characteristics (e.g., nutrient value) could support the growth of an infectious agent it contains and thus could increase the waste's risk.

Certain waste treatments reduce multiple hazards in one step. For example, incineration can destroy oxidizable organic chemicals and infectious agents, waste feed rates can be controlled to meet emission limits for volatile radionuclides, and radioactive ash can be disposed of as a dry radioactive waste. Likewise, some chemical treatment methods (e.g., those using bleach) both oxidize toxic chemicals and disinfect biological hazards. Such treatment could convert a chemical-radioactive-biological waste to a radioactive waste.

7.C.5 Future Trends in Management of Multihazardous Waste

Multihazardous waste is becoming the focus of much attention by regulatory agencies, as well as by the laboratories that must deal with it. The U.S. NRC relieved much of the laboratory mixed waste problem by allowing liquid scintillation fluid (LSF) with less than 1.85 kBq/g of ^{3}H or ^{14}C to be disposed of without regard to radioactivity. Thus ignitable LSF below this limit need not be managed as a mixed waste but only as a hazardous chemical waste. As explained above, although U.S. NRC policy has not established a *de minimis* level for other types of laboratory radioactive waste, licensees can often propose a license-specific *de minimis* level, below which mixed waste can be released for management as a chemical waste.

Regional EPA offices and state and local hazardous waste authorities differ in their regulation of storage and chemical treatment of mixed waste. Under certain conditions, regulators may allow such activities without a permit or may temporarily waive the storage or treatment permit requirements. Most regulators allow separation or treatment of chemical and radioactive components in the waste collection container or as part of a process without a RCRA permit.

Within the context of these changes, attention is being directed at promising technologies for the treatment of multihazardous waste. At this time, it cannot be determined if and when these technologies will be available, or if they will be developed for use in the laboratory or on a commercial scale. Nevertheless, they deserve careful attention, as do other approaches that will surely be developed in the near future, including the following:

- Ultraviolet peroxidation is being reviewed by the National Institutes of Health as a method to treat laboratory aqueous mixed waste containing low concentrations of organics. The process, though still under development, is expected to treat a wide range of organics and sterilize the waste.
- Wet oxidation and supercritical fluid oxidation (for aqueous solutions containing 1 to 10% organics) are being developed to destroy the chemical constituents in mixed waste.
- Biodegradation has been used successfully to treat soil, sludge, and other contaminated waste streams containing up to 1% organic waste. Laboratory-scale bioreactors are commercially available.
- Plasma torch, thermal desorption, molten salt pyrolysis, vitrification, and arc or hearth pyrolysis followed by incineration have all been used to treat mixed waste.

7.D PROCEDURES FOR THE LABORATORY-SCALE TREATMENT OF SURPLUS AND WASTE CHEMICALS

Concerns about environmental protection, bans on landfill disposal of waste, and limited access to sewer disposal have encouraged the development of strategies to reduce hazardous waste from laboratories. Many management methods are considered in earlier chapters of this book (see Chapter 4, section 4.B, and Chapter 5, section 5.B). The small-scale treatment and deactivation of products and by-products as part of the experiment plan is one approach that can be used to address the problem at the level of the actual generator, the laboratory worker. However, unless there is a significant reduction in risk by such action, there may be little benefit in carrying out a procedure that will simply produce another kind of waste with similar

risks and challenges for disposal. Furthermore, the question of what constitutes "legal" treatment within the laboratory is still unresolved.

Nevertheless, there is often merit for such in-laboratory treatment. Below are some procedures of general use at the laboratory scale. Additional procedures can be found in the earlier edition of this book (*Prudent Practices for Disposal of Chemicals from Laboratories*; NRC, 1983) and other books listed in the bibliography. More specific procedures for laboratory treatment are increasingly being included in the experimental sections of chemical journals and in publications such as *Organic Syntheses* and *Inorganic Syntheses*.

Safety must be the first consideration before undertaking any of the procedures below. The procedures presented here are intended to be carried out only by, or under the direct supervision of, a trained scientist or technologist who understands the chemistry and hazards involved. Appropriate personal protection should be used. With the exception of neutralization, the procedures are intended for application to small quantities, that is, not more than a few hundred grams. *Because risks tend to increase exponentially with scale, larger quantities should be treated only in small batches unless a qualified chemist has demonstrated that the procedure can be scaled up safely.* The generator must ensure that the procedure eliminates the regulated hazard before the products are disposed of as nonhazardous waste. In addition, if the procedure suggests disposal of the product into the sanitary sewer, this strategy must comply with local regulations.

(See Chapter 6, section 6.F, for further information on protective clothing and also Chapter 5, section 5.C.2.6.)

7.D.1 Acids and Bases

Neutralization of acids and bases (corrosives) is generally exempt from a RCRA treatment permit. However, because the products of the reaction are often disposed of in the sanitary sewer, it is important to ensure that hazardous waste such as toxic metal ions is not a part of the effluent.

In most laboratories, both waste acids and waste bases are generated, and so it is most economical to collect them separately and then neutralize one with the other. If additional acid or base is required, sulfuric or hydrochloric acid and sodium or magnesium hydroxide, respectively, can be used.

If the acid or base is highly concentrated, it is prudent to first dilute it with cold water (adding the acid or base to the water) to a concentration below 10%. Then the acid and base are mixed, and the additional water is slowly added when necessary to cool and dilute the neutralized product. The concentration of neutral salts disposed of in the sanitary sewer should generally be below 1%.

7.D.2 Organic Chemicals

7.D.2.1 Thiols and Sulfides

Small quantities of thiols (mercaptans) and sulfides can be destroyed by oxidation to a sulfonic acid with sodium hypochlorite. If other groups that can be oxidized by hypochlorite are also present, the quantity of this reagent used must be increased accordingly.

Procedure for oxidizing 0.1 mol of a liquid thiol:

$$RSH + 3OCl^- \rightarrow RSO_3H + 3Cl^-$$

Five hundred milliliters (0.4 mol, 25% excess) of commercial hypochlorite laundry bleach (5.25% sodium hypochlorite) is poured into a 5-L three-necked flask located in a fume hood. The flask is equipped with a stirrer, thermometer, and dropping funnel. The thiol (0.1 mol) is added dropwise to the stirred hypochlorite solution, initially at room temperature. A solid thiol can be added gradually through a neck of the flask or can be dissolved in tetrahydrofuran or other appropriate nonoxidizable solvent and the solution added to the hypochlorite. (The use of tetrahydrofuran introduces a flammable liquid that could alter the final disposal method.) Traces of thiol can be rinsed from the reagent bottle and dropping funnel with additional hypochlorite solution. Oxidation, accompanied by a rise in temperature and dissolution of the thiol, usually starts after a small amount of the thiol has been added. If the reaction has not started spontaneously after about 10% of the thiol has been added, addition is stopped and the mixture warmed to about 50 °C to initiate this reaction. Addition is resumed only after it is clear that oxidation is occurring. The temperature is maintained at 45 to 50 °C by adjusting the rate of addition and using an ice bath for cooling if necessary. Addition requires about 15 minutes. If the pH drops below 6 because of generation of the sulfonic acid, it may be necessary to add some sodium hydroxide or additional bleach because hypochlorite is destroyed under acidic conditions. Stirring is continued for 2 hours while the temperature gradually falls to room temperature. The mixture should be a clear solution, perhaps containing traces of oily by-products. The reaction mixture can usually be flushed down the drain with excess water. The unreacted laundry bleach need not be decomposed.

(Because sodium hypochlorite solutions deteriorate on storage, it is advisable to have relatively fresh material available. A 5.25% solution of sodium hypochlorite

has 25 g of active chlorine per liter. If determination of the active hypochlorite content is justified, it can be accomplished as follows. Ten milliliters of the sodium hypochlorite solution is diluted to 100.0 mL, and then 10.0 mL of this diluted reagent is added to a solution of 1 g of potassium iodide and 12.5 mL of 2 M acetic acid in 50 mL of distilled water. Using a starch solution as indicator, titrate the solution with 0.1 N sodium thiosulfate. One milliliter of titrant corresponds to 3.5 mg of active chlorine. A 5.25% solution of sodium hypochlorite requires approximately 7 mL of titrant.)

Calcium hypochlorite may be used as an alternative to sodium hypochlorite and requires a smaller volume of liquid. For 0.1 mol of thiol, 42 g (25% excess) of 65% calcium hypochlorite (technical grade) is stirred into 200 mL of water at room temperature. The hypochlorite soon dissolves, and the thiol is then added as in the above procedure.

Laboratory glassware, hands, and clothing contaminated with thiols can be deodorized by a solution of Diaperene®, a tetraalkylammonium salt used to deodorize containers in which soiled diapers have been washed.

Small amounts of sulfides, RSR', can be oxidized to sulfones (RSO₂R') to eliminate their disagreeable odors. The hypochlorite procedure used for thiols can be employed for this purpose, although the resulting sulfones are often water-insoluble and may have to be separated from the reaction mixture by filtration.

Small amounts of the inorganic sulfides, sodium sulfide or potassium sulfide, can be destroyed in aqueous solution by sodium or calcium hypochlorite using the procedure described for oxidizing thiols.

$$Na_2S + 4OCl^- \rightarrow Na_2SO_4 + 4Cl^-$$

7.D.2.2 Acyl Halides and Anhydrides

Acyl halides, sulfonyl halides, and anhydrides react readily with water, alcohols, and amines. They should never be allowed to come into contact with waste that contains such substances. Most compounds in this class can be hydrolyzed to water-soluble products of low toxicity.

Procedure for hydrolyzing 0.5 mol of RCOX, RSO₂X, or (RCO)₂O:

$$RCOX + 2NaOH \rightarrow RCO_2Na + NaX + H_2O$$

A 1-L three-necked flask equipped with a stirrer, dropping funnel, and thermometer is placed on a steam bath in a hood, and 600 mL of 2.5 M aqueous sodium hydroxide (1.5 mol, 50% excess) are poured into the flask. A few milliliters of the acid derivative are added dropwise with stirring. If the derivative is a solid, it can be added in small portions through a neck of the flask. If reaction occurs, as indicated by a rise in temperature and dissolution of the acid derivative, addition is continued at such a rate that the temperature does not rise above 45 °C. If the reaction is sluggish, as may be the case with less soluble compounds such as *p*-toluenesulfonyl chloride, the mixture is heated before adding any more acid derivative. When the initial added material has dissolved, the remainder is added dropwise. As soon as a clear solution is obtained, the mixture is cooled to room temperature, neutralized to about pH 7 with dilute hydrochloric or sulfuric acid, and washed down the drain with excess water.

7.D.2.3 Aldehydes

Many aldehydes are respiratory irritants, and some, such as formaldehyde and acrolein, are quite toxic. There is sometimes merit in oxidation of aldehydes to the corresponding carboxylic acids, which are usually less toxic and less volatile.

Procedure for permanganate oxidation of 0.1 mol of aldehyde:

$$3RCHO + 2KMnO_4 \rightarrow$$
$$2RCO_2K + RCO_2H + 2MnO_2 + H_2O$$

A mixture of 100 mL of water and 0.1 mol of aldehyde is stirred in a 1-L round-bottomed flask equipped with a thermometer, dropping funnel, stirrer, steam bath, and, if the aldehyde boils below 100 °C, a condenser. Approximately 30 mL of a solution of 12.6 g (0.08 mol, 20% excess) of potassium permanganate in 250 mL of water is added over a period of 10 minutes. If the temperature rises above 45 °C, the solution should be cooled. If this addition is not accompanied by a rise in temperature and loss of the purple permanganate color, the mixture is heated by the steam bath until a temperature is reached at which the color is discharged. The rest of the permanganate solution is added slowly at within 10 °C of this temperature. The temperature is then raised to 70 to 80 °C, and stirring continued for 1 hour or until the purple color has disappeared, whichever occurs first. The mixture is cooled to room temperature and acidified with 6 N sulfuric acid. **(CAUTION: Do not add concentrated sulfuric acid to permanganate solution because explosive manganese oxide (Mn_2O_7) may precipitate.)** Enough solid sodium hydrogen sulfite (at least 8.3 g, 0.08 mol) is added with stirring at 20 to 40 °C to reduce all the

manganese, as indicated by loss of purple color and dissolution of the solid manganese dioxide. The mixture is washed down the drain with a large volume of water.

If the aldehyde contains a carbon-carbon double bond, as in the case of the highly toxic acrolein, 4 mol (20% excess) of permanganate per mol of aldehyde is required to oxidize the alkene bond and the aldehyde group.

Formaldehyde is oxidized conveniently to formic acid and carbon dioxide by sodium hypochlorite. Thus 10 mL of formalin (37% formaldehyde) in 100 mL of water is stirred into 250 mL of hypochlorite laundry bleach (5.25% NaOCl) at room temperature and allowed to stand for 20 minutes before being flushed down the drain. This procedure is not recommended for other aliphatic aldehydes because it leads to chloro acids, which are more toxic and less biodegradable than corresponding unchlorinated acids.

7.D.2.4 Amines

Acidified potassium permanganate efficiently degrades aromatic amines. Diazotization followed by hypophosphorus acid protonation is a method for deamination of aromatic amines, but the procedure is more complex than oxidation.

Procedure for permanganate oxidation of 0.01 mol of aromatic amine:

A solution of 0.01 mol of aromatic amine in 3 L of 1.7 N sulfuric acid is prepared in a 5-L flask; 1 L of 0.2 M potassium permanganate is added, and the solution allowed to stand at room temperature for 8 hours. Excess permanganate is reduced by slow addition of solid sodium hydrogen sulfite until the purple color disappears. The mixture is then flushed down the drain.

7.D.2.5 Organic Peroxides and Hydroperoxides

(CAUTION: Peroxides are particularly dangerous. These procedures should be carried out only by knowledgeable laboratory workers.) Peroxides can be removed from a solvent by passing it through a column of basic activated alumina, by treating it with indicating Molecular Sieves®, or by reduction with ferrous sulfate. Although these procedures remove hydroperoxides, which are the principal hazardous contaminants of peroxide-forming solvents, they do not remove dialkyl peroxides, which may also be present in low concentrations. Commonly used peroxide reagents, such as acetyl peroxide, benzoyl peroxide, *t*-butyl hydroperoxide, and di-*t*-butyl peroxide, are less dangerous than the adventitious peroxides formed in solvents.

Removal of peroxides with alumina:

A 2 × 33 cm column filled with 80 g of 80-mesh basic activated alumina is usually sufficient to remove all peroxides from 100 to 400 mL of solvent, whether water-soluble or water-insoluble. After passage through the column, the solvent should be tested for peroxide content. Peroxides formed by air oxidation are usually decomposed by the alumina, not merely absorbed on it. However, for safety it is best to slurry the wet alumina with a dilute acidic solution of ferrous sulfate before it is discarded.

Removal of peroxides with Molecular Sieves®:

Reflux 100 mL of the solvent with 5 g of 4- to 8-mesh indicating activated 4A Molecular Sieves® for several hours under nitrogen. The sieves are separated from the solvent and require no further treatment because the peroxides are destroyed during their interaction with the sieves.

Removal of peroxides with ferrous sulfate:

$$ROOH + 2Fe^{2+} + 2H^+ \rightarrow ROH + 2Fe^{3+} + H_2O$$

A solution of 6 g of $FeSO_4 \cdot 7H_2O$, 6 mL of concentrated sulfuric acid, and 11 mL of water is stirred with l L of water-insoluble solvent until the solvent no longer gives a positive test for peroxides. Usually only a few minutes are required.

Diacyl peroxides can be destroyed by this reagent as well as by aqueous sodium hydrogen sulfite, sodium hydroxide, or ammonia. However, diacyl peroxides with low solubility in water, such as dibenzoyl peroxide, react very slowly. A better reagent is a solution of sodium iodide or potassium iodide in glacial acetic acid.

Procedure for destruction of diacyl peroxides:

$$(RCO_2)_2 + 2NaI \rightarrow 2RCO_2Na + I_2$$

For 0.01 mol of diacyl peroxide, 0.022 mol (10% excess) of sodium or potassium iodide is dissolved in 70 mL of glacial acetic acid, and the peroxide added gradually with stirring at room temperature. The solution is rapidly darkened by the formation of iodine. After a minimum of 30 minutes, the solution is washed down the drain with a large excess of water.

Most dialkyl peroxides (ROOR) do not react readily at room temperature with ferrous sulfate, iodide, ammonia, or the other reagents mentioned above. However, these peroxides can be destroyed by a modification of the iodide procedure.

Procedure for destruction of dialkyl peroxides:

One milliliter of 36% (w/v) hydrochloric acid is added to the above acetic acid/potassium iodide solution as an accelerator, followed by 0.01 mol of the dialkyl peroxide. The solution is heated to 90 to 100 °C on a steam bath over the course of 30 minutes and held at that temperature for 5 hours.

7.D.3 Inorganic Chemicals

7.D.3.1 Metal Hydrides

Most metal hydrides react violently with water with the evolution of hydrogen, which can form an explosive mixture with air. Some, such as lithium aluminum hydride, potassium hydride, and sodium hydride, are pyrophoric. Most can be decomposed by gradual addition of (in order of *decreasing* reactivity) methyl alcohol, ethyl alcohol, *n*-butyl alcohol, or *t*-butyl alcohol to a stirred, ice-cooled solution or suspension of the hydride in an inert liquid, such as diethyl ether, tetrahydrofuran, or toluene, under nitrogen in a three-necked flask. Although these procedures reduce the hazard and should be a part of any experimental procedure that uses reactive metal hydrides, the products from such deactivation may be hazardous waste that must be treated as such on disposal.

Hydrides commonly used in laboratories are lithium aluminum hydride, potassium hydride, sodium hydride, sodium borohydride, and calcium hydride. The following methods for their disposal demonstrate that the reactivity of metal hydrides varies considerably. Most hydrides can be decomposed safely by one of the four methods, but the properties of a given hydride must be well understood in order to select the most appropriate method. **(CAUTION: Most of the methods described below produce hydrogen gas, which can present an explosion hazard. The reaction should be carried out in a hood, behind a shield, and with proper safeguards to avoid exposure of the effluent gas to spark or flame. Any stirring device must be spark-proof.)**

Decomposition of lithium aluminum hydride:

Lithium aluminum hydride ($LiAlH_4$) can be purchased as a solid or as a solution in toluene, diethyl ether, tetrahydrofuran, or other ethers. Although dropwise addition of water to its solutions under nitrogen in a three-necked flask has frequently been used to decompose it, vigorous frothing often occurs. An alternative is to use 95% ethanol, which reacts less vigorously than water. A safer procedure is to decompose the hydride with ethyl acetate, because no hydrogen is formed.

$$2CH_3CO_2C_2H_5 + LiAlH_4 \rightarrow LiOC_2H_5 + Al(OC_2H_5)_3$$

To the hydride solution in a flask equipped with a stirrer, ethyl acetate is added slowly. The mixture sometimes becomes so viscous after the addition that stirring is difficult and additional solvent may be required. When the reaction with ethyl acetate has ceased, a saturated aqueous solution of ammonium chloride is added with stirring. The mixture separates into an organic layer and an aqueous layer containing inert inorganic solids. The upper, organic layer should be separated and disposed of as a flammable liquid. The lower, aqueous layer can often be disposed of in the sanitary sewer.

Decomposition of potassium or sodium hydride:

Potassium and sodium hydride (KH, NaH) in the dry state are pyrophoric, but they can be purchased as a relatively safe dispersion in mineral oil. Either form can be decomposed by adding enough dry hydrocarbon solvent (e.g., heptane) to reduce the hydride concentration below 5% and then adding excess *t*-butyl alcohol dropwise under nitrogen with stirring. Cold water is then added dropwise, and the resulting two layers are separated. The organic layer can be disposed of as a flammable liquid. The aqueous layer can often be neutralized and disposed of in the sanitary sewer.

Decomposition of sodium borohydride:

Sodium borohydride (NaBH$_4$) is so stable in water that a 12% aqueous solution stabilized with sodium hydroxide is sold commercially. In order to effect decomposition, the solid or aqueous solution is added to enough water to make the borohydride concentration less than 3%, and then excess equivalents of dilute aqueous acetic acid are added dropwise with stirring under nitrogen.

Decomposition of calcium hydride:

Calcium hydride (CaH$_2$), the least reactive of the materials discussed here, is purchased as a powder. It is decomposed by adding 25 mL of methyl alcohol per gram of hydride under nitrogen with stirring. When reaction has ceased, an equal volume of water is gradually added to the stirred slurry of calcium methoxide. The mixture is then neutralized with acid and disposed of in the sanitary sewer.

7.D.3.2   Inorganic Cyanides

Inorganic cyanides can be oxidized to cyanate using aqueous hypochlorite following a procedure similar to the oxidation of thiols. Hydrogen cyanide can be converted to sodium cyanide by neutralization with aqueous sodium hydroxide, and then oxidized.

Procedure for oxidation of cyanide:

$$NaCN + NaOCl \rightarrow NaOCN + NaCl$$

An aqueous solution of the cyanide salt in an ice-cooled, three-necked flask equipped with a stirrer, thermometer, and dropping funnel is cooled to 4 to 10 °C. A 50% excess of commercial hypochlorite laundry bleach containing 5.25% (0.75 M) sodium hypochlorite is added slowly with stirring while maintaining the low temperature. When the addition is complete and heat is no longer being evolved, the solution is allowed to warm to room temperature and stand for several hours. The mixture can then be washed down the drain with excess water. The same procedure can be applied to insoluble cyanides such as cuprous cyanide (though copper salts should not be disposed of in the sanitary sewer). In calculating the quantity of hypochlorite required, the experimenter should remember that additional equivalents may be needed if the metal ion can be oxidized to a higher valence state, as in the reaction,

$$2CuCN + 3OCl^- + H_2O \rightarrow$$
$$2Cu^{2+} + 2OCN^- + 2OH^- + 3Cl^-$$

A similar procedure can be used to destroy hydrogen cyanide, but precautions must be taken to avoid exposure to this very toxic gas. Hydrogen cyanide is dissolved in several volumes of ice water. Approximately 1 molar equivalent of aqueous sodium hydroxide is added at 4 to 10 °C to convert the hydrogen cyanide into its sodium salt, and then the procedure described above for sodium cyanide is followed. (**CAUTION: Sodium hydroxide or other bases, including sodium cyanide, must not be allowed to come into contact with liquid hydrogen cyanide because they may initiate a violent polymerization of the hydrogen cyanide.**)

This procedure also destroys soluble ferrocyanides and ferricyanides. Alternatively, these can be precipitated as the ferric or ferrous salt, respectively, for possible landfill disposal.

(See Chapter 6, section 6.D, for details on working with hazardous gases.)

7.D.3.3 Metal Azides

Heavy metal azides are notoriously explosive and should be handled by trained personnel. Silver azide (and also fulminate) can be generated from Tollens reagent, which is often found in undergraduate laboratories. Sodium azide is explosive only when heated to near its decomposition temperature (300 °C), but heating it should be avoided. Sodium azide should never be flushed down the drain. This practice has caused serious accidents because the azide can react with lead or copper in the drain lines to produce an azide that may explode. It can be destroyed by reaction with nitrous acid:

$$2NaN_3 + 2HNO_2 \rightarrow 3N_2 + 2NO + 2NaOH$$

Procedure for destruction of sodium azide:

The operation must be carried out in a hood because of the formation of toxic nitric oxide. An aqueous solution containing no more than 5% sodium azide is put into a three-necked flask equipped with a stirrer and a dropping funnel. Approximately 7 mL of 20% aqueous solution of sodium nitrite (40% excess) per gram of sodium azide is added with stirring. A 20% aqueous solution of sulfuric acid is then added gradually until the reaction mixture is acidic to litmus paper. **(CAUTION: The order of addition is essential. Poisonous, volatile hydrazoic acid (HN_3) will evolve if the acid is added before the nitrite.)** When the evolution of nitrogen oxides ceases, the acidic solution is tested with starch iodide paper. If it turns blue, excess nitrite is present, and the decomposition is complete. The reaction mixture is washed down the drain.

7.D.3.4 Alkali Metals

Alkali metals react violently with water, with common hydroxylic solvents, and with halogenated hydrocarbons. They should always be handled in the absence of these materials. The metals are usually destroyed by controlled reaction with an alcohol. The final aqueous alcoholic material can usually be disposed of in the sanitary sewer.

Procedure for destruction of alkali metals:

Waste sodium is readily destroyed with 95% ethanol. The procedure is carried out in a three-necked, round-bottomed flask equipped with a stirrer, dropping funnel, condenser, and heating mantle. Solid sodium should be cut into small pieces with a sharp knife while wet with a hydrocarbon, preferably mineral oil, so that the unoxidized surface is exposed. A dispersion of sodium in mineral oil can be treated directly. The flask is flushed with nitrogen and the pieces of sodium placed in it. Then 13 mL of 95% ethanol per gram of sodium are added at a rate that causes rapid refluxing. **(CAUTION: Hydrogen gas is evolved and can present an explosion hazard. The reaction should be carried out in a hood, behind a shield, and with proper safeguards (such as in Chapter 5, sections 5.G.4 and 5.G.5) to avoid exposing the effluent gas to spark or flame. Any stirring device must be spark-proof.)** Stirring is commenced as soon as enough ethanol has been added to make this possible. The mixture is stirred and heated under reflux until the sodium is dissolved. The heat source is removed, and an equal volume of water added at a rate that causes no more than mild refluxing. The solution is then cooled, neutralized with 6 M sulfuric or hydrochloric acid, and washed down the drain.

To destroy metallic potassium, the same procedure and precautions as for sodium are used, except that the less reactive *t*-butyl alcohol is used in the proportion of 21 mL/g of metal. **(CAUTION: Potassium metal can form explosive peroxides. Metal that has formed a yellow oxide coating from exposure to air should not be cut with a knife, even when wet with a hydrocarbon, because an explosion can be promoted.)** If the potassium is dissolving too slowly, a few percent of methanol can be added gradually to the refluxing *t*-butyl alcohol. Oxide-coated potassium sticks should be put directly into the flask and decomposed with *t*-butyl alcohol. The decomposition will require considerable time because of the low surface/volume ratio of the metal sticks.

Lithium metal can be treated by the same procedure, but using 30 mL of 95% ethanol per gram of lithium. The rate of dissolution is slower than that of sodium.

7.D.3.5 Metal Catalysts

Metal catalysts such as Raney nickel and other fine metal powders can be slurried into water; dilute hydrochloric acid is then added carefully until the solid dissolves. Depending on the metal and on local regulations, the solution can be discarded in the sanitary sewer or with other hazardous or nonhazardous solid waste. Precious metals should be recovered from this process.

7.D.3.6 Water-Reactive Metal Halides

Liquid halides, such as $TiCl_4$ and $SnCl_4$, can be added to well-stirred water in a round-bottomed flask cooled by an ice bath as necessary to keep the exothermic reaction under control. It is usually more convenient to add solid halides, such as $AlCl_3$ and $ZrCl_4$, to stirring water and crushed ice in a flask or beaker. The acidic solution can be neutralized and, depending on the metal and local regulations, discarded in the sanitary sewer or with other hazardous or nonhazardous solid waste.

7.D.3.7 Halides and Acid Halides of Nonmetals

Halides and acid halides such as PCl_3, PCl_5, $SiCl_4$, $SOCl_2$, SO_2Cl_2, and $POCl_3$ are water-reactive. The liquids can be hydrolyzed conveniently using 2.5 M sodium hydroxide by the procedure described earlier for acyl halides and anhydrides. These compounds are irritating to the skin and respiratory passages and, even more than most chemicals, require a good hood and skin protection when handling them. Moreover, PCl_3 may give off small amounts of highly toxic phosphine (PH_3) during hydrolysis.

Sulfur monochloride (S_2Cl_2) is a special case. It is hydrolyzed to a mixture of sodium sulfide and sodium sulfite, so that the hydrolyzate must be treated with hypochlorite as described earlier for sulfides before it can be flushed down the drain.

The solids of this class (e.g., PCl_5) tend to cake and fume in moist air and therefore are not conveniently hydrolyzed in a three-necked flask. It is preferable to add them to a 50% excess of 2.5 M sodium hydroxide solution in a beaker or wide-mouth flask equipped with a stirrer and half-filled with crushed ice. If the solid has not all dissolved by the time the ice has melted and the stirred mixture has reached room temperature, the reaction can be completed by heating on a steam bath, and then the acidic solution neutralized and disposed of in the sanitary sewer.

7.D.3.8 Inorganic Ions

Many inorganic wastes consist of a cation (metal or metalloid atom) and an anion (which may or may not contain a metalloid component). It is often helpful to examine the cationic and anionic parts of the substance separately to determine whether either possesses a hazard.

If a substance contains a "heavy metal," it is often assumed that it is highly toxic. While salts of some heavy metals, such as lead, thallium, and mercury, are highly toxic, those of others, such as gold and tantalum, are not. On the other hand, compounds of beryllium, a "light metal," are highly toxic. In Table 7.1, cations of metals and metalloids are listed alphabetically in two groups: those whose toxic properties as described in the toxicological literature present a significant haz-

ard, and those whose properties do not. The basis for separation is relative, and the separation does not imply that those in the second list are "nontoxic." Similarly, Table 7.2 lists anions according to their level of toxicity and other dangerous properties, such as strong oxidizing power (e.g., perchlorate), flammability (e.g., amide), water reactivity (e.g., hydride), and explosivity (e.g., azide).

Materials that pose a hazard because of significant radioactivity are outside the scope of this volume, although they may be chemically treated in a manner similar to the nonradioactive materials discussed in this chapter. Their handling and disposal are highly regulated in most countries. Low-level radioactive mixed waste is discussed in section 7.C above.

7.D.3.8.1 Chemicals in Which Neither the Cation Nor the Anion Presents a Significant Hazard

Chemicals in which neither the cation nor the anion presents a significant hazard consist of those chemicals composed of ions from the right-hand columns of Tables 7.1 and 7.2. Those that are soluble in water to the extent of a few percent can usually be disposed of in the sanitary sewer. Only laboratory quantities should be disposed of in this manner, and at least 100 parts of water per part of chemical should be used. Local regulations should be checked for possible restrictions. Dilute slurries of insoluble materials, such as calcium sulfate or aluminum oxide, also can be handled in this way, provided the material is finely divided and not contaminated with tar, which might clog the piping. Some incinerators can handle these chemicals. If time and space permit, dilute aqueous solutions can be boiled down or allowed to evaporate to leave only a sludge of the inorganic solid for landfill disposal. However, appropriate precautions, including the use of traps, must be considered to ensure that toxic or other prohibited materials are not released to the atmosphere.

An alternative procedure is to precipitate the metal ion by the agent recommended in Table 7.1. The precipitate can often be disposed of in a secure landfill. The most generally applicable procedure is to precipitate the cation as the hydroxide by adjusting the pH to the range shown in Table 7.3.

7.D.3.8.2 Precipitation of Cations as Their Hydroxides

Because the pH range for precipitation varies greatly among metal ions, it is important to control it carefully. The aqueous solution of the metal ion is adjusted to the recommended pH (Table 7.3) by addition of a solu-

TABLE 7.1 High- and Low-Toxicity Cations and Preferred Precipitants

High Toxic Hazard		Low Toxic Hazard	
Cation	Precipitant[a]	Cation	Precipitant[a]
Antimony	OH^-, S^{2-}	Aluminium	OH^-
Arsenic	S^{2-}	Bismuth	OH^-, S^{2-}
Barium	SO_4^{2-}, CO_3^{2-}	Calcium	SO_4^{2-}, CO_3^{2-}
Beryllium	OH^-	Cerium	OH^-
Cadmium	OH^-, S^{2-}	Cesium	
Chromium(III)[b]	OH^-	Copper[c]	OH^-, S^{2-}
Cobalt(II)[b]	OH^-, S^{2-}	Gold	OH^-, S^{2-}
Gallium	OH^-	Iron[c]	OH^-, S^{2-}
Germanium	OH^-, S^{2-}	Lanthanides	OH^-
Hafnium	OH^-	Lithium	
Indium	OH^-, S^{2-}	Magnesium	OH^-
Iridium[d]	OH^-, S^{2-}	Molybdenum(VI)[b,e]	
Lead	OH^-, S^{2-}	Niobium(V)	OH^-
Manganese(II)[b]	OH^-, S^{2-}	Palladium	OH^-, S^{2-}
Mercury	OH^-, S^{2-}	Potassium	
Nickel	OH^-, S^{2-}	Rubidium	
Osmium(IV)[b,f]	OH^-, S^{2-}	Scandium	OH^-
Platinum(II)[b]	OH^-, S^{2-}	Sodium	
Rhenium(VII)[b]	S^{2-}	Strontium	SO_4^{2-}, CO_3^{2-}
Rhodium(III)[b]	OH^-, S^{2-}	Tantalum	OH^-
Ruthenium(III)[b]	OH^-, S^{2-}	Tin	OH^-, S^{2-}
Selenium	S^{2-}	Titanium	OH^-
Silver[d]	Cl^-, OH^-, S^{2-}	Yttrium	OH^-
Tellurium	S^{2-}	Zinc[c]	OH^-, S^{2-}
Thallium	OH^-, S^{2-}	Zirconium	OH^-
Tungsten(VI)[b,e]			
Vanadium	OH^-, S^{2-}		

[a]Precipitants are listed in order of preference: OH^-, CO_3^{2-} = base (sodium hydroxide or sodium carbonate), S^{2-} = sulfide, SO_4^{2-} = sulfate, and Cl^- = chloride.

[b]The precipitant is for the indicated valence state.

[c]Very low maximum tolerance levels have been set for these low-toxicity ions in some countries, and large amounts should not be put into public sewer systems. The small amounts typically used in laboratories will not normally affect water supplies, although they may be prohibited by the local pubicly owned treatment works (POTW).

[d]Recovery of these rare and expensive metals may be economically favorable.

[e]These ions are best precipitated as calcium molybdate(VI) or calcium tungstate(VI).

[f]CAUTION: Osmium tetroxide, OsO_4, a volatile, extremely poisonous substance, is formed from almost any osmium compound under acid conditions in the presence of air. Reaction with corn oil or powdered milk will destroy it.

tion of 1 M sulfuric acid, or 1 M sodium hydroxide or carbonate. The pH can be determined over the range 1 through 10 by use of pH test paper.

The precipitate is separated by filtration, or as a heavy sludge by decantation, and packed for disposal. Some gelatinous hydroxides are difficult to filter. In such cases, heating the mixture close to 100 °C or stirring with diatomaceous earth, approximately 1 to 2 times the weight of the precipitate, often facilitates filtration.

As shown in Table 7.1, precipitants other than a base may be superior for some metal ions, such as sulfuric acid for calcium ion. For some ions, the hydroxide precipitate will redissolve at a high pH (Table 7.3). For a number of metal ions the use of sodium carbonate will result in precipitation of the metal carbonate or a mixture of hydroxide and carbonate.

7.D.3.8.3 Chemicals in Which the Cation Presents a Relatively High Hazard from Toxicity

In general, waste chemicals containing any of the cations listed as highly hazardous in Table 7.1 can be precipitated as their hydroxides or oxides. Alternatively, many can be precipitated as insoluble sulfides by treatment with sodium sulfide in neutral solution (Table 7.4). Several sulfides will redissolve in excess sulfide ion, and so it is important that the sulfide ion concentration be controlled by adjustment of the pH.

Precipitation as the hydroxide is achieved as described above. Precipitation as the sulfide is accomplished by adding a 1 M solution of sodium sulfide to the metal ion solution, and then adjusting the pH to

TABLE 7.2 High- and Low-Hazard Anions and Preferred Precipitants

High-Hazard Anions

Ion	Hazard Type[a]	Precipitant	Low-Hazard Anions
Aluminium hydride, AlH_4^-	F, W	—	Bisulfite, HSO_3^-
Amide, NH_2^-	F, E[b]	—	Borate, BO_3^{3-}, $B_4O_7^{2-}$
Arsenate, AsO_3^-, AsO_4^{3-}	T	Cu^{2+}, Fe^{2+}	Bromide, Br^-
Arsenite, AsO_2^-, AsO_3^{3-}	T	Pb^{2+}	Carbonate, CO_3^{2-}
Azide, N_3^-	E, T	—	Chloride, Cl^-
Borohydride, BH_4^-	F	—	Cyanate, OCN^-
Bromate, BrO_3^-	O, F, E	—	Hydroxide, OH^-
Chlorate, ClO_3^-	O, E	—	Iodide, I^-
Chromate, CrO_4^{2-}, $Cr_2O_7^{2-}$	T, O	c	Oxide, O^-
Cyanide, CN^-	T	—	Phosphate, PO_4^{3-}
Ferricyanide, $\{Fe(CN)_6\}^{3-}$	T	Fe^{2+}	Sulfate, SO_4^{2-}
Ferrocyanide, $\{FE(CN)_6\}^{4-}$	T	Fe^{3+}	Sulfite, SO_3^{2-}
Fluoride, F^-	T	Ca^{2+}	Thiocyanate, SCN^-
Hydride, H^-	F, W	—	
Hydroperoxide, O_2H^-	O, E	—	
Hydrosulfide, SH^-	T	—	
Hypochlorite, OCl^-	O	—	
Iodate, IO_3^-	O, E	—	
Nitrate, NO_3^-	O	—	
Nitrite, NO_2^-	T, O	—	
Perchlorate, ClO_4^-	O, E	—	
Permanganate, MnO_4^-	T, O	—	
Peroxide, O_2^{2-}	O, E	d	
Persulfate, $S_2O_8^{2-}$	O	—	
Selenate, SeO_4^{2-}	T	Pb^{2+}	
Selenide, Se^{2-}	T	Cu^{2+}	
Sulfide, S^{2-}	T	e	

[a]T = toxic; O = oxidant; F = flammable; E = explosive; and W = water reactive.
[b]Metal amides readily form explosive peroxides on exposure to air.
[c]Reduce and precipitate as Cr(III).
[d]Reduce and precipitate as Mn(II); see Table 7.1.
[e]See Table 7.4.

neutral with 1 M sulfuric acid. **(CAUTION: Avoid acidifying the mixture because hydrogen sulfide could be formed.)** The precipitate is separated by filtration or decantation and packed for disposal. Excess sulfide ion can be destroyed by the addition of hypochlorite to the clear aqueous solution.

The following ions are most commonly found as oxyanions and are not precipitated by base: As^{3+}, As^{5+}, Re^{7+}, Se^{4+}, Se^{6+}, Te^{4+}, and Te^{6+}. These elements can be precipitated from their oxyanions as the sulfides by the above procedure. Oxyanions of Mo^{6+} and W^{6+} can be precipitated as their calcium salts by the addition of calcium chloride. Some ions can be absorbed by passing their solutions over ion-exchange resins. The resins can be landfilled, and the effluent solutions poured down the drain.

Another class of compounds whose cations may not be precipitated by the addition of hydroxide ions are the most stable complexes of metal cations with Lewis bases, such as ammonia, amines, and tertiary phosphines. Because of the large number of these compounds and their wide range of properties, it is not possible to give a general procedure for separating the cations. In many cases, metal sulfides can be precipitated directly from aqueous solutions of the complexes by the addition of aqueous sodium sulfide. If a test-tube experiment shows that other measures are needed, the addition of hydrochloric acid to produce a slightly acidic solution will often decompose the complex by protonation of the basic ligand. Metal ions that form insoluble sulfides under acid conditions can then be precipitated by dropwise addition of aqueous sodium sulfide.

A third option for this waste is incineration, provided that the incinerator ash is to be sent to a secure landfill. Incineration to ash reduces the volume of

TABLE 7.3 pH Ranges for Precipitation of Metal Hydroxides and Oxides

pH:	1	2	3	4	5	6	7	8	9	10

Cation	Precipitation pH range
Ag^{1+}	9 → 1N
Al^{3+}	7–8
As^{3+}	Not precipitated (precipitate as sulfide)
As^{5+}	Not precipitated (precipitate as sulfide)
Au^{3+}	7–8
Be^{2+}	7–8
Bi^{3+}	7 → 1N
Cd^{2+}	7 → 1N
Co^{2+}	8 → 1N
Cr^{3+}	7 → 1N
Cu^{1+}	9 → 1N
Cu^{2+}	7 → 1N
Fe^{2+}	7 → 1N
Fe^{3+}	7 → 1N
Ga^{3+}	7–8
Ge^{4+}	6–8
Hf^{4+}	6–7
Hg^{1+}	8 → 1N
Hg^{2+}	8 → 1N
In^{3+}	6 → pH13
Ir^{4+}	6–8
Mg^{2+}	9 → 1N
Mn^{2+}	8 → 1N
Mn^{4+}	7 → 1N
Mo^{6+}	Not precipitated (precipitate as Ca salt)
Nb^{5+}	1 → 10
Ni^{2+}	8 → 1N
Os^{4+}	7–8
Pb^{2+}	7–8
Pd^{2+}	7–8
Pd^{4+}	7–8
Pt^{2+}	7–8
Re^{3+}	6 → 1N
Re^{7+}	Not precipitated (precipitate as sulfide)
Rh^{3+}	7–8
Ru^{3+}	7 → 1N
Sb^{3+}	7–8
Sb^{5+}	7–8
Sc^{3+}	8 → 1N
Se^{4+}	Not precipitated (precipitate as sulfide)
Se^{6+}	Not precipitated (precipitate as sulfide)
Sn^{2+}	7–8
Sn^{4+}	7–8
Ta^{5+}	1 → 10
Te^{4+}	Not precipitated (precipitate as sulfide)
Te^{6+}	Not precipitated (precipitate as sulfide)
Th^{4+}	7 → 1N
Ti^{3+}	8 → 1N
Ti^{4+}	8 → 1N
Tl^{3+}	9 → 1N
V^{4+}	7–8
V^{5+}	7–8
W^{6+}	Not precipitated (precipitate as Ca salt)
Zn^{2+}	8–9
Zr^{4+}	6–7

NOTE: Most metal ions are precipitated as hydroxides or oxides at high pH. However, many precipitates will redissolve in excess base. For this reason, it is necessary to control pH closely in a number of cases. This table shows the recommended pH range for precipitating many cations in their most common oxidation state. The notation "1 N" in the right-hand column indicates that the precipitate will not dissolve in 1 N sodium hydroxide (pH 14).

SOURCES: Erdey (1965) and Burns et al. (1981).

TABLE 7.4 Precipitation of Sulfides

Precipitated at pH 7	Not Precipitated at Low pH	Soluble Complex at High pH
Ag^+		
As^{3+a}		x
Au^{+a}		x
Bi^{3+}		
Cd^{2+}		
Co^{2+}	x	
Cr^{3+a}		
Cu^{2+}		
Fe^{2+a}	x	
Ge^{2+}		x
Hg^{2+}		x
In^{3+}	x	
Ir^{4+}		x
Mn^{2+a}	x	
Mo^{3+}		x
Ni^{2+}	x	
Os^{4+}		
Pb^{2+}		
Pd^{2+a}		
Pt^{2+a}		x
Re^{4+}		
Rh^{2+a}		
Ru^{4+}		
Sb^{3+a}		x
Se^{2+}		x
Sn^{2+}		x
Te^{4+}		x
Tl^{+a}	x	
V^{4+a}		
Zn^{2+}	x	

NOTE: Precipitation of ions listed without an x is usually not pH-dependent.

[a]Higher oxidation states of this ion are reduced by sulfide ion and precipitated as this sulfide.

SOURCE: Swift and Schaefer (1961).

waste going to a landfill. Waste that contains mercury, thallium, gallium, osmium, selenium, or arsenic should not be incinerated because volatile, toxic combustion products may be emitted.

7.D.3.8.4 Chemicals in Which an Anion Presents a Relatively High Hazard

The more common dangerous anions are listed in Table 7.2. Many of the comments made above about the disposal of dangerous cations apply to these anions. The hazard associated with some of these anions is their reactivity or potential to explode, which makes them unsuitable for landfill disposal. Most chemicals containing these anions can be incinerated, but strong oxidizing agents and hydrides should be introduced into the incinerator only in containers of

not more than a few hundred grams. Incinerator ash from anions of chromium or manganese should be transferred to a secure landfill.

Some of these anions can be precipitated as insoluble salts for landfill disposal, as indicated in Table 7.2. Small amounts of strong oxidizing agents, hydrides, cyanides, azides, metal amides, and soluble sulfides or fluorides can be converted into less hazardous substances in the laboratory before being disposed of. Suggested procedures are presented in the following paragraphs.

7.D.3.8.5 Procedure for Reduction of Oxidizing Salts

Hypochlorites, chlorates, bromates, iodates, periodates, inorganic peroxides and hydroperoxides, persulfates, chromates, molybdates, and permanganates can be reduced by sodium hydrogen sulfite. A dilute solution or suspension of a salt containing one of these anions has its pH reduced to less than 3 with sulfuric acid, and a 50% excess of aqueous sodium hydrogen sulfite is added gradually with stirring at room temperature. An increase in temperature indicates that the reaction is taking place. If the reaction does not start on addition of about 10% of the sodium hydrogen sulfite, a further reduction in pH may initiate it. Colored anions (e.g., permanganate and chromate) serve as their own indicators of completion of the reduction. The reduced mixtures can often be washed down the drain. However, if large amounts of permanganate have been reduced, it may be necessary to transfer the manganese dioxide to a secure landfill, possibly after a reduction in volume by concentration or precipitation. *Do not dispose of chromium salts in the sanitary sewer.*

Hydrogen peroxide can be reduced by the sodium hydrogen sulfite procedure or by ferrous sulfate as described earlier for organic hydroperoxides. However, it is usually acceptable to dilute it to a concentration of less than 3% and dispose of it in the sanitary sewer. Solutions with a hydrogen peroxide concentration greater than 30% should be handled with great care to avoid contact with reducing agents, including all organic materials, or with transition metal compounds, which can catalyze a violent reaction.

Concentrated perchloric acid (particularly when stronger than 60%) must be kept away from reducing agents, including weak ones such as ammonia, wood, paper, plastics, and all other organic substances, because it can react violently with them. Dilute perchloric acid is not reduced by common laboratory reducing agents such as sodium hydrogen sulfite, hydrogen sulfide, hydriodic acid, iron, or zinc. Perchloric acid is most easily disposed of by stirring it gradually into enough cold water to make its concentration less than 5%, neutralizing it with aqueous sodium hydroxide, and washing the solution down the drain with a large excess of water.

Nitrate is most dangerous in the form of concentrated nitric acid (70% or higher), which is a potent oxidizing agent for organic materials and all other reducing agents. It can also cause serious skin burns. Dilute aqueous nitric acid is not a dangerous oxidizing agent and is not easily reduced by common laboratory reducing agents. Dilute nitric acid should be neutralized with aqueous sodium hydroxide before disposal down the drain; concentrated nitric acid should be diluted carefully by adding it to about 10 volumes of water before neutralization. Metal nitrates are generally quite soluble in water. Those of the metals listed in Table 7.1 as having a low toxic hazard, as well as ammonium nitrate, should be kept separate from oil or other organic materials because on heating such a combination, fire or explosion can occur. Otherwise, these can be treated as chemicals that present no significant hazard.

Nitrites in aqueous solution can be destroyed by adding about 50% excess aqueous ammonia and acidifying with hydrochloric acid to pH 1:

$$HNO_2 + NH_3 \rightarrow N_2 + 2H_2O$$

8 Laboratory Facilities

8.A INTRODUCTION

Laboratory workers must understand how the facilities designed for chemical laboratories operate. They should be familiar with ventilation systems, environmental controls, fume hoods. and other exhaust devices associated with such equipment. In order to reduce accidents, the experimental work should be viewed in the context of the entire laboratory and its facilities, both for safety and efficiency.

Many laboratory operations depend on special utility requirements that have become common, such as "clean" electric power, high voltages, high-volume water or cooling, and special gas services. Larger and/or special exhaust devices are also increasingly necessary. The large scale of some experiments may require special laboratory configurations and hazard containment measures; thus, these factors need careful consideration during the experimental design stage. Further, enhanced measures to protect laboratory workers from exposure to potential dangers, coupled with steps to conserve energy and minimize waste, have triggered the implementation of sophisticated systems to provide a safe, comfortable, and cost-effective work environment. Individuals who wish to gain the full benefits that these systems offer must know how they work.

Inspection programs are an important component of maintaining both the physical infrastructure of the laboratory and good relationships with environmental health and safety support staff and facility engineering and maintenance staff. These experts can address specific questions about operating equipment that is peculiar to a particular laboratory and can give expert advice before specialized equipment is ordered.

Increasing demands are being placed on laboratory staff to work safely. As experiments have become increasingly elaborate, so has the infrastructure to support them. The laboratory worker should be kept abreast of changes in the working environment, should keep relations with environmental, engineering, and maintenance staff friendly, and should always confer with them about newly acquired equipment. Institutions must make every effort to maximize the quality of training in this regard and provide opportunities for workers to maintain a safe and successful workplace as changes are introduced in the laboratory.

8.B LABORATORY INSPECTION PROGRAMS

Good housekeeping practices supplemented by a program of periodic laboratory inspections will help to keep laboratory facilities and equipment in a safe operating condition. Inspections can safeguard the quality of the institution's laboratory safety program.

A variety of inspection protocols may be used, and the organization's management should select and participate in the design of the inspection program appropriate for that institution's unique needs. The program should embrace goals to:

- maintain laboratory facilities and equipment in a safe operating condition,
- provide a comfortable and safe working environment for all employees and the public, and
- ensure that all laboratory procedures and experiments are conducted in a safe and prudent manner.

These goals should be approached with a considerable degree of flexibility. The different types of inspections, the frequency with which they are conducted, and who conducts them should be considered. A discussion of items to inspect and several possible inspection protocols follows, but neither list is all-inclusive.

8.B.1 Items to Include in an Inspection Program

The following list is representative, not exhaustive. Depending on the laboratory and the type of work conducted in it, other items may also be targeted for inspection. A typical maintenance inspection might consider the following potential hazards:

- Keep water in drain traps, particularly for floor drains or sinks used infrequently. Vapors emitted from dry, unsealed drains may cause an explosive or flammable condition; such vapors are also the most common source of unexplained laboratory odors. Running water into a drain for 20 to 30 seconds is usually sufficient to fill a drain trap.
- Secure plastic or rubber hose connections. If proper clamps are not used, hoses can slip off fittings, causing serious floods and water damage. Backflow preventers often require periodic testing to comply with local building codes. Locking quick-disconnects are available for locations where lines often need to be disconnected.
- Routinely check for inadequate or defective wiring. Frayed cords, improper connections, inappropriate use of extension cords, and power and control wires routed in the same path are found most commonly. Circuit breakers and electrical protection devices should also be checked. These problems create the potential for fires and electrocution.
- Inspect for improper gas tubing, and faulty valve and regulator installations. These types of errors can be difficult to identify but are important because they pose significant hazards in systems where toxic or flammable gases are used. The typical items to inspect

for should include incorrect fittings (incompatible materials), installation errors (e.g., excessive tightness), and defective or missing regulators, flow controls, and monitoring devices. Also check for leaks in inert gas systems to avoid the cost for "lost" gases.

• Check proximity of flammable materials to any potential ignition sources. Open flames and devices that generate sparks should not be near flammables. Pay special attention to devices placed in fume hoods that do not meet National Electrical Safety Code (U.S. DOC, 1993) Division 1, Group C and D explosion-resistance specifications for electrical devices. Stirrers, hot plates, Variacs, heat tape, outlet strips, ovens (all types), refrigerators, flame sources (e.g., flame ionization detectors (FIDs) and atomic absorption spectrometers), and heat guns constitute the majority of devices that do not typically conform to these code requirements (see section 8.C.6.1).

• Check guards for rotating machinery and heating devices as protective measures for mechanical and thermal hazards. Test safety switches and emergency stops periodically. Inspect setups for unattended operation, shielding of high-pressure and vacuum equipment, and any other equipment hazards. Compliance with OSHA "lock-out/tag-out" regulations (29 CFR 1910.147) also needs to be verified.

• Periodically test and inspect emergency devices (e.g., safety showers and eyewash stations) and safety equipment (e.g., fire extinguishers, fire blankets, and first aid and spill control kits) to make sure they are functional. Inspectors need to verify that workers are using personal protection and safety equipment appropriately in their day-to-day work.

• Keep aisles and emergency exits free of obstacles. Unused supplies and equipment should be stored so as to avoid blocking exits.

• Carefully examine any documentation required by the institution. Such documentation, which should be made readily available, may include experiment plans, training plans and records, chemical and equipment hazard information, operating plans, and an up-to-date emergency evacuation plan. An emergency plan should always be prepared for the contingencies of ventilation failure (resulting from power failure, for example) and other emergencies, such as fire or explosion in a chemical fume hood.

8.B.2 Types of Inspection Programs: Who Conducts Them and What They Offer

The easiest inspection program to implement requires laboratory supervisors to inspect their own work space and equipment on a periodic basis. These individuals are the "first line of defense" for a program of safety excellence. They need to note items such as open containers, faulty faucets or valves, frayed wiring, broken apparatus, obstructions on floors or aisles, and unsafe clutter. They also need to follow through to make sure that any problems receive prompt resolution.

General equipment inspection should also be done fairly frequently. For certain types of equipment in constant use, such as gas chromatographs, daily inspections may be appropriate. Other types of equipment may need only weekly or monthly inspection, or inspection prior to use if operated infrequently.

The challenge for any inspection program is to keep laboratory workers continuously vigilant. Workers need positive encouragement to develop the habit of inspection and to adopt the philosophy that good housekeeping and maintenance for their work space protect them and may help them produce better results for their efforts. Incentive programs may stimulate workers to pay closer attention to the condition of their equipment and work space.

Probably the most traditional type of inspection is that conducted by the laboratory supervisor. This form of inspection presents an excellent opportunity to promote a culture of safety and prudence within an organization. The supervisor gains the opportunity to take a close look at the facilities and operations. He or she also can discuss with individual workers issues of interest or concern that may fall outside the scope of the actual inspection. Again, a constructive and positive approach to observed problems and issues will foster an attitude of cooperation and leadership with regard to safety. It can help build and reinforce a culture of teamwork and cooperation that has benefits far beyond protecting the people and physical facilities.

The International Loss Control Institute (ILCI), a private organization contracted to assess safety standards at industrial facilities, recommends a monthly inspection of industrial laboratory facilities and equipment by supervisory personnel. Some institutions may deem this frequency unnecessarily high. Supervisors should probably make inspections no less frequently than once per quarter.

More senior staff should be encouraged to inspect facilities periodically, too. In most cases, these inspections should occur at least annually. They offer the added benefit of providing a senior leader with a good overview of the condition of the facilities, the work conducted in the laboratories, who does what work, and how people feel about their work. In essence, these kinds of inspections supplement the normal practice of "management by walking around."

One of the most effective safety tools a larger institution can employ is periodic peer-level inspections. Usually, the people who fulfill this role work in the organization they serve, but not in the area being surveyed. Individuals may function on an ad hoc basis, or the institution may select specific individuals and confer on them various formal appellations such as "safety committee member." A peer inspection program has the intrinsic advantage of being perceived as less threatening than other forms of surveys or audits.

Peer inspections depend heavily on the knowledge and commitment of the people who conduct them. Individuals who volunteer or are selected to perform inspections for only a brief period of service may not learn enough about an operation or procedure to observe and comment constructively. On the other hand, people who receive involuntary appointments or who serve too long may not maintain the desired level of diligence. Peer-level inspectors generally perform best if they have some ongoing responsibility with regard to the safety program, such as consulting on experiment designs. They will be more familiar with the work and will conduct the review more effectively.

Having a high-quality peer-level inspection program may reduce the need for frequent inspections by supervisory personnel. However, peer inspections should not replace other inspections completely. Walk-throughs by the organization's leadership demonstrate commitment to the safety programs, which is key to their continuing success.

Another option is to have the organization's environmental health and safety staff conduct inspections. In a smaller organization, these types of inspections may adequately address the ILCI frequency recommendations. In larger institutions, the safety staff may be limited to semiannual or even annual walk-throughs. A more practical use of safety staff for inspections may be to target certain operations or experiments. Or the safety staff could focus on a particular type of inspection, such as safety equipment and systems. Finally, they could perform "audits" to check the work of other inspectors or look specifically at previous problem areas. It is important for the safety staff to address noted deficiencies with appropriate reminders and/or additional training. Punitive measures should be employed, but only for chronic offenders or deliberate problems that pose a serious potential hazard.

Safety staff are not the only nonlaboratory personnel who should conduct safety inspections. Facility engineers or maintenance personnel may add considerable value to safety inspection programs. It also gives them the opportunity to gain a better perspective on the laboratory work.

Other types of inspections and audits use individuals or groups outside the laboratory organization to conduct the survey. Inspections may be mandatory, such as annual site visits by EPA to hazardous waste generators, or they may be surprise inspections with 24 hours notice or less given prior to the visit. Organizations subject to such inspections must keep their programs and records up-to-date at all times. In fact, all organizations should strive for that goal. Any significant incident or accident within a facility will trigger one or more inspections and investigations by outside organizations. If the underlying safety programs are found to be sound, that factor may help limit negative findings and potential penalties.

Many different types of elective inspections or audits can be conducted by outside experts or organizations. They may inspect a particular facility, piece(s) of equipment, or procedure either during the pre-experiment design phase or during operations.

Tours, walk-throughs, and inspections by outside organizations offer the opportunity to build relationships with governmental agencies and the public. For example, an annual visit by the fire department serving a particular facility will acquaint personnel with the operations and the location of particular hazards. If these individuals are ever called into the facility to handle an emergency, their familiarity with the facility will give them a greater degree of effectiveness. During their walk-through, they may offer comments and suggestions for improvements. A relationship built over a period of time will help make this input positive and constructive.

If a pending operation or facility change may cause public attention and concern, an invitation targeted to specific people or groups may prevent problems. Holding public open houses from time to time can help build a spirit of support and trust. Many opportunities exist to apply this type of open approach to dealing with the public. An organization only needs to consider when to use it and what potential benefits may accrue.

> An industrial laboratory arranged an on-site visit for representatives from the Sierra Club and other local citizen action groups who had begun to raise objections to the approval of the site's RCRA Part B permit for a hazardous waste incinerator. Concerns about the noise, smoke, and smell from the incinerator were dispelled after the individuals stood right next to it and did not realize what it was and that it was operating at the time!

8.C LABORATORY VENTILATION

8.C.1 Laboratory Fume Hoods

Laboratory fume hoods are the most important components used to protect laboratory workers from exposure to hazardous chemicals and agents used in the laboratory. Functionally, a standard fume hood is a fire- and chemical-resistant enclosure with one opening (face) in the front with a movable window (sash) to allow user access into the interior. Large volumes of air are drawn through the face and out the top to contain and remove contaminants from the laboratory.

8.C.2 Fume Hood Face Velocity

The average velocity of the air drawn through the face of the hood is called the face velocity. It is generally calculated as the total volumetric exhaust flow rate for the hood, divided by the area of the open face, less an adjustment for hood air leakage. The face velocity of a hood greatly influences the ability of the hood to contain hazardous substances, that is, its containment efficiency. Face velocities that are too low or too high will reduce the containment efficiency of a fume hood. In most cases, the recommended face velocity is between 80 and 100 feet per minute (fpm). Face velocities between 100 and 120 fpm may be used for substances of very high toxicity or where outside influences adversely affect hood performance. However, energy costs to operate the fume hood are directly proportional to the face velocity. Face velocities approaching or exceeding 150 fpm should not be used, because they may cause turbulence around the periphery of the sash opening and actually reduce the capture efficiency of the fume hood.

Average face velocity is determined either by measuring individual points across the plane of the sash opening and calculating their average or by measuring the hood volume flow rate with a pitot tube in the exhaust duct and dividing this rate by the open face area. Containment may be verified by using one of the flow visualization techniques found in section 8.C.5 on fume hood testing. Once the acceptable average face velocity, minimum acceptable velocity, and maximum standard deviation of velocities have been determined for a hood, laboratory, facility, or site, they should be incorporated into the laboratory's Chemical Hygiene Plan.

8.C.3 Factors That Affect Fume Hood Performance

Tracer gas containment testing of fume hoods has revealed that air currents impinging on the face of a hood at a velocity exceeding 30 to 50% of the hood face velocity will reduce the containment efficiency of the hood by causing turbulence and interfering with the laminar flow of the air entering the hood. Thirty to fifty percent of a hood face velocity of 100 fpm, for example, is 30 to 50 fpm, which represents a *very* low velocity that can be produced in many ways. The rate of 20 fpm is considered to be still air because that is the velocity at which most people first begin to sense air movement.

8.C.3.1 Adjacency to Traffic

Most people walk at a velocity of approximately 250 fpm (about 3 miles per hour). Wakes or vortices form behind a person who is walking, and velocities in those vortices exceed 250 fpm. When a person walks in front of an open fume hood, the vortices can overcome the fume hood face velocity and pull contaminants out of the fume hood, into the vortex, and into the laboratory. Therefore, fume hoods should not be located on heavily traveled aisles, and those that are should be kept closed when not in use. Foot traffic near these hoods should be avoided, or special care should be taken.

8.C.3.2 Adjacency to Supply Air Diffusers

Air is supplied continuously to laboratories to replace the air exhausted from the fume hoods and other exhaust sources and to provide ventilation and temperature/humidity control. This air usually enters the laboratory through devices called supply air diffusers located in the ceiling. Velocities that can exceed 800 fpm are frequently encountered at the face of these diffusers. If air currents from these diffusers reach the face of a fume hood before they decay to 30 to 50% of the hood face velocity, they can cause the same effect as air currents produced by a person walking in front of the hood. Normally, the effect is not quite as pronounced as the traffic effect, but it occurs constantly, whereas the traffic effect is transient. Relocating the diffuser, replacing it with another type, or rebalancing the diffuser air volumes in the laboratory can alleviate this problem.

8.C.3.3 Adjacency to Windows and Doors

Exterior windows with movable sashes are not recommended in laboratories. Wind blowing through the windows and high-velocity vortices caused when doors open can strip contaminants out of the fume hoods and interfere with laboratory static pressure controls.

8.C.4 General Safe Operating Procedures for Fume Hoods

In addition to protecting the laboratory worker from toxic or unpleasant agents used in them, fume hoods can provide an effective containment device for accidental spills of chemicals. There should be at least one hood for every two workers in laboratories where most work involves hazardous chemicals, and the hoods should be large enough to provide each worker with at least 2.5 linear feet of working space at the face. If this amount of hood space is not available, other types of local ventilation should be provided, and special care should be exercised to monitor and restrict the use of hazardous substances.

8.C.4.1 Prevention of Intentional Release of Hazardous Substances into Fume Hoods

Fume hoods should be regarded as backup safety devices that can contain and exhaust toxic, offensive, or flammable materials when the containment of an experiment or procedure fails and vapors or dusts escape from the apparatus being used. Note the following:

- Just as you should never flush a laboratory waste down the drain, never intentionally send waste up the hood.
- Instead, fit all apparatus used in hoods with condensers, traps, or scrubbers to contain and collect waste solvents or toxic vapors or dusts.

QUICK GUIDE
GUIDELINES FOR MAXIMIZING HOOD EFFICIENCY

Many factors can compromise the efficiency of a hood operation. Most of these are avoidable; thus, it is important to be aware of all behavior that can, in some way, modify the hood and its capabilities. The following should always be considered when using a hood:

- Keep fume hood exhaust fans on at all times.
- If possible, position the fume hood sash so that work is performed by extending the arms under or around the sash, placing the head in front of the sash, and keeping the glass between the worker and the chemical source. The worker views the procedure through the glass, which will act as a primary barrier if a spill, splash, or explosion should occur.
- Avoid opening and closing the fume hood sash rapidly, and avoid swift arm and body movements in front of or inside the hood. These actions may increase turbulence and reduce the effectiveness of fume hood containment.
- Place chemical sources and apparatus at least 6 inches behind the face of the hood. In some laboratories, a colored stripe is painted on, or tape applied to, the hood work surface 6 inches back from the face to serve as a reminder. Quantitative fume hood containment tests reveal that the concentration of contaminant in the breathing zone can be 300 times higher from a source located at the front of the hood face than from a source placed at least 6 inches back. This concentration declines further as the source is moved farther toward the back of the hood.
- Place equipment as far to the back of the hood as practical without blocking the bottom baffle.
- Separate and elevate each instrument by using blocks or racks so that air can flow easily around all apparatus.
- Do not use large pieces of equipment in a hood, because they tend to cause dead spaces in the airflow and reduce the efficiency of the hood.
- If a large piece of equipment emits fumes or heat outside a fume hood, then have a special-purpose hood designed and installed to ventilate that particular device. This method of ventilation is much more efficient than placing the equipment in a fume hood, and it will consume much less air.
- Do not modify fume hoods in any way that adversely affects the hood performance. This includes adding, removing, or changing any of the fume hood components, such as baffles, sashes, airfoils, liners, and exhaust connections.

• Make sure that all highly toxic or offensive vapors are scrubbed or adsorbed before the exit gases are released into the hood exhaust system (see section 8.C.8.1 on fume hood scrubbers).

8.C.4.2   Fume Hood Performance Checks

It is necessary to check if the hoods are performing properly. Observe the following guidelines:

• Evaluate each hood before use and on a regular basis (preferably once a year) to verify that the face velocity meets the criteria specified for it in the laboratory's Chemical Hygiene Plan.
• Also verify the absence of excessive turbulence (see section 8.C.4.4 below).
• Make sure that a continuous monitoring device for adequate hood performance is present, and check it every time the hood is used.

(For further information, see section 8.C.5 on testing and verification.)

8.C.4.3   Housekeeping

Laboratory fume hoods and adjacent work areas should be kept clean and free of debris at all times. Solid objects and materials (such as paper) should be kept from entering the exhaust ducts of hoods, because they can lodge in the ducts or fans and adversely affect their operation. Also, the hood will have better airflow across its work surface if there are minimal numbers of bottles, beakers, and laboratory apparatus inside the hood; therefore, it is prudent to keep *unnecessary* equipment and glassware outside of the hood at all times and store all chemicals in approved storage cans, containers, or cabinets (not in the fume hood). Furthermore, it is best to keep the work space neat and clean in all laboratory operations, particularly those involving the use of hoods, so that any procedure or experiment can be undertaken without the possibility of disturbing, or even destroying, what is being done.

8.C.4.4   Sash Operation

Except when adjustments to the apparatus are being made, the hood should be kept closed, with vertical sashes down and horizontal sashes closed, to help prevent the spread of a fire, spill, or other hazard into the laboratory. Sliding sashes should not be removed from horizontal sliding sash hoods. The face opening of the hood should be kept small to improve the overall performance of the hood. If the face velocity becomes

excessive, the facility engineers should make adjustments or corrections.

For hoods without face velocity controls (see section 8.C.6.3.2), the sash should be positioned to produce the recommended face velocity, which often occurs only over a limited range of sash positions. This range should be determined and marked during fume hood testing. For hoods with face velocity controls, it is imperative to keep the sash closed when the hood is not in use.

8.C.4.5   Constant Operation of Fume Hoods

Although turning fume hoods off when not in use saves energy, keeping them on at all times is safer, especially for fume hoods connected directly to a single fan. Because most laboratory facilities are under negative pressure, air may be drawn backward through the nonoperating fan, down the duct, and into the laboratory unless an ultralow-leakage backdraft damper is used in the duct. If the air is cold, it may freeze liquids in the hood. Fume hood ducts are rarely insulated; therefore, condensation and ice may form in cold weather. When the fume hood is turned on again and the duct temperature rises, the ice will melt, and water will run down the ductwork, drip into the hood, and possibly react with chemicals in the hood.

Fume hoods connected to a common exhaust manifold offer an advantage. The main exhaust system will rarely be shut down; hence, positive ventilation is available to each hood on the system at all times. In a constant air volume (CAV) system (see section 8.C.6.3.1), "shutoff" dampers to each hood can be installed, allowing passage of enough air to prevent fumes from leaking out of the fume hoods and into the laboratory when the sash is closed. It is prudent to allow 10 to 20% of the full volume of the hood flow to be drawn through the hood in the off position to prevent excessive corrosion.

8.C.5   Testing and Verification

All fume hoods should be tested, before they leave the manufacturer, by using ASHRAE/ANSI standard 110, Methods of Testing Performance of Laboratory Fume Hoods. The hood should pass the low- and high-volume smoke challenges with no leakage or flow reversals and have a control level of 0.05 parts per million (ppm) or less on the tracer gas test. ASHRAE/ANSI 110 testing of fume hoods after installation in their final location by trained personnel is highly recommended. The control level of tracer gas for an "as installed" or "as used" test via the ASHRAE/ANSI 110 method should not exceed 0.1 ppm. Periodic performance testing consisting of a face velocity analysis and flow visualization using smoke tubes, bombs, or fog generators should be performed annually. Laboratory workers should request a fume hood performance evaluation any time there is a change in any aspect of the ventilation system. Thus, changes in the total volume of supply air, changes in the locations of supply air diffusers, or the addition of other auxiliary local ventilation devices (e.g., more hoods, vented cabinets, and snorkels) all call for reevaluation of the performance of all hoods in the laboratory.

The ASHRAE/ANSI 110 test is the most practical way to determine fume hood capture efficiency quantitatively. The test includes several components, which may be used together or separately, including face velocity testing, flow visualization, face velocity controller response testing, and tracer gas containment testing. Performance should be evaluated against the design specifications for uniform airflow across the hood face as well as for the total exhaust air volume. Equally important is the evaluation of operator exposure. The first step in the evaluation of hood performance is the use of a smoke tube or similar device to determine that the hood is on and exhausting air. The second step is to measure the velocity of the airflow at the face of the hood. The third step is to determine the uniformity of air delivery to the hood face by making a series of face velocity measurements taken in a grid pattern.

Traditional hand-held instruments are subject to probe movement and positioning errors as well as reading errors owing to the optimistic bias of the investigator. Also, the traditional method yields only a snapshot of the velocity data, and no measure of variation over time is possible. To overcome this limitation, it is recommended that velocity data be taken while using a velocity transducer connected to a data acquisition system and read continuously by a computer for approximately 30 seconds at each traverse point. If the transducer is fixed in place, using a ring stand or similar apparatus, and is properly positioned and oriented, this method can overcome the errors and drawbacks associated with the traditional method. The variation in data for a traverse point can then be used as an indicator of turbulence, an important additional performance indicator that has been almost completely overlooked in the past. If the standard deviation of the average velocity profile at each point exceeds 20% of the mean, or the average standard deviation of velocities at each traverse point (turbulence) exceeds 15% of the mean face velocity, corrections should be made by adjusting the interior hood baffles and, if necessary, by altering the path of the supply air flowing into the room. Most laboratory hoods are equipped with a baffle that has movable slot openings at both the top and the bottom, which should be moved until the airflow is essentially uniform. Larger hoods may require additional slots in the baffle to achieve uniform airflow across the hood face. These adjustments should be made by an experienced laboratory ventilation engineer or technician using proper instrumentation.

The total volume of air exhausted by a hood is the sum of the face volume (average face velocity times face area of the hood) plus air leakage, which averages about 5 to 15% of the face volume. If the hood and the general ventilating system are properly designed, face velocities in the range of 80 to 100 fpm will provide a laminar flow of air over the work surface and sides of the hood. Higher face velocities (150 fpm or more), which exhaust the general laboratory air at a greater rate, both waste energy and are likely to degrade hood performance by creating air turbulence at the hood face and within the hood, causing vapors to spill out into the laboratory.

Because a substantial amount of energy is required to supply tempered supply air to even a small hood, the use of hoods to store bottles of toxic or corrosive chemicals is a very wasteful practice, which can also, as noted above, seriously impair the effectiveness of the hood as a local ventilation device. Thus, it is preferable to provide separate vented cabinets for the storage of toxic or corrosive chemicals. The amount of air exhausted by such cabinets is much less than that exhausted by a properly operating hood. (Also see section 8.C.4.)

Perhaps the most meaningful (but also the most time-consuming and expensive) method for evaluating hood performance is to measure worker exposure while the hood is being used for its intended purpose. By using commercial personal air-sampling devices that can be worn by the hood user, worker exposure (both excursion peak and time-weighted average) can be measured by using standard industrial hygiene techniques. The criterion for evaluating the hood should be the desired performance (i.e., does the hood contain vapors and gases at the desired worker-exposure level?). A sufficient number of measurements should be made to define a statistically significant maximum exposure based on worst-case operating conditions. Direct-reading instruments are available for determining the short-term concentration excursions that may occur in laboratory hood use.

8.C.6 Fume Hood Design and Construction

When specifying a laboratory fume hood for use in a particular activity, the laboratory worker should be aware of all the design features of the hood. Assistance from an industrial hygienist, ventilation engineer, or laboratory consultant is recommended when deciding to purchase a fume hood.

8.C.6.1   General Design Recommendations

Laboratory fume hoods and the associated exhaust ducts should be constructed of nonflammable materials. They should be equipped with either vertical or horizontal sashes that can be closed. The glass within the sash should be either laminated safety glass that is at least 7/32 inch thick or other equally safe material that will not shatter if there is an explosion within the hood. The utility control valves, electrical receptacles, and other fixtures should be located outside the hood to minimize the need to reach within the hood proper. Other specifications regarding the construction materials, plumbing requirements, and interior design will vary, depending on the intended use of the hood. (See Chapter 6, sections 6.C.1.1 and 6.C.1.2.)

Although hoods are most commonly used to control concentrations of toxic vapors, they can also serve to dilute and exhaust flammable vapors. Although theoretically possible, it is extremely unlikely (even under most worst-case scenarios) that the concentration of flammable vapors will reach the lower explosive limit (LEL) in the exhaust duct. However, somewhere between the source and the exhaust outlet of the hood, the concentration will pass through the upper explosive limit (UEL) and the LEL before being fully diluted at the outlet. Both the hood designer and the user should recognize this hazard and eliminate possible sources of ignition within the hood and its ductwork if there is a potential for explosion. The use of duct sprinklers or other suppression methods in laboratory fume ductwork is not necessary, or desirable, in the vast majority of situations.

8.C.6.2   Special Design Features

There have been two major improvements in fume hood design—airfoils and baffles—since the fume hood was invented. Both features should be included on any new fume hoods that are purchased.

Airfoils built into the fume hood at the bottom and sides of the sash opening significantly reduce boundary turbulence and improve capture performance. All fume hoods purchased should be fitted with airfoils.

When air is drawn through a hood without a baffle (see Figure 8.1), most of the air is drawn through the upper part of the opening, producing an uneven velocity distribution across the face opening. When baffles are installed, the velocity distribution is greatly improved. All fume hoods should have baffles. Adjustable baffles can improve hood performance and are desirable if the adjustments are made by an experienced industrial hygienist, consultant, or hood technician.

8.C.6.3   Fume Hood Airflow Types

The first fume hoods were simply boxes that were open on one side and connected to an exhaust duct. Since they were first introduced, many variations on this basic design have been made. Six of the major variants in fume hood airflow design are listed below with their characteristics. Conventional hoods are the most common and include benchtop, distillation, and walk-in hoods of the constant air volume (CAV), variable air volume (VAV), bypass and non-bypass variety, with or without airfoils. Auxiliary air hoods and ductless fume hoods are not considered "conventional" and are used less often. Laboratory workers should know what kind of hood they are using and what its advantages and limitations are.

8.C.6.3.1   Constant Air Volume (CAV) Hoods

A constant air volume (CAV) fume hood draws a constant exhaust volume through the hood regardless of sash position. Because the volume is constant, the face velocity varies inversely with the sash position. The fume hood volume should be adjusted to achieve the proper face velocity at the desired working height of the sash, and then the hood should be operated at this height. (See section 8.C.4.)

8.C.6.3.2   Variable Air Volume (VAV) Hoods

A variable air volume (VAV) fume hood, also known as a constant velocity hood, is any hood that has been fitted with a face velocity control, which varies the amount of air exhausted from the fume hood in response to the sash opening to maintain a constant face velocity. In addition to providing an acceptable face velocity over a relatively large sash opening (compared to a CAV hood), VAV hoods also provide significant energy savings by reducing the flow rate from the hood when it is closed. These types of hoods are usually of the non-bypass design to reduce air volume (see below).

8.C.6.3.3   Non-Bypass Hoods

A non-bypass hood has only one major opening through which the air may pass into the hood, that is, the sash opening. The airflow pattern of this type of hood is shown in Figure 8.2. A CAV non-bypass hood

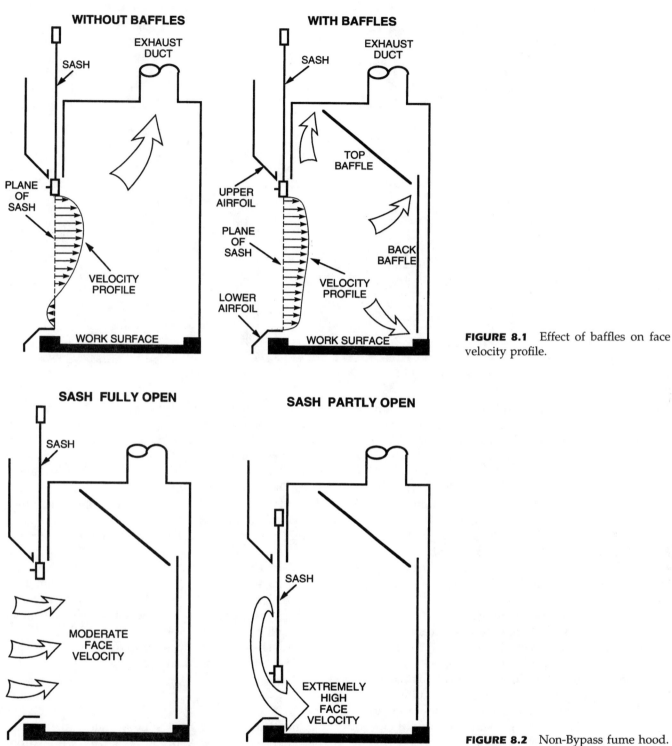

FIGURE 8.1 Effect of baffles on face velocity profile.

FIGURE 8.2 Non-Bypass fume hood.

has the undesirable characteristic of producing very large face velocities at small sash openings. As the sash is lowered, face velocities may exceed 1,000 fpm near the bottom. Face velocities are limited by the leakage of the hood through cracks and under the airfoil and by the increasing pressure drop as the sash is closed.

A common misconception is that the volume of air exhausted by this type of hood decreases when the sash is closed. Although the pressure drop through the hood increases slightly as the sash is closed, no appreciable change in volume occurs. This does not mean that these fume hoods should not be closed when

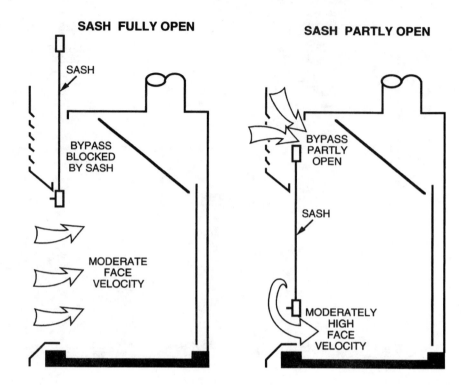

FIGURE 8.3 Bypass fume hood.

not in use, however. A closed hood provides a primary barrier to the spread of a fire or chemical release.

Many workers are reluctant to close their CAV non-bypass hoods because of the increase in air velocity and noise that occurs when the sash is lowered. This high-velocity air jet sweeping over the work surface often disturbs gravimetric measurements, causes undesired cooling of heated vessels and glassware, and can blow sample trays, gloves, and paper towels to the back of the hood, where they may be drawn into the exhaust system. Care should be exercised to prevent materials from entering the exhaust system. They can lodge in the ductwork, reducing airflow to the hood, or can be conveyed through the system and drawn into the exhaust fan and damage the fan or cause sparks. Variable-volume non-bypass hoods avoid these undesirable features by maintaining the velocity at a constant rate except at very low sash positions, where it may increase to a few hundred feet per minute. Hoods with horizontal sashes are usually of the non-bypass variety, because this arrangement is more difficult to connect to the bypass mechanism.

8.C.6.3.4   Bypass Hoods

A bypass fume hood is shown in Figure 8.3. It is similar to the non-bypass design but has an opening above the sash through which air may pass at low sash positions. Because the opening is usually 20 to 30% of the maximum open area of the sash, this hood will still exhibit the increasing velocity characteristic of the non-bypass hood as the sash is lowered. But the face velocity stops increasing as the sash is lowered to the position where the bypass opening is exposed by the falling sash. The terminal face velocity of these types of hoods depends on the bypass area but is usually in the range of 300 to 500 fpm—significantly higher than the recommended operating face velocity. Therefore, the air volume for bypass hoods should also be adjusted to achieve the desired face velocity at the desired sash height, and the hood should be operated at this position. This arrangement is usually found in combination with a vertical sash, because this is the simplest arrangement for opening the bypass. Varieties are available for horizontal sashes, but the bypass mechanisms are complicated and may cause maintenance problems.

8.C.6.3.5   Auxiliary Air Hoods

Quantitative tracer gas testing of many auxiliary air fume hoods has revealed that, even when adjusted properly and with the supply air properly conditioned, significantly higher worker exposure to the materials used in the hood may occur than with conventional (non–auxiliary air) hoods. Auxiliary air hoods should not be purchased for new installations, and existing auxiliary air hoods should be replaced or modified to eliminate the supply air feature of the hood. This feature causes a disturbance of the velocity profile and leakage of fumes from the hood into the worker's breathing zone.

The auxiliary air fume hood was developed in the 1970s primarily to reduce laboratory energy consumption. It is a combination of a bypass fume hood and a supply air diffuser located at the top of the sash. These hoods were intended to introduce unconditioned or tempered air, as much as 70% of the air exhausted from the hood, directly to the front of the hood. Ideally, this unconditioned air bypasses the laboratory and significantly reduces air conditioning and heating costs in the laboratory. In practice, however, many problems are caused by introducing unconditioned or slightly conditioned air above the sash, all of which may produce a loss of containment.

8.C.6.3.6 Ductless Fume Hoods

Ductless fume hoods are ventilated enclosures that have their own fan, which draws air out of the hood and through filters and ultimately recirculates it into the laboratory. The filters are designed to trap vapors generated in the hood and exhaust "clean" air back into the laboratory. These hoods usually employ activated carbon filters. The collection efficiency of the filters decreases over time. Ductless fume hoods have *extremely* limited applications and should be used *only* where the hazard is very low, where the access to the hood and the chemicals used in the hood are carefully controlled, and under the supervision of a laboratory supervisor who is familiar with the serious limitations and potentially hazardous characteristics of these devices. If these limitations cannot be accommodated, then this type of device should not be used.

8.C.7 Fume Hood Configurations

8.C.7.1 Benchtop Hoods

As the name implies, a benchtop fume hood sits on a laboratory bench with the work surface at bench height. These hoods can be of the CAV or VAV variety and can be a bypass or non-bypass design. The sash can be a vertical rising or a horizontal sliding type, or a combination of the two. Normally, the work surface is dished or has a raised lip around the periphery to contain spills in the hood. Sinks in hoods are not recommended because they encourage laboratory workers to dispose of chemicals in them. If they must be used, to drain cooling water from a condenser, for instance, then they should be fitted with a standpipe to prevent chemical spills from entering the drain. The condenser water drain can be run into the standpipe.

Spills will be caught in the cupsink by the standpipe for later cleanup and disposal. A typical benchtop fume hood is shown in Figure 8.4.

8.C.7.2 Distillation (Knee-high or Low-boy) Hoods

The distillation hood is similar to the benchtop hood except that the work surface is closer to the floor to allow more vertical space inside the hood for tall apparatus such as distillation columns. A typical distillation hood is shown in Figure 8.5.

8.C.7.3 Walk-in Hoods

A walk-in hood stands on the floor of the laboratory and is used for very tall or large apparatus. The sash type can be either horizontal or double- or triple-hung vertical. These hoods are also usually of the non-bypass type. The word "walk-in" is really a misnomer. One should never actually walk into a fume hood when it is operating and contains hazardous chemicals. Once past the plane of the sash, the worker is inside the hood with the chemicals. If the worker is required to enter the hood during operations where hazardous chemicals are present, personal protective equipment appropriate for the hazard should be worn. This may include respirators, goggles, rubber gloves, boots, suits, and self-contained breathing apparatus. A typical walk-in hood is shown in Figure 8.6.

8.C.7.4 California Hoods and Ventilated Enclosures

The California hood is a ventilated enclosure with a movable sash on more than one side. These hoods can usually be accessed through a horizontal sliding sash from the front and rear. They may also have a sash on the ends. Their configuration precludes the use of baffles and airfoils and therefore may not provide a suitable face velocity distribution across their many openings.

A ventilated enclosure is any site-fabricated hood designed primarily for containing processes such as scale-up or pilot plant equipment. Most do not have baffles or airfoils, and most designs have not had the rigorous testing and design refinement that conventional mass-produced fume hoods enjoy. Both the California hood and the ventilated enclosure are designed primarily to contain, but not capture, fumes like a conventional fume hood. Working at the opening of the devices, even when the plane of the opening has not been broken, may expose the worker to higher concentrations of hazardous materials than if a conventional hood were used.

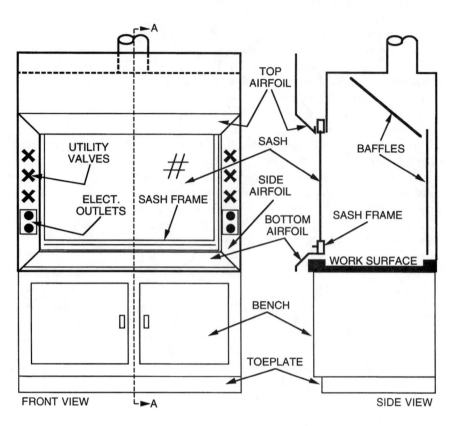

FIGURE 8.4 Benchtop fume hood.

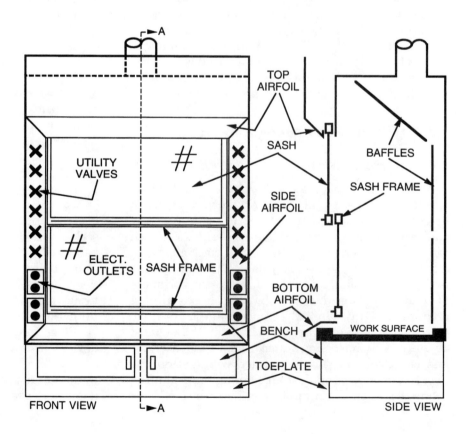

FIGURE 8.5 Distillation fume hood.

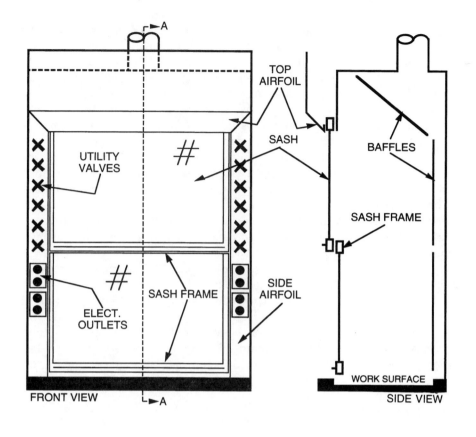

FRONT VIEW   SIDE VIEW   **FIGURE 8.6** "Walk-in" fume hood.

8.C.7.5 Perchloric Acid Hoods

The perchloric acid hood, with its associated duct-work, exhaust fan, and support systems, is designed especially for use with perchloric acid and other materials that can deposit shock-sensitive crystalline materials in the hood and exhaust system. These materials become pyrophoric when they dry or dehydrate (see also Chapter 5, section 5.G.6). Special water spray systems are employed to wash down all interior surfaces of the hood, duct, fan, and stack, and special drains are necessary to handle the effluent from the wash-down. The hood liner and work surface are usually stainless steel and are sealed by welding all seams. Perchloric acid hoods have drains in the work surface. Water spray heads are usually installed in the top of the hood, behind the baffles, and in the hood interior. The water spray should be turned on whenever perchloric acid is being heated in the fume hood. Welded or flanged and gasketed fittings to provide air- and water-tight connections are recommended. The duct-work, fan, and stack should be fabricated of plastic, glass, or stainless steel and should be fitted with spray heads approximately every 10 feet on vertical runs and at each change in direction. Horizontal runs should be avoided because they inhibit drain-down, and the spray action is not as effective on the top and sides of the duct. The washdown piping, located outside, should be protected from freezing. A three-way valve on the water supply piping that allows it to drain when not in use is helpful. Care should be used when routing the water lines to prevent the creation of traps that will retain water. Special operating procedures should be written to cover the washdown procedure for these types of hoods. The exhaust from a perchloric acid hood should not be manifolded with that from other types of fume hoods.

8.C.7.6 Radioisotope Hoods

Hoods used for work with radioactive sources or materials should be designed so that they can be decontaminated completely on a regular basis. A usual feature is a one-piece, stainless steel, welded liner with smooth, coved corners, which can be cleaned easily and completely. The superstructure of radioisotope hoods is usually made stronger than that of a conventional hood in order to support lead bricks and other shielding that may be required in the hood. Special treatment of the exhaust from radioisotope hoods may be required by government agencies to prevent the release of radioactive material into the environment. This usually involves the use of high-efficiency particulate air (HEPA) filters (see section 8.C.8.4).

8.C.8 Fume Hood Exhaust Treatment

Until now, treatment of fume hood exhausts has been limited. Because effluent quantities and concentrations are relatively low compared to those of other industrial air emissions sources, their removal is technologically challenging. And the chemistry for a given fume hood effluent can be difficult to predict and may change over time.

Nevertheless, legislation and regulations increasingly recognize that certain materials exhausted by a fume hood may be sufficiently hazardous that they can no longer be expelled directly into the air. Therefore, the practice of removing these materials from hood exhaust streams will become increasingly more prevalent.

8.C.8.1 Fume Hood Scrubbers and Contaminant Removal Systems

A number of technologies are evolving for treating fume hood exhaust by means of fume hood scrubbers and containment removal systems. Whenever possible, experiments involving such materials should be designed so that the toxic materials are collected in traps or scrubbers rather than being released into the hood. If for some reason this is impossible, then HEPA filters are recommended for highly toxic particulates. Liquid scrubbers may also be used to remove particulates, vapors, and gases from the exhaust system. None of these methods, however, is completely effective, and all trade an air pollution problem for a solid or liquid waste disposal problem. Incineration may be the ultimate method for destroying combustible compounds in exhaust air, but adequate temperature and dwell time are required to ensure complete combustion (see section 8.C.9).

Incinerators require considerable capital to build and energy to operate; hence, other methods should be studied before resorting to their use. The optimal system for collecting or destroying toxic materials in exhaust air must be determined on a case-by-case basis. In all cases, such treatment of exhaust air should be considered only if it is not practical to pass the gases or vapors through a scrubber or adsorption train before they enter the exhaust airstream. Also, if an exhaust system treatment device is added to an existing fume hood, the impact on the fan and other exhaust system components must be carefully evaluated. These devices require significant additional energy to overcome the pressure drop they add to the system. (See also Chapter 7, section 7.B.6.1.)

8.C.8.2 Liquid Scrubbers

A fume hood scrubber is a laboratory-scale version of a typical packed-bed liquid scrubber used for indus-

trial air pollution control. Figure 8.7 shows a schematic of a typical fume hood scrubber.

Contaminated air from the fume hood enters the unit and passes through the packed bed, liquid spray section, and mist eliminator, and then into the exhaust system for release up the stack. The air and the scrubbing liquor pass in a countercurrent fashion for efficient gas-liquid contact. The scrubbing liquor is recirculated from the sump and back to the top of the system using a pump. Water-soluble gases, vapors, and aerosols are dissolved into the scrubbing liquor. Particulates are also captured quite effectively by this type of scrubber. Removal efficiencies for most water-soluble acid- and base-laden airstreams are usually between 95 and 98%.

Scrubber units are typically configured vertically and are located next to the fume hood as shown in Figure 8.7. They are also produced in a top mount version, in which the packing, spray manifold, and mist eliminator sections are located on top of the hood and the sump and liquid handling portion are underneath the hood for a compact arrangement taking up no more floor area than the hood itself.

8.C.8.3 Other Gas-Phase Filters

There is another basic type of gas-phase filtration available for fume hoods in addition to liquid scrubbers. These are "inert" adsorbents and chemically active adsorbents. The "inert" variety includes activated carbon, activated alumina, and Molecular Sieves®. These substances typically come in bulk form for use in a deep bed and are available also as cartridges and as panels for use in housings similar to particulate filter housings. They are usually manufactured in the form of beads, but they may take many forms. The beads are porous and have extremely large surface areas with sites onto which gas and vapor molecules are trapped or adsorbed as they pass through. Chemically active adsorbents are simply inert adsorbents impregnated with a strong oxidizer, such as potassium permanganate (purple media), which reacts with and destroys the organic vapors. Although there are other oxidizers targeted to specific compounds, the permanganates are the most popular. Adsorbents can handle hundreds of different compounds, including most volatile organic components (VOCs), but also have an affinity for harmless species such as water vapor.

As the air passes through the adsorbent bed, gases are removed in a section of the bed. (For this discussion, "gas" means gases and vapors.) As the bed loads with gases, and if the adsorbent is not regenerated or replaced, eventually contaminants will break through the end of the bed. After breakthrough occurs, gases will pass through the bed at higher and higher concen-

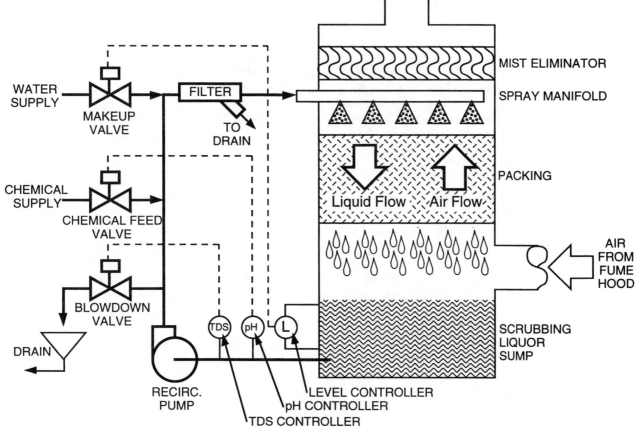

FIGURE 8.7 Typical fume hood scrubber schematic.

trations at a steady state until the upstream and downstream levels are almost identical. To prevent breakthrough, the adsorbent must be either changed or regenerated on a regular basis. Downstream monitoring to detect breakthrough or sampling of the media to determine the remaining capacity of the bed should be performed regularly.

An undesirable characteristic of these types of scrubbers is that if high concentrations of organics or hydrocarbons are carried into the bed, as would occur if a liquid were spilled inside the hood, a large exotherm will occur in the reaction zone of the bed. This exotherm may cause a fire in the scrubber. These devices and other downstream devices such as particulate filters should be located to minimize the effects of a fire, should one occur. Fires can start in these devices at surprisingly low temperatures due to the catalytic action of the adsorbent matrix. Therefore, such devices should be used and operated with care.

8.C.8.4 Particulate Filters

Air from fume hoods and biological safety cabinets in which radioactive or biologically active materials are used should be properly filtered to remove these agents so that they are not released into the atmosphere. Other hazardous particulates may require this type of treatment as well. The most popular method of accomplishing this removal is by using a HEPA filter bank. These HEPA filters trap 99.97% of all particulates greater than 3 microns in diameter. These systems must be specified, purchased, and installed so that the filters can be changed without exposing the worker or the environment to the agents trapped in the filter. Sterilizing the filter bank is prudent before changing filters that may contain etiologic agents.

The "bag-in, bag-out" method of replacing filters is a popular way to prevent worker exposure. This method separates the contaminated filter and housing from the worker and the environment by using a special plastic barrier "bag" and special procedures to prevent exposure to or release of the hazardous agent.

8.C.9 Thermal Oxidizers and Incinerators

Thermal oxidizers and incinerators are extremely expensive to purchase, install, operate, and maintain. However, they are one of the most effective methods of handling toxic and etiologic agents. The operational aspects of these devices are beyond the scope of this book. Also, their application to fume hoods has historically been rare. When considering this method of pollution control, an expert should be called to assist.

8.C.10 Other Local Exhaust Systems

There are many types of laboratory equipment and apparatus that generate vapors and gases but should not be used in a conventional fume hood. Some examples are gas chromatographs, atomic absorption spectrophotometers, mixers, vacuum pumps, and ovens. If the vapors or gases emitted by this type of equipment are hazardous or noxious, or if it is undesirable to release them into the laboratory because of odor or heat, then they should be contained and removed using local exhaust equipment. Local capture equipment and systems should be designed only by an experienced engineer or industrial hygienist. Also, users of these devices must have proper training, or they may be ineffectively used.

Whether the emission source is a vacuum-pump discharge vent, a gas chromatograph exit port, or the top of a fractional distillation column, the local exhaust requirements are similar. The total airflow should be high enough to transport the volume of gases or vapors being emitted, and the capture velocity should be sufficient to collect the gases or vapors.

Despite limitations, specific ventilation capture systems provide effective control of emissions of toxic vapors or dusts if they are installed and used correctly. A separate, dedicated exhaust system is recommended. The capture system should not be attached to an existing hood duct unless fan capacity is increased and airflow to both hoods is properly balanced. One important consideration is the effect that such added local exhaust systems will have on the ventilation for the rest of the laboratory. Each additional capture hood will be a new exhaust port in the laboratory and will compete with the existing exhaust sources for supply air.

Downdraft ventilation has been used effectively to contain dusts and other dense particulates and high concentrations of heavy vapors that, because of their density, tend to fall. Such systems require special engineering considerations to ensure that the particulates are transported in the airstream. Here again, a ventilation engineer or industrial hygienist should be consulted if this type of system is deemed suitable for a particular laboratory operation.

8.C.10.1 Elephant Trunks

An elephant trunk, or snorkel, is a piece of flexible duct or hose connected to an exhaust system. It cannot effectively capture contaminants that are farther than about one-half a diameter from the end of the hose. Elephant trunks are particularly effective for capturing discharges from gas chromatographs, pipe nipples, and pieces of tubing if the hose is placed directly on top of the discharge with the end of the discharge protruding to the hose. In this case, the volume flow rate of the hose must be at least 110 to 150% of the flow rate of the discharge.

The capture velocity is approximately 8.5% of the face velocity at a distance equal to the diameter of the local exhaust opening. Thus, a 3-inch-diameter snorkel or elephant trunk having a face velocity of 150 fpm will have a capture velocity of only 11 fpm at a distance of 3 inches from the opening. Because the air movement velocity is typically at least 20 fpm, capture of vapors emitted at 3 inches from the snorkel will be incomplete. However, vapors emitted at distances of 2 inches or less from the snorkel opening may be captured completely under these conditions.

8.C.10.2 Slot Hoods

Slot hoods are specially designed industrial ventilation hoods intended to capture contaminants generated according to a specific rate, distance in front of the hood, and release velocity for specific ambient airflow. In general, if designed properly, these hoods are more effective and operate using much less air than either elephant trunks or canopy hoods. In order to be effective, however, the geometry, flow rate, and static pressure must all be correct. Typical slot hoods are shown in Figure 8.8. Each type has different capture characteristics and applications. If the laboratory worker believes that one of these devices is necessary, then a qualified ventilation engineer should be called to design the hood and exhaust system.

8.C.10.3 Canopy Hoods

The canopy hood is not only the most common local exhaust system but also probably the most misunderstood piece of industrial ventilation equipment. It is estimated by industrial ventilation experts that as many as

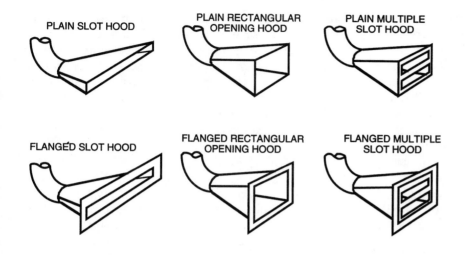

FIGURE 8.8 Typical specific ventilation hoods.

95% of the canopy hoods in use (other than in homes and restaurants) are misapplied *and* ineffective. The capture range of a canopy hood is extremely limited, and a large volume of air is needed for it to operate effectively. Thus, a canopy hood works best when thermal or buoyant forces exist that will move the contaminant up to the hood capture zone (a few inches below the opening). However, because canopy hoods are generally placed well above a contaminant source so that workers can operate underneath them, they draw contaminants past the worker's breathing zone and then into the exhaust system. If a canopy hood exists in a laboratory, it should be used only for nonhazardous service, such as capturing heated air or water vapor.

8.C.10.4 *Flammable-Liquid Storage Cabinets*

Flammable and combustible liquids should be stored only in approved flammable-liquid storage cabinets, not in a fume hood, out on the benches, or in a nonapproved storage cabinet.

Whether or not flammable-liquid storage cabinets should be ventilated is a matter for debate. One view is that all such cabinets should be vented by using an approved exhaust system, because it reduces the concentration of flammable vapors below the lower explosive limit inside the cabinet. A properly designed cabinet ventilation system will do this under most circumstances and results in a situation in which no fuel is rich enough in vapor to support combustion. However, there is still liquid in the cabinet and a source of fresh air provided by the ventilation system. All that is needed is an ignition source. The other view is that in most circumstances flammable-liquid storage cabinets should not be ventilated, because the cabinet is designed to extinguish a fire by depriving it of oxygen and ventilation defeats this purpose.

Both opinions are valid, depending on the conditions. Ventilation is prudent when the liquids stored in the cabinet are highly toxic or extremely odoriferous. Particularly odoriferous substances such as mercaptans have such a low odor threshold that even with meticulous housekeeping the odors persist; hence, ventilation may be desired.

If a ventilated flammable-liquid storage cabinet is used under a fume hood, it should not be vented into the fume hood above it. It should have a separate exhaust duct connected to the exhaust system. Fires occur most frequently in fume hoods. Fire from a fume hood may propagate into a flammable-liquid storage cabinet directly vented into the hood.

If a special-flammable storage cabinet ventilation system is installed, it should employ an AMCA-C type spark-resistant fan and an explosion-proof motor. Most fractional horsepower fans commonly used for this purpose do not meet this criterion and should not be used. If the building has a common fume hood exhaust system, then it is appropriate to hook a flammable-storage cabinet up to it if it must be ventilated.

8.C.11 General Laboratory Ventilation and Environmental Control Systems

General ventilation systems control the quantity and quality of the air supplied to and exhausted from the laboratory. The general ventilation system should ensure that the laboratory air is continuously replaced so that concentrations of odoriferous or toxic substances do not increase during the workday and are not recirculated from laboratory to laboratory.

Exhaust systems fall into two main categories: general and specific. General systems serve the laboratory as a whole, and include devices such as fume hoods and snorkels, as codes and good design practices allow.

Specific systems serve isotope hoods, perchloric acid hoods, or other high-hazard sources that require isolation from the general laboratory exhaust systems.

A general ventilation system that gives 6 to 12 room air changes per hour is normally adequate. More airflow may be required to cool laboratories with high internal heat loads, such as those with analytical equipment, or to service laboratories with large specific exhaust system requirements. In all cases, air should flow from the offices, corridors, and support spaces into the laboratories. All air from chemical laboratories should be exhausted outdoors and not recirculated. Thus, the air pressure in chemical laboratories should be negative with respect to the rest of the building unless the laboratory is also a clean room (see section 8.E.2). The outside air intakes for a laboratory building should be in a location that reduces the possibility of re-entrainment of laboratory exhaust or contaminants from other sources such as waste disposal areas and loading docks. (See Chapter 7, section 7.B.6.4, for further information.)

Although the supply system itself provides dilution of toxic gases, vapors, aerosols, and dust, it gives only modest protection, especially if these impurities are released into the laboratory in any significant quantity. Operations that can release these toxins, such as running reactions, heating or evaporating solvents, and transfer of chemicals from one container to another, should normally be performed in a fume hood. Toxic substances should be stored in cabinets fitted with an exhaust device. Likewise, laboratory apparatus that may discharge toxic vapors, such as vacuum pump exhausts, gas chromatograph exit ports, liquid chromatographs, and distillation columns, should vent to an exhaust device such as a snorkel.

The steady increase in the cost of energy in recent years, coupled with a greater awareness of the risks associated with the use of chemicals in the laboratory, has caused a conflict between the desire to minimize the costs of heating, cooling, humidifying, and dehumidifying laboratory air and the need to provide laboratory workers with adequate ventilation. However, cost considerations should never take precedence over ensuring that workers are protected from hazardous concentrations of airborne toxic substances.

8.C.11.1 Supply Systems

Well-designed laboratory air supply systems approach the ideal condition of laminar airflow, directing clean incoming air over laboratory personnel and sweeping contaminated air away from their breathing zone. Ventilation systems with well-designed diffusers that optimize "complete mixing" may also be satisfactory. Usually, several carefully selected supply air diffusers are used in the laboratory. Air entry through perforated ceiling panels may also successfully provide uniform airflow, but proper air distribution above the plenum is required. The plenum, diffuser, or perforated ceiling panels must be kept free of obstructions in order for the supply system to function properly.

8.C.11.2 Constant Air Volume

Constant air volume (CAV) air supply systems are the traditional design standard for laboratories. This method assumes constant exhaust and supply airflow rates through the laboratory. Although such systems are the easiest to design, and in some cases the easiest to operate, they have significant drawbacks due to their high energy consumption and limited flexibility. Classical CAV design assumes that all fume hoods operate 24 hours/day, 7 days/week, and at constant maximum volume. Adding, changing, or removing fume hoods or other exhaust sources for CAV systems requires rebalancing the entire system to accommodate the changes. Most CAV systems in operation today are unbalanced and operate under significant negative pressure. These conditions are caused by the inherent inflexibility of this design type, coupled with the addition of fume hoods not originally planned.

8.C.11.3 Variable Air Volume

Variable air volume (VAV) laboratories are rapidly replacing traditional CAV laboratories as the design standard. These systems are based on fume hoods with face velocity controls. As the users operate the fume hoods, the exhaust volume from the laboratory changes and the supply air volume must adapt to maintain a volume balance and room pressure control. An experienced laboratory ventilation engineer must be consulted to design these systems, because the systems and controls are complex and must be designed, sized, and matched so they operate effectively together.

8.C.12 Exhaust Systems

8.C.12.1 Individual Hood Fans

In traditional exhaust systems, each fume hood has its own exhaust fan. This arrangement has the following disadvantages and advantages: There is no way to dilute the fume hood effluent before release. The possibility of cross-contamination from one fume hood discharge to another is eliminated. Providing redundancy and emergency power for this arrangement is difficult and expensive; however, a fan failure will affect only one fume hood. The potential to use diversity is limited, as is the potential to use VAV controls. The potential to treat individual fume hood exhaust (as opposed to treating all fume hood exhaust) is excellent. For the same reason, the potential to recover heat from individual fans is almost nonexistent. The maintenance requirement for these systems is considerable, because they contain many pieces of equipment and have many roof penetrations, which can cause leakage problems. The mechanical (shaft) space requirements, initial cost, and operating cost are higher than for alternative systems, such as manifolded systems.

8.C.12.2 Manifolded (Common Header) Systems

For compatible exhaust streams, providing a common, manifolded exhaust system is an attractive design alternative to individual hood fans. This design is chosen increasingly for new laboratory buildings and is compatible with VAV systems. Manifolded systems have the following advantages and disadvantages: The potential for mixing and dilution of high concentrations of contaminants from a single fume hood by the air exhaust from all the other fume hoods on the system is excellent. The cross-contamination potential from one hood to another is minimal. The potential to provide redundancy of exhaust fans and/or provide emergency power to these systems is excellent. Conversely, the effects of a fan failure are widespread and serious; hence, redundancy is required in most cases. The potential to take advantage of VAV diversity and flow variation is also excellent, as is the ability to oversize the system for future expansion and flexibility. The ability to treat individual exhausts is retained by using new in-line liquid scrubber technologies. The maintenance, operating, and initial costs of these systems are all lower than for individual hood fan systems, and these systems require fewer roof penetrations. The heat recovery potential for these systems is maximized by collecting all the exhaust sources into a common duct.

8.C.12.3 Hybrid Exhaust Systems

Certain types of fume hoods and exhaust sources, such as perchloric acid hoods, should not be manifolded with other types of fume hood exhausts. In large buildings where the designer wishes to take advantage of the benefits of manifolded exhaust systems but

wishes to isolate a few exhaust streams, a combination, or hybrid, of these two types of systems is usually the most prudent and cost-effective alternative.

8.C.12.4 Exhaust Stacks

Proper stack design and placement are an extremely important aspect of good exhaust system design. Recirculation of contaminated air from the fume hood exhaust system into the fresh air supply of the facility or adjacent facilities may occur if stacks are not provided or if they are not designed properly to force the contaminated exhaust airstream up and into the prevailing wind stream. Stack design should take into account building aerodynamics, local terrain, nearby structures, and local meteorological information. An experienced laboratory consultant or an expert in atmospheric dispersion should be consulted to design exhaust stacks for a laboratory facility.

8.D ROOM PRESSURE CONTROL SYSTEMS

Laboratories and clean rooms usually require that a differential pressure be maintained between them and adjoining nonlaboratory spaces. This requirement may come from code considerations or from the intended use of the space. For example, NFPA Standard 45 states that "laboratory work units and laboratory work areas in which hazardous chemicals are being used shall be maintained at an air-pressure that is negative relative to the corridors or adjacent non-laboratory areas . . ." (NFPA, 1991d). This rule helps to prevent the migration of fire, smoke, and chemical releases from the laboratory space. Laboratories containing radiation hazards or biohazards may also be required by government agencies to maintain a negative pressure in order to contain these hazards. Clean rooms, on the other hand, are normally operated at a positive static pressure to prevent infiltration of particulates. (See sections 8.E.2 and 8.E.3 below for further information.)

8.E SPECIAL SYSTEMS

8.E.1. Glove Boxes

Glove boxes are usually small units that have multiple openings in which arm-length rubber gloves are mounted. The operator works inside the box by using these gloves. Construction materials vary widely, depending on the intended use. Clear plastic is frequently used, because it allows visibility of the work area and is easily cleaned.

Glove boxes generally operate under negative pressure, so that any air leakage is into the box. If the material being used is sufficiently toxic to require the use of an isolation system, the exhaust air will require special treatment before release into the regular exhaust system. These small units have a low airflow; therefore, scrubbing or adsorption (or both) can be easily accomplished.

Some glove boxes operate under positive pressure. These boxes are commonly used for experiments for which protection from atmospheric moisture or oxygen is desired. If positive-pressure glove boxes are used with highly toxic materials, they should be thoroughly tested for leaks before each use. Also, a method to monitor the integrity of the system (such as a shut-off valve or a pressure gauge designed into it) is required.

8.E.2 Clean Rooms

Clean rooms are special laboratories or work spaces in which large volumes of air are supplied through HEPA filters to reduce the particulates present in the room. Several classifications of clean rooms are commonly used. Clean room classifications refer to the number of particles larger than 0.5 microns in size per cubic foot of volume. Unfiltered ambient air has approximately 500,000 to 1,000,000 particles per cubic foot. Certain pharmaceutical, microbiological, optical, and microelectronic facilities require clean rooms of differing classifications from Class 10,000 to Class 10 or lower. Special construction materials and techniques, air handling equipment, filters, garments, and procedures are required, depending on the cleanliness level of the facility. A laboratory consultant or expert in clean room operation should be consulted before a clean room is worked in or built.

8.E.3 Environmental Rooms and Special Testing Laboratories

Environmental rooms, either refrigeration cold rooms or warm rooms, for growth of organisms and cells, are designed and built to be closed air circulation systems. Thus, the release of any toxic substance into these rooms poses potential dangers. Their contained atmosphere creates significant potential for the formation of aerosols and for cross-contamination of research projects. These problems should be controlled by preventing the release of aerosols or gases into the room. Special ventilation systems can be designed, but they will almost always degrade the temperature and humidity stability of the room. Special environmentally controlled cabinets are available to condition or store smaller quantities of materials at a much lower cost than in an environmental room.

Because environmental rooms have contained atmospheres, people who work inside them must be able to escape rapidly. Doors for these rooms should have magnetic latches (preferable) or breakaway handles to allow easy escape. These rooms should have emergency lighting so that a person will not be confined in the dark if the main power fails.

As is the case for other refrigerators, volatile flammable solvents should not be used in cold rooms (see Chapter 6, section 6.C.3.1). The exposed motors for the circulation fans can serve as a source of ignition and initiate an explosion. The use of volatile acids should also be avoided in these rooms, because such acids can corrode the cooling coils in the refrigeration system, which can lead to the development of leaks of refrigerants. Other asphyxiants such as nitrogen gas should also be avoided in enclosed spaces. Oxygen monitors and flammable gas detectors are recommended when the possibility of a low oxygen or flammable atmosphere in the room exists.

8.E.4 Biological Safety Cabinets and Biosafety Facilities

Biological safety cabinets (BSCs) are common containment and protection devices used in laboratories working with biological agents. BSCs and other facilities in which viable organisms are handled require special construction and operating procedures to protect workers and the environment. Conventional laboratory fume hoods should never be used to contain biological hazards. *Biosafety in Microbiological and Biomedical Laboratories* (U.S. DHHS, 1993) and *Biosafety in the Laboratory: Prudent Practices for the Handling and Disposal of Infectious Materials* (NRC, 1989) give detailed information on this subject.

8.F MAINTENANCE OF VENTILATION SYSTEMS

Even the best-engineered and most carefully installed ventilation system requires routine maintenance. Blocked or plugged air intakes and exhausts, as well as control system calibration and operation, can alter the performance of the total ventilation system. Filters become loaded, belts loosen, bearings require lubrication, motors need attention, ducts corrode, and minor components fail. These malfunctions, individually or collectively, can affect overall ventilation performance.

Facility-related environmental controls and safety systems, including fume hoods and room pressure controls, fire and smoke alarms, and special alarms and monitors for gases, should thus be inspected and maintained on a regular basis.

Each laboratory should be evaluated periodically for the quality and quantity of its general ventilation and any time a change is made, either to the general ventilation system for the building or to some aspect of local ventilation within the laboratory. The size of a room and its geometry, coupled with the velocity and volume of supply air, determine its air patterns. Airflow paths into and within a room can be determined by observing smoke patterns. Convenient sources of smoke for this purpose are the commercial smoke tubes available from local safety and laboratory supply companies. If the general laboratory ventilation is satisfactory, the movement of supply air from corridors and other diffusers into the laboratory and out through hoods and/or other exhaust sources should be relatively uniform. There should be no areas where air remains static or areas that have unusually high airflow velocities. If stagnant areas are found, a ventilation engineer should be consulted, and appropriate changes should be made to supply or exhaust sources to correct the deficiencies.

The number of air changes per hour within a laboratory can be estimated by dividing the total volume of the laboratory (in cubic feet) by the rate at which exhaust air is removed (in cubic feet per minute). For each exhaust port (e.g., hoods), the product of the face area (in square feet) and the average face velocity (in linear feet per minute) will give the exhaust rate for that source (in cubic feet per minute). The sum of these rates for all exhaust sources in the laboratory yields the total rate at which air is being exhausted from the laboratory. The rate at which air is exhausted from the laboratory facility should equal the rate at which supply air is introduced to the building. Thus, decreasing the flow rate of supply air (perhaps to conserve energy) decreases the number of air changes per hour in the laboratory, the face velocities of the hoods, and the capture velocities of all other local ventilation systems.

Airflows are usually measured with thermal anemometers or velometers. These instruments are available from safety supply companies or laboratory supply houses. The proper calibration and use of these instruments and the evaluation of the data are a separate discipline. An industrial hygienist or a ventilation engineer should be consulted whenever serious ventilation problems are suspected or when decisions on appropriate changes to a ventilation system are needed to achieve a proper balance of supply and exhaust air.

All ventilation systems should have a device that readily permits the user to monitor whether the total system and its essential components are functioning properly. Manometer, pressure gauges, and other devices that measure the static pressure in the air ducts are sometimes used to reduce the need to manually measure airflow. "Telltales" and other similar simple devices can also serve as indicators of airflow. The need for and the type of monitoring device should be determined on a case-by-case basis. If the substance of interest has excellent warning properties and the consequence of overexposure is minimal, the system will need less stringent control than if the substance is highly toxic or has poor warning properties.

9 Governmental Regulation of Laboratories

9.A INTRODUCTION

Recent years have seen a proliferation in the number of laws, regulations, and ordinances, federal, state, and local, that affect laboratories. This body of law is vast, complex, and intricate in its details and interrelationships. Both the law and its application vary from state to state, among federal regulatory agency regional offices, and among local jurisdictions. The individual researcher or laboratory worker cannot possibly be familiar with all of these regulations, but it is important that there be a strong institutional capacity, usually in a specialized office of environmental health and safety professionals, that is familiar with the details of these rules and can act as a resource for the researcher. In those smaller institutions that may not have such a specialized office, a researcher or an assigned individual, perhaps from the chemistry department, should seek advice directly from the regulatory agencies, from knowledgeable environmental health and safety professionals from other institutions, or from private environmental health and safety professionals and consultants.

In addition, all researchers should be familiar with the principal provisions and concepts of the most important laws and regulations that affect laboratories across the country. The two most important are the Occupational Safety and Health Administration's regulation, *Occupational Exposure to Hazardous Chemicals in Laboratories* (the OSHA Laboratory Standard) and the Resource Conservation and Recovery Act (RCRA), under which the Environmental Protection Agency (EPA) regulates hazardous waste. Because of its importance, the text of the OSHA Laboratory Standard is reprinted in Appendix A. Researchers, laboratory workers, and, in particular, laboratory managers and supervisors should read and understand these regulations.

9.B RISK AND REGULATION

Within the scientific research community, ambivalence about environmental health and safety regulations has raised much attention. The ethical conduct of science encourages excellence in scientific discovery while upholding societal values. Not surprisingly, highly trained chemists and other scientists, while fully supporting prudent and safe practices in the laboratory, sometimes feel that regulatory requirements do not distinguish scientifically among the varying levels of risk presented by some substances and procedures. To some extent this is true, but carried to an extreme, this perception can preclude a sufficient understanding of the rationale and realities that underlie environmental health and safety laws and regulations.

The basic laws—the Occupational Safety and Health Act, governing worker safety and health, the Resource Conservation and Recovery Act, governing the safe generation, storage, transport, and disposal of hazardous chemical waste, and the Clean Air Act and Federal Water Pollution Control Act, protecting public health and the environment—have been conceived soundly. These and other laws reflect congressional, state, and local legislative concern about the problems they address and generally enjoy strong public support. These and other relevant laws, along with their associated regulations, are listed in Table 9.1.

The research chemist might ask, "Why do I have to label all the chemicals in my laboratory? I know what's in those bottles." "Why can't I pour this chemical down the drain? It's such a small quantity that it will be diluted to harmlessness." "Why must I fill out detailed labels for the waste containers in my laboratory? Who needs all this paperwork?" and so forth.

The regulator would answer, "Yes, but if your chemicals aren't labeled, new workers in the laboratory won't know what is in those containers. If they don't know the contents, how can they properly use, store, or dispose of the unlabeled chemicals? Also, what happens if there's a fire or other emergency? The emergency crew responding won't know what's in your laboratory." Or, "If everyone poured just a little bit of that chemical down the drain, it could, cumulatively, seriously degrade water quality in sensitive rivers, lakes, or streams."

This debate is complicated by the fact that it is a virtual impossibility for environmental health and safety regulators to weigh *every* risk precisely. To attempt chemical-by-chemical regulation of the thousands of known, and unknown, chemicals would be so onerous and time-consuming as to leave many serious hazards unregulated. Consequently, the regulators have tried to strike a balance, regulating by class of risk, for example, ignitability, corrosivity, reactivity, and toxicity, and relying on employers to exercise reason and prudence in caring for the safety of their employees, as in OSHA's "General Duty" clause and the Hazard Communication Standard and the Laboratory Standard.

Those managing and working in laboratories should also recognize that violation of environmental health and safety laws and regulations not only may pose unnecessary risks to those in the laboratory and the surrounding community, but also can result in serious collateral consequences—imposition of civil penalties (fines of up to $25,000 per day per violation), as well

TABLE 9.1 Federal Laws and Regulations Affecting Laboratories

Law/Regulation	Citation	Purpose	Comments
Occupational Safety and Health Act (OSHA)	29 USC 651 et seq.	Worker protection	
Occupational Exposure to Hazardous Chemicals in Laboratories (Laboratory Standard)	29 CFR 1910.1450	Laboratory worker protection from chemical use	
Hazard Communication Standard	29 CFR 1910.1200	General worker protection from chemical use	
Occupational Exposure to Bloodborne Pathogens	29 CFR 1910.1030	Worker protecton from exposure to bloodborne pathogens	
Air Contaminants	29 CFR 1910.1000-1050	Standards for exposure to hazardous chemicals	
Hazardous Waste Operations and Emergency Response (HAZWOPER)	29 CFR 1910.120 and 40 CFR 311	Worker protection during hazardous waste cleanup	Applies to state and local government employees not covered by OSHA
Hazardous Materials	29 CFR 1910.101-111	Protection against hazards of compressed gases, flammable and combustible liquids, explosives, anhydrous ammonia	See also Uniform Fire Code (UFC) and National Fire Protection Association (NFPA) standards
Personal Protective Equipment	29 CFR 1910.132-138	Head, hand, foot, eye, face, and respiratory tract protection	See also American National Standards Institute (ANSI) standards
Medical Services and First Aid	29 CFR 1910.151	Provision of medical services, first aid equipment, and facilities for quick drenching and flushing of eyes	
Control of Hazardous Energy (Lock out/Tag out)	29 CFR 1910.147	Worker protection from electrical and other stored energy hazards	
Machinery and Machine Guarding	29 CFR 1910.211-219	Worker protection from mechanical hazards	
Ionizing Radiation	29 CFR 1910.96	Worker protection from radiation	See also Atomic Energy Act
General Duty Clause	29 USC 654 5(a) and (b)	Assurance of workplace free from recognized hazards that are causing or likely to cause serious physical harm	
Resource Conservation and Recovery Act (RCRA)	42 USC 6901 et seq.	Protection of human health and environment	
Hazardous Waste Management	40 CFR 260-272	"Cradle-to-grave" control of chemical waste	
Underground Storage Tanks	40 CFR 280	Protection against groundwater and soil contamination	

(continued on next page)

TABLE 9.1   Federal Laws and Regulations Affecting Laboratories *(continued)*

Law/Regulation	Citation	Purpose	Comments
Comprehensive Environmental Response, Compensation, and Liability Act (CERCLA)	42 USC 9601 et seq.	Remediation of past chemical disposal sites and assignment of liability	Also known as "Superfund" law
Superfund Amendments and Reauthorization Act (SARA)	42 USC 9601 et seq. 42 USC 11000 et seq.	Planning for emergencies and reporting of hazardous materials	Title III also known as Community Right-to-Know Act
National Contingency Plan	40 CFR 300-302	Cleanup requirements for spills and disposal sites	
Emergency Planning and Notification	40 CFR 355	Requirements for reporting of extremely hazardous materials and unplanned releases	Applies to all chemical users
Hazardous Chemical Reporting: Community Right-to-Know Act (SARA 311-312)	40 CFR 370	Requirements for reporting of hazardous chemicals in use	Exempts hazardous chemicals used in research laboratories (states may vary)
Toxic Chemical Release Reporting (SARA 313)	40 CFR 372	Requirements for reporting of chemical releases	Applies only to manufacturing facilities (states may vary)
Toxic Substances Control Act (TSCA)	15 USC 2601 et seq.	Protection of human health and the environment by requiring testing and necessary restrictions on use of certain chemical substances	Collection and development of information on chemicals
Reporting and Recordkeeping Requirements	40 CFR 704	One provision exempts users of small quantities solely for research and development (R&D)	Must follow R&D exemption requirements
Significant Adverse Reaction	40 CFR 717	Record of new allegation that chemical substance or mixture caused significant adverse effect for health or the environment	TSCA 8(c)
Premanufacture Notification (PMN)	40 CFR 720	Premanufacture notification (PMN) for chemical not on TSCA Inventory	Requires notification of EPA before manufacture or import of new chemical substance
Technically Qualified Individual (TQI)	40 CFR 720.3(ee)	Definition of technically qualified individual (TQI) by background, understanding of risks, responsibilities, and legal requirements	Follow TQI requirements with R&D
TSCA Exemption for Research and Development (R&D)	40 CFR 720.36	Exemption for R&D from PMN if chemical substance not on TSCA Inventory is manufactured or imported only in small quantities solely for R&D	Follow R&D exemption requirements including labeling and MSDS information
Polychlorinated Biphenyls (PCBs)	40 CFR 761	Prohibitions against PCBs in manufacturing, processing, distribution in commerce, and use prohibitions	Permits certain limited laboratory use of PCBs

(continued on facing page)

TABLE 9.1 Federal Laws and Regulations Affecting Laboratories *(continued)*

Law/Regulation	Citation	Purpose	Comments
Clean Air Act (CAA)	42 USC 7401 et seq.	Protection of air quality and human health	
Clean Air Act Amendments of 1990 (CAAA)	42 USC 7409 et seq.	Expansion of air quality protection	Requires development of specific rules for laboratories
National Emission Standards for Hazardous Air Pollutants (NESHAP)	40 CFR 70	Control of air pollutant emissions	
Montreal Protocol: Protection of Stratospheric Ozone	40 CFR 82	Control of emission of ozone-depleting compounds	Severely limits use of certain chlorofluorocarbons
Federal Water Pollution Control Act (FWPCA)	33 USC 1251 et seq.	Improvement and protection of water quality	
Criteria and Standards for the National Pollutant Discharge Elimination System (NPDES)	40 CFR 125	Control of discharge to public waters	
General Pretreatment Regulations for Existing and New Sources of Pollution	40 CFR 403	Control of discharge of pollutants to public treatment works	Implemented by local sewer authorities
Hazardous Materials Transportation Act (HMTA)	49 USC 1801 et seq.	Control of movement of hazardous materials	
Hazardous Material Regulations	49 CFR 100-199	Regulation of packaging, labeling, placarding, and transporting	
Hazardous Materials Training Requirements	49 CFR 172.700-704	Assurance of training for all persons involved in transportation of hazardous materials	Also known as HM126F
Atomic Energy Act (AEA) Energy Reorganization Act (ERA)	42 USC 2073 et seq. 42 USC 5841 et seq.	Establish standards for protection against radiation hazards	See also OSHA, Ionizing Radiation
Standards for Protection Against Radiation; Licenses	10 CFR 20 10 CFR 30-35	Establish exposure limits and license conditions	Rules promulgated by Nuclear Regulatory Commission
National Environmental Policy Act (NEPA)	42 USC 4321 et seq.	Ensure consideration of all environmental effects	
Requirements of the Council on Environmental Quality	40 CFR 6 and 1506	Indicate requirements for Environmental Impact Statement (EIS)	

as criminal penalties. Perhaps of general importance, violations can erode community confidence in an institution's seriousness of purpose in safeguarding the environment and complying with the law. It should be recognized that "prudent practice" involves not only scientific prudence but also prudent behavior in terms of the risks of violation of law or regulation, that is, the risk of adverse publicity for the institution and the risk of damaging the important trust and support of the community for the overall academic enterprise.

It is also prudent for institutions that handle chemicals in laboratories to participate in the regulatory process so that regulators will understand the impact that proposed rules can have on the laboratory environment. Because environmental regulations, particularly RCRA regulations, have not yet, in general, recognized the unique nature of the laboratory setting, some rules may be unnecessarily onerous for laboratories, without producing the expected increase in environmental protection. The best way to influence the regulatory process is through dialogue with the regulators. This dialogue, particularly at the state and local levels, can take place directly or in collaboration with the institution's environmental health and safety or governmental relations office. Also, professional associations, such as the American Chemical Society, the American Industrial Hygiene Association, the American Conference of Governmental Industrial Hygienists, and the American Institute of Chemical Engineers, as well as trade associations such as the Chemical Manufacturers Association, regularly comment on proposed regulations, especially proposed federal regulations (which, by law, require solicitation of comment from interested parties). Participation in the regulatory process through such groups is encouraged.

A brief description of the federal legislative and regulatory processes may be helpful. Laws are a product of legislative activity. Legislation is usually proposed by senators and representatives to achieve a desired result, for example, improved employee safety or environmental protection. Proposed laws are often known by their Senate or House file numbers, for example, S.xxx or H.R.xxx. Copies of proposed laws can be obtained, even at this early stage in the process, by requesting them from local offices of House or Senate members. Sponsors of proposed legislation are open to comment from the public. Once a law is passed, it is known by its Public Law number, for example, P.L. 94-580, Resource Conservation and Recovery Act (RCRA). It is published in the *United States Code* and is referenced by volume and chapter number; 42 USC 6901 et seq. is the citation for RCRA.

When a law is passed, it is assigned to an administrative unit (agency or department) for development of rules and regulations that will implement the purpose of the legislation. The two major agencies involved in regulation of chemicals are the Occupational Safety and Health Administration (OSHA) and the Environmental Protection Agency (EPA). Proposed regulations are published in the *Federal Register*, a daily publication of federal agency activities. A public comment period and perhaps public hearings are specified, during which all affected parties have an opportunity to present their support for or concerns with the regulations as proposed. This is the second significant opportunity for involvement in the regulatory process. Final rules are published in the *Federal Register* and in the *Code of Federal Regulations* (CFR), which is updated annually to include all changes during the previous year. Rules in the CFR are referenced by title and part number; 40 CFR 260-272 is the citation for RCRA's hazardous waste rules.

9.C THE OSHA LABORATORY STANDARD: OCCUPATIONAL EXPOSURE TO HAZARDOUS CHEMICALS IN LABORATORIES

Beginning in the early 1970s, groups and individuals representing laboratories contended that the then-existing OSHA standards, designed for exposure conditions in an industrial production setting, were inappropriate for the very different exposure conditions in laboratories. As a result of these concerns, OSHA in 1981 undertook the development of a special regulatory regime for laboratories. The Laboratory Standard, promulgated in 1990, was the result (see Appendix A). In its Laboratory Standard, OSHA refers to the National Research Council's (1981) *Prudent Practices for Handling Hazardous Chemicals in Laboratories* as "nonmandatory . . . guidance to assist employers in the development of the Chemical Hygiene Plan." It is anticipated that the present edition, *Prudent Practices in the Laboratory: Handling and Disposal of Chemicals*, will likewise be referenced.

9.C.1 The Chemical Hygiene Plan

The centerpiece of the Laboratory Standard is the Chemical Hygiene Plan. This is a written plan developed by the employer (e.g., university or research organization) and has the following major elements:

- employee information and training about the hazards of chemicals in the work area, including how to detect their presence or release, work practices and how to use protective equipment, and emergency response procedures;
- the circumstances under which a particular laboratory operation requires prior approval from the employer;
- standard operating procedures for work with hazardous chemicals;
- criteria for use of control measures, such as engineering controls or personal protection equipment;
- measures to ensure proper operation of fume hoods and other protective equipment;
- provisions for additional employee protection for work with "select carcinogens" (as defined in the Laboratory Standard) and for reproductive toxins or substances that have a high degree of acute toxicity;
- provisions for medical consultations and examinations for employees; and
- designation of a chemical hygiene officer.

In many large institutions, the environmental health and safety department has developed a *generic* Chemical Hygiene Plan, but the plan must be modified to include detailed protections that are specific to each laboratory and its workers. This approach allows considerable flexibility in achieving the performance-based goals of the Laboratory Standard. Model Chemical Hygiene Plans are available from the OSHA consultation service, from the American Chemical Society, and from some professional associations or commercial sources.

9.C.2 Relation of the OSHA Laboratory Standard to Other OSHA Standards

Several points about the OSHA Laboratory Standard deserve special mention. The intention of the standard is to supersede existing OSHA *health* standards, but other OSHA rules on topics not specifically addressed in the standard remain applicable. The so-called "general duty" clause of the Occupational Safety and Health Act, which requires an employer to "furnish to each of his employees . . . a place of employment . . . free from recognized hazards that are likely to cause death or serious physical harm . . . " and requires an employee to "comply with occupational safety and health standards and all rules . . . issued pursuant to this chapter which are applicable to his own actions and conduct" continues to be applicable and, indeed, is one of the most commonly cited sections in cases of alleged OSHA violations. Other OSHA standards relating to possible eye or skin contact must continue to be observed. There are dozens of chemicals in this category. They are listed in 29 CFR 1910 as well as in specific standards following Section 1910.1000, such as the vinyl chloride standard, 29 CFR 1910.1017, which prohibits direct contact with liquid vinyl chloride.

Other OSHA standards setting forth permissible exposure limits (PELs) apply to the extent that they require limiting exposures to below the PEL and, where the PEL or "action level" is routinely exceeded, the Laboratory Standard's provisions require exposure monitoring and medical surveillance. The requirements for exposure monitoring and medical surveillance are found in Appendix A, sections (d) and (g) of the Laboratory Standard.

9.C.3 PELs and TLVs

OSHA's permissible exposure limits were directly adopted, in 1970, from the similar list of threshold limit values (TLVs), a nonregulatory consensus document prepared by the American Conference of Governmental Industrial Hygienists (ACGIH). Quoting the TLV booklet (ACGIH, 1994), "Threshold Limit Values (TLVs) refer to airborne concentrations of substances and represent conditions under which it is believed that nearly all workers may be repeatedly exposed day after day without adverse health affects." TLVs are average concentrations over a normal 8-hour work day and a 40-hour work week. The existence of a "time-weighted average (TWA) exposure" approach to control of airborne contaminants indicates that exposures may be somewhat higher or lower than the average at various times of the day, which is typical of work with chemicals.

ACGIH also recommends, for some compounds, a short-term exposure limit (STEL), which establishes a safe exposure limit of no more than four 15-minute periods a day. STELs are published only for compounds where toxic effects have been reported from high short-term exposures in either humans or animals. STELs typically are no more than 25 to 200% above the associated TLV. Some compounds have a "C" preceding the numerical TLV designation. This indicates that the TLV is a ceiling level concentration that should not be exceeded during any part of the working exposure. Ceiling limits have generally been applied to compounds that are fast acting. For compounds that include neither a STEL nor a C notation, a limit on the upper level of exposure should still be imposed. According to the TLV booklet, "Excursions in worker exposure levels may exceed 3 times the TLV-TWA for no more than a total of 30 minutes during a work day, and under no circumstances should they exceed 5 times the TLV-TWA, provided that the TLV-TWA is not exceeded."

The "action level" is an OSHA regulatory term occurring in a few substance-specific regulations. It is an airborne concentration (lower than the associated PEL) that, when exceeded, requires certain activities, such as exposure monitoring or medical surveillance. If the researcher is working with one of the listed substances, it is important to understand when air monitoring and/or medical surveillance are required.

9.C.4 Particularly Hazardous Substances

There are special provisions in the Laboratory Standard regarding work with "particularly hazardous substances," a term that includes "select carcinogens," "reproductive toxins," and "substances with a high degree of acute toxicity." A select carcinogen is defined in the standard as any substance (1) regulated by OSHA as a carcinogen, (2) listed as "known to be a carcinogen" in the *Annual Report on Carcinogens* published by the National Toxicology Program (NTP), (3) listed under Group 1 ("carcinogenic to humans") by the *International Agency for Research on Cancer (IARC) Monographs*, or (4) *in certain cases*, listed in either Group

2A or 2B by IARC or under the category "reasonably anticipated to be carcinogens" by NTP. A category (4) substance is considered a select carcinogen only if it causes statistically significant tumor incidence in experimental animals in accordance with any of the following criteria: (1) after inhalation exposure of 6 to 7 hours per day, 5 days per week, for a significant portion of a lifetime to dosages of less than 10 mg/m^3; (2) after repeated skin application of less than 300 mg/kg of body weight per week; or (3) after oral dosages of less than 50 mg/kg of body weight per day.

"Reproductive toxins" are defined as those chemicals that affect reproductive capabilities, including chromosomal damage (mutations) and effects on fetuses (teratogenesis). Chemicals with a "high degree of toxicity" also require special provisions for worker health. Although "select carcinogens" are specifically identified through reference to other publications, "reproductive toxins" and chemicals with a "high degree of acute toxicity" are not specified further, which has made it difficult to apply these categories. Some institutions have chosen to adopt the OSHA Hazard Communication Standard definition of "highly toxic" (LD$_{50}$ < 50 mg/kg oral dose) as a workable definition of "high degree of acute toxicity." There is very little agreement on how to determine "reproductive toxins."

It is important to understand that the OSHA PELs and substance-specific standards do not include *all* hazardous chemicals. It is the laboratory manager's responsibility under the Laboratory Standard and the "general duty" clause to apply scientific knowledge in safeguarding workers against risks, even though there may be no specifically applicable OSHA standard.

The OSHA-mandated special provisions for work with carcinogens, reproductive toxins, and substances that have a high degree of acute toxicity include consideration of "designated areas," use of containment devices, special handling of contaminated waste, and decontamination procedures. The OSHA requirement is for evaluation, assessment, and implementation of these special controls, when appropriate. These special provisions are to be included in the Chemical Hygiene Plan.

The Laboratory Standard also requires quite detailed recordkeeping, particularly with regard to records of exposure monitoring and medical surveillance, in those circumstances where exposure limits are exceeded or where work with especially hazardous substances is conducted.

9.C.5 Protection of Other Personnel in Laboratories

OSHA standards apply only to "employees" of laboratory facilities. In many cases, students are not employees within the scope of the Occupational Safety and Health Act, but both moral and legal considerations suggest that colleges and universities provide the same protections to students as are provided to all employees regularly working in the laboratory.

Custodial and maintenance staff who service the laboratory continue to be governed by other OSHA standards, particularly the Hazard Communication Standard, which sets forth the information, training, and health and safety protections required to be provided to nonlaboratory employees.

9.C.6 Federal Versus State Regulations

Enforcement of the Laboratory Standard, as well as other OSHA standards, may be a shared responsibility of the federal government and of state occupational safety and health programs. Under Section 18 of the Occupational Safety and Health Act, individual states may be authorized by federal OSHA to administer the act if they adopt a plan for development and enforcement of standards that is "at least as effective as the Federal standards." These states are known as "state-plan" states. In states that do not administer their own occupational health and safety programs, federal OSHA is the regulator, covering all nonpublic employers. State-plan states have generally included public employees in their regulatory approach. What this means is that a given institution may be subject to (1) the federal Laboratory Standard, enforced by federal OSHA, (2) a state Laboratory Standard, enforced by state OSHA, or (3) if a public institution is not subject to OSHA regulation, state public institution health and safety regulations enforced by a state agency. The environmental health and safety office at each institution should have a copy of the applicable standard.

Of the violations of the Laboratory Standard issued by OSHA, many have been for failure to have a Chemical Hygiene Plan or for a missing element in the plan. Another commonly cited violation is failure to meet the "Employee Information and Training" requirements of the Laboratory Standard. It is likely that OSHA enforcement of the Laboratory Standard will increase in the future, as state and federal OSHA inspectors focus more on laboratory activities and the implementation of this still relatively new standard.

9.C.7 Laboratory Standard Versus Hazard Communication Standard

It is important to understand the distinction between the Laboratory Standard and the Hazard Communication Standard. As noted above, the Laboratory Standard is intended, with limited exceptions, to be *the*

OSHA standard governing employees who routinely work in laboratories. The Hazard Communication Standard, on the other hand, applies to all nonlaboratory operations "where chemicals are either used, distributed or are produced for use or distribution." The obvious difficulty is that workers in maintenance shops, even if in a laboratory building, would be covered by the Hazard Communication Standard, not the Laboratory Standard. The requirements of the Hazard Communication Standard are, in certain respects, more demanding than those of the Laboratory Standard. For example, the Hazard Communication Standard requires that *each* container of hazardous chemicals used by the employee be labeled clearly with the identity of the chemical and appropriate hazard warnings, whereas the Laboratory Standard requires only that employers "ensure that labels on incoming containers of hazardous chemicals are not removed or defaced." The Hazard Communication Standard further requires that copies of Material Safety Data Sheets (MSDSs) for *each* hazardous chemical be readily accessible to employees, whereas the Laboratory Standard requires only that employers "maintain MSDSs that are received with incoming shipments, and ensure that they are readily accessible. . . ." Many institutions, faced with the difficulty of designing environmental health and safety programs that meet the requirements of the Laboratory Standard *and* the requirements of the Hazard Communication Standard, have opted to follow the requirements of the Hazard Communication Standard for *all* workplaces, laboratory and nonlaboratory, while additionally adopting and implementing the Chemical Hygiene Plan requirements of the Laboratory Standard as they apply to laboratories. Careful comparison of the two standards should be made when designing an environmental health and safety program.

9.D THE RESOURCE CONSERVATION AND RECOVERY ACT

The Resource Conservation and Recovery Act (RCRA) was enacted by Congress in 1976 to address the problem of waste disposal and reduction. Subtitle C of that act established a system for controlling hazardous waste from generation to disposal. This is often referred to as the "cradle to grave" system. The cradle, however, is the point at which the hazardous material first becomes a "hazardous waste," not when it is first received in a laboratory. Under RCRA, the Environmental Protection Agency is given great responsibilities in promulgating detailed regulations governing the generation, transport, treatment, storage, and disposal of hazardous waste. RCRA and EPA regulations

were written with a focus on industrial-scale generation of hazardous waste, but, with very limited exceptions, they also apply to laboratories that use chemicals.

9.D.1 Definition of a Generator

It is important to understand who is a generator under RCRA. A generator is "any person, *by site* [emphasis added], whose act or process produces hazardous waste. . . ." There are three categories of generator: (1) Large-quantity generators are those whose facilities generate 1,000 kg or more per month (about five 55-gallon drums of hazardous waste) or over 1 kg of "acutely hazardous waste" per month. By this measure, most large research institutions, including the larger universities, are large-quantity generators. (2) Small-quantity generators, to whom special rules apply, generate more than 100 but less than 1,000 kg of hazardous waste per month (and accumulate less than 6,000 kg at any one time) and less than 1 kg of "acutely hazardous waste" per month (and accumulate less than 1 kg at any one time). (3) A conditionally exempt small-quantity generator generates 100 kg or less of hazardous waste per month and less than 1 kg of "acutely hazardous waste." The special requirements applicable to conditionally exempt small-quantity generators can be found in 40 CFR 261.5.

"Individual generation site" is defined by RCRA regulation as a contiguous site at or on which hazardous waste is generated. Because most colleges, universities, and research institutions are located in one geographic site, they are viewed as a single generator under the definition above, and the entire facility has a single EPA generator identification number. *This means that each individual laboratory generating waste is not a "generator" within the terms of RCRA, but, as a part of the overall institutional "generator," each laboratory must comply with generator requirements.* Obviously, multisite institutions are required to have separate EPA identification numbers for each site.

In determining whether a site is an "individual generation site," some EPA regional offices have been quite strict in relying on the distinction of whether the site has any public roads, the reason being RCRA's objective of not allowing unregulated transport of hazardous waste on public ways. In the definitions section of the RCRA regulations, "on-site" is defined as "the same or geographically contiguous property which may be divided by public or private right-of-way, provided the entrance and exit between the properties is at a crossroads intersection, and access is by *crossing as opposed to going along* [emphasis added] the right-of-way." The significance of this interpretation is that hazardous waste that is being transported can be sent only to a permitted treatment, storage, and disposal facility (TSDF). It cannot be moved legally to an unpermitted holding facility if such movement requires transport along a public road, even if the public road and the receiving location are within the boundaries of an institution. This has caused serious management problems for some multibuilding institutions and companies.

Generators must obtain an EPA identification number, prepare the waste for transport, follow accumulation and storage requirements, manifest hazardous waste, and adhere to detailed recordkeeping and reporting requirements. Although conditionally exempt small-quantity generators are partially exempt from these requirements, they must still

- identify their waste to determine whether it is hazardous,
- not accumulate more than 1,000 kg of hazardous waste, and
- treat or dispose of the waste on site, or ensure that the waste is sent to a permitted TSDF or a recycling facility.

Also, generators producing more than 1 kg in a calendar month of "acute hazardous waste" (see below) are subject to full regulation under RCRA.

While some industrial research laboratories and a few large universities have on-site EPA-permitted TSDFs, most colleges and smaller universities and research institutions do not. Their hazardous waste is shipped off-site, treated, stored, and disposed of at commercial EPA-permitted TSDFs. The process and requirements for EPA-permitting of a TSDF are very complex, involving construction of costly facilities, detailed operational plans, and specialized staff.

It is important to note that state classification of generators may be different from the classifications outlined above. Some states regulate all generators of hazardous waste with no exemptions, and some states classify generators by waste type rather than by volume.

9.D.2 Definition of Hazardous Waste

RCRA defines "hazardous waste" as solid waste that, "because of its quantity, concentration, or physical, chemical, or infectious characteristics may: 1) cause, or significantly contribute to an increase in mortality or an increase in serious irreversible, or incapacitating reversible, illness or 2) pose a substantial present or potential hazard to human health or the environment when improperly treated, stored, transported, or disposed of, or otherwise managed."

A "solid waste" under RCRA need not be solid! "Solid waste" is defined as "any . . . discarded material, including solid, liquid, semisolid, or contained gaseous material. . . ." The term "discarded" includes any material that is abandoned, recycled, or "inherently wastelike." The term "hazardous waste" means any solid waste (as defined above) that:

- exhibits any of the characteristics of a hazardous waste (i.e., ignitability, corrosivity, reactivity, or toxicity as determined by EPA's Toxicity Characteristic Leaching Procedure (TCLP) test);
- has been listed as a hazardous waste by EPA regulation;
- is a mixture containing a listed hazardous waste and a nonhazardous solid waste; or
- is a waste derived from the treatment, storage, or disposal of a listed hazardous waste.

Certain otherwise hazardous wastes are excluded from regulation. These include samples sent for testing, household waste, agricultural waste (not including pesticides), oil and gas production waste, and others not generally associated with laboratory activities.

The RCRA regulations contain detailed provisions and testing procedures for determining if any particular chemical waste, not specifically listed in the regulations, is a "hazardous waste" within the RCRA definition.

"Acute hazardous wastes" are those listed as such in 40 CFR 261.31 (the FO20-27 series, the dioxin precursors) and 261.33 (e), the P list.

9.D.3 Accumulation Times and Amounts

Under RCRA, a large-quantity generator may accumulate hazardous waste for up to 90 days without a special EPA permit, providing

- the waste container is in good condition,
- the container material or liner is compatible with the waste contained,
- the container is kept closed except when actually adding or removing waste, and
- the container is properly handled and stored.

To accumulate waste for a longer period, an EPA permit as a "storage facility" is required. Some larger institutions have a central EPA-permitted TSDF, and longer storage times are allowed in accordance with the regulations and the permit conditions applicable to that institution.

9.D.4 Satellite Accumulation

One RCRA regulation of particular relevance to academic and research laboratories is the provision that a generator may accumulate up to 55 gallons of hazardous waste, or 1 quart of acutely hazardous waste, "at or near any point of generation where wastes initially accumulate which is under the control of the operator of the process generating the waste. . . ." Under this provision, which is recognized by some but not all state RCRA-enforcement agencies, individual laboratories, or "satellite" locations as they are sometimes termed, may be allowed to accumulate and store hazardous waste, but the waste must be dated and removed within 3 calendar days after reaching the 55-gallon limit. However, safe practice and the requirements of state and local regulation should guide laboratory managers in adhering to reasonable limits on accumulation volume and time. Prudent practice is to limit accumulation time to no more than 1 year (preferably less), and 55 gallons may be too large a quantity for a laboratory location, because of space limitations, fire code limitations, and potential spill or release hazards. When waste accumulates for long periods, laboratory workers tend to forget about it. *Hazardous waste, even in small quantities, should never be forgotten.*

9.D.5 Drain Disposal of Hazardous Waste

Only a few years ago, it was common practice to dispose of many laboratory wastes down the drain. Today, the indiscriminate disposal to the sanitary sewer of laboratory chemicals is not acceptable. Most laboratory drain systems are connected to sanitary sewer systems, and their effluent will eventually go to a sewage treatment plant. Some chemicals can interfere with the proper functioning of sewage treatment facilities or affect particularly sensitive bodies of water into which the chemical is discharged. In the laboratory drain system itself, some chemicals can create hazards of fire, explosion, or local air pollution. Others can corrode the drain system.

It is essential to recognize that the characteristics of wastewater treatment plants vary from one locality to another. These factors and local regulations govern what types and concentrations of chemicals can be disposed of. While RCRA regulations do exempt mixtures of hazardous waste and domestic sewage from hazardous waste regulation, local regulations of drain disposal are often more restrictive. While drain disposal of some *nonhazardous* chemicals may be acceptable, drain disposal of *hazardous* chemicals is permissible only under carefully prescribed circumstances. Some institutions, responding to strict local controls and concerns, have, as a matter of policy, simply adopted a conservative course and banned the drain disposal of *any* laboratory chemicals.

If the institution, after appropriate consideration, determines that some drain disposal should be permitted, the following general points of guidance should be adhered to:

- Use drain disposal only if the drain system flows to a wastewater treatment plant, and not into a septic tank system or a storm sewer system that flows directly into surface water.
- Make sure that the chemical to be disposed of is compatible with other materials being disposed of, and compatible with the piping material.
- Monitor sewer disposal of laboratory waste by individual workers or students for adherence to guidelines on types of chemicals, quantities and rates, and flushing procedures.
- Remember that only those compounds that are reasonably soluble in water are suitable for drain disposal.

(Chapter 7 contains more detailed guidance on these and other points regarding drain disposal.)

Under the Federal Water Pollution Control Act, all direct dischargers, that is, those who discharge effluents directly into rivers, streams, or other bodies of water, are required to have a National Pollutant Discharge Elimination Systems (NPDES) permit. Indirect dischargers, that is, those who discharge effluents into publicly owned sewage treatment works, are not required to have a permit, but must be subject to discharge conditions set by the local wastewater treatment authority. Because many of these local sanitary sewer and water treatment districts have their own, more restrictive, requirements, there should be *no* drain disposal of hazardous materials without checking with local authorities.

9.D.6 Empty Containers

A container or inner liner of a container that contained hazardous waste is "empty" under RCRA regulations if all waste has been removed by standard practice and no more than 2.5 cm (1 inch) of residue, or 3 percent by weight of containers less than 110 gallons, remains. These should be taken as upper limits, and in practice all "capturable" quantities of the substance should be removed. As an "empty" container, it is no longer subject to RCRA regulation. It should be noted, however, that if the container held *acute* hazardous waste, triple rinsing or equivalent measures are required before the container is "empty" within the RCRA regulations. Indeed, rinsing with water or a detergent solution is prudent for *all* hazardous waste containers. Rinsates resulting from the cleaning of empty containers that contained acutely hazardous waste are themselves hazardous waste and must be disposed of accordingly. Rinsates resulting from cleaning of hazardous waste containers may or may not be hazardous waste, depending on whether the waste in the container was listed in 40 CFR 261, Subpart D, or in all cases on whether they fall within the RCRA hazardous waste characteristics of ignitability, corrosivity, reactivity, or toxicity (40 CFR 261, Subpart C). It should also be noted that the regulations regarding empty containers apply to all hazardous waste containers, not just those containing laboratory chemicals. Paint cans, insecticide containers, cleaning supply bottles, and so forth, are also covered. Some institutions view it as more convenient and/or more prudent simply to handle all empty chemical containers from laboratories as hazardous waste and dispose of them accordingly. Others find it feasible to clean the containers as required, remove labels, dispose of the rinsate properly, and then dispose of or recycle the empty containers along with ordinary nonhazardous waste. Obviously, the latter course can substantially reduce the volume of reportable hazardous waste quantities generated as well as reduce the total hazardous waste disposal costs.

9.D.7 Land Disposal of Hazardous Waste

In the 1984 RCRA amendments, Congress required EPA to prohibit land disposal of hazardous waste unless it was processed using EPA-developed treatment standards that were protective of human health and the environment. As a result, EPA established certain prohibitions on land disposal and also established treatment standards. The net effect of these regulations is *a virtual total ban on direct land disposal* of hazardous waste. The detailed EPA regulations may be found at 40 CFR 261. There are provisions for special handling and treatment of certain wastes disposed of in Lab Packs.

9.D.8 In-laboratory Treatment

"Treatment" is very broadly defined by RCRA (40 CFR 260.10) to include "any method, technique, or process, including neutralization, designed to change the physical, chemical, or biological character or composition of any hazardous waste. . . ." Basically, RCRA prohibits any treatment without a permit, with only very limited exceptions. Those exceptions are as follows:

• Treatability studies, that is, investigation of new methods of treating or detoxifying hazardous waste, in which the quantities of hazardous waste treated are under certain specified limits. While a permit is not required, there are significant recordkeeping and reporting requirements.

• Treatment procedures in laboratories where the treatment procedure is *part of the experiment*. While not specifically addressed in the RCRA regulations, a rationale for this exception is that the material has not been declared a "waste" subject to RCRA regulation.

• "Closed loop" treatment processes where "only tank storage is involved and the entire process through completion of reclamation is closed by being entirely connected with pipes" are allowed. For most laboratories, such a closed loop system is not feasible for treatment.

• Elementary neutralization of waste that is hazardous only because of the characteristic of corrosivity.

• Treatment under certain circumstances by conditionally exempt small-quantity generators.

For most laboratories, especially academic research laboratories, the obtaining of a special EPA treatment permit is not feasible because of the great expense involved in the application process, and the detailed recordkeeping, analytical, and other requirements attendant upon a permitted treatment operation. Some states have adopted a "permit-by-rule" regulatory approach, allowing categorical or blanket permitting of certain small-scale treatment methods. Because of the regulatory complexities governing in-laboratory treatment and the differing interpretations applied by RCRA enforcement agencies, it is recommended that the researcher contact, either directly or through the institution's environmental health and safety and legal offices, the cognizant EPA office prior to undertaking in-laboratory treatment.

9.D.9 Waste Minimization Under RCRA

One of the goals of RCRA is to reduce or eliminate the generation of hazardous waste. EPA uses the term "waste minimization" to mean the reduction of hazardous waste generated prior to any treatment, storage, or disposal of the waste. It is defined as any source reduction or recycling activity that results in either reduction of the total volume of hazardous waste or reduction of the toxicity of the hazardous waste. RCRA requires large-quantity generators to identify in their biennial report to EPA (or the state) what has actually been achieved as a result of efforts undertaken to reduce the volume and toxicity of waste. In addition, generators are required to certify on the manifest accompanying off-site shipment of waste that they have a waste minimization program in place. Interim guidelines for waste minimization programs have been published by EPA (U.S. EPA, 1990).

9.D.10 Transportation of Chemicals and Hazardous Waste

Transportation of both hazardous materials and hazardous waste is primarily regulated by the Department of Transportation (DOT), under authority of the Hazardous Materials Transportation Act. The RCRA rules include some additional requirements for transportation of hazardous waste.

The DOT regulations applicable to transport of laboratory chemicals include those governing packaging, labeling, marking, placarding, and reporting of discharges. A transporter is defined as any person engaged in the *off-site* transportation of hazardous materials or waste. These regulations apply not only to those who actually transport, but also to those who *initiate or receive* hazardous waste shipments. Those who prepare hazardous materials for transportation must also meet certain training requirements. These requirements have been recently promulgated under the Hazardous Materials Transportation Uniform Safety Act. For institutions whose laboratory operations are at a single site, transportation within that site is not regulated, as long as that transport involves no travel along public ways. Most institutions, however, have developed policies for on-site transportation covering labeling, segregation of incompatibles, containment and double containment, and other necessary safeguards to prevent accidental release to the environment or injury to persons during transportation, even if not required by governmental regulation.

If off-site transportation of any hazardous material, including shipments of small samples to colleagues, is contemplated, the laboratory worker should consult the institution's environmental health and safety office and/or the EPA/DOT regulations.

9.D.11 Underground Storage Tanks

Subtitle I of RCRA was enacted to control and prevent leaks from underground storage tanks storing hazardous substances, including petroleum products. It is not uncommon to discover underground storage tanks that may have been installed many years ago in or near laboratory buildings. Inventorying of these tanks and establishing systems for leak detection, recordkeeping, reporting, cleanup or other corrective action, and ultimately closure, are required by RCRA regulations. All new tanks installed must meet EPA and state design specifications for the tanks themselves and associated piping and cathodic protection systems. All underground storage tanks must comply with the regulations for new tanks by December 22, 1998.

9.E THE CLEAN AIR ACT

The Clean Air Act (CAA) regulates emissions into the air. Under the 1990 amendments to the CAA, emissions of sulfur dioxide, volatile organic compounds (VOCs), hazardous air pollutants (HAPs), and ozone-depleting chemicals (ODCs) are being more rigorously regulated. Institutions with large research laboratory operations will be affected by these rules, given their use of a variety of volatile chemicals and solvents that result in VOC or HAP emissions. Laboratory managers should work closely with the institution's office of environmental health and safety in addressing issues of CAA regulatory application and compliance. Under the structure of the CAA, states and local air quality districts are responsible with EPA for developing emissions standards for local areas. These standards may vary. For example, areas with serious air pollution problems may develop stricter emissions controls.

Colleges, universities, or other institutions with research laboratories that have the potential to emit one or more of the listed hazardous air pollutants in amounts greater than 10 tons/year of a single hazardous air pollutant or 25 tons/year of total hazardous air pollutants will be considered a major source and covered by the regulations. The "tons per year" calculation is based on total potential emissions by facilities in a contiguous area and under common control, including such sources as power plants and boilers, and, in a few states or localities, fume hoods. Thus, many of the larger research institutions will be affected by these new standards. EPA will be establishing emission standards based on "maximum achievable control technologies" for each source category. The 1990 amendments also require EPA to establish a special source category for research or laboratory facilities. EPA has not yet issued specific laboratory-directed regulations, but, as noted above, laboratories that are part of a large university or research institution will be regulated as part of that combined major source.

Laboratories should also be aware of regulatory provisions relating to stratospheric-ozone-depleting substances. The list of such substances can be found at 40 CFR 82, Appendixes A and B to Subpart A. The list includes as "Class I" substances most common freons, carbon tetrachloride, and methyl chloroform. The regulations prohibit the movement in interstate commerce of these substances and other listed ozone-depleting substances.

It is clear, however, that CAA enforcement, which has, up to now, largely focused on industrial emissions, will focus increasingly on emissions from universities and other research institutions. Laboratory researchers should work closely with the institution's environmental health and safety staff and/or with the local air quality authorities in monitoring and achieving compliance with these new CAA standards. Laboratory managers should also participate, along with their institution's environmental regulatory specialists, with the regulatory agencies in the development of workable rules for the laboratory special source category.

9.F SARA TITLE III, COMMUNITY RIGHT-TO-KNOW AND EMERGENCY NOTIFICATION AND RESPONSE

Under Title III of SARA, the Superfund Amendments and Reauthorization Act, facilities that use hazardous chemicals in their operations must maintain the Material Safety Data Sheets (MSDSs) required under OSHA's Hazard Communication Standard, submit copies of the MSDSs, provide inventories of hazardous chemicals, and report accidental releases to emergency planning authorities. The basic rationale for these regulations is twofold; the community's "right to know" what hazardous materials are present in facilities in their community, and the need for emergency response authorities and local fire departments to know what substances are being used or stored in case they are required to respond to a fire, explosion, release, or other emergency. "Hazardous chemical," however, is defined to exclude any chemical "to the extent it is used in a research laboratory or hospital or other medical facility under the direct supervision of a technically qualified individual." It is important to note, however, that (1) *nonlaboratory* uses of hazardous chemicals are not excluded, and (2) state or local laws relating to community right-to-know or emergency notification and response may, regardless of SARA exemptions, require chemical inventories or release notification for research laboratories.

Two other aspects of SARA Title III deserve mention. They are the emergency planning notification requirement and the release notification requirement. The emergency planning notification regulation requires that any institution that has an EPA-listed "extremely hazardous substance" on site in greater than EPA-specified "threshold planning quantities" must notify the state emergency response authorities. The quantity limits are based on the total quantity of the hazardous chemical present at the facility rather than in an individual laboratory.

The release notification provisions require that an institution notify state and community authorities in the event of a release into the environment of a "hazardous substance" or an "extremely hazardous substance" in excess of EPA-established "reportable quantities."

Another important aspect of SARA is the requirement for training of those who respond to emergencies involving hazardous materials. Although the training requirements are contained in OSHA regulations, Congress has stipulated that they should apply to all employers in all states. Researchers should contact their institution's environmental health and safety and/or local emergency response agencies to ascertain locally applicable requirements for notification and training.

9.G THE TOXIC SUBSTANCES CONTROL ACT

The Toxic Substances Control Act (TSCA) is intended to control new or existing chemicals that may present unreasonable risks to human health or the environment. In 1976, Congress enacted this statute to fill gaps in chemical control not covered by other laws and agencies, such as the Federal Insecticide, Fungicide and Rodenticide Act (FIFRA), the Food and Drug Administration (FDA), and the Occupational Safety and Health Administration (OSHA). TSCA is not intended to overlap other laws that already regulate specified chemical uses. TSCA authorizes EPA to administer and enforce the rules it develops under TSCA.

TSCA's numerous sections deal with various aspects of chemical control:

- reporting and recordkeeping requirements of chemical manufacturers and processors,
- establishment of an inventory of existing chemicals in U.S. commerce (the TSCA Inventory), and
- requirements for premanufacture notification (PMN) of new chemicals to EPA.

TSCA also gives EPA authority to

- require chemical testing,
- ban or otherwise control chemicals in commerce,
- control polychlorinated biphenyls (PCBs),
- enforce and set penalties for violations, and
- provide protection for confidential business information submitted to EPA.

TSCA's provisions have a major impact on chemical manufacturers and their associated research and development (R&D) laboratories. Primary regulations that can affect R&D laboratories are the R&D exemption from the PMN and "significant adverse reactions" in TSCA 8(c).

9.G.1 Research and Development (R&D) Exemption from the Premanufacture Notification (PMN)

Before a company can manufacture or import a new chemical (a chemical not listed on the TSCA Inventory), it must file a PMN with EPA and allow the agency a specified period of time to assess the risks to health or the environment. Failure to do so can result in penalties as high as $25,000 per day of violation. It should be noted that importation is the same as manufacturing under TSCA.

There is an exemption from the PMN requirement for chemicals used for R&D, or in a use regulated by another agency (such as solely for medical use regulated by FDA), but EPA still considers R&D use to be a commercial activity. There are significant personnel notification and recordkeeping requirements under this exemption (R&D Exemption: 40 CFR 720.3, 720.36, and 720.78).

The following is a summary of TSCA rules concerning R&D exemption requirements:

1. Operation under the R&D exemption requires notification in writing to those who may be exposed to the exempt substance. However, R&D conducted solely in R&D laboratories results in minimal requirements. New substances for which no hazard information is available need only carry the appropriate labels.

2. All containers of R&D-exempt material or of mixtures containing the material should have an experimental material label and a TSCA R&D label. While EPA does not specify the language, the following is an example of a TSCA R&D label:

This material is not listed on the TSCA Inventory. It should be used for research and development purposes only under the direct supervision of a technically qualified individual.

3. On any Material Safety Data Sheet (MSDS) in the "composition, information on ingredients" section for an R&D-exempt material, words to the following effect should be indicated:

This material is for R&D evaluation only. It can only be used for R&D evaluations until PMN review by EPA is completed. If this material is used in plants or non-R&D locations for R&D evaluation, its use must be supervised by a technically qualified individual. Review all sections of this MSDS prior to use.

4. If an R&D-exempt material leaves a "prudent laboratory practices" area—e.g., if a sample is sent to a customer or the material is used in a *pilot plant* or *manufacturing plant*—you must provide written notification to those exposed to the substance.

- Your notification must inform users of *any* information in your possession or control on the risk to health from such exposure, or must state that no information is known about the toxicity of the substance.

- You should request a review of health studies or EPA actions on the R&D material by your laboratory's toxicologist and maintain a record of the response. The adequacy of the notification(s) is the responsibility of the manufacturer or importer, and copies of such notice(s) must be kept for 5 years.

5. A checklist for R&D samples sent to plants or outside processors would include:

- shipping the material with
  —appropriate experimental material and TSCA R&D container labels affixed, and
  —an enclosed copy of the MSDS.
- providing a letter to each recipient that indicates
  —any known hazard information,
  —that the substance is solely for R&D purposes, and
  —that use of the substance requires supervision by a technically qualified individual.
- maintaining a record of
  —names and addresses of sample recipients,
  —sample identification code,
  —amount of material, and
  —purpose of the sample.

Copies of the written notifications provided (i.e., the letter, the MSDS, and copies of all labels affixed to sample containers) should be maintained in a file for 5 years (40 CFR 720.78, Recordkeeping).

6. EPA did not set exact quantities for R&D exemption volumes. The EPA standard is "not greater than reasonably necessary for such (R&D) purposes." Let that serve as your standard for R&D-stage activities, with the understanding that you are in the best position to determine what reasonable R&D volumes are, given the technology and product development requirements for making and processing a particular material for your R&D needs. Note that stockpiling of R&D material for later manufacturing is prohibited if the material is not on the TSCA Inventory.

9.G.2 Recordkeeping Requirements for Significant Adverse Reaction Allegations—TSCA 8(c)

One subpart of Section 8 of TSCA can affect R&D laboratories. TSCA 8(c) is the requirement to keep records of allegations of significant adverse effects of chemicals. The allegations need not be proved to be subject to this recordkeeping requirement. If a worker becomes aware of an allegation of a significant adverse reaction to a chemical (e.g., a skin rash, an allergic reaction, respiratory effects), he or she should contact the supervisor or the health, safety, and environmental office in the organization. This office or another responsible office must maintain a written record of the allegation in its files. TSCA requires that this record be kept for 5 years if made by someone outside the organization, and for 30 years if made by an employee.

9.G.3 Regulations Covering Polychlorinated Biphenyls (PCBs)

One set of regulations under TSCA speaks directly to research facilities, colleges, and universities, however. Those are the regulations governing polychlorinated biphenyls (PCBs) and monochlorobiphenyl. Essentially, PCB manufacturing in concentrations of more than 50 ppm is banned, and PCB use in any concentration is banned unless used in a "totally enclosed manner." There are exceptions to these conditions. Two of possible relevance to researchers are those that allow the use of "small quantities for research and development" and use "as an immersion oil in microscopy." Because of the stringency of regulation, researchers contemplating working with PCBs should contact their institution's environmental health and safety office or the regional EPA office to determine any applicable requirements.

The most common use of PCBs is as dielectric fluid in transformers, capacitors, and other electrical equipment. Although PCB-containing equipment is no longer manufactured, PCB-containing transformers are still in use. PCBs are often found in transformers and capacitors of laboratory equipment, especially high-voltage transformers. Such items should be tested and appropriate steps taken before continued use or disposal. Any leaking oil from a transformer should be checked for PCB content.

Due to the bioaccumulation of PCBs, EPA has issued very strict regulations governing the use, servicing, and disposal of PCB-containing transformers. While most PCB-containing transformers may continue to be used for the remainder of their useful lives, they must be inventoried, inspected frequently for leaks, and serviced in accordance with stringent EPA procedures, and they are subject to recordkeeping and reporting requirements. EPA offices have been very rigorous in the enforcement of the PCB regulations, and large fines have been imposed on some institutions. Researchers should notify their institution's environmental health and safety office of any transformers or capacitors that might contain PCBs.

9.H REGULATION OF LABORATORY DESIGN AND CONSTRUCTION

Laboratory design and construction are regulated mainly by state and local laws that incorporate, by reference, generally accepted standard practices set out in various uniform codes, such as the Uniform Building Code, the Uniform Fire Code, and the National Fire Protection Association standards. For laboratory buildings where hazardous chemicals are stored or used, there are detailed requirements covering such things as spill control, drainage, containment, ventilation, emergency power, special controls for hazardous gases, fire prevention, and building height.

In addition, OSHA standards affect some key laboratory design and construction issues, for example, eyewashes, safety showers, and special ventilation requirements. Other consensus standards prepared by organizations such as the American National Standards Institute (ANSI) and the American Society of Heating, Refrigeration, and Air Conditioning Engineers (ASHRAE) are relevant to laboratory design. It is not uncommon for various codes and consensus standards to be incorporated into state or federal regulations.

Bibliography

A. M. Best Company. Best's Safety Directory. Revised annually. Morristown, N.J.: A. M. Best Company.

Amdur, M. O., J. Doull, and C. D. Klaassen, Eds. 1991. Casarett and Doull's Toxicology: The Basic Science of Poisons, 4th ed. New York: Pergamon Press.

American Chemical Society (ACS). Chemical & Engineering News (Letters to the Editor). Washington, D.C.: ACS.

American Chemical Society (ACS). 1991. Design of Safe Chemical Laboratories: Suggested References, 2nd ed. Committee on Chemical Safety. Washington, D.C.: ACS.

American Chemical Society (ACS). 1993. Less Is Better: Laboratory Chemical Management for Waste Reduction, 2nd ed. Task Force on Laboratory Waste Management, Department of Government Relations and Science Policy. Washington, D.C.: ACS.

American Chemical Society (ACS). 1994. Laboratory Waste Management: A Guidebook. Task Force on Laboratory Waste Management, Department of Government Relations and Science Policy. Washington, D.C.: ACS.

American Chemical Society (ACS). 1995. A Model Chemical Hygiene Plan for High Schools. Committee on Chemical Safety. Washington, D.C.: ACS.

American Chemical Society (ACS). 1995. Safety in Academic Chemistry Laboratories, 6th ed. Committee on Chemical Safety. Washington, D.C.: ACS.

American Chemical Society (ACS). 1995. Understanding Chemical Hazards: A Guide for Students. Task Force on Occupational Health and Safety. Washington, D.C.: ACS.

American Chemical Society (ACS). Chemical Health and Safety. Bimonthly publication of the American Chemical Society. Washington, D.C.: ACS.

American Conference of Governmental Industrial Hygienists (ACGIH). 1978. Industrial Ventilation, 15th ed. Cincinnati, Ohio: ACGIH.

American Conference of Governmental Industrial Hygienists (ACGIH). 1991. Documentation of the Threshold Limit Values and Biological Exposure Indices, 6th ed. Cincinnati, Ohio: ACGIH.

American Conference of Governmental Industrial Hygienists (ACGIH). 1994. Threshold Limit Values for Chemical Substances and Physical Agents and Biological Exposure Indices, 1994–1995. Cincinnati, Ohio: ACGIH.

American Industrial Hygiene Association (AIHA). 1989. Odor Thresholds for Chemicals with Established Occupational Health Standards. Fairfax, Va: AIHA.

American Industrial Hygiene Association (AIHA). 1992. Laboratory Ventilation. ANSI/AIHA Z9.5-1992. Fairfax, Va.: AIHA.

American National Standards Institute (ANSI). 1989. Practices for Occupational and Educational Eye and Face Protection: ANSI Z87.1-1989. New York: ANSI.

American National Standards Institute (ANSI). 1991. Supplement to Practices for Occupational and Educational Eye and Face Protection: ANSI Z87.1A-1991. New York: ANSI.

American National Standards Institute (ANSI). 1990. Emergency Eyewash and Shower Equipment: ANSI Z358.1-1990. New York: ANSI.

American National Standards Institute (ANSI). 1992. Practices for Respiratory Protection: ANSI Z88.2-1992. New York: ANSI.

American National Standards Institute (ANSI). 1993. Safe Use of Lasers: ANSI Z136.1-1993. New York: ANSI.

American National Standards Institute (ANSI). 1994. Hazardous Industrial Chemicals—Precautionary Labeling: ANSI Z129.1-1994. New York: ANSI.

American Red Cross. 1987. Adult CPR. Washington, D.C.: American Red Cross.

American Red Cross. 1987. Standard First Aid and Personal Safety, 3rd ed. Washington, D.C.: American Red Cross.

American Red Cross. 1995. Adult CPR. Washington, D.C.: American Red Cross.

American Society for Testing and Materials (ASTM). 1965. Fire and Explosion Hazards of Peroxy Compounds. ASTM Special Technical Publication 394. Philadelphia: ASTM.

American Society for Testing and Materials (ASTM). 1988. Standard Test Methods for Chemical Permeability: ASTM Standard F-739. Philadelphia: ASTM.

American Society of Heating, Refrigeration, and Air Conditioning Engineers (ASHRAE). 1978. Laboratories, Chapter 14 in Applications Handbook. Atlanta: ASHRAE.

American Society of Heating, Refrigeration, and Air Conditioning Engineers (ASHRAE). 1985. Methods of Testing Performance of Laboratory Fume Hoods: Standard 110-1985. Atlanta: ASHRAE.

American Society of Mechanical Engineers (ASME). 1992. Boiler and Pressure Vessel Code. Section VII, B31.1. PTC 25.3, NQA-1. Fairfield, N.J.: ASME Press.

Armour, Margaret-Ann. 1991. Hazardous Laboratory Chemicals Disposal Guide. Boca Raton, Fla.: CRC Press.

Armour, Margaret-Ann, Lois M. Weir, and L. Gordon. 1981. Hazardous Chemicals: Information and Disposal. Edmonton, Alta., Canada: University of Alberta.

Ashbrook, Peter C., and Malcolm M. Renfrew, Eds. 1991. Safe Laboratories: Principles and Practices for Design and Remodeling. Chelsea, Mich.: Lewis Publishers.

Barton, John, and Richard Rogers, Eds. 1993. Chemical Reaction Hazards: A Guide. Rugby, Warwickshire, England: Institution of Chemical Engineers.

Beyler, R. E., and V. K. Myers. 1982. What every chemist should know about teratogens. Journal of Chemical Education 759:59.

Blank, M. 1993. Biological effects of electromagnetic fields. Bioelectrochemistry and Bioenergetics 32:203–210.

Braker, W., and A. L. Mossman. 1980. Matheson Gas Data Book. Lyndhurst, N.J.: Matheson Gas Products.

Braker, W., A. L. Mossman, and D. Siegel. 1988. Effects of Exposure to Toxic Gases: First Aid and Medical Treatment, 3rd ed. Secaucus, N.J.: Matheson Gas Products.

Bretherick, L., Ed. 1986. Hazards in the Chemical Laboratory, 4th ed. Cambridge, United Kingdom: Royal Society of Chemistry.

Bretherick, L. 1990. Bretherick's Handbook of Reactive Chemical Hazards, 4th ed. London: Butterworth.

Bright, F. V., and M. E. McNally, Eds. 1992. Supercritical Fluid Technology: Theoretical and Applied Approaches in Analytical Chemistry. American Chemical Society (ACS) Symposium Series, No. 488. Washington, D.C.: ACS.

Brill, T. B., and K. James. 1993. Kinetics and mechanisms of thermal decomposition of nitroaromatic explosives. Chemistry Reviews 93:2667–2692.

Bruker Instruments. 1992. Site Planning Guide for Superconducting NMR Systems. Bellerica, Mass.: Bruker.

Budinger, T. F. 1992. Emerging nuclear magnetic resonance technologies: Health and Safety. Annals of the New York Academy of Sciences 649:1–18.

Bulloff, J. J. 1991. Improving Safety in the Chemical Laboratory, 2nd ed. New York: John Wiley & Sons.

Burns, D. T., A. Townsend, and A. H. Carter, 1981. Inorganic Reaction Chemistry, Vol 2. New York: Ellis Horwood.

Campbell, Monica, and William Glenn. 1982. Profit from Pollution Prevention. Toronto, Ontario: Pollution Probe Foundation.

Caplan, K. J., and G. W. Knutson. 1977. The Effect of Room Air Challenge in the Efficiency of Laboratory Fume Hoods. ASHRAE Transactions 83, Part 1.

Castegnaro, M., and E. B. Sansone. 1986. Chemical Carcinogens: Some Guidelines for Handling and Disposal in the Laboratory. New York: Springer-Verlag.

Center for Labor Education and Research. 1992. Emergency Responder Training Manual for the Hazardous Materials Technician. Lori P. Andrews, Ed. New York: Van Nostrand Reinhold.

Chamberlin, R. I., and J. E. Leahy. 1978. Laboratory Fume Hood Standards. Contract No. 68-01-4661. Facilities Engineering and Real Properties Branch. Washington, D.C.: Environmental Protection Agency.

Clansky, K. B., Ed. 1987. Chemical Guide to the OSHA Hazard Communication Standard. Burlingame, Calif.: Roytech.

Clansky, K. B., Ed. 1989. Suspect Chemicals Sourcebook: Guide to Industrial Chemicals Covered Under Major Federal Regulatory and Advisory Programs. Burlingame, Calif.: Roytech.

Clayton, G. D., and F. E. Clayton, Eds. 1994. Patty's Industrial Hygiene and Toxicology, 4th ed. Volume 2, Part C, Toxicology. New York: Wiley Interscience.

Compressed Gas Association. 1990. Handbook of Compressed Gases, 3rd ed. New York: Compressed Gas Association.

Costner, Pat, and Joe Thornton. 1990. Playing with Fire: Hazardous Waste Incineration. Washington, D.C.: Greenpeace.

DiBerardinis, I. J., et al. 1993. Guidelines for Laboratory Design: Health and Safety Considerations, 2nd ed. New York: Wiley-Interscience.

Dixon, Lloyd S., Deborah S. Drezner, and James K. Hammitt. 1993. Private-Sector Cleanup Expenditures and Transaction Costs at 18 Superfund Sites. Santa Monica, Calif.: Rand.

Dreisbach, R. H., and W. O. Robertson. 1987. Handbook of Poisoning: Diagnosis and Treatment, 12th ed. Los Altos, Calif.: Appleton and Lange.

Erdey, L., 1965. Gravimetric Analysis, Part II. New York: Pergamon Press.

Farris, C. A., and P. T. Anastas. 1993. Alternative synthetic design for pollution prevention. Initiatives at the U.S. Environmental Protection Agency. A symposium presented at the 206th American Chemical Society meeting (August), Chicago, Ill.

Fawcett, H. H., Ed. 1988. Hazardous and Toxic Materials: Safe Handling and Disposal, 2nd ed. New York: Wiley-Interscience.

Fischer, Kenneth E. 1985. Contracts to dispose of laboratory waste. Journal of Chemical Education A118(April):62.

Fischer, Kenneth E. 1989. Certifications for professional hazardous materials and waste management. Journal of Chemical Education A112–A114 (April):66.

Fortuna, Richard C., and David J. Lennett. 1987. Hazardous Waste Regulation, The New Era: An Analysis and Guide to RCRA and the 1984 Amendments. New York: McGraw-Hill.

Freeman, E., Ed. 1990. Hazardous Waste Minimization. New York: McGraw-Hill.

Fuller, F. H., and A. W. Etchells. 1979. Safe Operations with the 0.3 m/s (60 fpm) Laboratory Hood. Journal No. 49. Atlanta: American Society of Heating, Refrigeration, and Air Conditioning Engineers.

Fuscaldo, A. A., B. J. Erlick, and B. Hindman. 1980. Laboratory Safety: Theory and Practice. New York: Academic Press.

Goldman, Benjamin A., James A. Hulme, and Cameron Johnson. 1986. Hazardous Waste Management: Reducing the Risk. Washington, D.C.: Island Press.

Gosselin, Robert E., Roger P. Smith, and Harold C. Hodge. 1984. Clinical Toxicology of Commercial Products, 5th ed. Baltimore, Md.: Williams & Wilkins.

Government Institutes, Inc. 1988. Environmental Health and Safety Manager's Handbook. Rockville, Md.: Government Institutes.

Hamstead, A. C. 1964. Ind. Eng. Chem. 56(6):37.

Hathaway, G. J., N. H. Proctor, J. P. Hughes, and M. L. Fischman, Eds. 1991. Proctor and Hughes' Chemical Hazards of the Workplace, 3rd ed. New York: Van Nostrand Reinhold.

Hazardous Materials Information Center. 1986. Hazard Classification Systems: Comparative Guide to Definitions and Labels. Middletown, Conn.: Inter/Face.

Hazardous Waste Task Force. 1987. Hazardous Waste Management at Educational Institutions. Washington, D.C.: National Association of College and University Business Officers.

Hileman, B. 1993. Health effects of electromagnetic fields remain unresolved. Chemical & Engineering News (November 8):15–29.

Hitchings, Dale T. 1994. Laboratory space pressurization control systems. ASHRAE Journal 36 (February 1):36.

Hitchings, Dale T., and R. S. Shull. 1993. Measuring and calculating laboratory exhaust diversity—Three case studies. ASHRAE Transactions 99, Part 2.

Inorganic Syntheses. New York: John Wiley & Sons.

International Agency for Research on Cancer (IARC). 1982. IARC Monographs on the Evaluation of Carcinogenic Risk of Chemicals to Humans: Supplement No. 3. Ann Arbor, Mich.: Books on Demand.

International Union of Pure and Applied Chemistry (IUPAC) and World Health Organization (WHO). 1992. Chemical Safety Matters. Cambridge, United Kingdom: Cambridge University Press.

Kaufman, J. A., Ed. 1990. Waste Disposal in Academic Institutions. Chelsea, Mich.: Lewis Publishers.

Kletz, T. 1988. Learning from Accidents in Industry. London: Butterworth.

Laboratory Safety & Environmental Management. Bimonthly newsletter. The Target Group, Inc., 1907 W. Burbank Blvd., Burbank, CA 91506.

Lenga, Robert E., Ed. 1988. Sigma-Aldrich Library of Chemical Safety Data, 2d ed. Milwaukee, Wis.: Sigma-Aldrich Chemical Company.

Lenga, Robert E., Ed. 1993. Sigma-Aldrich Library of Regulatory and Safety Data. Milwaukee, Wis.: Aldrich Chemical Company.

Lewis, Richard J., Sr. 1991. Reproductively Active Chemicals: A Reference Guide. New York: Van Nostrand Reinhold.

Lewis, Richard J., Sr. 1992. Sax's Dangerous Properties of Industrial Materials, 8th ed. New York: Van Nostrand Reinhold.

Lewis, Richard J., Sr. 1993. Hazardous Chemicals Desk Reference, 3rd ed. New York: Van Nostrand Reinhold.

Lewis, Richard J., Sr. 1993. Hawley's Condensed Chemical Dictionary, 12th ed. New York: Van Nostrand Reinhold.

Lunn, G., and E. B. Sansone. 1990. Destruction of Hazardous Chemicals in the Laboratory. New York: John Wiley & Sons.

Luxon, S. G., Ed. 1992. Hazards in the Chemical Laboratory, 5th ed. Cambridge, United Kingdom: Royal Society of Chemistry.

Matheson Gas Products. 1983. Guide to Safe Handling of Compressed Gases. Secaucus, N.J.: Matheson Gas Products.

Mayo, D. D. W., S. S. Butcher, R. M. Pike, C. M. Foote, J. R. Hotham, and D. S. Page. 1986. Microscale Organic Laboratory. New York: Wiley.

McHugh, Mark A., and Val J. Krukonis. 1994. Supercritical Fluid Extraction: Principles and Practice, 2nd ed. Boston: Butterworth-Heinemann.

McKusick, B. C. 1984. Procedures for laboratory destruction of chemicals. Journal of Chemical Education A152:61.

Mikell, W. G., and W. C. Drinkard. 1984. Good practices for hood use. Journal of Chemical Education A13:61.

Mikell, W. G., and F. H. Fuller. 1988. Good practices for safe hood operation. Journal of Chemical Education A36:65.

Mikell, W. G., and L. R. Hobbs. 1981. Laboratory hood studies. Journal of Chemical Education A165:58.

National Fire Protection Association (NFPA). 1975. Hazardous Chemicals Data. Quincy, Mass.: NFPA.

National Fire Protection Association (NFPA). 1975. Manual of Hazardous Chemical Reactions: NFPA Manual 491M. Quincy, Mass.: NFPA.

National Fire Protection Association (NFPA). 1986. NFPA Directory. Quincy, Mass.: NFPA.

National Fire Protection Association (NFPA). 1990. Flammable and Combustible Liquids Code Handbook. 4th ed. Quincy, Mass.: NFPA.

National Fire Protection Association (NFPA). 1991a. National Electrical Code. National Fire Codes, Triennial, 5 Volumes. Quincy, Mass.: NFPA.

National Fire Protection Association (NFPA). 1991b. Fire Protection Guide to Hazardous Materials, 10th ed. Quincy, Mass.: NFPA.

National Fire Protection Association (NFPA). 1991c. Flammable and Combustible Liquids. NFPA Standard 30. Quincy, Mass.: NFPA.

National Fire Protection Association (NFPA). 1991d. Fire Protection for Laboratories Using Chemicals. NFPA Standard 45. Quincy, Mass.: NFPA.

National Fire Protection Association (NFPA). 1992. National Fire Codes, 1992, 12 Volumes. Quincy, Mass.: NFPA.

National Fire Protection Association (NFPA). 1993. National Electrical Code Handbook. Triennial. Quincy, Mass.: NFPA.

National Fire Protection Association (NFPA). 1994. Fire Protection for Laboratories Using Chemicals, 2nd ed. Quincy, Mass.: NFPA.

National Institutes of Health (NIH), U.S. Department of Health and Human Services (DHHS). 1981. NIH Guidelines for the Laboratory Use of Chemical Carcinogens. Washington, D.C.: U.S. Government Printing Office.

National Research Council (NRC). 1981. Prudent Practices for Handling Hazardous Chemicals in Laboratories. Washington, D.C.: National Academy Press.

National Research Council (NRC). 1983. Prudent Practices for Disposal of Chemicals from Laboratories. Washington, D.C.: National Academy Press.

National Research Council (NRC). 1989. Biosafety in the Laboratory: Prudent Practices for the Handling and Disposal of Infectious Materials. Washington, D.C.: National Academy Press.

Noyes, Robert, Ed. 1992. Handbook of Leak, Spill, and Accidental Release Prevention Techniques. Park Ridge, N.J.: Noyes Publications.

Organic Syntheses. New York: John Wiley & Sons.

Patnaik, P. A. 1992. A Comprehensive Guide to the Hazardous Properties of Chemical Substances. New York: Van Nostrand Reinhold.

Peacock, R. D. 1993. Journal of Vacuum Science and Technology A11:1627–1630.

Persson, B. R. R., and F. Stahlberg. 1989. Health and Safety of Clinical NMR Examinations. Boca Raton, Fla.: CRC Press.

Peterson, J. E. 1959. An approach to a rational method of recommending face velocities for laboratory hoods. American Industrial Hygienists Association Journal 20:259.

Peterson, J. E. 1963. Laboratory fume hoods and their exhaust systems. Air Conditioning, Heating and Ventilation 60:63.

Phifer, R. W., and W. R. McTigue, Jr. 1988. Handbook of Hazardous Waste Management for Small Quantity Generators. Chelsea, Mich.: Lewis Publishers.

Pike, R. M., Z. Szafran, and M. M. Singh. 1992. Microscale chemistry: A vision for the future. Presented at the 203rd meeting of the American Chemical Society, Symposium on Pollution Prevention and Waste Minimization in Laboratories (April), San Francisco, Calif.

Pine, Stanley H. 1988. Laboratory safety and emergency preparedness. Journal of Chemical Education 65:A98.

Pine, Stanley H. 1994. Safety lessons from an earthquake zone. Chemical Health and Safety. Volume 10. Washington, D.C.: American Chemical Society.

Pipitone, D. A., Ed. 1991. Safe Storage of Laboratory Chemicals, 2nd ed. New York: John Wiley & Sons.

Pitt, M. J. 1984. Some thoughts on temporary labels in the laboratory. Journal of Chemical Education A231-A232:61.

Pitt, M. J., and E. Pitt. 1985. Handbook of Laboratory Waste Disposal: A Practical Manual. New York: Halsted.

Purchase, R., Ed. 1994. The Laboratory Environment. Special Publication No. 136. Cambridge, United Kingdom: Royal Society of Chemistry.

Reese, K. M., Ed. 1981. Teaching Chemistry to Physically Handicapped Students. Washington, D.C.: American Chemical Society.

Ross, L. 1993. New center promotes microscale chemistry to cut wastes. Chemical & Engineering News (August 9):21.

Royal Society of Chemistry. 1989–1992. Chemical Safety Data Sheets, 5 Volumes. Cambridge, United Kingdom: Royal Society of Chemistry.

Shapiro, J. 1990. Radiation Protection—A Guide for Scientists and Physicians, 3rd ed. Cambridge, Mass.: Harvard University Press.

Shepard, Thomas H. 1992. Catalog of Teratogenic Agents, 7th ed. Baltimore, Md.: Johns Hopkins University Press.

Sittig, Marshall. 1991. Handbook of Toxic and Hazardous Chemicals and Carcinogens, 3rd ed. Park Ridge, N.J.: Noyes Publications.

Snow, J. T., Ed. 1982. Handling of Carcinogens and Hazardous Compounds. San Diego, Calif.: Calbiochem-Behring.

Stecher, Paul G., Ed. 1968. The Merck Index: An Encyclopedia of Chemicals and Drugs. 8th ed. Rahway, N.J.: Merck.

Steere, Norman V., Ed. 1968. Safety in the Chemical Laboratory. Easton, Pa.: Division of Chemical Education of the American Chemical Society.

Stimson, James A., Jeffrey J. Kimmel, and Sara Thurin Rollin. 1993. Guide to Environmental Laws: From Premanufacture to Disposal. Washington, D.C.: Bureau of National Affairs.

Stricoff, R. Scott, and Douglas B. Walters. 1990. Laboratory Health and Safety Handbook: A Guide for the Preparation of a Chemical Hygiene Plan. New York: Wiley.

Swanson, A. B., and N. V. Steere. 1981. Safety considerations for physically handicapped individuals in the chemistry laboratory. Journal of Chemical Education 234:58.

Swift, E. H., and W. P. Shaefer. 1961. Qualitative Elemental Analysis. San Francisco: Freeman.

Szafran, Z., R. M. Pike, and M. M. Singh. 1991. Microscale Inorganic Experiments. New York: John Wiley & Sons.

3M. 3M Guide to Laboratory Practices. Technical Council. Health, Safety and Environment Committee. St. Paul, Minn.: 3M Center.

Tuma, L. D. 1991. Identification of process hazards using thermal analytical techniques. Thermochimica Acta 192:121–128.

U.S. Congress. 1971. Atomic Energy Act of 1946 and Amendments. Joint Committee on Atomic Energy. Washington, D.C.: U. S. Government Printing Office.

U.S. Congress, Senate. 1978. Resource Conservation and Recovery Act, C4-4170. Committee on Environmental and Public Works. Subcommittee on Resource Protection. Washington, D.C.: U.S. Government Printing Office.

U.S. Department of Commerce and National Institute of Standards and Technology. 1993. National Electrical Safety Code. Washington, D.C.: U.S. Government Printing Office.

U.S. Department of Health, Education, and Welfare (DHEW), Public Health Service, Center for Disease Control, National Institute for Occupational Safety and Health. 1977. Carcinogens—Regulation and Control: A Management Guide to Carcinogens. Publication No. 77-205. Washington, D.C.: U.S. Government Printing Office.

U.S. Department of Health, Education, and Welfare (DHEW), Public Health Service, Center for Disease Control, National Institute for Occupational Safety and Health. 1977. Carcinogens—Regulation and Control: Working with Carcinogens: A Guide to Good Health Practices. Publication No. 77-206. Washington, D.C.: U.S. Government Printing Office.

U.S. Department of Health and Human Services (DHHS), Public Health Service, Center for Disease Control, National Institute for Occupational Safety and Health, 1981. Occupational Health Guidelines for Chemical Hazards, 3 Volumes, Mackison, F. W., R. S. Stricoff, and L. J. Partridge, Eds. DHHS (NIOSH) Publication No. 81-123. Washington, D.C.: U.S. Government Printing Office.

U.S. Department of Health and Human Services (DHHS), Public Health Service, Centers for Disease Control, National Institute for Occupational Safety and Health. 1988. Supplement to Occupational Health Guidelines for Chemical Hazards. DHHS (NIOSH) Publication No. 89-104. Washington, D.C.: U.S. Government Printing Office.

U.S. Department of Health and Human Services (DHHS), Public Health Service, Centers for Disease Control, National Institute for Occupational Safety and Health. 1990. NIOSH Pocket Guide to Chemical Hazards. Mackison, F. W., R. S. Stricoff, and L. J. Partridge, eds. DHHS (NIOSH) Publication No. 90-117. Washington, D.C.: U.S. Government Printing Office.

U.S. Department of Health and Human Services (DHHS), Public Health Service, Centers for Disease Control, National Toxicology Program. 1991. Annual Report on Carcinogens, 6th ed. Summary from the National Toxicology Program. Research Triangle Park, N.C.: Public Information Office.

U.S. Department of Health and Human Services (DHHS), Public Health Service, Centers for Disease Control and Prevention, and National Institutes of Health. 1993. Biosafety in Microbiological and Biomedical Laboratories, 3rd ed. Jonathan Richmond and Robert McKinney, Eds. Washington, D.C.: U.S. Government Printing Office.

U.S. Department of Health and Human Services (DHHS), Public Health Service, National Institute for Occupational Safety and Health. Registry of Toxic Effects of Chemical Substances. Revised annually. Washington, D.C.: U.S. Government Printing Office.

U.S. Department of Health and Human Services (DHHS), Public Health Service, Centers for Disease Control and Prevention, National Institute for Occupational Safety and Health. 1994. Applications Manual for the Revised NIOSH Lifting Equation. DHHS (NIOSH) Publication No. 94-110. Washington, D.C.: U.S. Government Printing Office.

U.S. Department of Labor. Occupational Safety and Health Administration (OSHA). 1987. Chemical Hazard Communication. OSHA Publication No. 3084. Washington, D.C.: U.S. Government Printing Office.

U.S. Department of Labor. Occupational Safety and Health Administration. 1990. Occupational Exposures to Pollution Prevention: Selected Hospital Waste Streams. Washington, D.C.: U.S. Government Printing Office.

U.S. Department of Labor. Occupational Safety and Health Administration. 1991. Chemical Information Manual, 2nd ed. Rockville, Md.: Government Institutes.

U.S. Department of Transportation (DOT). 1993. Emergency Response Guidebook: A Guidebook for First Responders During the Initial Phase of a Hazardous Materials Incident. Washington, D.C.: U.S. Government Printing Office.

U.S. Environmental Protection Agency (EPA). Office of Research and Development and Municipal Environmental Research Laboratory. 1980. A Method of Determining the Compatibility of Hazardous Wastes. Cincinnati, Ohio: U.S. Environmental Protection Agency.

U.S. Environmental Protection Agency (EPA). 1986. Understanding the Small Quantity Generator Hazardous Waste Rules: A Handbook for Small Business. EPA Publication No. 530-SW-86-019. Washington, D.C.: EPA.

U.S. Environmental Protection Agency (EPA). 1990. Guides to Pollution Prevention: Research and Educational Institutions. Washington, D.C.: U.S. Government Printing Office.

U.S. National Committee on Radiation Protection and Measurements. 1960. Protection Against Radiation from Sealed Gamma Sources. National Bureau of Standards Handbook 73. Washington, D.C.

U.S. Office of Technology Assessment. 1985. Reproductive Health Hazards in the Workplace. Washington, D.C.: U.S. Government Printing Office.

U.S. Office of Technology Assessment. 1989. Low-Level Radioactive Waste Policy Act of 1980. United States Code, Volume 42, sections 2021 (b) to 2021 (j). Washington, D.C.: U.S. Government Printing Office.

Wagner, Travis. 1990. The Hazardous Waste Q&A. New York: Van Nostrand Reinhold.

Wagner, Travis. 1991. The Complete Guide to the Hazardous Waste Regulations, 2nd ed. New York: Van Nostrand Reinhold.

Walters, D. B., Ed. 1980. Safe Handling of Chemical Carcinogens, Mutagens, Teratogens, and Highly Toxic Substances. Ann Arbor, Mich.: Ann Arbor Science.

Wayda, A. L., and M. Y. Darensbourg. 1987. Experimental Organometallic Chemistry, A Practicum in Synthesis and Characterization. Washington, D.C.: American Chemical Society.

Weiss, G., Ed. 1986. Hazardous Chemicals Data Book, 2nd ed. Park Ridge, N.J.: Noyes Data Corporation.

Wexler, P., Ed. 1981. Information Resources in Toxicology. New York: Elsevier.

Williamson, K. L., 1989. Macroscale and Microscale Organic Experiments. Lexington, Mass.: D. C. Heath.

Woodside, Gayle, and Dianna S. Kocurek. 1994. Resources and References: Hazardous Waste and Hazardous Materials Management. Park Ridge, N.J.: Noyes Publications.

Young, J. A., Ed. 1991. Improving Safety in the Chemical Laboratory: A Practical Guide, 2nd ed. New York: John Wiley & Sons.

Young, J. A., W. R. Kingsley, and G. H. Wahl. 1990. Developing a Chemical Hygiene Plan. Washington, D.C.: American Chemical Society.

APPENDIXES

Appendix A: OSHA Laboratory Standard

29 CFR 1910.1450—Occupational Exposure to Hazardous Chemicals in Laboratories

(a) Scope and application.

(1) This section shall apply to all employers engaged in the laboratory use of hazardous chemicals as defined below.

(2) Where this section applies, it shall supersede, for laboratories, the requirements of all other OSHA health standards in 29 CFR part 1910, subpart Z, except as follows:

(i) For any OSHA health standard, only the requirement to limit employee exposure to the specific permissible exposure limit shall apply for laboratories, unless that particular standard states otherwise or unless the conditions of paragraph (a)(2)(iii) of this section apply.

(ii) Prohibition of eye and skin contact where specified by any OSHA health standard shall be observed.

(iii) Where the action level (or in the absence of an action level, the permissible exposure limit) is routinely exceeded for an OSHA regulated substance with exposure monitoring and medical surveillance requirements paragraphs (d) and (g)(1)(ii) of this section shall apply.

(3) This section shall not apply to:

(i) Uses of hazardous chemicals which do not meet the definition of laboratory use, and in such cases, the employer shall comply with the relevant standard in 29 CFR part 1910, subpart Z, even if such use occurs in a laboratory.

(ii) Laboratory uses of hazardous chemicals which provide no potential for employee exposure. Examples of such conditions might include:

(A) Procedures using chemically-impregnated test media such as Dip-and-Read tests where a reagent strip is dipped into the specimen to be tested and the results are interpreted by comparing the color reaction to a color chart supplied by the manufacturer of the test strip; and

(B) Commercially prepared kits such as those used in performing pregnancy tests in which all of the reagents needed to conduct the test are contained in the kit.

(b) Definitions—"Action level" means a concentration designated in 29 CFR part 1910 for a specific substance, calculated as an eight (8)-hour time-weighted average, which initiates certain required activities such as exposure monitoring and medical surveillance.

"Assistant Secretary" means the Assistant Secretary of Labor for Occupational Safety and Health, U.S. Department of Labor, or designee. "Carcinogen" (see "select carcinogen").

"Chemical Hygiene Officer" means an employee who is designated by the employer, and who is qualified by training or experience, to provide technical guidance in the development and implementation of the provisions of the Chemical Hygiene Plan. This definition is not intended to place limitations on the position description or job classification that the designated individual shall hold within the employer's organizational structure.

"Chemical Hygiene Plan" means a written program developed and implemented by the employer which sets forth procedures, equipment, personal protective equipment and work practices that (i) are capable of protecting employees from the health hazards presented by hazardous chemicals used in that particular workplace and (ii) meets the requirements of paragraph (e) of this section. "Combustible liquid" means any liquid having a flashpoint at or above 100 deg. F (37.8 deg. C), but below 200 deg. F (93.3 deg. C), except any mixture having components with flashpoints of 200 deg. F (93.3 deg. C), or higher, the total volume of which make up 99 percent or more of the total volume of the mixture.

"Compressed gas" means: (i) A gas or mixture of gases having, in a container, an absolute pressure exceeding 40 psi at 70 deg. F (21.1 deg. C); or (ii) A gas or mixture of gases having, in a container, an absolute pressure exceeding 104 psi at 130 deg. F (54.4 deg. C) regardless of the pressure at 70 deg. F (21.1 deg. C); or (iii) A liquid having a vapor pressure exceeding 40 psi at 100 deg. F (37.8 deg. C) as determined by ASTM D-323-72.

"Designated area" means an area which may be used for work with "select carcinogens," reproductive toxins or substances which have a high degree of acute toxicity. A designated area may be the entire laboratory, such as a laboratory hood.

"Emergency" means any occurrence such as, but not limited to, equipment failure, rupture of containers or failure of control equipment which results in an uncontrolled release of a hazardous chemical into the workplace.

"Employee" means an individual employed in a laboratory workplace who may be exposed to hazardous chemicals in the course of his or her assignments.

"Explosive" means a chemical that causes a sudden, almost instantaneous release of pressure, gas, and heat when subjected to sudden shock, pressure, or high temperature.

"Flammable" means a chemical that falls into one of the following categories:

(i) "Aerosol, flammable" means an aerosol that, when tested by the method described in 16 CFR 1500.45, yields a flame protection exceeding 18 inches at full valve opening, or a flashback (a flame extending back to the valve) at any degree of valve opening;

(ii) "Gas, flammable" means: (A) A gas that, at ambient temperature and pressure, forms a flammable mixture with air at a concentration of 13 percent by volume or less; or (B) A gas that, at ambient temperature and pressure, forms a range of flammable mixtures with air wider than 12 percent by volume, regardless of the lower limit.

(iii) "Liquid, flammable" means any liquid having a flashpoint below 100 deg F (37.8 deg. C), except any mixture having components with flashpoints of 100 deg. C or higher, the total of which makes up 99 percent or more of the total volume of the mixture.

(iv) "Solid, flammable" means a solid, other than a blasting agent or explosive as defined in 1910.109(a), that is liable to cause fire through friction, absorption of moisture, spontaneous chemical change, or retained heat from manufacturing or processing, or which can be ignited readily and when ignited burns so vigorously and persistently as to create a serious hazard. A chemical shall be considered to be a flammable solid if, when tested by the method described in 16 CFR 1500.44, it ignites and burns with a self-sustained flame at a rate greater than one-tenth of an inch per second along its major axis.

"Flashpoint" means the minimum temperature at which a liquid gives off a vapor in sufficient concentration to ignite when tested as follows:

(i) Tagliabue Closed Tester (See American National Standard Method of Test for Flash Point by Tag Closed Tester, Z11.24-1979 (ASTM D 56-79))—for liquids with a viscosity of less than 45 Saybolt Universal Seconds (SUS) at 100 deg. F (37.8 deg. C), that do not contain suspended solids and do not have a tendency to form a surface film under test; or

(ii) Pensky-Martens Closed Tester (See American National Standard Method of Test for Flashpoint by Pensky-Martens Closed Tester, Z11.7-1979 (ASTM D 93-79))—for liquids with a viscosity equal to or greater than 45 SUS at 100 deg. F (37.8 deg. C), or that contain suspended solids, or that have a tendency to form a surface film under test; or

(iii) Setaflash Closed Tester (see American National Standard Method of Test for Flash Point by Setaflash Closed Tester (ASTM D 3278-78)). Organic peroxides, which undergo autoaccelerating thermal decomposition, are excluded from any of the flashpoint determination methods specified above.

"Hazardous chemical" means a chemical for which there is statistically significant evidence based on at least one study conducted in accordance with established scientific principles that acute or chronic health effects may occur in exposed employees. The term "health hazard" includes chemicals which are carcinogens, toxic or highly toxic agents, reproductive toxins, irritants, corrosives, sensitizers, hepatotoxins, nephrotoxins, neurotoxins, agents which act on the hematopoietic systems, and agents which damage the lungs, skin, eyes, or mucous membranes. Appendices A and B of the Hazard Communication Standard (29 CFR 1910.1200) provide further guidance in defining the scope of health hazards and determining whether or not a chemical is to be considered hazardous for purposes of this standard.

"Laboratory" means a facility where the "laboratory use of hazardous chemicals" occurs. It is a workplace where relatively small quantities of hazardous chemicals are used on a non-production basis.

"Laboratory scale" means work with substances in which the containers used for reactions, transfers, and other handling of substances are designed to be easily and safely manipulated by one person.

"Laboratory scale" excludes those workplaces whose function is to produce commercial quantities of materials. "Laboratory-type hood" means a device located in a laboratory, enclosure on five sides with a movable sash or fixed partial enclosed on the remaining side; constructed and maintained to draw air from the laboratory and to prevent or minimize the escape of air contaminants into the laboratory; and allows chemical manipulations to be conducted in the enclosure without insertion of any portion of the employee's body other than hands and arms. Walk-in hoods with adjustable sashes meet the above definition provided that the sashes are adjusted during use so that the airflow and the exhaust of air contaminants are not compromised and employees do not work inside the enclosure during the release of airborne hazardous chemicals.

"Laboratory use of hazardous chemicals" means handling or use of such chemicals in which all of the following conditions are met:

(i) Chemical manipulations are carried out on a "laboratory scale;"

(ii) Multiple chemical procedures or chemicals are used;

(iii) The procedures involved are not part of a production process, nor in any way simulate a production process; and

(iv) "Protective laboratory practices and equipment" are available and in common use to minimize the potential for employee exposure to hazardous chemicals.

"Medical consultation" means a consultation which takes place between an employee and a licensed physician for the purpose of determining what medical

examinations or procedures, if any, are appropriate in cases where a significant exposure to a hazardous chemical may have taken place.

"Organic peroxide" means an organic compound that contains the bivalent —O—O— structure and which may be considered to be a structural derivative of hydrogen peroxide where one or both of the hydrogen atoms have been replaced by an organic radical.

"Oxidizer" means a chemical other than a blasting agent or explosive as defined in 1910.109(a), that initiates or promotes combustion in other materials, thereby causing fire either of itself or through the release of oxygen or other gases.

"Physical hazard" means a chemical for which there is scientifically valid evidence that it is a combustible liquid, a compressed gas, explosive, flammable, an organic peroxide, an oxidizer pyrophoric, unstable (reactive) or water-reactive.

"Protective laboratory practices and equipment" means those laboratory procedures, practices and equipment accepted by laboratory health and safety experts as effective, or that the employer can show to be effective, in minimizing the potential for employee exposure to hazardous chemicals.

"Reproductive toxins" means chemicals which affect the reproductive chemicals which affect the reproductive capabilities including chromosomal damage (mutations) and effects on fetuses (teratogenesis).

"Select carcinogen" means any substance which meets one of the following criteria:

(i) It is regulated by OSHA as a carcinogen; or

(ii) It is listed under the category, "known to be carcinogens," in the Annual Report on Carcinogens published by the National Toxicology Program (NTP) (latest edition); or

(iii) It is listed under Group 1 ("carcinogenic to humans") by the International Agency for Research on Cancer Monographs (IARC) (latest editions); or

(iv) It is listed in either Group 2A or 2B by IARC or under the category, "reasonably anticipated to be carcinogens" by NTP, and causes statistically significant tumor incidence in experimental animals in accordance with any of the following criteria: (A) After inhalation exposure of 6-7 hours per day, 5 days per week, for a significant portion of a lifetime to dosages of less than 10 mg/m(3); (B) After repeated skin application of less than 300 (mg/kg of body weight) per week; or (C) After oral dosages of less than 50 mg/kg of body weight per day.

"Unstable (reactive)" means a chemical which in the pure state, or as produced or transported, will vigorously polymerize, decompose, condense, or will become self-reactive under conditions of shocks, pressure or temperature. "Water-reactive" means a chemical that reacts with water to release a gas that is either flammable or presents a health hazard.

(c) Permissible exposure limits. For laboratory uses of OSHA regulated substances, the employer shall assure that laboratory employees' exposures to such substances do not exceed the permissible exposure limits specified in 29 CFR part 1910, subpart Z.

(d) Employee exposure determination

(1) Initial monitoring. The employer shall measure the employee's exposure to any substance regulated by a standard which requires monitoring if there is reason to believe that exposure levels for that substance routinely exceed the action level (or in the absence of an action level, the PEL).

(2) Periodic monitoring. If the initial monitoring prescribed by paragraph (d)(1) of this section discloses employee exposure over the action level (or in the absence of an action level, the PEL), the employer shall immediately comply with the exposure monitoring provisions of the relevant standard.

(3) Termination of monitoring. Monitoring may be terminated in accordance with the relevant standard.

(4) Employee notification of monitoring results. The employer shall, within 15 working days after the receipt of any monitoring results, notify the employee of these results in writing either individually or by posting results in an appropriate location that is accessible to employees.

(e) Chemical hygiene plan—General. (Appendix A of this section is non-mandatory but provides guidance to assist employers in the development of the Chemical Hygiene Plan.)

(1) Where hazardous chemicals as defined by this standard are used in the workplace, the employer shall develop and carry out the provisions of a written Chemical Hygiene Plan which is:

(i) Capable of protecting employees from health hazards associated with hazardous chemicals in that laboratory and

(ii) Capable of keeping exposures below the limits specified in paragraph (c) of this section.

(2) The Chemical Hygiene Plan shall be readily available to employees, employee representatives and, upon request, to the Assistant Secretary.

(3) The Chemical Hygiene Plan shall include each of the following elements and shall indicate specific measures that the employer will take to ensure laboratory employee protection:

(i) Standard operating procedures relevant to safety and health considerations to be followed when laboratory work involves the use of hazardous chemicals;

(ii) Criteria that the employer will use to determine and implement control measures to reduce employee exposure to hazardous chemicals including engineering controls, the use of personal protective equipment and hygiene practices; particular attention shall be given to the selection of control measures for chemicals that are known to be extremely hazardous;

(iii) A requirement that fume hoods and other protective equipment are functioning properly and specific measures that shall be taken to ensure proper and adequate performance of such equipment;

(iv) Provisions for employee information and training as prescribed in paragraph (f) of this section;

(v) The circumstances under which a particular laboratory operation, procedure or activity shall require prior approval from the employer or the employer's designee before implementation;

(vi) Provisions for medical consultation and medical examinations in accordance with paragraph (g) of this section;

(vii) Designation of personnel responsible for implementation of the Chemical Hygiene Plan including the assignment of a Chemical Hygiene Officer, and, if appropriate, establishment of a Chemical Hygiene Committee; and

(viii) Provisions for additional employee protection for work with particularly hazardous substances. These include "select carcinogens," reproductive toxins and substances which have a high degree of acute toxicity. Specific consideration shall be given to the following provisions which shall be included where appropriate:

(A) Establishment of a designated area;

(B) Use of containment devices such as fume hoods or glove boxes;

(C) Procedures for safe removal of contaminated waste; and

(D) Decontamination procedures.

(4) The employer shall review and evaluate the effectiveness of the Chemical Hygiene Plan at least annually and update it as necessary.

(f) Employee information and training.

(1) The employer shall provide employees with information and training to ensure that they are apprised of the hazards of chemicals present in their work area.

(2) Such information shall be provided at the time of an employee's initial assignment to a work area where hazardous chemicals are present and prior to assignments involving new exposure situations. The frequency of refresher information and training shall be determined by the employer.

(3) Information. Employees shall be informed of:

(i) The contents of this standard and its appendices which shall be made available to employees;

(ii) the location and availability of the employer's Chemical Hygiene Plan;

(iii) The permissible exposure limits for OSHA regulated substances or recommended exposure limits for other hazardous chemicals where there is no applicable OSHA standard;

(iv) Signs and symptoms associated with exposures to hazardous chemicals used in the laboratory; and

(v) The location and availability of known reference material on the hazards, safe handling, storage and disposal of hazardous chemicals found in the laboratory including, but not limited to, Material Safety Data Sheets received from the chemical supplier.

(4) Training.

(i) Employee training shall include:

(A) Methods and observations that may be used to detect the presence or release of a hazardous chemical (such as monitoring conducted by the employer, continuous monitoring devices, visual appearance or odor of hazardous chemicals when being released, etc.);

(B) The physical and health hazards of chemicals in the work area; and

(C) The measures employees can take to protect themselves from these hazards, including specific procedures the employer has implemented to protect employees from exposure to hazardous chemicals, such as appropriate work practices, emergency procedures, and personal protective equipment to be used.

(ii) The employee shall be trained on the applicable details of the employer's written Chemical Hygiene Plan.

(g) Medical consultation and medical examinations.

(1) The employer shall provide all employees who work with hazardous chemicals an opportunity to receive medical attention, including any follow-up examinations which the examining physician determines to be necessary, under the following circumstances:

(i) Whenever an employee develops signs or symptoms associated with a hazardous chemical to which the employee may have been exposed in the laboratory, the employee shall be provided an opportunity to receive an appropriate medical examination.

(ii) Where exposure monitoring reveals an exposure level routinely above the action level (or in the absence of an action level, the PEL) for an OSHA regulated substance for which there are exposure monitoring and medical surveillance requirements, medical surveillance shall be established for the affected employee as prescribed by the particular standard.

(iii) Whenever an event takes place in the work area such as a spill, leak, explosion or other occurrence resulting in the likelihood of a hazardous exposure, the affected employee shall be provided an opportunity for a medical consultation. Such consultation shall be for the purpose of determining the need for a medical examination.

(2) All medical examinations and consultations shall be performed by or under the direct supervision of a licensed physician and shall be provided without cost to the employee, without loss of pay and at a reasonable time and place.

(3) Information provided to the physician. The employer shall provide the following information to the physician:

(i) The identity of the hazardous chemical(s) to which the employee may have been exposed;

(ii) A description of the conditions under which the exposure occurred including quantitative exposure data, if available; and

(iii) A description of the signs and symptoms of exposure that the employee is experiencing, if any.

(4) Physician's written opinion.

(i) For examination or consultation required under this standard, the employer shall obtain a written opinion from the examining physician which shall include the following:

(A) Any recommendation for further medical follow-up;

(B) The results of the medical examination and any associated tests;

(C) Any medical condition which may be revealed in the course of the examination which may place the employee at increased risk as a result of exposure to a hazardous workplace; and

(D) A statement that the employee has been informed by the physician of the results of the consultation or medical examination and any medical condition that may require further examination or treatment.

(ii) The written opinion shall not reveal specific findings of diagnoses unrelated to occupational exposure.

(h) Hazard identification.

(1) With respect to labels and material safety data sheets:

(i) Employers shall ensure that labels on incoming containers of hazardous chemicals are not removed or defaced.

(ii) Employers shall maintain any material safety data sheets that are received with incoming shipments of hazardous chemicals, and ensure that they are readily accessible to laboratory employees.

(2) The following provisions shall apply to chemical substances developed in the laboratory:

(i) If the composition of the chemical substance which is produced exclusively for the laboratory's use is known, the employer shall determine if it is a hazardous chemical as defined in paragraph (b) of this section. If the chemical is determined to be hazardous, the employer shall provide appropriate training as required under paragraph (f) of this section.

(ii) If the chemical produced is a byproduct whose composition is not known, the employer shall assume that the substance is hazardous and shall implement paragraph (e) of this section.

(iii) If the chemical substance is produced for another user outside of the laboratory, the employer shall comply with the Hazard Communication Standard (29 CFR 1910.120) including the requirements for preparation of material safety data sheets and labeling.

(i) Use of respirators. Where the use of respirators is necessary to maintain exposure below permissible exposure limits, the employer shall provide, at no cost to the employee, the proper respiratory equipment. Respirators shall be selected and used in accordance with the requirements of 29 CFR 1910.134.

(j) Recordkeeping.

(1) The employer shall establish and maintain for each employee an accurate record of any measurements taken to monitor employee exposures and any medical consultation and examinations including tests or written opinions required by this standard.

(2) The employer shall assure that such records are kept, transferred, and made available in accordance with 29 CFR 1910.20.

(k) Dates.

(1) Effective date. This section shall become effective May 1, 1990.

(2) Start-up dates.

(i) Employers shall have developed and implemented a written Chemical Hygiene Plan no later than January 31, 1991.

(ii) Paragraph (a)(2) of this section shall not take effect until the employer has developed and implemented a written Chemical Hygiene Plan.

(l) Appendices. The information contained in the appendices is not intended, by itself, to create any additional obligations not otherwise imposed or to detract from any existing obligation.

Appendix A to 1910.1450—National Research Council Recommendations Concerning Chemical Hygiene in Laboratories (Non-Mandatory)

Table of Contents

Foreword

Corresponding Sections of the Standard and This Appendix

A. General Principles

1. Minimize All Chemical Exposures
2. Avoid Underestimation of Risk
3. Provide Adequate Ventilation
4. Institute a Chemical Hygiene Program
5. Observe the PELs and TLVs

B. Responsibilities

1. Chief Executive Officer
2. Supervisor of Administrative Unit
3. Chemical Hygiene Officer
4. Laboratory Supervisor
5. Project Director
6. Laboratory Worker

C. The Laboratory Facility

1. Design
2. Maintenance
3. Usage
4. Ventilation

D. Components of the Chemical Hygiene Plan

1. Basic Rules and Procedures
2. Chemical Procurement, Distribution, and Storage
3. Environmental Monitoring
4. Housekeeping, Maintenance and Inspections
5. Medical Program
6. Personal Protective Apparel and Equipment
7. Records
8. Signs and Labels
9. Spills and Accidents
10. Training and Information
11. Waste Disposal

E. General Procedures for Working with Chemicals

1. General Rules for All Laboratory Work with Chemicals
2. Allergens and Embryotoxins
3. Chemicals of Moderate Chronic or High Acute Toxicity
4. Chemicals of High Chronic Toxicity
5. Animal Work with Chemicals of High Chronic Toxicity

F. Safety Recommendations

G. Material Safety Data Sheets

Foreword

As guidance for each employer's development of an appropriate laboratory Chemical Hygiene Plan, the following non-mandatory recommendations are provided. They were extracted from "Prudent Practices for Handling Hazardous Chemicals in Laboratories" (referred to below as "Prudent Practices"), which was published in 1981 by the National Research Council and is available from the National Academy Press, 2101 Constitution Ave., NW, Washington DC 20418.

"Prudent Practices" is cited because of its wide distribution and acceptance and because of its preparation by members of the laboratory community through the sponsorship of the National Research Council. However, none of the recommendations given here will modify any requirements of the laboratory standard. This appendix merely presents pertinent recommendations from "Prudent Practices," organized into a form convenient for quick reference during operation of a laboratory facility and during development and application of a Chemical Hygiene Plan. Users of this appendix should consult "Prudent Practices" for a more extended presentation and justification for each recommendation.

"Prudent Practices" deals with both safety and chemical hazards while the laboratory standard is concerned primarily with chemical hazards. Therefore, only those recommendations directed primarily toward control of toxic exposures are cited in this appendix, with the term "chemical hygiene" being substituted for the word "safety." However, since conditions producing or threatening physical injury often pose toxic risks as well, page references concerning major categories of safety hazards in the laboratory are given in section F.

The recommendations from "Prudent Practices" have been paraphrased, combined, or otherwise reorganized, and headings have been added. However, their sense has not been changed.

Corresponding Sections of the Standard and This Appendix

The following table is given for the convenience of those who are developing a Chemical Hygiene Plan which will satisfy the requirements of paragraph (e) of the standard. It indicates those sections of this appendix which are most pertinent to each of the sections of paragraph (e) and related paragraphs.

Paragraph and topic in laboratory standard	Relevant appendix section
(e)(3)(i) Standard operating procedures for handling toxic chemicals.	C, D, E
(e)(3)(ii) Criteria to be used for implementation of measures to reduce exposures.	D
(e)(3)(iii) Fume hood performance.	C4b
(e)(3)(iv) Employee information and training (including emergency procedures).	D10, D9
(e)(3)(v) Requirements for prior approval of laboratory activities.	E2b, E4b
(e)(3)(vi) Medical consultation and medical examinations.	D5, E4f
(e)(3)(vii) Chemical hygiene responsibilities.	B
(e)(3)(viii) Special precautions for work with particularly hazardous substances.	E2, E3, E4

In this appendix, those recommendations directed primarily at administrators and supervisors are given in sections A-D. Those recommendations of primary concern to employees who are actually handling laboratory chemicals are given in section E. (References to page numbers in "Prudent Practices" are given in parentheses.)

A. General Principles for Work with Laboratory Chemicals

In addition to the more detailed recommendations listed below in sections B-E, "Prudent Practices" expresses certain general principles, including the following:

1. It is prudent to minimize all chemical exposures. Because few laboratory chemicals are without hazards, general precautions for handling all laboratory chemicals should be adopted, rather than specific guidelines for particular chemicals (2,10). Skin contact with chemicals should be avoided as a cardinal rule (198).

2. Avoid underestimation of risk. Even for substances of no known significant hazard, exposure should be minimized; for work with substances which present special hazards, special precautions should be taken (10, 37, 38). One should assume that any mixture will be more toxic than its most toxic component (30, 103) and that all substances of unknown toxicity are toxic (3, 34).

3. Provide adequate ventilation. The best way to prevent exposure to airborne substances is to prevent their escape into the working atmosphere by use of hoods and other ventilation devices (32, 198).

4. Institute a chemical hygiene program. A mandatory chemical hygiene program designed to minimize exposures is needed; it should be a regular, continuing effort, not merely a standby or short-term activity (6,11). Its recommendations should be followed in academic teaching laboratories as well as by full-time laboratory workers (13).

5. Observe the PELs, TLVs. The Permissible Exposure Limits of OSHA and the Threshold Limit Values of the American Conference of Governmental Industrial Hygienists should not be exceeded (13).

B. Chemical Hygiene Responsibilities

Responsibility for chemical hygiene rests at all levels (6, 11, 21) including the:

1. Chief executive officer, who has ultimate responsibility for chemical hygiene within the institution and must, with other administrators, provide continuing support for institutional chemical hygiene (7, 11).

2. Supervisor of the department or other administrative unit, who is responsible for chemical hygiene in that unit (7).

3. Chemical hygiene officer(s), whose appointment is essential (7) and who must:

(a) Work with administrators and other employees to develop and

(b) Monitor procurement, use, and disposal of chemicals used in the lab (8);

(c) See that appropriate audits are maintained (8);

(d) Help project directors develop precautions and adequate facilities (10);

(e) Know the current legal requirements concerning regulated substances (50); and

(f) Seek ways to improve the chemical hygiene program (8, 11).

4. Laboratory supervisor, who has overall responsibility for chemical hygiene in the laboratory (21) including responsibility to:

(a) Ensure that workers know and follow the chemical hygiene rules, that protective equipment is available and in working order, and that appropriate training has been provided (21, 22);

(b) Provide regular, formal chemical hygiene and housekeeping inspections including routine inspections of emergency equipment (21, 171);

(c) Know the current legal requirements concerning regulated substances (50, 231);

(d) Determine the required levels of protective apparel and equipment (156, 160, 162); and

(e) Ensure that facilities and training for use of any material being ordered are adequate (215).

5. Project director or director of other specific operation, who has primary responsibility for chemical hygiene procedures for that operation (7).

6. Laboratory worker, who is responsible for:

(a) Planning and conducting each operation in accordance with the institutional chemical hygiene procedures (7, 21, 22, 230); and

(b) Developing good personal chemical hygiene habits (22).

C. The Laboratory Facility

1. Design. The laboratory facility should have:

(a) An appropriate general ventilation system (see C4 below) with air intakes and exhausts located so as to avoid intake of contaminated air (194);

(b) Adequate, well-ventilated stockrooms/storerooms (218, 219);

(c) Laboratory hoods and sinks (12, 162);

(d) Other safety equipment including eyewash fountains and drench showers (162, 169); and

(e) Arrangements for waste disposal (12, 240).

2. Maintenance. Chemical-hygiene-related equipment (hoods, incinerator, etc.) should undergo continual appraisal and be modified if inadequate (11, 12).

3. Usage. The work conducted (10) and its scale (12) must be appropriate to the physical facilities available and, especially, to the quality of ventilation (13).

4. Ventilation—(a) General laboratory ventilation. This system should: Provide a source of air for breathing and for input to local ventilation devices (199); it should not be relied on for protection from toxic substances released into the laboratory (198); ensure that laboratory air is continually replaced, preventing increase of air concentrations of toxic substances during the working day (194); direct air flow into the laboratory from non-laboratory areas and out to the exterior of the building (194).

(b) Hoods. A laboratory hood with 2.5 linear feet of hood space per person should be provided for every 2 workers if they spend most of their time working with chemicals (199); each hood should have a continuous monitoring device to allow convenient confirmation of adequate hood performance before use (200, 209). If this is not possible, work with substances of unknown toxicity should be avoided (13) or other types of local ventilation devices should be provided (199). See pp. 201-206 for a discussion of hood design, construction, and evaluation.

(c) Other local ventilation devices. Ventilated storage cabinets, canopy hoods, snorkels, etc. should be provided as needed (199). Each canopy hood and snorkel should have a separate exhaust duct (207).

(d) Special ventilation areas. Exhaust air from glove boxes and isolation rooms should be passed through scrubbers or other treatment before release into the regular exhaust system (208). Cold rooms and warm rooms should have provisions for rapid escape and for escape in the event of electrical failure (209).

(e) Modifications. Any alteration of the ventilation system should be made only if thorough testing indicates that worker protection from airborne toxic substances will continue to be adequate (12, 193, 204).

(f) Performance. Rate: 4-12 room air changes/hour is normally adequate general ventilation if local exhaust systems such as hoods are used as the primary method of control (194).

(g) Quality. General air flow should not be turbulent and should be relatively uniform throughout the laboratory, with no high velocity or static areas (194, 195); airflow into and within the hood should not be excessively turbulent (200); hood face velocity should be adequate (typically 60-100 lfm) (200, 204).

(h) Evaluation. Quality and quantity of ventilation should be evaluated on installation (202), regularly monitored (at least every 3 months) (6, 12, 14, 195), and reevaluated whenever a change in local ventilation devices is made (12, 195, 207). See pp. 195-198 for methods of evaluation and for calculation of estimated airborne contaminant concentrations.

D. Components of the Chemical Hygiene Plan

1. Basic Rules and Procedures
(Recommendations for these are given in section E, below.)

2. Chemical Procurement, Distribution, and Storage
(a) Procurement. Before a substance is received, information on proper handling, storage, and disposal should be known to those who will be involved (215, 216). No container should be accepted without an adequate identifying label (216). Preferably, all substances should be received in a central location (216).

(b) Stockrooms/storerooms. Toxic substances should be segregated in a well-identified area with local exhaust ventilation (221). Chemicals which are highly toxic (227) or other chemicals whose containers have been opened should be in unbreakable secondary containers (219). Stored chemicals should be examined periodically (at least annually) for replacement, deterioration, and container integrity (218-19). Stockrooms/storerooms should not be used as preparation or repackaging areas, should be open during normal working hours, and should be controlled by one person (219).

(c) Distribution. When chemicals are hand carried, the container should be placed in an outside container or bucket. Freight-only elevators should be used if possible (223).

(d) Laboratory storage. Amounts permitted should be as small as practical. Storage on bench tops and in hoods is inadvisable. Exposure to heat or direct sunlight should be avoided. Periodic inventories should be conducted, with unneeded items being discarded or returned to the storeroom/stockroom (225-6, 229).

3. Environmental Monitoring
Regular instrumental monitoring of airborne concentrations is not usually justified or practical in laboratories but may be appropriate when testing or redesigning hoods or other ventilation devices (12) or when a highly toxic substance is stored or used regularly (e.g., 3 times/week) (13).

4. Housekeeping, Maintenance, and Inspections
(a) Cleaning. Floors should be cleaned regularly (24).

(b) Inspections. Formal housekeeping and chemical hygiene inspections should be held at least quarterly (6, 21) for units which have frequent personnel changes and semiannually for others; informal inspections should be continual (21).

(c) Maintenance. Eye wash fountains should be inspected at intervals of not less than 3 months (6). Respirators for routine use should be inspected periodically by the laboratory supervisor (169). Other safety equipment should be inspected regularly (e.g., every 3-6 months) (6, 24, 171). Procedures to prevent restarting of out-of-service equipment should be established (25).

(d) Passageways. Stairways and hallways should not be used as storage areas (24). Access to exits, emergency equipment, and utility controls should never be blocked (24).

5. Medical Program
(a) Compliance with regulations. Regular medical surveillance should be established to the extent required by regulations (12).

(b) Routine surveillance. Anyone whose work involves regular and frequent handling of toxicologically significant quantities of a chemical should consult a qualified physician to determine on an individual basis whether a regular schedule of medical surveillance is desirable (11, 50).

(c) First aid. Personnel trained in first aid should be available during working hours and an emergency room with medical personnel should be nearby (173). See pp. 176-178 for description of some emergency first aid procedures.

6. Protective Apparel and Equipment
These should include for each laboratory:
(a) Protective apparel compatible with the required degree of protection for substances being handled (158-161);

(b) An easily accessible drench-type safety shower (162, 169);

(c) An eyewash fountain (162)

(d) A fire extinguisher (162-164);

(e) Respiratory protection (164-9), fire alarm and telephone for emergency use (162) should be available nearby; and

(f) Other items designated by the laboratory supervisor (156, 160).

7. Records
(a) Accident records should be written and retained (174).

(b) Chemical Hygiene Plan records should document that the facilities and precautions were compatible with current knowledge and regulations (7).

(c) Inventory and usage records for high-risk substances should be kept as specified in section E3e below.

(d) Medical records should be retained by the institution in accordance with the requirements of state and federal regulations (12).

8. Signs and Labels

Prominent signs and labels of the following types should be posted:

(a) Emergency telephone numbers of emergency personnel/facilities, supervisors, and laboratory workers (28);

(b) Identity labels, showing contents of containers (including waste receptacles) and associated hazards (27, 48);

(c) Location signs for safety showers, eyewash stations, other safety and first aid equipment, exits (27) and areas where food and beverage consumption and storage are permitted (24); and

(d) Warnings at areas or equipment where special or unusual hazards exist (27).

9. Spills and Accidents

(a) A written emergency plan should be established and communicated to all personnel; it should include procedures for ventilation failure (200), evacuation, medical care, reporting, and drills (172).

(b) There should be an alarm system to alert people in all parts of the facility including isolation areas such as cold rooms (172).

(c) A spill control policy should be developed and should include consideration of prevention, containment, cleanup, and reporting (175).

(d) All accidents or near accidents should be carefully analyzed with the results distributed to all who might benefit (8, 28).

10. Information and Training Program

(a) Aim: To assure that all individuals at risk are adequately informed about the work in the laboratory, its risks, and what to do if an accident occurs (5, 15).

(b) Emergency and Personal Protection Training: Every laboratory worker should know the location and proper use of available protective apparel and equipment (154, 169). Some of the full-time personnel of the laboratory should be trained in the proper use of emergency equipment and procedures (6). Such training as well as first aid instruction should be available to (154) and encouraged for (176) everyone who might need it.

(c) Receiving and stockroom/storeroom personnel should know about hazards, handling equipment, protective apparel, and relevant regulations (217).

(d) Frequency of Training: The training and education program should be a regular, continuing activity—not simply an annual presentation (15).

(e) Literature/Consultation: Literature and consult-ing advice concerning chemical hygiene should be readily available to laboratory personnel, who should be encouraged to use these information resources (14).

11. Waste Disposal Program

(a) Aim: To assure that minimal harm to people, other organisms, and the environment will result from the disposal of waste laboratory chemicals (5).

(b) Content (14, 232, 233, 240): The waste disposal program should specify how waste is to be collected, segregated, stored, and transported and include consideration of what materials can be incinerated. Transport from the institution must be in accordance with DOT regulations (244).

(c) Discarding Chemical Stocks: Unlabeled containers of chemicals and solutions should undergo prompt disposal; if partially used, they should not be opened (24, 27). Before a worker's employment in the laboratory ends, chemicals for which that person was responsible should be discarded or returned to storage (226).

(d) Frequency of Disposal: Waste should be removed from laboratories to a central waste storage area at least once per week and from the central waste storage area at regular intervals (14).

(e) Method of Disposal: Incineration in an environmentally acceptable manner is the most practical disposal method for combustible laboratory waste (14, 238, 241). Indiscriminate disposal by pouring waste chemicals down the drain (14, 231, 242) or adding them to mixed refuse for landfill burial is unacceptable (14). Hoods should not be used as a means of disposal for volatile chemicals (40, 200). Disposal by recycling (233, 243) or chemical decontamination (40, 230) should be used when possible.

E. Basic Rules and Procedures for Working with Chemicals

The Chemical Hygiene Plan should require that laboratory workers know and follow its rules and procedures. In addition to the procedures of the sub programs mentioned above, these should include the rules listed below.

1. General Rules

The following should be used for essentially all laboratory work with chemicals:

(a) Accidents and Spills—Eye Contact: Promptly flush eyes with water for a prolonged period (15 minutes) and seek medical attention (33, 172).

Ingestion: Encourage the victim to drink large amounts of water (178).

Skin Contact: Promptly flush the affected area with water (33, 172, 178) and remove any contaminated clothing (172, 178). If symptoms persist after washing, seek medical attention (33). Clean-up. Promptly clean up spills, using appropriate protective apparel and equipment and proper disposal (24, 33). See pp. 233-237 for specific clean-up recommendations.

(b) Avoidance of "routine" exposure: Develop and encourage safe habits (23); avoid unnecessary exposure to chemicals by any route (23). Do not smell or taste chemicals (32). Vent apparatus which may discharge toxic chemicals (vacuum pumps, distillation columns, etc.) into local exhaust devices (199). Inspect gloves (157) and test glove boxes (208) before use. Do not allow release of toxic substances in cold rooms and warm rooms, since these have contained recirculated atmospheres (209).

(c) Choice of chemicals: Use only those chemicals for which the quality of the available ventilation system is appropriate (13).

(d) Eating, smoking, etc.: Avoid eating, drinking, smoking, gum chewing, or application of cosmetics in areas where laboratory chemicals are present (22, 24, 32, 40); wash hands before conducting these activities (23, 24). Avoid storage, handling, or consumption of food or beverages in storage areas, refrigerators, glassware or utensils which are also used for laboratory operations (23, 24, 226).

(e) Equipment and glassware: Handle and store laboratory glassware with care to avoid damage; do not use damaged glassware (25). Use extra care with Dewar flasks and other evacuated glass apparatus; shield or wrap them to contain chemicals and fragments should implosion occur (25). Use equipment only for its designed purpose (23, 26).

(f) Exiting: Wash areas of exposed skin well before leaving the laboratory (23).

(g) Horseplay: Avoid practical jokes or other behavior which might confuse, startle or distract another worker (23).

(h) Mouth suction: Do not use mouth suction for pipeting or starting a siphon (23, 32).

(i) Personal apparel: Confine long hair and loose clothing (23, 158). Wear shoes at all times in the laboratory but do not wear sandals, perforated shoes, or sneakers (158).

(j) Personal housekeeping: Keep the work area clean and uncluttered, with chemicals and equipment being properly labeled and stored; clean up the work area on completion of an operation or at the end of each day (24).

(k) Personal protection: Assure that appropriate eye protection (154-156) is worn by all persons, including visitors, where chemicals are stored or handled (22, 23, 33, 154). Wear appropriate gloves when the potential for contact with toxic materials exists (157); inspect the gloves before each use, wash them before removal,

and replace them periodically (157). (A table of resistance to chemicals of common glove materials is given on p. 159). Use appropriate (164-168) respiratory equipment when air contaminant concentrations are not sufficiently restricted by engineering controls (164-5), inspecting the respirator before use (169). Use any other protective and emergency apparel and equipment as appropriate (22, 157-162). Avoid use of contact lenses in the laboratory unless necessary; if they are used, inform supervisor so special precautions can be taken (155). Remove laboratory coats immediately on significant contamination (161).

(l) Planning: Seek information and advice about hazards (7), plan appropriate protective procedures, and plan positioning of equipment before beginning any new operation (22, 23).

(m) Unattended operations: Leave lights on, place an appropriate sign on the door, and provide for containment of toxic substances in the event of failure of a utility service (such as cooling water) to an unattended operation (27, 128).

(n) Use of hood: Use the hood for operations which might result in release of toxic chemical vapors or dust (198-9). As a rule of thumb, use a hood or other local ventilation device when working with any appreciably volatile substance with a TLV of less than 50 ppm (13). Confirm adequate hood performance before use; keep hood closed at all times except when adjustments within the hood are being made (200); keep materials stored in hoods to a minimum and do not allow them to block vents or air flow (200). Leave the hood "on" when it is not in active use if toxic substances are stored in it or if it is uncertain whether adequate general laboratory ventilation will be maintained when it is "off" (200).

(o) Vigilance: Be alert to unsafe conditions and see that they are corrected when detected (22).

(p) Waste disposal: Assure that the plan for each laboratory operation includes plans and training for waste disposal (230). Deposit chemical waste in appropriately labeled receptacles and follow all other waste disposal procedures of the Chemical Hygiene Plan (22, 24). Do not discharge to the sewer concentrated acids or bases (231); highly toxic, malodorous, or lachrymatory substances (231); or any substances which might interfere with the biological activity of waste water treatment plants, create fire or explosion hazards, cause structural damage or obstruct flow (242).

(q) Working alone: Avoid working alone in a building; do not work alone in a laboratory if the procedures being conducted are hazardous (28).

2. Working with Allergens and Embryotoxins

(a) Allergens (examples: diazomethane, isocyanates, bichromates): Wear suitable gloves to prevent hand contact with allergens or substances of unknown allergenic activity (35).

(b) Embryotoxins (34-5) (examples: organomercurials, lead compounds, formamide): If you are a woman of childbearing age, handle these substances only in a hood whose satisfactory performance has been confirmed, using appropriate protective apparel (especially gloves) to prevent skin contact. Review each use of these materials with the research supervisor and review continuing uses annually or whenever a procedural change is made. Store these substances, properly labeled, in an adequately ventilated area in an unbreakable secondary container. Notify supervisors of all incidents of exposure or spills; consult a qualified physician when appropriate.

3. Work with Chemicals of Moderate Chronic or High Acute Toxicity

Examples: diisopropylfluorophosphate (41), hydrofluoric acid (43), hydrogen cyanide (45).

Supplemental rules to be followed in addition to those mentioned above (Procedure B of "Prudent Practices", pp. 39-41):

(a) Aim: To minimize exposure to these toxic substances by any route using all reasonable precautions (39).

(b) Applicability: These precautions are appropriate for substances with moderate chronic or high acute toxicity used in significant quantities (39).

(c) Location: Use and store these substances only in areas of restricted access with special warning signs (40, 229). Always use a hood (previously evaluated to confirm adequate performance with a face velocity of at least 60 linear feet per minute) (40) or other containment device for procedures which may result in the generation of aerosols or vapors containing the substance (39); trap released vapors to prevent their discharge with the hood exhaust (40).

(d) Personal protection: Always avoid skin contact by use of gloves and long sleeves (and other protective apparel as appropriate) (39). Always wash hands and arms immediately after working with these materials (40).

(e) Records: Maintain records of the amounts of these materials on hand, amounts used, and the names of the workers involved (40, 229).

(f) Prevention of spills and accidents: Be prepared for accidents and spills (41). Assure that at least 2 people are present at all times if a compound in use is highly toxic or of unknown toxicity (39). Store breakable containers of these substances in chemically resistant trays; also work and mount apparatus above such trays or cover work and storage surfaces with removable, absorbent, plastic backed paper (40). If a major spill occurs outside the hood, evacuate the area; assure

that cleanup personnel wear suitable protective apparel and equipment (41).

(g) Waste: Thoroughly decontaminate or incinerate contaminated clothing or shoes (41). If possible, chemically decontaminate by chemical conversion (40). Store contaminated waste in closed, suitably labeled, impervious containers (for liquids, in glass or plastic bottles half-filled with vermiculite) (40).

4. Work with Chemicals of High Chronic Toxicity

Examples: dimethylmercury and nickel carbonyl (48), benzo-a-pyrene (51), N-nitrosodiethylamine (54), other human carcinogens or substances with high carcinogenic potency in animals (38).

Further supplemental rules to be followed, in addition to all those mentioned above, for work with substances of known high chronic toxicity (in quantities above a few milligrams to a few grams, depending on the substance) (47). (Procedure A of "Prudent Practices" pp. 47-50.)

(a) Access: Conduct all transfers and work with these substances in a "controlled area": a restricted access hood, glove box, or portion of a lab, designated for use of highly toxic substances, for which all people with access are aware of the substances being used and necessary precautions (48).

(b) Approvals: Prepare a plan for use and disposal of these materials and obtain the approval of the laboratory supervisor (48).

(c) Non-contamination/Decontamination: Protect vacuum pumps against contamination by scrubbers or HEPA filters and vent them into the hood (49). Decontaminate vacuum pumps or other contaminated equipment, including glassware, in the hood before removing them from the controlled area (49, 50). Decontaminate the controlled area before normal work is resumed there (50).

(d) Exiting: On leaving a controlled area, remove any protective apparel (placing it in an appropriate, labeled container) and thoroughly wash hands, forearms, face, and neck (49).

(e) Housekeeping: Use a wet mop or a vacuum cleaner equipped with a HEPA filter instead of dry sweeping if the toxic substance was a dry powder (50).

(f) Medical surveillance: If using toxicologically significant quantities of such a substance on a regular basis (e.g., 3 times per week), consult a qualified physician concerning desirability of regular medical surveillance (50).

(g) Records: Keep accurate records of the amounts of these substances stored (229) and used, the dates of use, and names of users (48).

(h) Signs and labels: Assure that the controlled area is conspicuously marked with warning and restricted access signs (49) and that all containers of these substances are appropriately labeled with identity and warning labels (48).

(i) Spills: Assure that contingency plans, equipment, and materials to minimize exposures of people and property in case of accident are available (233-4).

(j) Storage: Store containers of these chemicals only in a ventilated, limited access (48, 227, 229) area in appropriately labeled, unbreakable, chemically resistant, secondary containers (48, 229).

(k) Glove boxes: For a negative pressure glove box, ventilation rate must be at least 2 volume changes/ hour and pressure at least 0.5 inches of water (48). For a positive pressure glove box, thoroughly check for leaks before each use (49). In either case, trap the exit gases or filter them through a HEPA filter and then release them into the hood (49).

(l) Waste: Use chemical decontamination whenever possible; ensure that containers of contaminated waste (including washings from contaminated flasks) are transferred from the controlled area in a secondary container under the supervision of authorized personnel (49, 50, 233).

5. Animal Work with Chemicals of High Chronic Toxicity

(a) Access: For large scale studies, special facilities with restricted access are preferable (56).

(b) Administration of the toxic substance: When possible, administer the substance by injection or gavage instead of in the diet. If administration is in the diet, use a caging system under negative pressure or under laminar air flow directed toward HEPA filters (56).

(c) Aerosol suppression: Devise procedures which minimize formation and dispersal of contaminated aerosols, including those from food, urine, and feces (e.g., use HEPA filtered vacuum equipment for cleaning, moisten contaminated bedding before removal from the cage, mix diets in closed containers in a hood) (55, 56).

(d) Personal protection: When working in the animal room, wear plastic or rubber gloves, fully buttoned laboratory coat or jumpsuit and, if needed because of incomplete suppression of aerosols, other apparel and equipment (shoe and head coverings, respirator) (56).

(e) Waste disposal: Dispose of contaminated animal tissues and excreta by incineration if the available incinerator can convert the contaminant to non-toxic products (238); otherwise, package the waste appropriately for burial in an EPA-approved site (239).

F. Safety Recommendations

The above recommendations from "Prudent Practices" do not include those which are directed primar-

ily toward prevention of physical injury rather than toxic exposure. However, failure of precautions against injury will often have the secondary effect of causing toxic exposures. Therefore, we list below page references for recommendations concerning some of the major categories of safety hazards which also have implications for chemical hygiene:

1. Corrosive agents: (35-6)
2. Electrically powered laboratory apparatus: (179-92)
3. Fires, explosions: (26, 57-74, 162-4, 174-5, 219-20, 226-7)
4. Low temperature procedures: (26, 88)
5. Pressurized and vacuum operations (including use of compressed gas cylinders): (27, 75-101)

G. Material Safety Data Sheets

Material safety data sheets are presented in "Prudent Practices" for the chemicals listed below. (Asterisks denote that comprehensive material safety data sheets are provided.)

*Acetyl peroxide (105)
*Acrolein (106)
*Acrylonitrile
 Ammonia (anhydrous) (91)
*Aniline (109)
*Benzene (110)
*Benzo[a]pyrene (112)
*Bis(chloromethyl) ether (113)
 Boron trichloride (91)
 Boron trifluoride (92)
 Bromine (114)
*Tert-butyl hydroperoxide (148)
*Carbon disulfide (116)
 Carbon monoxide (92)
*Carbon tetrachloride (118)
*Chlorine (119)
 Chlorine trifluoride (94)
*Chloroform (121)
 Chloromethane (93)
*Diethyl ether (122)
 Diisopropyl fluorophosphate (41)
*Dimethylformamide (123)
*Dimethyl sulfate (125)
*Dioxane (126)
*Ethylene dibromide (128)
*Fluorine (95)
*Formaldehyde (130)
*Hydrazine and salts (132)
 Hydrofluoric acid (43)
 Hydrogen bromide (98)

Hydrogen chloride (98)
*Hydrogen cyanide (133)
*Hydrogen sulfide (135)
Mercury and compounds (52)
*Methanol (137)
*Morpholine (138)
*Nickel carbonyl (99)
*Nitrobenzene (139)
Nitrogen dioxide (100)
N-nitrosodiethylamine (54)
*Peracetic acid (141)
*Phenol (142)
*Phosgene (143)
*Pyridine (144)
*Sodium azide (145)
*Sodium cyanide (147)
Sulfur dioxide (101)
*Trichloroethylene (149)
*Vinyl chloride (150)

29 CFR 1910.1450 App. B References (Non-Mandatory)

Appendix B to 1910.1450—References (Non-Mandatory)

The following references are provided to assist the employer in the development of a Chemical Hygiene Plan. The materials listed below are offered as non-mandatory guidance. References listed here do not imply specific endorsement of a book, opinion, technique, policy or a specific solution for a safety or health problem. Other references not listed here may better meet the needs of a specific laboratory.

(a) MATERIALS FOR THE DEVELOPMENT OF THE CHEMICAL HYGIENE PLAN:
1. American Chemical Society, Safety in Academic Chemistry Laboratories, 4th edition, 1985.
2. Fawcett, H.H. and W.S. Wood, Safety and Accident Prevention in Chemical Operations, 2nd edition, Wiley-Interscience, New York, 1982.
3. Flury, Patricia A., Environmental Health and Safety in the Hospital Laboratory, Charles C. Thomas Publisher, Springfield, IL, 1978.
4. Green, Michael E. and Turk, Amos, Safety in Working with Chemicals, Macmillan Publishing Co., NY, 1978.
5. Kaufman, James A., Laboratory Safety Guidelines, Dow Chemical Co., Box 1713, Midland, MI 48640, 1977.
6. National Institutes of Health, NIH Guidelines for the Laboratory Use of Chemical Carcinogens, NIH Pub. No. 81-2385, GPO, Washington, DC 20402, 1981.
7. National Research Council, Prudent Practices for Disposal of Chemicals from Laboratories, National Academy Press, Washington, DC, 1983.

8. National Research Council, Prudent Practices for Handling Hazardous Chemicals in Laboratories, National Academy Press, Washington, DC, 1981.
9. Renfrew, Malcolm, Ed., Safety in the Chemical Laboratory, Vol. IV, J. Chem. Ed., American Chemical Society, Easlon, PA, 1981.
10. Steere, Norman V., Ed., Safety in the Chemical Laboratory, J. Chem. Ed. American Chemical Society, Easlon, PA, 18042, Vol. I, 1967, Vol. II, 1971, Vol. III, 1974.
11. Steere, Norman V., Handbook of Laboratory Safety, the Chemical Rubber Company, Cleveland, OH, 1971.
12. Young, Jay A., Ed., Improving Safety in the Chemical Laboratory, John Wiley & Sons, Inc., New York, 1987.

(b) HAZARDOUS SUBSTANCES INFORMATION:
1. American Conference of Governmental Industrial Hygienists, Threshold Limit Values for Chemical Substances and Physical Agents in the Workroom Environment with Intended Changes, 6500 Glenway Avenue, Bldg. D-7, Cincinnati, OH 45211-4438.
2. Annual Report on Carcinogens, National Toxicology Program U.S. Department of Health and Human Services, Public Health Service, U.S. Government Printing Office, Washington, DC (latest edition).
3. Best Company, Best Safety Directory, Vols. I and II, Oldwick, NJ, 1981.
4. Bretherick, L., Handbook of Reactive Chemical Hazards, 2nd edition, Butterworths, London, 1979.
5. Bretherick, L., Hazards in the Chemical Laboratory, 3rd edition, Royal Society of Chemistry, London, 1986.
6. Code of Federal Regulations, 29 CFR part 1910 subpart Z. U.S. Govt. Printing Office, Washington, DC 20402 (latest edition).
7. IARC Monographs on the Evaluation of the Carcinogenic Risk of Chemicals to Man, World Health Organization Publications Center, 49 Sheridan Avenue, Albany, New York 12210 (latest editions).
8. NIOSH/OSHA Pocket Guide to Chemical Hazards. NIOSH Pub. No. 85-114, U.S. Government Printing Office, Washington, DC, 1985 (or latest edition).
9. Occupational Health Guidelines, NIOSH/OSHA. NIOSH Pub. No. 81-123, U.S. Government Printing Office, Washington, DC, 1981.
10. Patty, F.A., Industrial Hygiene and Toxicology, John Wiley & Sons, Inc., New York, NY (Five Volumes).
11. Registry of Toxic Effects of Chemical Substances, U.S. Department of Health and Human Services, Public Health Service, Centers for Disease Control, National Institute for Occupational Safety and Health, Revised Annually, for sale from Superintendent of Documents, U.S. Govt. Printing Office, Washington, DC 20402.
12. The Merck Index: An Encyclopedia of Chemicals and Drugs. Merck and Company Inc., Rahway, NJ, 1976 (or latest edition).
13. Sax, N.I. Dangerous Properties of Industrial Materials, 5th edition, Van Nostrand Reinhold, NY, 1979.
14. Sittig, Marshall, Handbook of Toxic and Hazardous Chemicals, Noyes Publications, Park Ridge, NJ, 1981.

(c) INFORMATION ON VENTILATION:

1. American Conference of Governmental Industrial Hygienists. Industrial Ventilation (latest edition), 6500 Glenway Avenue, Bldg. D-7, Cincinnati, OH 45211-4438.

2. American National Standards Institute, Inc. American National Standards Fundamentals Governing the Design and Operation of Local Exhaust Systems ANSI Z 9.2-1979, American National Standards Institute, NY 1979.

3. Imad, A.P. and Watson, C.L. Ventilation Index: An Easy Way to Decide about Hazardous Liquids, Professional Safety, pp. 15-18, April 1980.

4. National Fire Protection Association, Fire Protection for Laboratories Using Chemicals NFPA-45, 1982. Safety Standard for Laboratories in Health Related Institutions, NFPA, 56c, 1980. Fire Protection Guide on Hazardous Materials, 7th edition, 1978. National Fire Protection Association, Batterymarch Park, Quincy, MA 02269.

5. Scientific Apparatus Makers Association (SAMA), Standard for Laboratory Fume Hoods, SAMA LF7-1980, 1101 16th Street, NW, Washington, DC 20036.

(d) INFORMATION ON AVAILABILITY OF REFERENCED MATERIAL:

1. American National Standards Institute (ANSI), 1430 Broadway, New York, NY 10018.

2. American Society for Testing and Materials (ASTM), 1916 Race Street, Philadelphia, PA 19103.

(Approved by the Office of Management and Budget under control number 1218-0131)

[55 FR 3327, Jan. 31, 1990]

Appendix B: Laboratory Chemical Safety Summaries

This appendix presents Laboratory Chemical Safety Summaries (LCSSs) for 88 substances commonly encountered in laboratories. These summaries have been prepared in accord with the general and comprehensive approach to experiment planning and risk assessment that is outlined in Chapters 2 ("Prudent Planning of Experiments") and 3 ("Evaluating Hazards and Assessing Risks in the Laboratory") of this volume, and they should be used only by individuals familiar with the content of those chapters. The scope of coverage and degree of detail provided in these summaries should be appropriate for prudent experiment planning in most commonly encountered laboratory situations. Each summary includes chemical and toxicological information derived from the various secondary sources discussed in Chapter 3, as well as from Material Safety Data Sheets (MSDSs).

The committee encourages the dissemination of these summaries as a means of promoting the prudent use of hazardous chemicals in laboratory work. It anticipates that these summaries will also serve as models for the preparation of additional LCSSs for chemicals not included in this appendix. In fact, *the committee recommends that laboratory workers routinely prepare new LCSSs for unfamiliar substances as part of the risk assessment they should carry out for each experiment as outlined at the conclusion of Chapter 3.*

The preparation and use of Laboratory Chemical Safety Summaries as described here are consistent with the Chemical Hygiene Plans required for every laboratory under the OSHA Laboratory Standard. Thus, the identification of substances that meet the OSHA criteria for "particularly hazardous substances" or "select carcinogens" should be facilitated by the use of these summaries.

LIMITATIONS OF LCSSs

All users of Laboratory Chemical Safety Summaries should understand their limitations. In each summary, the content of the section on toxicity is dependent on the quality of the information available. For some chemicals the description of toxicity hazards is based on extensive experience with human exposure, while in other cases this discussion is based on limited data from animal tests. If a substance meets the OSHA definition of a "select carcinogen" (based on current information), that fact is noted here. The discussion of toxic effects has been written so as to be comprehensible to the average laboratory worker, with full knowledge that the use of plain language may lead to a lack of precision in the description of toxic effects. The section on reactivity and incompatibility summarizes only those items that are likely to be encountered in normal laboratory use and should not be considered comprehensive. If more extensive information is required for any of the categories of information given in these summaries, the sources listed in Chapter 3 should be consulted. In addition, OSHA regulations (Standards—29 CFR) are now available on the WorldWide-Web, as are further links to safety and health information: http://www.osha.gov/safhlth.html.

These summaries should be used only by laboratory workers with general training in the safe handling of chemicals. LCSSs are intended to be used in conjunction with Chapters 3 through 7 of this volume, and these summaries make frequent reference to the contents of those chapters. The information in these summaries has been selected for its relevance to the *laboratory use* of chemicals. In particular, the listing of chemicals and toxicological hazards is not intended to be a comprehensive review of the literature for a given substance. These summaries do not contain information on

- household or nonlaboratory use of a chemical;
- commercial, manufacturing, or other large-scale use of chemicals;
- consequences of abuse of a chemical by deliberate ingestion, inhalation, or injection;
- environmental effects of release, disposal, or incineration of a chemical; or
- shipment or transportation of a chemical in accordance with applicable laws and regulations.

The information contained in these summaries is believed to be accurate at the time of publication of this volume. A recent MSDS should be consulted for updated information, especially on exposure limits.

PREPARATION OF NEW LCSSs

All of the information required for the preparation of new LCSSs should be available in the sources discussed in Chapter 3. The following directions should be helpful in preparing specific sections of new LCSSs:

- *Odor:* Information on odor and odor thresholds can be found in NIOSH Guidelines, Royal Society Chemical Safety Data Sheets, and in the AIHA publication "Odor Thresholds for Chemicals with Established Occupational Health Standards."
- *Toxicity data:* LD_{50} and LC_{50} values can be found in MSDSs and other sources listed in Chapter 3. Exposure limits are included in MSDSs and are listed in the ACGIH Threshold Limit Value booklet, which is updated annually.
- *Major hazards:* This section should provide key words indicating only the most important potential hazards associated with the title substance.
- *Toxicity:* The first paragraph of this section should discuss acute toxicity hazards using plain language. Symptoms of exposure by inhalation, skin contact, eye contact, and ingestion should be separately described, and the degree of hazard of the substance should be identified as "high," "moderate," or "low," as discussed in Chapter 3. The paragraph should indicate whether there are adequate warning properties for the substance. The second paragraph should address chronic toxicity. For potential carcinogens, whether the substance is classified as an OSHA "select carcinogen" should be indicated.
- *Flammability and explosibility:* This section should indicate the NFPA rating for the substance, explosion limits, toxic substances that may be produced in a fire, and the type of fire extinguisher appropriate for fighting fires.
- *Storage and handling:* This section should make reference to the appropriate sections of Chapter 5 and should also highlight any special procedures of particular importance in work with the title substance.

SECTIONS INCLUDED IN LCSSs

Each of the 88 LCSSs supplied in this appendix includes those of the following sections that apply to the title substance:

- Substance,
- Formula,
- Physical properties,
- Odor,
- Vapor density,
- Vapor pressure,
- Flash point,
- Autoignition temperature,
- Toxicity data,
- Major hazards,
- Toxicity,
- Flammability and explosibility,
- Reactivity and incompatibility,
- Storage and handling,
- Accidents,
- Disposal.

LABORATORY CHEMICAL SAFETY SUMMARIES

Acetaldehyde
Acetic Acid
Acetone
Acetonitrile
Acetylene
Acrolein
Acrylamide
Acrylonitrile
Aluminum Trichloride
Ammonia (Anhydrous)
Ammonium Hydroxide
Aniline
Arsine
Benzene
Boron Trifluoride
Bromine
Tert-Butyl Hydroperoxide
Butyllithiums
Carbon Disulfide
Carbon Monoxide
Carbon Tetrachloride
Chlorine
Chloroform
Chloromethyl Methyl Ether (and Related Compounds)
Chromium Trioxide and Other Chromium(VI) Salts
Cyanogen Bromide
Diazomethane
Diborane
Dichloromethane
Diethyl Ether
Diethylnitrosamine (and Related Nitrosamines)
Dimethyl Sulfate
Dimethyl Sulfoxide
Dimethylformamide
Dioxane
Ethanol
Ethidium Bromide
Ethyl Acetate
Ethylene Dibromide
Ethylene Oxide
Fluorides (Inorganic)
Fluorine
Formaldehyde
Hexamethylphosphoramide
Hexane (and Related Hydrocarbons)

Hydrazine
Hydrobromic Acid and Hydrogen Bromide
Hydrochloric Acid and Hydrogen Chloride
Hydrogen
Hydrogen Cyanide
Hydrogen Fluoride and Hydrofluoric Acid
Hydrogen Peroxide
Hydrogen Sulfide
Iodine
Lead and Its Inorganic Compounds
Lithium Aluminum Hydride
Mercury
Methanol
Methyl Ethyl Ketone
Methyl Iodide
Nickel Carbonyl
Nitric Acid
Nitrogen Dioxide
Osmium Tetroxide
Oxygen
Ozone
Palladium on Carbon
Peracetic Acid (and Related Percarboxylic Acids)
Perchloric Acid (and Inorganic Perchlorates)
Phenol
Phosgene
Phosphorus
Potassium
Potassium Hydride and Sodium Hydride
Pyridine
Silver and Its Compounds
Sodium
Sodium Azide
Sodium Cyanide and Potassium Cyanide
Sodium Hydroxide and Potassium Hydroxide
Sulfur Dioxide
Sulfuric Acid
Tetrahydrofuran
Toluene
Toluene Diisocyanate
Trifluoroacetic Acid
Trimethylaluminum (and Related Organoaluminum Compounds)
Trimethyltin Chloride (and Other Organotin Compounds)

LABORATORY CHEMICAL SAFETY SUMMARY: ACETALDEHYDE

Substance	Acetaldehyde (Ethanal, acetic aldehyde) CAS 75-07-0
Formula	CH_3CHO
Physical Properties	Colorless liquid bp 21 °C, mp −124 °C Miscible with water
Odor	Pungent, fruity odor detectable at 0.0068 to 1000 ppm (mean = 0.067 ppm)
Vapor Density	1.52 (air = 1.0)
Vapor Pressure	740 mmHg at 20 °C
Flash Point	−38 °C
Autoignition Temperature	185 °C

Toxicity Data

LD_{50} oral (rat)	661 mg/kg
LC_{50} inhal (rat)	20,550 ppm (37,000 mg/m^3; 30 min)
PEL (OSHA)	200 ppm (360 mg/m^3)
TLV-TWA (ACGIH)	100 ppm (180 mg/m^3)
STEL (ACGIH)	150 ppm (270 mg/m^3)

Major Hazards Highly flammable liquid; irritating to the eyes and respiratory system.

Toxicity The acute toxicity of acetaldehyde is low by inhalation and moderate by ingestion. Exposure to acetaldehyde by inhalation is irritating to the respiratory tract and mucous membranes; this substance is a narcotic and can cause central nervous system depression. Ingestion of acetaldehyde may cause severe irritation of the digestive tract leading to nausea, vomiting, headache, and liver damage. Acetaldehyde causes irritation and burning upon skin contact and is a severe eye irritant.

Acetaldehyde has caused nasal tumors in rats exposed by inhalation and is listed by IARC in Group 2B ("possible human carcinogen"). It is not classified as a "select carcinogen" according to the criteria of the OSHA Laboratory Standard. Acetaldehyde is mutagenic and has been shown to be a reproductive toxin in animals. Acetaldehyde is formed by metabolism of ethanol, and chronic exposure can produce symptoms similar to alcoholism.

Flammability and Explosibility Acetaldehyde is a dangerous fire hazard (NFPA rating = 4) owing to its volatility and low autoignition temperature. Its vapor is explosive in the concentration range 4 to 66% in air and may be ignited by hot surfaces such as hot plates or lightbulbs, or by static

(continued on facing page)

electricity discharges. The vapor is heavier than air and may travel a considerable distance to an ignition source and "flash back." Carbon dioxide or dry chemical extinguishers should be used to fight acetaldehyde fires.

Reactivity and Incompatibility

Acetaldehyde is a reactive substance and on storage in the presence of air may undergo oxidation to form explosive peroxides. It may also polymerize violently when in contact with strong acids or trace metals such as iron. Acetaldehyde may undergo violent reactions with acid chlorides, anhydrides, amines, hydrogen cyanide, and hydrogen sulfide.

Storage and Handling

Acetaldehyde should be handled in the laboratory using the "basic prudent practices" described in Chapter 5.C, supplemented by the additional precautions for dealing with extremely flammable substances (Chapter 5.F). In particular, acetaldehyde should be used only in areas free of ignition sources, and quantities greater than 1 liter should be stored in tightly sealed metal containers in areas separate from oxidizers. Acetaldehyde should always be stored under an inert atmosphere of nitrogen or argon to prevent autoxidation.

Accidents

In the event of skin contact, immediately wash with soap and water and remove contaminated clothing. In case of eye contact, promptly wash with copious amounts of water for 15 min (lifting upper and lower lids occasionally) and obtain medical attention. If acetaldehyde is ingested, obtain medical attention immediately. If large amounts of this compound are inhaled, move the person to fresh air and seek medical attention at once.

In the event of a spill, remove all ignition sources, soak up the acetaldehyde with a spill pillow or absorbent material, place in an appropriate container, and dispose of properly. Alternatively, acetaldehyde spills may be neutralized with sodium bisulfite solution before cleanup. Respiratory protection may be necessary in the event of a large spill or release in a confined area.

Disposal

Excess acetaldehyde and waste material containing this substance should be placed in an appropriate container, clearly labeled, and handled according to your institution's waste disposal guidelines. For more information on disposal procedures, see Chapter 7 of this volume.

The information in this LCSS has been compiled by a committee of the National Research Council from literature sources and Material Safety Data Sheets and is believed to be accurate as of July 1994. This summary is intended for use by trained laboratory personnel in conjunction with the NRC report *Prudent Practices in the Laboratory: Handling and Disposal of Chemicals*. This LCSS presents a concise summary of safety information that should be adequate for most laboratory uses of the title substance, but in some cases it may be advisable to consult more comprehensive references. This information should not be used as a guide to the nonlaboratory use of this chemical.

LABORATORY CHEMICAL SAFETY SUMMARY: ACETIC ACID

Substance	Acetic acid (Ethanoic acid) CAS 64-19-7
Formula	CH_3COOH
Physical Properties	Colorless liquid bp 118 °C, mp 17 °C Miscible in water (100 g/100 mL)
Odor	Strong, pungent, vinegar-like odor detectable at 0.2 to 1.0 ppm
Vapor Density	2.1 (air = 1.0)
Vapor Pressure	11 mmHg at 20 °C
Flash Point	39 °C
Autoignition Temperature	426 °C

Toxicity Data

LD_{50} oral (rat)	3310 mg/kg
LD_{50} skin (rabbit)	1060 mg/kg
LC_{50} inhal (mice)	5620 ppm (1 h)
PEL (OSHA)	10 ppm (25 mg/m^3)
TLV-TWA (ACGIH)	10 ppm (25 mg/m^3)
STEL (ACGIH)	15 ppm (37 mg/m^3)

Major Hazards

Corrosive to the skin and eyes; vapor or mist is very irritating and can be destructive to the eyes, mucous membranes, and respiratory system; ingestion causes internal irritation and severe injury.

Toxicity

The acute toxicity of acetic acid is low. The immediate toxic effects of acetic acid are due to its corrosive action and dehydration of tissues with which it comes in contact. A 10% aqueous solution of acetic acid produced mild or no irritation on guinea pig skin. At 25 to 50%, generally severe irritation results. In the eye, a 4 to 10% solution will produce immediate pain and sometimes injury to the cornea. Acetic acid solutions of 80% or greater concentration can cause serious burns of the skin and eyes. Acetic acid is slightly toxic by inhalation; exposure to 50 ppm is extremely irritating to the eyes, nose, and throat.

Acetic acid has not been found to be carcinogenic or to show reproductive or developmental toxicity in humans.

Flammability and Explosibility

Acetic acid is a combustible substance (NFPA rating = 2). Heating can release vapors that can be ignited. Vapors or gases may travel considerable distances to ignition source

(continued on facing page)

and "flash back." Acetic acid vapor forms explosive mixtures with air at concentrations of 4 to 16% (by volume). Carbon dioxide or dry chemical extinguishers should be used for acetic acid fires.

Reactivity and Incompatibility	Contact with strong oxidizers may cause fire.
Storage and Handling	Acetic acid should be handled in the laboratory using the "basic prudent practices" described in Chapter 5.C. In particular, acetic acid should be used only in areas free of ignition sources, and quantities greater than 1 liter should be stored in tightly sealed metal containers in areas separate from oxidizers.
Accidents	In the event of skin contact, immediately wash with soap and water and remove contaminated clothing. In case of eye contact, promptly wash with copious amounts of water for 15 min (lifting upper and lower lids occasionally) and obtain medical attention. If acetic acid is ingested, obtain medical attention immediately. If large amounts of this compound are inhaled, move the person to fresh air and seek medical attention at once.

In the event of a spill, remove all ignition sources, soak up the acetic acid with a spill pillow or absorbent material, place in an appropriate container, and dispose of properly. Cleaned-up material is a RCRA Hazardous Waste. Respiratory protection may be necessary in the event of a large spill or release in a confined area. |
| **Disposal** | Excess acetic acid and waste material containing this substance should be placed in a covered metal container, clearly labeled, and handled according to your institution's waste disposal guidelines. For more information on disposal procedures, see Chapter 7 of this volume. |

LABORATORY CHEMICAL SAFETY SUMMARY: ACETONE

Substance	Acetone (2-Propanone) CAS 67-64-1
Formula	CH_3COCH_3
Physical Properties	Colorless liquid bp 56 °C, mp −94 °C Miscible with water
Odor	Characteristic pungent odor detectable at 33 to 700 ppm (mean = 130 ppm)
Vapor Density	2.0 (air = 1.0)
Vapor Pressure	180 mmHg at 20 °C
Flash Point	−18 °C
Autoignition Temperature	465 °C

Toxicity Data

LD_{50} oral (rat)	5800 mg/kg
LD_{50} skin (rabbit)	20,000 mg/kg
LC_{50} inhal (rat)	50,100 mg/m^3
PEL (OSHA)	1000 ppm (2400 mg/m^3)
TLV-TWA (ACGIH)	750 ppm
STEL (ACGIH)	1000 ppm (2400 mg/m^3)

Major Hazards Highly flammable.

Toxicity The acute toxicity of acetone is low. Acetone is primarily a central nervous system depressant at high concentrations (greater than 12,000 ppm). Unacclimated volunteers exposed to 500 ppm acetone experienced eye and nasal irritation, but it has been reported that 1000 ppm for an 8-hour day produced no effects other than slight transient irritation to eyes, nose, and throat. Therefore there are good warning properties for those unaccustomed to working with acetone; however, frequent use of acetone seems to cause accommodation to its slight irritating properties. Acetone is practically nontoxic by ingestion. A case of a man swallowing 200 mL of acetone resulted in his becoming stuporous after 1 hour and then comatose; he regained consciousness 12 hour later. Acetone is slightly irritating to the skin, and prolonged contact may cause dermatitis. Liquid acetone produces moderate transient eye irritation.

Acetone has not been found to be carcinogenic in animal tests or to have effects on reproduction or fertility.

(continued on facing page)

The information in this LCSS has been compiled by a committee of the National Research Council from literature sources and Material Safety Data Sheets and is believed to be accurate as of July 1994. This summary is intended for use by trained laboratory personnel in conjunction with the NRC report *Prudent Practices in the Laboratory: Handling and Disposal of Chemicals*. This LCSS presents a concise summary of safety information that should be adequate for most laboratory uses of the title substance, but in some cases it may be advisable to consult more comprehensive references. This information should not be used as a guide to the nonlaboratory use of this chemical.

Flammability and Explosibility

Acetone is extremely flammable (NFPA rating = 3), and its vapor can travel a considerable distance to an ignition source and "flash back." Acetone vapor forms explosive mixtures with air at concentrations of 2 to 13% (by volume). Carbon dioxide or dry chemical extinguishers should be used for acetone fires.

Reactivity and Incompatibility

Fires and/or explosions may result from the reaction of acetone with strong oxidizing agents (e.g., chromium trioxide) and very strong bases (e.g., potassium *t*-butoxide).

Storage and Handling

Acetone should be handled in the laboratory using the "basic prudent practices" described in Chapter 5.C, supplemented by the additional precautions for dealing with extremely flammable substances (Chapter 5.F). In particular, acetone should be used only in areas free of ignition sources, and quantities greater than 1 liter should be stored in tightly sealed metal containers in areas separate from oxidizers.

Accidents

In the event of skin contact, immediately wash with soap and water and remove contaminated clothing. In case of eye contact, promptly wash with copious amounts of water for 15 min (lifting upper and lower lids occasionally) and obtain medical attention. If acetone is ingested, obtain medical attention immediately. If large amounts of this compound are inhaled, move the person to fresh air and seek medical attention at once.

In the event of a spill, remove all ignition sources, soak up the acetone with a spill pillow or absorbent material, place in an appropriate container, and dispose of properly. Respiratory protection may be necessary in the event of a large spill or release in a confined area.

Disposal

Excess acetone and waste material containing this substance should be placed in an appropriate container, clearly labeled, and handled according to your institution's waste disposal guidelines. For more information on disposal procedures, see Chapter 7 of this volume.

LABORATORY CHEMICAL SAFETY SUMMARY: ACETONITRILE

Substance	Acetonitrile (Methyl cyanide, cyanomethane) CAS 75-05-8
Formula	$H_3C-C{\equiv}N$
Physical Properties	Colorless liquid bp 82 °C, mp -46°C Miscible with water (>100 g/100 mL)
Odor	Aromatic ether-like odor detectable at 40 ppm
Vapor Density	1.42 (air = 1.0)
Vapor Pressure	73 mmHg at 20 °C
Flash Point	6 °C
Autoignition Temperature	524 °C

Toxicity Data

LD_{50} oral (rat)	2730 mg/kg
LD_{50} skin (rabbit)	1250 mg/kg
LC_{50} inhal (rat)	7551 ppm (8 h)
PEL (OSHA)	40 ppm (70 mg/m^3)
STEL (OSHA)	60 ppm (105 mg/m^3)
TLV-TWA (ACGIH)	40 ppm (70 mg/m^3)
STEL (ACGIH)	60 ppm (105 mg/m^3)

Major Hazards Flammable liquid and vapor; liquid severely irritates the eyes.

Toxicity Acetonitrile is slightly toxic by acute exposure through oral intake, skin contact, and inhalation. However, acetonitrile can be converted by the body to cyanide. Symptoms of exposure include weakness, flushing, headache, difficult and/or rapid breathing, nausea, vomiting, diarrhea, blue-gray discoloration of the skin and lips (due to a lack of oxygen), stupor, and loss of consciousness. Acetonitrile is severely irritating to the eyes and slightly irritating to the skin. Prolonged contact can lead to absorption through the skin and more intense irritation. Acetonitrile is regarded as having adequate warning properties.

Acetonitrile is not mutagenic in bacterial and animal cells and has not been found to be a carcinogen in humans. Single high-dose exposure in animals during pregnancy produced birth defects possibly due to the liberation of cyanide. Multiple oral doses during pregnancy did not produce birth defects. Repeated exposure in animals produced adverse lung effects.

(continued on facing page)

Flammability and Explosibility

Acetonitrile is a flammable liquid (NFPA rating = 3), and its vapor can travel a considerable distance to an ignition source and "flash back." Acetonitrile vapor forms explosive mixtures with air at concentrations of 4 to 16% (by volume). Hazardous gases produced in a fire include hydrogen cyanide, carbon monoxide, carbon dioxide, and oxides of nitrogen. Carbon dioxide or dry chemical extinguishers should be used for acetonitrile fires.

Reactivity and Incompatibility

Contact of acetonitrile with strong oxidizers can result in violent reactions. Acetonitrile hydrolyzes on exposure to strong acids and bases. It is incompatible with reducing agents and alkali metals and may attack plastics, rubber, and some coatings.

Storage and Handling

Acetonitrile should be handled in the laboratory using the "basic prudent practices" described in Chapter 5.C, supplemented by the additional precautions for dealing with highly flammable substances (Chapter 5.F). In particular, acetonitrile should be used only in areas free of ignition sources, and quantities greater than 1 liter should be stored in tightly sealed metal containers in areas separate from oxidizers.

Accidents

In the event of skin contact, immediately wash with soap and water and remove contaminated clothing. In case of eye contact, promptly wash with copious amounts of water for 15 min (lifting upper and lower lids occasionally) and obtain medical attention. If acetonitrile is ingested, obtain medical attention immediately. If large amounts of this compound are inhaled, move the person to fresh air and seek medical attention at once.

In the event of a spill, remove all ignition sources, soak up the acetonitrile with a spill pillow or absorbent material, place in an appropriate container, and dispose of properly. Evacuation and cleanup using respiratory and skin protection may be necessary in the event of a large spill or release in a confined area.

Disposal

Excess acetonitrile and waste material containing this substance should be placed in an appropriate container, clearly labeled, and handled according to your institution's waste disposal guidelines. For more information on disposal procedures, see Chapter 7 of this volume.

LABORATORY CHEMICAL SAFETY SUMMARY: ACETYLENE

Substance	Acetylene (Ethyne; welding gas) CAS 74-86-2
Formula	$HC{\equiv}CH$
Physical Properties	Colorless gas bp $-84\ ^{\circ}C$ (sublimes), mp $-82\ ^{\circ}C$ Slightly soluble in water (0.106 g/100 mL)
Odor	Odorless, although garlic-like or "gassy" odor often detectable because of trace impurities
Vapor Density	0.91 (air = 1.0)
Vapor Pressure	3.04×10^4 mmHg (~40 atmospheres) at 16.8 $^{\circ}C$
Flash Point	$-18\ ^{\circ}C$
Autoignition Temperature	305 $^{\circ}C$
Toxicity Data	LC_{50} inhal (rat) simple asphyxiant (>500,000 ppm) TLV-TWA (ACGIH) simple asphyxiant
Major Hazards	Extremely flammable gas; simple asphyxiant.
Toxicity	Acetylene is relatively nontoxic and has been used as an anesthetic. Inhalation of acetylene can be hazardous because of its action as a simple asphyxiant. Concentrations of about 10% in air cause slight intoxication, and levels of 20% in air may produce headaches and labored breathing. At higher concentrations (33% and above), acetylene acts as a narcotic, causing unconsciousness in 7 min or less, with rapid and full recovery normally seen on removal from exposure of less than several hours. Concentrations of acetylene above 50% in air can cause death by asphyxiation within 5 min. Commercially available acetylene may contain highly toxic impurities, including phosphine, arsine, and hydrogen sulfide; the presence of these impurities must be considered in setting acceptable exposure levels to acetylene. For example, the concentration of acetylene containing 95 ppm of phosphine impurity (which has a TLV of 0.3 ppm) should not exceed 3160 ppm to stay within the TLV for phosphine. There is no evidence that acetylene is a human carcinogen or reproductive toxin.
Flammability and Explosibility	Acetylene is a highly flammable gas and forms explosive mixtures with air over an unusually wide range of concentrations (2 to 80%). Acetylene can polymerize exothermically, leading to deflagration. With a very high positive free energy of formation, acetylene

(continued on facing page)

The information in this LCSS has been compiled by a committee of the National Research Council from literature sources and Material Safety Data Sheets and is believed to be accurate as of July 1994. This summary is intended for use by trained laboratory personnel in conjunction with the NRC report *Prudent Practices in the Laboratory: Handling and Disposal of Chemicals.* This LCSS presents a concise summary of safety information that should be adequate for most laboratory uses of the title substance, but in some cases it may be advisable to consult more comprehensive references. This information should not be used as a guide to the nonlaboratory use of this chemical.

is thermodynamically unstable and is sensitive to shock and pressure. Its stability is enhanced by the presence of small amounts of other compounds such as methane, and acetylene in cylinders is relatively safe to handle because it is dissolved in acetone. Acetylene fires can be fought with carbon dioxide, dry chemical, and halon extinguishers; firefighting is greatly facilitated by shutting off the gas supply.

Reactivity and Incompatibility

Acetylene forms highly unstable acetylides with many metals, including copper, brass, mercury, potassium, silver, and gold. The dry acetylides are sensitive, powerful explosives. Acetylene may react violently with fluorine and other halogens (chlorine, bromine, iodine) and forms explosive compounds on contact with nitric acid.

Storage and Handling

Acetylene should be handled in the laboratory using the "basic prudent practices" described in Chapter 5.C, supplemented by the additional precautions for dealing with extremely flammable substances (Chapter 5.F) and compressed gases (Chapter 5.H). In particular, acetylene should be used only in well-ventilated areas free of ignition sources. Acetylene is supplied in specially designed steel cylinders containing acetone and an inert material. More than 1 liter of the gas should never be stored in other containers. Brass or copper tubing, valves, or fittings should never be allowed to come in contact with acetylene. If acetylene must be purified, it should be passed through concentrated H_2SO_4 and NaOH (do not use activated carbon).

Accidents

If large amounts of this compound are inhaled, move the person to fresh air and seek medical attention at once.

In the event of an acetylene leak, shut down and remove all ignition sources and ventilate the area at once to prevent flammable mixtures from forming. Carefully remove cylinders with slow leaks to remote outdoor locations. Limit access to an affected area. Respiratory protection may be necessary in the event of a large release or a leak in a confined area.

Disposal

Excess acetylene should be returned to the vendor for disposal; disposal should not be attempted in the laboratory. Excess acetylene should be vented from reaction flasks, tubing, etc., rather than scrubbed with strong base to avoid the formation of acetylides. For more information on disposal procedures, see Chapter 7 of this volume.

LABORATORY CHEMICAL SAFETY SUMMARY: ACROLEIN

Substance	Acrolein (Acrylaldehyde, acrylic aldehyde, 2-propenal) CAS 107-02-8
Formula	$H_2C = CHCHO$
Physical Properties	Colorless to yellow liquid bp 53 °C, mp −87 °C Highly soluble in water (21 g/100 mL)
Odor	Pungent, lacrimatory, intensely irritating odor detectable at 0.02 to 0.4 ppm
Vapor Density	1.9 (air = 1.0)
Vapor Pressure	210 mmHg at 20 °C
Flash Point	−26 °C
Autoignition Temperature	234 °C

Toxicity Data

LD_{50} oral (rat)	42 to 46 mg/kg
LD_{50} skin (rabbit)	562 mg/kg
LC_{50} inhal (rat)	300 mg/m³ (30 min)
PEL (OSHA)	0.1 ppm (0.25 mg/m³)
STEL (OSHA)	0.3 ppm (0.69 mg/m³)
TLV-TWA (ACGIH)	0.1 ppm (0.23 mg/m³)
STEL (ACGIH)	0.3 ppm (0.69 mg/m³)

Major Hazards

Highly toxic; causes severe irritation and corrosion of skin, eyes, nose, and respiratory system; highly flammable; may polymerize violently upon loss or removal of inhibitor or initiation by chemical agents.

Toxicity

Acrolein is a highly toxic and corrosive substance. Inhalation of acrolein can cause moderate to severe eye, nose, and respiratory system irritation after a few minutes of exposure to concentrations as low as 0.25 ppm. Higher concentrations can cause immediate and/or delayed lung injury including pulmonary edema and respiratory insufficiency; fatal reactions have occurred upon exposure to as little as 10 ppm. This substance is a powerful lacrimator, and eye contact with acrolein liquid or vapor can cause severe burns. Skin contact can cause severe redness, swelling, burns with blistering, and corrosion. Acrolein can be absorbed through the skin, leading to systemic effects including delayed pulmonary edema. Ingestion of acrolein can cause gastrointestinal distress, pulmonary congestion, and edema. Acrolein has been reported to be a weak skin sensitizer in some individuals. This substance is regarded as having adequate warning properties.

(continued on facing page)

Acrolein is mutagenic in bacteria but did not cause increased tumor incidence in animals exposed chronically by injection or inhalation. Administration to pregnant rats caused malformations and lethality to embryos. Chronic exposure to as little as 0.21 ppm acrolein caused inflammatory changes in lungs, liver, kidneys, and brains of experimental animals.

Flammability and Explosibility

Acrolein is a highly flammable liquid (NFPA rating = 3) and its vapor can travel a considerable distance and "flash back." Acrolein vapor forms explosive mixtures with air at concentrations of 2.8 to 31% (by volume). Carbon dioxide or dry chemical extinguishers should be used for acrolein fires.

Reactivity and Incompatibility

Acrolein can polymerize violently upon exposure to heat (temperatures above 50 °C), light, or various chemical initiators such as amines, bases, and acids. Commercial acrolein contains an inhibitor such as hydroquinone; samples from which the inhibitor has been removed (e.g., by distillation) are extremely hazardous.

Storage and Handling

Because of its corrosivity, flammability, and high acute toxicity, acrolein should be handled using the "basic prudent practices" of Chapter 5.C, supplemented by the additional precautions for work with compounds of high toxicity (Chapter 5.D) and extremely flammable substances (Chapter 5.F). In particular, work with acrolein should be conducted in a fume hood to prevent exposure by inhalation, and splash goggles and butyl rubber gloves should be worn at all times to prevent eye and skin contact. Acrolein should be used only in areas free of ignition sources. Containers of acrolein should be stored in secondary containers in areas separate from amines, oxidizers, acids, and bases.

Accidents

In the event of skin contact, immediately wash with soap and water and remove contaminated clothing. In case of eye contact, promptly wash with copious amounts of water for 15 min (lifting upper and lower lids occasionally) and obtain medical attention. If acrolein is ingested, obtain medical attention immediately. If acrolein is inhaled, move the person to fresh air and seek medical attention at once, since immediate or delayed respiratory injury may result.

In the event of a spill, remove all ignition sources, soak up the acrolein with a spill pillow or absorbent material, place in an appropriate container, and dispose of properly. Respiratory protection should be employed owing to the risk of severe eye, nose, and respiratory injury.

Disposal

Excess acrolein and waste material containing this substance should be placed in an appropriate container, clearly labeled, and handled according to your institution's waste disposal guidelines. For more information on disposal procedures, see Chapter 7 of this volume.

Substance	Acrylamide (2-Propeneamide, vinyl amide) CAS 79-06-1
Formula	$H_2C = CH–CONH_2$
Physical Properties	Colorless crystals bp 125 °C (25 mmHg), mp 85 °C Soluble in water (216 g/100 mL)
Odor	Odorless solid

Toxicity Data

LD_{50} oral (rat)	124 mg/kg
LD_{50} skin (rat)	400 mg/kg
PEL (OSHA)	0.3 mg/m^3—skin
TLV-TWA (ACGIH)	0.03 mg/m^3—skin

Major Hazards

Suspected human carcinogen (OSHA "select carcinogen") and neurotoxin.

Toxicity

The acute toxicity of acrylamide is moderate by ingestion or skin contact. Skin exposure leads to redness and peeling of the skin of the palms. Aqueous acrylamide solutions cause eye irritation; exposure to a 50% solution of acrylamide caused slight corneal injury and slight conjunctival irritation, which healed in 8 days.

The chronic toxicity of acrylamide is high. Repeated exposure to ~2 mg/kg per day may result in neurotoxic effects, including unsteadiness, muscle weakness, and numbness in the feet (leading to paralysis of the legs), numbness in the hands, slurred speech, vertigo, and fatigue. Exposure to slightly higher repeated doses in animal studies has induced multisite cancers and reproductive effects, including abortion, reduced fertility, and mutagenicity. Acrylamide is listed in IARC Group 2B ("possible human carcinogen") and is classified as a "select carcinogen" under the criteria of the OSHA Laboratory Standard.

Flammability and Explosibility

The volatility of acrylamide is low (0.03 mmHg at 40 °C), and it does not pose a significant flammability hazard.

Reactivity and Incompatibility

May polymerize violently on strong heating or exposure to strong base. Acrylamide may react violently with strong oxidizers.

Storage and Handling

Because of its carcinogenicity and neurotoxicity, acrylamide should be handled using the "basic prudent practices" of Chapter 5.C, supplemented by the additional precautions for work with compounds of high chronic toxicity (Chapter 5.D). In particular, this substance should be handled only when wearing appropriate impermeable gloves to prevent skin contact, and all operations that have the potential of producing acrylamide dusts or aerosols of solutions should be conducted in a fume hood to prevent exposure by inhalation.

(continued on facing page)

Accidents

In the event of skin contact, immediately wash with soap and water and remove contaminated clothing. In case of eye contact, promptly wash with copious amounts of water for 15 min (lifting upper and lower lids occasionally) and obtain medical attention. If acrylamide is ingested, obtain medical attention immediately. If large amounts of acrylamide dust are inhaled, move the person to fresh air and seek medical attention at once.

In the event of a spill, mix acrylamide with an absorbent material (avoid raising dust), place in an appropriate container, and dispose of properly. Evacuation and cleanup using respiratory protection may be necessary in the event of a large spill or release in a confined area.

Disposal

Excess acrylamide and waste material containing this substance should be placed in an appropriate container, clearly labeled, and handled according to your institution's waste disposal guidelines. For more information on disposal procedures, see Chapter 7 of this volume.

LABORATORY CHEMICAL SAFETY SUMMARY: ACRYLONITRILE

Substance	Acrylonitrile (Vinyl cyanide, 2-propenenitrile, cyanoethylene, ACN) CAS 107-13-1
Formula	$H_2C=CH-C\equiv N$
Physical Properties	Colorless liquid bp 77 °C, mp −82 °C Moderately soluble in water (7.3 g/100 mL)
Odor	Mild pyridine-like odor at 2 to 22 ppm
Vapor Density	1.83 (air = 1.0)
Vapor Pressure	100 mmHg at 22.8 °C
Flash Point	−1 °C
Autoignition Temperature	481 °C

Toxicity Data

LD_{50} oral (rat)	78 mg/kg
LD_{50} skin (rabbit)	250 mg/kg
LC_{50} inhal (rat)	425 ppm (4 h)
PEL (OSHA)	2 ppm
TLV-TWA (ACGIH)	2 ppm—skin

Major Hazards

Probable human carcinogen (OSHA "select carcinogen"); moderate acute toxicity; highly flammable.

Toxicity

Acrylonitrile is classified as moderately toxic by acute exposure through oral intake, skin contact, and inhalation. Symptoms of exposure include weakness, lightheadedness, diarrhea, nausea, and vomiting. Acrylonitrile is severely irritating to the eyes and mildly irritating to the skin; prolonged contact with the skin can lead to burns.

Acrylonitrile is mutagenic in bacterial and mammalian cell cultures and embryotoxic/teratogenic in rats at levels that produce maternal toxicity. Acrylonitrile is carcinogenic in rats and is regulated by OSHA as a carcinogen (29 CFR 1910.1045). Acrylonitrile is listed in IARC Group 2A ("probable human carcinogen") and is classified as a "select carcinogen" under the criteria of the OSHA Laboratory Standard.

Flammability and Explosibility

Highly flammable liquid (NFPA rating = 3). Vapor forms explosive mixtures with air at concentrations of 3 to 17% (by volume). Hazardous gases produced in fire include hydrogen cyanide, carbon monoxide, and oxides of nitrogen. Carbon dioxide or dry chemical extinguishers should be used to fight acrylonitrile fires.

(continued on facing page)

Reactivity and Incompatibility

Violent reaction may occur on exposure to strong acids and bases, amines, strong oxidants, copper, and bromine. Violent polymerization can be initiated by heat, light, strong bases, peroxides, and azo compounds.

Storage and Handling

Because of its carcinogenicity and flammability, acrylonitrile should be handled using the "basic prudent practices" of Chapter 5.C, supplemented by the additional precautions for work with compounds of high chronic toxicity (Chapter 5.D) and extremely flammable substances (Chapter 5.F). In particular, work with acrylonitrile should be conducted in a fume hood to prevent exposure by inhalation, and splash goggles and impermeable gloves should be worn at all times to prevent eye and skin contact. Acrylonitrile should be used only in areas free of ignition sources. Containers of acrylonitrile should be stored in secondary containers in the dark in areas separate from oxidizers and bases.

Accidents

In the event of skin contact, immediately wash with soap and water and remove contaminated clothing. In case of eye contact, promptly wash with copious amounts of water for 15 min (lifting upper and lower lids occasionally) and obtain medical attention. If acrylonitrile is ingested, obtain medical attention immediately. If large amounts of this compound are inhaled, move the person to fresh air and seek medical attention at once.

In the event of a spill, remove all ignition sources, soak up the acrylonitrile with a spill pillow or absorbent material, place in an appropriate container, and dispose of properly. Evacuation and cleanup using respiratory protection may be necessary in the event of a large spill or release in a confined area.

Disposal

Excess acrylonitrile and waste material containing this substance should be placed in an appropriate container, clearly labeled, and handled according to your institution's waste disposal guidelines. For more information on disposal procedures, see Chapter 7 of this volume.

LABORATORY CHEMICAL SAFETY SUMMARY: ALUMINUM TRICHLORIDE

Substance

Aluminum trichloride
(Aluminum chloride, trichloroaluminum)
CAS 7446-70-0

Formula

$AlCl_3$

Physical Properties

White crystalline solid
Sublimes at 181 °C
Reacts violently with water (90 g/100 mL)

Odor

Hydrogen chloride odor detectable when exposed to moist air

Vapor Pressure

1 mmHg at 100 °C

Toxicity Data

LD_{50} oral (rat)	3730 mg/kg
LD_{50} skin (rabbit)	>2 g/kg
TLV-TWA (ACGIH)	2 mg(Al)/m^3

Major Hazards

Highly corrosive solid that reacts with water to form hydrochloric acid.

Toxicity

Aluminum chloride is strongly irritating and highly corrosive to the skin, eyes, and mucous membranes owing to its reaction with water to form hydrochloric acid. It is slightly toxic by ingestion but can cause severe burns to the mouth and digestive tract until hydrolyzed in the stomach. Inhalation of aluminum trichloride dust, vapor, or its hydrolysis products can result in severe damage to the tissues of the respiratory tract and can lead to shortness of breath, wheezing, coughing, and headache; inhalation of large amounts may lead to respiratory tract spasms and pulmonary edema and can be fatal. Skin and eye contact with aluminum chloride can cause severe burns.

Aluminum chloride may cause allergic skin reactions. Long-term exposure can cause damage to lungs. In some animal tests, aluminum chloride has shown developmental and reproductive toxicity. Aluminum chloride has not been found to be carcinogenic in humans.

Flammability and Explosibility

Aluminum chloride is not flammable but reacts violently with water, so fires involving this substance should be extinguished with carbon dioxide or dry chemicals. Toxic fumes (HCl and reaction products) can be released during fires.

Reactivity and Incompatibility

Anhydrous aluminum chloride reacts violently with water to produce HCl and a great deal of heat. Aluminum chloride reacts violently on heating with nitrobenzene and may react violently or explosively with ethylene oxide, organic azides, organic perchlorates, and sodium borohydride. In the presence of moisture, this substance is highly corrosive to most metals.

(continued on facing page)

Storage and Handling

Aluminum chloride should be handled in the laboratory using the "basic prudent practices" described in Chapter 5.C. In particular, work with this substance should be conducted in a fume hood, and impermeable gloves should be worn at all times when handling AlCl$_3$. Aluminum chloride should be stored in sealed containers under an inert atmosphere in a cool, dry place. Care should be taken in opening containers of this compound because of the possibility of the buildup of HCl vapor from hydrolysis with traces of moisture.

Accidents

In the event of skin contact, immediately wash with soap and water and remove contaminated clothing. In case of eye contact, promptly wash with copious amounts of water for 15 min (lifting upper and lower lids occasionally) and obtain medical attention. If aluminum chloride is ingested, wash the mouth with water and seek immediate medical attention. If large amounts of this compound (or the HCl vapor generated from its contact with water) are inhaled, move the person to fresh air and seek medical attention at once.

In the event of a spill, sweep up the aluminum chloride (avoid raising dust) and collect in a bag while wearing appropriate protective clothing. Small spills may be collected and carefully hydrolyzed with a large excess of cold water, neutralized with base, and disposed of properly. Respiratory protection may be necessary in the event of a large spill or release in a confined area.

Disposal

Small excess amounts of aluminum chloride and waste material containing this substance should be cautiously added to a large stirred excess of water, neutralized, and filtered. The insoluble solids should be placed in an appropriate container, clearly labeled, and handled according to your institution's waste disposal guidelines. The neutral aqueous solution should be flushed down a drain with plenty of water. For more information on disposal procedures, see Chapter 7 of this volume.

LABORATORY CHEMICAL SAFETY SUMMARY: AMMONIA (ANHYDROUS)

Substance	Ammonia (anhydrous) CAS 7664-41-7
Formula	NH_3
Physical Properties	Colorless gas bp -33 °C, mp -78 °C Highly soluble in water (89.9g/100 mL at 0 °C)
Odor	Intense pungent odor detectable at 17 ppm
Vapor Density	0.59 (air = 1.0)
Vapor Pressure	8.71 atm at 21 °C
Autoignition Temperature	690 °C

Toxicity Data

LD_{50} oral (rat)	350 mg/kg
LC_{50} inhal (rat)	2000 ppm (4 h)
PEL (OSHA)	35 ppm (27 mg/m³)
TLV-TWA (ACGIH)	25 ppm (17 mg/m³)
STEL (ACGIH)	35 ppm (27 mg/m³)

Major Hazards   Extremely irritating and corrosive to the eyes, skin, and respiratory tract.

Toxicity   Ammonia gas is extremely corrosive and irritating to the skin, eyes, nose, and respiratory tract. Exposure by inhalation causes irritation of the nose, throat, and mucous membranes. Lacrimation and irritation begin at 130 to 200 ppm, and exposure at 3000 ppm is intolerable. Exposure to high concentrations (above approximately 2500 ppm) is life threatening, causing severe damage to the respiratory tract, resulting in bronchitis, chemical pneumonitis, and pulmonary edema, which can be fatal. Eye contact with ammonia vapor is severely irritating, and exposure of the eyes to liquid ammonia or mists can result in serious damage, which may result in permanent eye injury and blindness. Skin contact with ammonia vapor, mists, and liquid can cause severe irritation and burns; contact with the liquid results in cryogenic burns as well. Ingestion of liquid ammonia burns the tissues, causing severe abdominal pain, nausea, vomiting, and collapse and can be fatal. Ammonia gas is regarded as having adequate warning properties.

Ammonia has not been found to be carcinogenic or to show reproductive or developmental toxicity in humans. Chronic exposure to ammonia can cause respiratory irritation and damage.

(continued on facing page)

Flammability and Explosibility

Ammonia vapor is slightly flammable (NFPA rating = 1) and ignites only with difficulty. Ammonia forms explosive mixtures with air in the range 16 to 25%. Water, carbon dioxide, or dry chemical extinguishers should be used for ammonia fires.

Reactivity and Incompatibility

Ammonia vapors or solutions can react with compounds of silver, gold, and mercury to produce unstable and highly explosive products. Do not use a mercury manometer for measuring ammonia gas pressure. Highly explosive nitrogen halides can form in reactions of ammonia with halogens, hypohalites, and similar compounds. Violent reactions can occur with oxidizing agents such as chromium trioxide, hydrogen peroxide, nitric acid, sodium and potassium nitrate, chlorite, chlorate, and bromate salts.

Storage and Handling

Anhydrous ammonia should be handled in the laboratory using the "basic prudent practices" described in Chapter 5.C, supplemented by the procedures for work with compressed gases discussed in Chapter 5.H. All work with ammonia should be conducted in a fume hood to prevent exposure by inhalation, and splash goggles and impermeable gloves should be worn at all times to prevent eye and skin contact. Cylinders of ammonia should be stored in locations appropriate for compressed gas storage and separated from incompatible compounds such as acids, halogens, and oxidizers. Only steel valves and fittings should be used on ammonia containers; copper, silver, and zinc should not be permitted to come into contact with ammonia.

Accidents

Prompt medical attention is required in all cases of overexposure to ammonia. In the event of skin contact, immediately wash with soap and water and remove contaminated clothing. In case of eye contact, promptly wash with copious amounts of water for 15 min (lifting upper and lower lids occasionally) and obtain medical attention. If ammonia is ingested, obtain medical attention immediately. If large amounts of ammonia are inhaled, move the person to fresh air and seek medical attention at once.

In the event of a gaseous ammonia leak, shut down all ignition sources and ventilate the area at once to prevent flammable mixtures from forming. Carefully remove cylinders with slow leaks to a fume hood or remote outdoor location. Limit access to an affected area. Respiratory protection may be necessary in the event of a large release in a confined area.

Disposal

Cylinders containing excess ammonia should be returned to the manufacturer. Ammonia gas may also be dissolved in water for neutralization and disposal. In some localities, aqueous ammonia may be disposed of down the drain after appropriate neutralization and dilution. If ammonia gas is directed into water, precautions should be taken to prevent the suckback of water into the ammonia-containing vessel or cylinder. Dissolution of ammonia in water is accompanied by heat evolution. In a fume hood, the diluted ammonia solution should be neutralized with a nonoxidizing strong acid such as HCl. The resulting solution can be discharged to the sanitary sewer. If drain disposal is not permitted, the aqueous ammonia should be placed in an appropriate container, clearly labeled, and handled according to your institution's waste disposal guidelines. For more information on disposal procedures, see Chapter 7 of this volume.

LABORATORY CHEMICAL SAFETY SUMMARY: AMMONIUM HYDROXIDE

Substance	Ammonium hydroxide (Aqua ammonia, ammonia) CAS 1336-21-6
Formula	28 to 30% NH_3 in H_2O
Physical Properties	Colorless liquid bp: unstable above 27.8 °C, mp −71.7 °C Concentrated ammonium hydroxide is a 29% solution of NH_3 in H_2O.
Odor	Strong pungent ammonia odor detectable at 17 ppm
Vapor Density	0.59 for anhydrous NH_3 (air = 1.0)
Vapor Pressure	115 mmHg at 20 °C for 29% solution
Autoignition Temperature	690 °C (for ammonia)

Toxicity Data

LD_{50} oral (rat)	350 mg/kg
PEL (OSHA)	35 ppm (27 mg/m^3)
TLV-TWA (ACGIH)	25 ppm (17 mg/m^3)
STEL (ACGIH)	35 ppm (27 mg/m^3)

Major Hazards Highly corrosive to the eyes, skin, and mucous membranes.

Toxicity Ammonia solutions are extremely corrosive and irritating to the skin, eyes, and mucous membranes. Exposure by inhalation can cause irritation of the nose, throat, and mucous membranes. Exposure to high concentrations of ammonia vapor (above approximately 2500 ppm) is life threatening, causing severe damage to the respiratory tract and resulting in bronchitis, chemical pneumonitis, and pulmonary edema, which can be fatal. Eye contact with ammonia vapor is severely irritating, and exposure of the eyes to ammonium hydroxide can result in serious damage and may cause permanent eye injury and blindness. Skin contact can result in severe irritation and burns; contact with the liquid results in cryogenic burns as well. Ingestion of ammonium hydroxide burns the mouth, throat, and gastrointestinal tract and can lead to severe abdominal pain, nausea, vomiting, and collapse.

Ammonium hydroxide has not been found to be carcinogenic or to show reproductive or developmental toxicity in humans. Chronic exposure to ammonia can cause respiratory irritation and damage.

Flammability and Explosibility Ammonia vapor is slightly flammable (NFPA rating = 1) and ignites only with difficulty. Ammonia forms explosive mixtures with air in the range 16 to 25%. Water, carbon dioxide, or dry chemical extinguishers should be used for ammonia fires.

(continued on facing page)

The information in this LCSS has been compiled by a committee of the National Research Council from literature sources and Material Safety Data Sheets and is believed to be accurate as of July 1994. This summary is intended for use by trained laboratory personnel in conjunction with the NRC report *Prudent Practices in the Laboratory: Handling and Disposal of Chemicals*. This LCSS presents a concise summary of safety information that should be adequate for most laboratory uses of the title substance, but in some cases it may be advisable to consult more comprehensive references. This information should not be used as a guide to the nonlaboratory use of this chemical.

Reactivity and Incompatibility

Highly explosive nitrogen halides will form in reactions with halogens, hypohalites, and similar compounds. Reaction with certain gold, mercury, and silver compounds may form explosive products. Violent reactions can occur with oxidizing agents such as chromium trioxide, hydrogen peroxide, nitric acid, chlorite, chlorate, and bromate salts. Exothermic and violent reactions may occur if concentrated ammonium hydroxide solution is mixed with strong acids, acidic metal and nonmetal halides, and oxyhalides.

Storage and Handling

Ammonium hydroxide should be handled in the laboratory using the "basic prudent practices" described in Chapter 5.C. All work with this substance should be conducted in a fume hood to prevent exposure by inhalation, and splash goggles and impermeable gloves should be worn at all times to prevent eye and skin contact. Containers should be tightly sealed to prevent escape of vapor and should be stored in a cool area separate from halogens, acids, and oxidizers. Containers stored in warm locations may build up dangerous internal pressures of ammonia gas.

Accidents

In the event of skin contact, immediately wash with soap and water and remove contaminated clothing. In case of eye contact, promptly wash with copious amounts of water for 15 min (lifting upper and lower lids occasionally) and obtain medical attention. If ammonium hydroxide is ingested, obtain medical attention immediately. If large amounts of ammonia are inhaled, move the person to fresh air and seek medical attention at once.

In the event of a spill, soak up ammonium hydroxide with a spill pillow or absorbent material, place in an appropriate container, and dispose of properly. Alternatively, flood the spill with water to dilute the ammonia before cleanup. Boric, citric, and similar powdered acids are good granular neutralizing spill cleanup materials. Respiratory protection may be necessary in the event of a large spill or release in a confined area.

Disposal

In some localities, ammonium hydroxide may be disposed of down the drain after appropriate neutralization and dilution. In a fume hood, the concentrated solution should be diluted with water to about 4% concentration in a suitably large container, and neutralized with a nonoxidizing strong acid such as HCl. The resulting solution can be discharged to the sanitary sewer. If neutralization and drain disposal are not permitted, excess ammonium hydroxide and waste material containing this substance should be placed in an appropriate container, clearly labeled, and handled according to your institution's waste disposal guidelines. For more information on disposal procedures, see Chapter 7 of this volume.

LABORATORY CHEMICAL SAFETY SUMMARY: ANILINE

Substance	Aniline (Phenylamine; aminobenzene) CAS 62-53-3
Formula	$C_6H_5NH_2$
Physical Properties	Colorless, oily liquid; darkens to brown on exposure to air and light bp 184 °C, mp −6 °C Moderately soluble in water (3.5 g/100 mL at 20 °C)
Odor	Sweet, amine-like odor detectable at 0.6 to 10 ppm
Vapor Density	3.2 (air = 1.0)
Vapor Pressure	0.7 mmHg at 20 °C
Flash Point	70 °C
Autoignition Temperature	615 °C

Toxicity Data

LD_{50} oral (rat)	250 mg/kg
LD_{50} skin (rabbit)	820 mg/kg
LC_{50} inhal (rat)	478 ppm
PEL (OSHA)	5 ppm (19 mg/m³)—skin
TLV-TWA (ACGIH)	2 ppm (7.6 mg/m³)—skin

Major Hazards
Moderately toxic if swallowed, inhaled, or absorbed through the skin; causes skin and eye irritation.

Toxicity
Aniline is a moderate skin irritant, a moderate to severe eye irritant, and a skin sensitizer in animals. Aniline is moderately toxic via inhalation and ingestion. Symptoms of exposure (which may be delayed up to 4 hours) include headache, weakness, dizziness, nausea, difficulty breathing, and unconsciousness. Exposure to aniline results in the formation of methemoglobin and can thus interfere with the ability of the blood to transport oxygen. Effects from exposure at levels near the lethal dose include hypoactivity, tremors, convulsions, liver and kidney effects, and cyanosis.

Aniline has not been found to be a carcinogen or reproductive toxin in humans. Some tests in rats demonstrate carcinogenic activity. However, other tests in which mice, guinea pigs, and rabbits were treated by various routes of administration gave negative results. Aniline produced developmental toxicity only at maternally toxic dose levels but did not have a selective toxicity for the fetus. It produces genetic damage in animals and in mammalian cell cultures but not in bacterial cell cultures.

(continued on facing page)

The information in this LCSS has been compiled by a committee of the National Research Council from literature sources and Material Safety Data Sheets and is believed to be accurate as of July 1994. This summary is intended for use by trained laboratory personnel in conjunction with the NRC report *Prudent Practices in the Laboratory: Handling and Disposal of Chemicals*. This LCSS presents a concise summary of safety information that should be adequate for most laboratory uses of the title substance, but in some cases it may be advisable to consult more comprehensive references. This information should not be used as a guide to the nonlaboratory use of this chemical.

Flammability and Explosibility	Aniline is a combustible liquid (NFPA rating = 2). Smoke from a fire involving aniline may contain toxic nitrogen oxides and aniline vapor. Toxic aniline vapors are given off at high temperatures and form explosive mixtures in air. Carbon dioxide or dry chemical extinguishers should be used to fight aniline fires.
Reactivity and Incompatibility	Reacts violently with strong oxidizing agents, including nitric acid, peroxides, and ozone.
Storage and Handling	Aniline should be handled in the laboratory using the "basic prudent practices" described in Chapter 5.C. In particular, aniline should only be used in areas free of ignition sources, and quantities greater than 1 liter should be stored in tightly sealed metal containers in areas separate from oxidizers.
Accidents	In the event of skin contact, immediately wash with soap and water and remove contaminated clothing. In case of eye contact, promptly wash with copious amounts of water for 15 min (lifting upper and lower lids occasionally) and obtain medical attention. If aniline is ingested, obtain medical attention immediately. If inhaled, move the person to fresh air and seek medical attention at once.
	In the event of a spill, remove all ignition sources, soak up the aniline with a spill pillow or absorbent material, place in a covered metal container, label clearly, and dispose of properly. Respiratory protection may be necessary in the event of a large spill or release in a confined area.
Disposal	Excess aniline and waste material containing this substance should be placed in a covered metal container, clearly labeled, and handled according to your institution's waste disposal guidelines. For more information on disposal procedures, see Chapter 7 of this volume.

LABORATORY CHEMICAL SAFETY SUMMARY: ARSINE

Substance	Arsine (Arsenic hydride, arsenic trihydride, hydrogen arsenide) CAS 7784-42-1
Formula	AsH_3
Physical Properties	Colorless gas bp $-62\,°C$, mp $-117\,°C$ Slightly soluble in water (0.07 g/100 mL at 20 °C)
Odor	Garlic-like odor detectable at 0.5 to 1 ppm
Vapor Density	2.7 (air = 1.0)
Vapor Pressure	>760 mmHg at 20 °C
Flash Point	$< -62\,°C$
Autoignition Temperature	Not established. Decomposes at 232 to 300 °C to form elemental arsenic and hydrogen.

Toxicity Data

LC_{LO} inhal (rat)	94 ppm (300 mg/m³; 15 min)
PEL (OSHA)	0.05 ppm (0.2 mg/m³)
TLV-TWA (ACGIH)	0.05 ppm (0.16 mg/m³)

Major Hazards

Extremely toxic gas that destroys red blood cells and can cause widespread organ injury and death.

Toxicity

The acute toxicity of arsine by inhalation is extremely high. This substance is a powerful systemic toxin with a strong affinity for the hemoglobin in the blood, causing hemolysis. Acute inhalation of arsine can cause the breakdown of red blood cells and hemoglobin, impairment of kidney function, damage to the liver and heart, electroencephalogram abnormality, hemolytic anemia, and death due to kidney or heart failure. Symptoms may be delayed for several hours, particularly if very low concentrations have been inhaled. Symptoms of exposure to arsine may include headache, malaise, weakness, dizziness, breathing difficulty, abdominal pain, nausea, vomiting, jaundice, dark red (bloody) urine followed by absence of urination, pulmonary edema, and coma. Exposure to a concentration of 5 to 10 ppm in air for several minutes may be hazardous to human health. The minimum amount of arsine detectable by odor is about 0.5 ppm; since the permissible exposure limit is 0.05 ppm, arsine does not have adequate warning properties to avoid overexposure.

In cases where the amount of inhaled arsine is insufficient to produce acute effects, or where small quantities are inhaled over prolonged periods, destruction of red blood cells

(continued on facing page)

will occur. The only symptoms noted may be general tiredness, pallor, breathlessness on exertion, and palpitations as would be expected with severe secondary anemia. The carcinogenicity of arsine in humans has not been established; however, arsenic and certain inorganic arsenic compounds are recognized human carcinogens.

Flammability and Explosibility

Arsine is flammable in air, having a lower explosion limit (LEL) of 5.8%. The upper limit has not been determined. Combustion products (arsenic trioxide and water) are less toxic than arsine itself. In the event of an arsine fire, stop the flow of gas if possible without risk of harmful exposure and let the fire burn itself out.

Reactivity and Incompatibility

Arsine is a strong reducing agent and reacts violently with oxidizing agents such as fluorine, chlorine, nitric acid, and nitrogen trichloride.

Storage and Handling

Because of its high acute toxicity, arsine should be handled using the "basic prudent practices" of Chapter 5.C, supplemented by the additional practices for work with compounds of high toxicity (Chapter 5.D), flammability (Chapter 5.F), and for work with compressed gases (Chapter 5.H). In particular, cylinders of arsine should be stored and used in a continuously ventilated gas cabinet or fume hood. Local fire codes should be reviewed for limitations on quantity and storage requirements. Carbon steel, stainless steel, Monel®, and Hastelloy®C are preferred materials for handling arsine; brass and aluminum should be avoided. Kel-F® and Teflon® are preferred gasket materials; Viton® and Nylon® are acceptable.

Accidents

In the event of a release of arsine, the area should be evacuated immediately. Regard anyone exposed to arsine as having inhaled a potentially toxic dose. Rescue of an affected individual requires appropriate respiratory protection. Remove exposed individuals to an uncontaminated area and seek immediate emergency medical help. Keep victim warm, quiet, and at rest; provide assisted respiration if breathing has stopped.

To respond to a release, use appropriate protective equipment and clothing. Positive pressure air-supplied respiratory protection is required. Close cylinder valve and ventilate area. Remove cylinder to a fume hood or remote area if it cannot be shut off.

Emergency response and rescue procedures should be in place before beginning work with arsine. Local rescue assistance may be needed and should be prearranged.

Disposal

Excess arsine should be returned to the manufacturer, according to your institution's waste disposal guidelines. For more information on disposal procedures, see Chapter 7 of this volume.

LABORATORY CHEMICAL SAFETY SUMMARY: BENZENE

Substance	Benzene (Benzol) CAS 71-43-2
Formula	C_6H_6
Physical Properties	Colorless liquid bp 80.1 °C, mp 5.5 °C Slightly soluble in water (0.18 g/100 mL)
Odor	"Paint-thinner-like" odor detectable at 12 ppm
Vapor Density	2.7 (air = 1.0)
Vapor Pressure	75 mmHg at 20 °C
Flash Point	-11.1 °C
Autoignition Temperature	560 °C

Toxicity Data

LD_{50} oral (rat)	930 mg/kg
LC_{50} inhal (rat)	10,000 ppm (7 h)
PEL (OSHA)	1 ppm (3.2 mg/m^3)
TLV-TWA (ACGIH)	10 ppm (32 mg/m^3)
STEL (ACGIH)	5 ppm (16 mg/m^3)

Major Hazards   Highly flammable; chronic toxin affecting the blood-forming organs; OSHA "select carcinogen."

Toxicity   The acute toxicity of benzene is low. Inhalation of benzene can cause dizziness, euphoria, giddiness, headache, nausea, drowsiness, and weakness. Benzene can cause moderate irritation to skin and severe irritation to eyes and mucous membranes. Benzene readily penetrates the skin to cause the same toxic effects as inhalation or ingestion.

The chronic toxicity of benzene is significant. Exposure to benzene affects the blood and blood-forming organs such as the bone marrow, causing irreversible injury; blood disorders including anemia and leukemia may result. The symptoms of chronic benzene exposure may include fatigue, nervousness, irritability, blurred vision, and labored breathing. Benzene is regulated by OSHA as a carcinogen (Standard 1910.1028) and is listed in IARC Group 1 ("carcinogenic to humans"). This substance is classified as a "select carcinogen" under the criteria of the OSHA Laboratory Standard.

Flammability and Explosibility   Benzene is a highly flammable liquid (NFPA rating = 3), and its vapors may travel a considerable distance to a source of ignition and "flash back." Vapor-air mixtures are

(continued on facing page)

explosive above the flash point. Carbon dioxide and dry chemical extinguishers should be used to fight benzene fires.

Reactivity and Incompatibility

Fire and explosion hazard with strong oxidizers such as chlorine, oxygen, and bromine (in the presence of certain catalysts such as iron) and with strong acids.

Storage and Handling

Because of its carcinogenicity and flammability, benzene should be handled using the "basic prudent practices" of Chapter 5.C, supplemented by the additional precautions for work with compounds of high chronic toxicity (Chapter 5.D) and extremely flammable substances (Chapter 5.F). In particular, work with benzene should be conducted in a fume hood to prevent exposure by inhalation, and splash goggles and impermeable gloves should be worn at all times to prevent eye and skin contact. Benzene should be used only in areas free of ignition sources.

Accidents

In the event of skin contact, immediately wash with soap and water and remove contaminated clothing. In case of eye contact, promptly wash with copious amounts of water for 15 min (lifting upper and lower lids occasionally) and obtain medical attention. If benzene is ingested, obtain medical attention immediately. If large amounts of this compound are inhaled, move the person to fresh air and seek medical attention at once.

In the event of a spill, remove all ignition sources, soak up the benzene with a spill pillow or absorbent material, place in an appropriate container, and dispose of properly. Respiratory protection should be employed during spill cleanup.

Disposal

Excess benzene and waste material containing this substance should be placed in an appropriate container, clearly labeled, and handled according to your institution's waste disposal guidelines. For more information on disposal procedures, see Chapter 7 of this volume.

LABORATORY CHEMICAL SAFETY SUMMARY: BORON TRIFLUORIDE

Substance	Boron trifluoride (Boron fluoride, trifluoroborane) CAS 7637-07-2
Formula	BF_3
Physical Properties	Colorless gas, fumes in moist air bp -100 °C, mp -127 °C Highly soluble in cold water (332 g/100 mL)
Odor	Pungent odor detectable at 1.5 ppm
Vapor Density	2.4 (air = 1.0)
Vapor Pressure	>1 mmHg at 20 °C
Flash Point	Noncombustible

Toxicity Data

LC_{50} inhal (rat)	387 ppm (1070 mg/m^3; 1 h)
PEL (OSHA)	1 ppm (3 mg/m^3; ceiling)
TLV (ACGIH)	1 ppm (3 mg/m^3; ceiling)

Major Hazards

Highly corrosive to skin, eyes, and mucous membranes; reacts violently with water to form highly toxic HF.

Toxicity

Boron trifluoride (and organic complexes such as BF_3-etherate) are extremely corrosive substances that are destructive to all tissues of the body. Upon contact with moisture in the skin and other tissues, these compounds react to form hydrofluoric acid and fluoroboric acid, which cause severe burns. Boron trifluoride gas is extremely irritating to the skin, eyes, and mucous membranes. Inhalation of boron trifluoride can cause severe irritation and burning of the respiratory tract, difficulty breathing, and possibly respiratory failure and death. Exposure of the eyes to BF_3 can cause severe burns and blindness. This compound is not considered to have adequate warning properties.

Boron trifluoride has not been found to be carcinogenic or to show reproductive or developmental toxicity in humans. Chronic exposure to boron trifluoride gas can cause respiratory irritation and damage.

Flammability and Explosibility

Boron trifluoride gas is noncombustible. Water should not be used to extinguish any fire in which boron trifluoride is present. Dry chemical powder should be used for fires involving organic complexes of boron trifluoride.

Reactivity and Incompatibility

Boron trifluoride reacts violently with water and alkali and alkaline earth metals such as sodium, potassium, and calcium. It may react exothermically with alkyl nitrates, ethylene oxide, and butadiene.

(continued on facing page)

Storage and Handling

Boron trifluoride should be handled in the laboratory using the "basic prudent practices" described in Chapter 5.C, supplemented in the case of work with gaseous boron trifluoride with the procedures of Chapter 5.H. All work with boron trifluoride should be conducted in a fume hood to prevent exposure by inhalation, and splash goggles and impermeable gloves should be worn to prevent eye and skin contact. Cylinders of boron trifluoride should be stored in locations appropriate for compressed gas storage and separated from alkali metals, alkaline earth metals, and other incompatible substances. Solutions of boron trifluoride should be stored in tightly sealed containers under an inert atmosphere in secondary containers.

Accidents

In the event of skin contact, immediately wash with soap and water and remove contaminated clothing. In case of eye contact, promptly wash with copious amounts of water for 15 min (lifting upper and lower lids occasionally) and obtain medical attention. If boron trifluoride is inhaled, move the person to fresh air and seek medical attention at once. If this compound is ingested, obtain medical attention immediately.

In the event of accidental release of boron trifluoride gas, evacuate the area, and if the cause of the release is a leaking cylinder, remove the cylinder to a fume hood or open area if it is possible to do so safely. Positive pressure air-supplied respiratory protection and protective clothing may be necessary to deal with a leaking cylinder of boron trifluoride, and emergency response personnel should be notified.

Disposal

Cylinders containing excess boron trifluoride should be returned to the manufacturer. Solutions of boron trifluoride should be labeled and disposed of according to your institution's disposal guidelines. For more information on disposal procedures, see Chapter 7 of this volume.

LABORATORY CHEMICAL SAFETY SUMMARY: BROMINE

Substance	Bromine CAS 7726-95-6
Formula	Br_2
Physical Properties	Dark red-brown liquid bp 59 °C, mp −7 °C Slightly soluble in water (3.5 g/100 mL)
Odor	Odor can be detected at concentrations as low as 0.05 ppm; exposure to concentrations below 1 ppm causes lacrimation.
Vapor Density	5.5 (air = 1.0)
Vapor Pressure	175 mmHg at 20 °C
Flash Point	Noncombustible

Toxicity Data

LD_{50} oral (rat)	2600 mg/kg
LC_{50} inhal (rat)	2700 mg/m^3
PEL (OSHA)	0.1 ppm
TLV-TWA (ACGIH)	0.1 ppm (0.7 mg/m^3)
STEL (ACGIH)	0.3 ppm (2 mg/m$^{3)}$

Major Hazards

Highly corrosive to skin and eyes; moderately toxic via inhalation; reacts violently with readily oxidized substances.

Toxicity

Bromine is highly corrosive to the skin, causing irritation and destruction with blister formation. If bromine is not removed from the skin immediately, deep-seated ulcers develop, which heal slowly. Severely painful and destructive eye burns may result from contact with either liquid or concentrated vapors of bromine. Bromine is a moderately toxic substance via inhalation. There are good warning properties for bromine: lacrimation begins at ~1 ppm, and 50 ppm is highly irritating to humans. A short exposure (minutes) to 1000 ppm would likely be fatal for humans. Vapor exposures can cause irritation and damage to the upper and lower respiratory tract (nose, throat, and lungs) to varying degrees depending on the concentration. If exposure is sufficiently high, it will cause pulmonary edema, which could lead to death. Other reported symptoms of overexposure include coughing, tightness of chest, nosebleed, headache, and dizziness, followed after some hours by abdominal pain, diarrhea, and a measles-like rash on the trunk and extremities.

Animal studies on the chronic toxicity of bromine revealed disturbances in the respiratory, nervous, and endocrine systems after exposure to 0.2 ppm for 4 months; similar exposure to 0.02 ppm did not produce any adverse effects.

(continued on facing page)

Flammability and Explosibility   Bromine alone is a noncombustible substance (NFPA rating = 0).

Reactivity and Incompatibility   Bromine reacts violently with easily oxidized substances, including many organic compounds and a number of metals. Explosions have been reported to occur, for example, on addition of bromine to methanol, acetaldehyde, and DMF. Fires and/or explosions may result from the reactions of bromine with hydrogen, acetylene, ammonia, aluminum, mercury, sodium, potassium, and phosphorus.

Storage and Handling   Bromine should be handled in the laboratory using the "basic prudent practices" described in Chapter 5.C. In particular, work with bromine should be conducted in a fume hood to prevent exposure by inhalation, and splash goggles and rubber gloves should be worn at all times when handling this corrosive substance. Containers of bromine should be stored at room temperature in a secondary container separately from readily oxidizable substances.

Accidents   In the event of skin contact, immediately wash with soap and water and remove contaminated clothing. If irritation or burns develop, seek medical attention. In case of eye contact, promptly wash with copious amounts of water and obtain medical attention. If bromine is ingested, give the person large amounts of milk or water to dilute the bromine (do *not* attempt to induce vomiting) and obtain medical attention immediately. If a significant amount of bromine is inhaled, move the person to fresh air and seek medical attention at once.

Treat small spills of bromine with sodium thiosulfate and an inert absorbent, place in an appropriate container, and dispose of properly. Large spills may require evacuation of the area and cleanup using full protective equipment.

Disposal   Excess bromine and waste material containing this substance should be placed in an appropriate container, clearly labeled, and handled according to your institution's waste disposal guidelines. Special care should be taken not to mix bromine with incompatible waste materials. For more information on disposal procedures, see Chapter 7 of this volume.

Substance	*tert*-Butyl hydroperoxide (and related organic peroxides) (TBHP; 2-hydroperoxy-2-methylpropane) CAS 75-91-2
Formula	$(CH_3)_3COOH$
Physical Properties	Colorless liquid Commercially available as 70 and 90% aqueous solutions and as "anhydrous solutions" in hydrocarbon solvents (e.g., decane) 70% aq TBHP: bp 96 °C, mp −3 °C Moderately soluble in water
Odor	Not available
Vapor Pressure	62 mmHg at 45 °C
Flash Point	27 to 54 °C
Autoignition Temperature	Self-accelerating decomposition at 88 to 93 °C

Toxicity Data

LD_{50} oral (rat)	406 mg/kg
LD_{50} skin (rabbit)	460 mg/kg
LC_{50} inhal (rat)	500 ppm (4 h)

Major Hazards	Highly reactive oxidizing agent; sensitive to heat and shock; eye and skin irritant.
Toxicity	Moderately toxic by inhalation and ingestion and severely irritating to the eyes and skin. *t*-Butyl hydroperoxide has not been found to be carcinogenic or to show reproductive or developmental toxicity in humans.
Flammability and Explosibility	*tert*-Butyl hydroperoxide is a flammable liquid and a highly reactive oxidizing agent. Pure TBHP is shock sensitive and may explode on heating. Carbon dioxide or dry chemical extinguishers should be used for fires involving *tert*-butyl hydroperoxide.
Reactivity and Incompatibility	*tert*-Butyl hydroperoxide and concentrated aqueous solutions of TBHP react violently with traces of acid and the salts of certain metals, including, in particular, manganese, iron, and cobalt. Mixing anhydrous *tert*-butyl hydroperoxide with organic and readily oxidized substances can cause ignition and explosion. TBHP can initiate polymerization of certain olefins.
Storage and Handling	*tert*-Butyl hydroperoxide should be handled in the laboratory using the "basic prudent practices" described in Chapter 5.C supplemented by the additional precautions for work with reactive and explosive substances (Chapter 5.G). In particular, *tert*-butyl hydroperox-

(continued on facing page)

ide should be stored in the dark at room temperature (do not refrigerate) separately from oxidizable compounds, flammable substances, and acids. Reactions involving this substance should be carried out behind a safety shield.

Accidents

In the event of skin contact, immediately wash with soap and water and remove contaminated clothing. In case of eye contact, promptly wash with copious amounts of water for 15 min (lifting upper and lower lids occasionally) and obtain medical attention. If *tert*-butyl hydroperoxide is inhaled or ingested, obtain medical attention immediately.

In the event of a spill, remove all ignition sources, soak up the *tert*-butyl hydroperoxide with a spill pillow or noncombustible absorbent material, place in an appropriate container, and dispose of properly. Respiratory protection may be necessary in the event of a large spill or release in a confined area. Cleanup of anhydrous *tert*-butyl hydroperoxide and concentrated solutions requires special precautions and should be carried out by trained personnel working from behind a body shield.

Disposal

Excess *tert*-butyl hydroperoxide and waste material containing this substance should be placed in an appropriate container, clearly labeled, and handled according to your institution's waste disposal guidelines. For more information on disposal procedures, see Chapter 7 of this volume.

LABORATORY CHEMICAL SAFETY SUMMARY: BUTYLLITHIUMS

Substance	Butyllithiums (and related alkyl lithium reagents) n-butyllithium: CAS 109-72-8 s-butyllithium (1-methylpropyllithium): CAS 598-30-1 t-butyllithium (1,1-dimethylethyllithium): CAS 594-19-4
Formula	C_4H_9Li
Physical Properties	Usually supplied and handled as solutions in ether or hydrocarbon solvents
Odor	Odor of the solvent
Toxicity Data	There is little toxicity data available for the butyllithiums; for data on ether and hydrocarbon solvents, see the appropriate LCSSs.
Major Hazards	Highly reactive; violent reactions may occur on exposure to water, CO_2 and other materials; may ignite spontaneously on exposure to air; highly corrosive to the skin and eyes.
Toxicity	Solutions of the butyllithiums are corrosive to the skin, eyes, and mucous membranes. Reaction with water generates highly corrosive lithium alkoxides and lithium hydroxide.
Flammability and Explosibility	The risk of fire or explosion on exposure of butyllithium solutions to the atmosphere depends on the identity of the organolithium compound, the nature of the solvent, the concentration of the solution, and the humidity. t-Butyllithium solutions are the most pyrophoric and may ignite spontaneously on exposure to air. Dilute solutions (1.6 M, 15% or less) of n-butyllithium in hydrocarbon solvents, although highly flammable, have a low degree of pyrophoricity and do not spontaneously ignite. Under normal laboratory conditions (25 °C, relative humidity of 70% or less), solutions of ~20% concentration will usually not ignite spontaneously on exposure to air. More concentrated solutions of n-butyllithium (50 to 80%) are most dangerous and will immediately ignite on exposure to air. Contact with water or moist materials can lead to fires and explosions, and the butyllithiums also react violently with oxygen.
Reactivity and Incompatibility	The butyllithiums are extremely reactive organometallic compounds. Violent explosions occur on contact with water with ignition of the solvent and of the butane produced. t-Butyllithium will ignite spontaneously in air. The butyllithiums ignite on contact with water, carbon dioxide, and halogenated hydrocarbons. The butyllithiums are incompatible with acids, halogenated hydrocarbons, alcohols, and many other classes of organic compounds.
Storage and Handling	Butyllithium solutions should be handled in the laboratory using the "basic prudent practices" described in Chapter 5.C, supplemented by the additional precautions for work with flammable (Chapter 5.F) and reactive (Chapter 5.G) substances. In particular, butyllithium should be stored and handled in areas free of ignition sources, and containers of butyllithium should be stored under an inert atmosphere. Work with butyllithium

(continued on facing page)

The information in this LCSS has been compiled by a committee of the National Research Council from literature sources and Material Safety Data Sheets and is believed to be accurate as of July 1994. This summary is intended for use by trained laboratory personnel in conjunction with the NRC report *Prudent Practices in the Laboratory: Handling and Disposal of Chemicals*. This LCSS presents a concise summary of safety information that should be adequate for most laboratory uses of the title substance, but in some cases it may be advisable to consult more comprehensive references. This information should not be used as a guide to the nonlaboratory use of this chemical.

should be conducted in a fume hood under an inert gas such as nitrogen or argon. Safety glasses, impermeable gloves, and a fire-retardant laboratory coat are required.

Accidents

In the event of skin contact, immediately wash with soap and water and remove contaminated clothing. In case of eye contact, promptly wash with copious amounts of water for 15 min (lifting upper and lower lids occasionally) and obtain medical attention. If butyllithium solution is ingested, obtain medical attention immediately. If large amounts of butyllithium solution are inhaled, move the person to fresh air and seek medical attention at once.

In the event of a spill, remove all ignition sources, and allow the butyllithium to react with atmospheric moisture. Carefully treat the residue with water, soak up with a spill pillow or absorbent material, place in an appropriate container, and dispose of properly. Respiratory protection may be necessary in the event of a large spill or release in a confined area.

Disposal

Excess butyllithium solution can be destroyed by dilution with hydrocarbon solvent to a concentration of approximately 5 wt %, followed by gradual addition to water with vigorous stirring under an inert atmosphere. Alternatively, the butyllithium solution can be slowly poured (transfer by cannula for *s*- or *t*-butyllithium) into a plastic tub or other container of powdered dry ice.

The residues from the above procedures and excess butyllithium should be placed in an appropriate container, clearly labeled, and handled according to your institution's waste disposal guidelines. For more information on disposal procedures, see Chapter 7 of this volume.

LABORATORY CHEMICAL SAFETY SUMMARY: CARBON DISULFIDE

Substance	Carbon disulfide (Carbon bisulfide) CAS 75-15-0
Formula	CS_2
Physical Properties	Colorless liquid bp 46 °C, mp −111 °C Slightly soluble in water (0.22 g/100 mL)
Odor	Cabbage-like odor detectable at 0.016 to 0.42 ppm (mean = 0.2 ppm)
Vapor Density	2.6 (air = 1.0)
Vapor Pressure	300 mmHg at 20 °C
Flash Point	−30 °C
Autoignition Temperature	90 °C

Toxicity Data

LD_{50} oral (rat)	3188 mg/kg
LC_{50} inhal (rat)	25,000 mg/m^3 (2 h)
STEL (OSHA)	12 ppm (36 mg/m^3)—skin
PEL (OSHA)	4 ppm (12 mg/m^3)
TLV-TWA (ACGIH)	10 ppm (31 mg/m^3)—skin

Major Hazards   Extremely flammable, volatile liquid; vapors are readily ignited by hot surfaces.

Toxicity   Carbon disulfide is only slightly toxic to laboratory animals by inhalation or ingestion, but its toxicity is relatively greater in humans. Exposure to 5000 ppm of carbon disulfide for 15 min can be fatal to humans. CS_2 may also exert its toxic effects after absorption through skin. By all routes of exposure, carbon disulfide affects the central nervous system. Overexposure to CS_2 may cause headache, dizziness, fatigue, muscle weakness, numbness, nervousness, or psychological disturbances. Contact of the liquid or high concentrations of CS_2 vapor with the eyes may cause irritation. Skin contact can also cause rash or skin irritation. Carbon disulfide is regarded as a substance with good warning properties.

Chronic exposure to relatively high concentrations of carbon disulfide may cause the central nervous system effects described above. In addition, chronic overexposure to carbon disulfide causes increased atherosclerosis, leading to risk of cardiovascular disease. Prolonged exposure of female workers to low concentrations of carbon disulfide has been associated with birth defects in offspring; exposure limit values provide little margin of safety for risk of developmental effects. Carbon disulfide has not been found to be a carcinogen in humans.

(continued on facing page)

The information in this LCSS has been compiled by a committee of the National Research Council from literature sources and Material Safety Data Sheets and is believed to be accurate as of July 1994. This summary is intended for use by trained laboratory personnel in conjunction with the NRC report *Prudent Practices in the Laboratory: Handling and Disposal of Chemicals*. This LCSS presents a concise summary of safety information that should be adequate for most laboratory uses of the title substance, but in some cases it may be advisable to consult more comprehensive references. This information should not be used as a guide to the nonlaboratory use of this chemical.

Flammability and Explosibility

Carbon disulfide is extremely flammable and is a dangerous fire hazard (NFPA rating = 3). It is has a high vapor pressure and extremely low autoignition temperature. Its vapor is heavier than air and can travel a considerable distance to a source of ignition and flash back. The vapor forms explosive mixtures in air at concentrations of 1.3 to 50%. Carbon disulfide can be ignited by hot surfaces such as steam baths that would ordinarily not constitute an ignition source for other flammable vapors. Rust (iron oxide) may increase the likelihood of ignition by hot surfaces. Carbon disulfide fires should be extinguished with CO_2 or dry chemical extinguishers.

Reactivity and Incompatibility

Reactions of alkali metals with carbon disulfide may cause explosions. Carbon disulfide reacts violently with metal azides.

Storage and Handling

Carbon disulfide should be handled in the laboratory using the "basic prudent practices" described in Chapter 5.C, supplemented by additional precautions for dealing with extremely flammable substances (Chapter 5.F). In particular, carbon disulfide should be used only in areas free of ignition sources (including hot plates, incandescent lightbulbs, and steam baths), and this substance should be stored in tightly sealed metal containers in areas separate from oxidizers.

Accidents

In the event of skin contact, immediately wash with soap and water and remove contaminated clothing. In case of eye contact, promptly wash with copious amounts of water for 15 min (lifting upper and lower lids occasionally) and obtain medical attention. If carbon disulfide is ingested, obtain medical attention immediately. If large amounts of this compound are inhaled, move the person to fresh air and seek medical attention at once.

In the event of a spill, take care to remove all ignition sources, soak up the carbon disulfide with a spill pillow or absorbent material, place in an appropriate container, and dispose of properly, taking appropriate precautions because of the extreme flammability of the liquid and vapor. Respiratory protection may be necessary in the event of a large spill or release in a confined area.

Disposal

Excess carbon disulfide and waste material containing this substance should be placed in an appropriate container, clearly labeled, and handled according to your institution's waste disposal guidelines. For more information on disposal procedures, see Chapter 7 of this volume.

LABORATORY CHEMICAL SAFETY SUMMARY: CARBON MONOXIDE

Substance   Carbon monoxide
                       (Carbonic oxide, monoxide)
                       CAS 630-08-0

Formula   CO

Physical Properties Colorless gas
                       bp −191.5 °C, mp −205 °C
                       Slightly soluble in water (0.004 g/100 mL at 20 °C)

Odor   Odorless gas

Vapor Density   0.97 (air = 1.0)

Vapor Pressure   >760 mmHg at 20 °C

Flash Point   < −191 °C

**Autoignition
Temperature**   609 °C

Toxicity Data LC_{50} inhal (rat)   1807 ppm (2065 mg/m³; 4 h)
                       LC_{LO} inhal (man)   4000 ppm (4570 mg/m³; 30 min)
                       PEL (OSHA)   50 ppm (55 mg/m³)
                       TLV-TWA (ACGIH)   25 ppm (29 mg/m³)

Major Hazards   Moderately toxic gas with no warning properties; decreases the ability of the blood to
                       carry oxygen to the tissues.

Toxicity   The acute toxicity of carbon monoxide by inhalation is moderate. Carbon monoxide is a
                       chemical asphyxiant that exerts its effects by combining preferentially with hemoglobin,
                       the oxygen-transport pigment of the blood, thereby excluding oxygen. Symptoms of
                       exposure to CO at 500 to 1000 ppm include headache, palpitations, dizziness, weakness,
                       confusion, and nausea. Loss of consciousness and death may result from exposure to
                       concentrations of 4000 ppm and higher; high concentrations may be rapidly fatal without
                       producing significant warning symptoms. Exposure to this gas may aggravate heart and
                       artery disease and may cause chest pain in individuals with preexisting heart disease.
                       Pregnant women are more susceptible to the effects of carbon monoxide exposure. Since
                       carbon monoxide is odorless, colorless, and tasteless, it has no warning properties, and
                       unanticipated overexposure to this highly dangerous gas can readily occur.

                       Carbon monoxide has not been found to be carcinogenic in humans. This substance has
                       shown developmental toxicity in animal tests. Chronic exposures to carbon monoxide at
                       levels around 50 ppm are thought by some investigators to have a negative impact on

(continued on facing page)

the results of behavioral tests such as time discrimination, visual vigilance, choice response tests, visual evoked responses, and visual discrimination thresholds.

Flammability and Explosibility

Carbon monoxide is a flammable gas. It forms explosive mixtures with air in the range of 12.5 to 74% by volume.

Reactivity and Incompatibility

Carbon monoxide is a reducing agent; it reacts violently with strong oxidizers. It undergoes violent reactions with many interhalogen compounds such as ClF_3, BrF_3, and BrF_5. CO reacts with many metals to form metal carbonyls, some of which may explode on heating, and reduces many metal oxides exothermically. Carbon monoxide reacts with sodium and with potassium to form explosive products that are sensitive to shock, heat, and contact with water.

Storage and Handling

Because of its toxic, flammable, and gaseous nature, carbon monoxide should be handled using the "basic prudent practices" of Chapter 5.C, supplemented by the additional precautions for work with flammable compounds (Chapter 5.F) and for work at high pressure (Chapter 5.H). In particular, cylinders of carbon monoxide should be stored and used in a continuously ventilated gas cabinet or fume hood. Local fire codes should be reviewed for limitations on quantity and storage requirements.

Accidents

In the event of a release of carbon monoxide, evacuate the area immediately. Rescue of an affected individual requires appropriate respiratory protection. Remove exposed individual to an uncontaminated area and seek immediate emergency help. Keep victim warm, quiet, and at rest and provide assisted respiration if breathing has stopped.

To respond to a release, use appropriate protective equipment and clothing. Positive pressure air-supplied respiratory protection is required. Close cylinder valve and ventilate area. Remove cylinder to a fume hood or remote area if it cannot be shut off.

Disposal

Excess carbon monoxide should be returned to the manufacturer, according to your institution's waste disposal guidelines. For more information on disposal procedures, see Chapter 7 of this volume.

The information in this LCSS has been compiled by a committee of the National Research Council from literature sources and Material Safety Data Sheets and is believed to be accurate as of July 1994. This summary is intended for use by trained laboratory personnel in conjunction with the NRC report *Prudent Practices in the Laboratory: Handling and Disposal of Chemicals*. This LCSS presents a concise summary of safety information that should be adequate for most laboratory uses of the title substance, but in some cases it may be advisable to consult more comprehensive references. This information should not be used as a guide to the nonlaboratory use of this chemical.

LABORATORY CHEMICAL SAFETY SUMMARY: CARBON TETRACHLORIDE

Substance	Carbon tetrachloride (Tetrachloromethane) CAS 56-23-5
Formula	CCl_4
Physical Properties	Colorless liquid bp 77 °C, mp −23 °C Insoluble in water (0.05 g/100 mL)
Odor	Ethereal, sweet, pungent odor detectable at 140 to 584 ppm (mean = 252 ppm)
Vapor Density	5.3 (air = 1.0)
Vapor Pressure	91 mmHg at 20 °C
Flash Point	Noncombustible

Toxicity Data

LD_{50} oral (rat)	2350 mg/kg
LD_{50} skin (rabbit)	>20 g/kg
LC_{50} inhal (rat)	8000 ppm (4 h)
PEL (OSHA)	2 ppm (13 mg/m^3)
TLV-TWA (ACGIH)	5 ppm (32.5 mg/m^3)—skin
STEL (ACGIH)	10 ppm (65 mg/m^3)

Major Hazards   Low to moderate acute toxicity; harmful to the liver, kidneys, and central nervous system.

Toxicity   The acute toxicity of carbon tetrachloride is low to moderate. Inhalation of carbon tetrachloride can produce symptoms such as dizziness, headache, fatigue, nausea, vomiting, stupor, and diarrhea. This substance is a depressant of the central nervous system, and inhalation of high concentrations causes damage to the liver, heart, and kidneys. Exposure to 1000 to 2000 ppm for 30 to 60 min can be fatal to humans. Ingestion of carbon tetrachloride leads to similar toxic effects, and swallowing as little as 4 mL can be lethal. Carbon tetrachloride irritates the skin, and prolonged contact may cause dryness and cracking. This substance is also slowly absorbed through the skin. Carbon tetrachloride liquid and vapor are also irritating to the eyes. The odor of carbon tetrachloride does not provide adequate warning of the presence of harmful concentrations.

Carbon tetrachloride shows carcinogenic effects in animal studies and is listed by IARC in Group 2B ("possible human carcinogen"). It is not classified as a "select carcinogen" according to the criteria of the OSHA Laboratory Standard. Prolonged or repeated exposure to this substance may result in liver and kidney damage. There is some evidence from animal studies that carbon tetrachloride may be a developmental and reproductive toxin in both males and females.

(continued on facing page)

Flammability and Explosibility

Carbon tetrachloride is noncombustible. Exposure to fire or high temperatures may lead to formation of phosgene, a highly toxic gas.

Reactivity and Incompatibility

Carbon tetrachloride may react explosively with reactive metals such as the alkali metals, aluminum, magnesium, and zinc. It can also react violently with boron and silicon hydrides, and upon heating with DMF.

Storage and Handling

Carbon tetrachloride should be handled in the laboratory using the "basic prudent practices" described in Chapter 5.C.

Accidents

In the event of skin contact, immediately wash with soap and water and remove contaminated clothing. In case of eye contact, promptly wash with copious amounts of water for 15 min (lifting upper and lower lids occasionally) and obtain medical attention. If carbon tetrachloride is ingested, obtain medical attention immediately. If large amounts of this compound are inhaled, move the person to fresh air and seek medical attention at once.

In the event of a spill, soak up carbon tetrachloride with a spill pillow or absorbent material, place in an appropriate container, and dispose of properly. Respiratory protection may be necessary in the event of a large spill or release in a confined area.

Disposal

Excess carbon tetrachloride and waste material containing this substance should be placed in an appropriate container, clearly labeled, and handled according to your institution's waste disposal guidelines. For more information on disposal procedures, see Chapter 7 of this volume.

LABORATORY CHEMICAL SAFETY SUMMARY: CHLORINE

Substance	Chlorine CAS 7782-50-5
Formula	Cl_2
Physical Properties	Greenish colored gas or amber liquid bp -34.1 °C, mp -101 °C Slightly soluble in water (0.7 g/100 mL)
Odor	Highly pungent, bleach-like odor detectable at 0.02 to 3.4 ppm (mean = 0.08 ppm)
Vapor Density	2.4 (air = 1.0)
Vapor Pressure	4800 mmHg at 20 °C

Toxicity Data

LC_{50} inhal (rat)	293 ppm (879 mg/m^3; 1 h)
PEL (OSHA)	1.0 ppm (3 mg/m^3)
TLV-TWA (ACGIH)	0.5 ppm (1.5 mg/m^3)
STEL (ACGIH)	1 ppm (2.9 mg/m^3)

Major Hazards

Highly irritating and corrosive to the eyes, skin, and respiratory tract; reacts violently with readily oxidized substances.

Toxicity

Chlorine is a severe irritant of the eyes, skin, and mucous membranes. Inhalation may cause coughing, choking, nausea, vomiting, headache, dizziness, difficulty breathing, and delayed pulmonary edema, which can be fatal. Exposure to ~500 ppm for 30 min may be fatal, and 1000 ppm can be lethal after a few breaths. Chlorine is highly irritating to the eyes and skin; exposure to 3 to 8 ppm causes stinging and burning of the eyes, and contact with liquid chlorine or high concentrations of the vapor can cause severe burns. Chlorine can be detected by its odor below the permissible limit; however, because of olfactory fatigue, odor may not always provide adequate warning of the presence of harmful concentrations of this substance.

Chronic exposures in animals up to 2.5 ppm for 2 years caused effects only in the upper respiratory tract, primarily the nose. Higher concentrations or repeated exposure has caused corrosion of the teeth. There is no evidence for carcinogenicity or reproductive or developmental toxicity of chlorine in humans.

Flammability and Explosibility

Chlorine is noncombustible but is a strong oxidizer and will support combustion of most flammable substances.

Reactivity and Incompatibility

Chlorine reacts violently or explosively with a wide range of substances, including hydrogen, acetylene, many hydrocarbons in the presence of light, ammonia, reactive metals, and metal hydrides and related compounds, including diborane, silane, and phosphine.

(continued on facing page)

Storage and Handling

Chlorine should be handled in the laboratory using the "basic prudent practices" described in Chapter 5.C, supplemented by the procedures for work with compressed gases discussed in Chapter 5.H. All work with chlorine should be conducted in a fume hood to prevent exposure by inhalation, and splash goggles and impermeable gloves should be worn at all times to prevent eye and skin contact. Cylinders of chlorine should be stored in locations appropriate for compressed gas storage and separated from incompatible compounds such as hydrogen, acetylene, ammonia, and flammable materials.

Accidents

In the event of skin contact, immediately wash with soap and water and remove contaminated clothing. In case of eye contact, promptly wash with copious amounts of water for 15 min (lifting upper and lower lids occasionally) and obtain medical attention. If chlorine is inhaled, move the person to fresh air and seek medical attention at once.

In case of accidental release of chlorine gas, such as from a leaking cylinder or associated apparatus, evacuate the area and eliminate the source of the leak if this can be done safely. Full-face supplied-air respiratory protection and protective clothing may be required to deal with a chlorine release. Cylinders with slow leaks should be carefully removed to a fume hood or remote outdoor locations. Chlorine leaks may be detected by passing a rag dampened with aqueous ammonia over the suspected valve or fitting. White fumes indicate escaping chlorine gas.

Disposal

Excess chlorine in cylinders should be returned to the manufacturer for disposal. For more information on disposal procedures, see Chapter 7 of this volume.

Substance	Chloroform (Trichloromethane) CAS 67-66-3
Formula	$CHCl_3$
Physical Properties	Colorless liquid bp 61 °C, mp −63.5 °C Slightly soluble in water (0.8 g/100 mL)
Odor	Ethereal, sweet odor detectable at 133 to 276 ppm (mean = 192 ppm)
Vapor Density	4.1 (air = 1.0)
Vapor Pressure	160 mmHg at 20 °C
Flash Point	Noncombustible

Toxicity Data

LD_{50} oral (rat)	908 mg/kg
LD_{50} skin (rabbit)	>20 g/kg
LC_{50} inhal (rat)	9937 ppm (47,702 mg/m^3; 4 h)
PEL (OSHA)	50 ppm (240 mg/m^3; ceiling)
TLV-TWA (ACGIH)	10 ppm (48 mg/m^3)

Major Hazards

Low acute toxicity; skin and eye irritant.

Toxicity

The acute toxicity of chloroform is low by all routes of exposure. Inhalation can cause dizziness, headache, drowsiness, and nausea, and at higher concentrations, disorientation, delirium, and unconsciousness. Inhalation of high concentrations may also cause liver and kidney damage. Exposure to 25,000 ppm for 5 min can be fatal to humans. Ingestion of chloroform can cause severe burning of the mouth and throat, chest pain, and vomiting. Chloroform is irritating to the skin and eyes, and liquid splashed in the eyes can cause burning pain and reversible corneal injury. Olfactory fatigue occurs on exposure to chloroform vapor, and it is not regarded as a substance with adequate warning properties.

Chloroform shows carcinogenic effects in animal studies and is listed by IARC in Group 2B ("possible human carcinogen"). It is not classified as a "select carcinogen" according to the criteria of the OSHA Laboratory Standard. Prolonged or repeated exposure to this substance may result in liver and kidney injury. There is some evidence from animal studies that chloroform is a developmental and reproductive toxin.

Flammability and Explosibility

Chloroform is noncombustible. Exposure to fire or high temperatures may lead to formation of phosgene, a highly toxic gas.

(continued on facing page)

Reactivity and Incompatibility	Chloroform reacts violently with alkali metals such as sodium and potassium, with a mixture of acetone and base, and with a number of strong bases such as potassium and sodium hydroxide, potassium *t*-butoxide, sodium methoxide, and sodium hydride. Chloroform reacts explosively with fluorine and dinitrogen tetroxide.
Storage and Handling	Chloroform should be handled in the laboratory using the "basic prudent practices" described in Chapter 5.C. In the presence of light, chloroform undergoes autoxidation to generate phosgene; this can be minimized by storing this substance in the dark under nitrogen. Commercial samples of chloroform frequently contain 0.5 to 1% ethanol as a stabilizer.
Accidents	In the event of skin contact, immediately wash with soap and water and remove contaminated clothing. In case of eye contact, promptly wash with copious amounts of water for 15 min (lifting upper and lower lids occasionally) and obtain medical attention. If chloroform is ingested, obtain medical attention immediately. If large amounts of this compound are inhaled, move the person to fresh air and seek medical attention at once.
	In the event of a spill, soak up chloroform with a spill pillow or absorbent material, place in an appropriate container, and dispose of properly. Respiratory protection may be necessary in the event of a large spill or release in a confined area.
Disposal	Excess chloroform and waste material containing this substance should be placed in an appropriate container, clearly labeled, and handled according to your institution's waste disposal guidelines. For more information on disposal procedures, see Chapter 7 of this volume.

Substance	Chloromethyl methyl ether (Methyl chloromethyl ether; CMME) CAS 107-30-2
Formula	$ClCH_2OCH_3$
Physical Properties	Colorless liquid bp 55 to 59 °C, mp -104 °C Hydrolyzes in water
Odor	Similar to HCl
Vapor Pressure	260 mmHg at 20 °C
Flash Point	15 °C
Autoignition Temperature	Not available

Toxicity Data

LC_{50} inhal (rat)	55 ppm (180 mg/m^3; 7 h)
LD_{50} oral (rat)	817 mg/kg

Major Hazards OSHA "select carcinogen"; highly irritating to the eyes, skin, and respiratory tract.

Toxicity The acute toxicity of chloromethyl methyl ether is moderate to high. Inhalation of the vapor is severely irritating to the eyes, skin, nose, and respiratory tract, and causes sore throat, fever, chills, and difficulty breathing. Exposure to high concentrations can lead to delayed pulmonary edema, which can be fatal. Eye or skin contact with the liquid can result in severe and painful burns. Ingestion of this substance may lead to severe burns of the mouth and stomach and can be fatal.

Chloromethyl methyl ether is regulated by OSHA as a carcinogen (29 CFR 1910.1006) and is listed in IARC Group 1 ("carcinogenic to humans"). This substance is classified as a "select carcinogen" under the criteria of the OSHA Laboratory Standard. Note also that some commercial samples of chloromethyl methyl ether contain up to 7% of highly carcinogenic bis(chloromethyl) ether. Hydrolysis of chloromethyl methyl ether produces HCl and formaldehyde, which can recombine to form bis(chloromethyl) ether. No information is available on the reproductive and developmental toxicity of chloromethyl methyl ether. Odor does not provide adequate warning of the harmful presence of this carcinogenic substance.

Flammability and Explosibility Chloromethyl methyl ether is highly flammable. Fires involving this substance should be extinguished with carbon dioxide or dry chemical extinguishers.

(continued on facing page)

The information in this LCSS has been compiled by a committee of the National Research Council from literature sources and Material Safety Data Sheets and is believed to be accurate as of July 1994. This summary is intended for use by trained laboratory personnel in conjunction with the NRC report *Prudent Practices in the Laboratory: Handling and Disposal of Chemicals*. This LCSS presents a concise summary of safety information that should be adequate for most laboratory uses of the title substance, but in some cases it may be advisable to consult more comprehensive references. This information should not be used as a guide to the nonlaboratory use of this chemical.

Reactivity and Incompatibility

Chloromethyl methyl ether decomposes in water to form HCl and formaldehyde, and reacts readily with oxidizing agents.

Storage and Handling

Because of its carcinogenicity, chloromethyl methyl ether should be handled using the "basic prudent practices" of Chapter 5.C, supplemented by the additional precautions for work with compounds of high chronic toxicity (Chapter 5.D) and high flammability (Chapter 5.F). In particular, work with this substance should be conducted in a fume hood to prevent exposure by inhalation, and appropriate impermeable gloves and splash goggles should be worn at all times to prevent skin and eye contact. Chloromethyl methyl ether is also highly flammable and should be used only in areas free of ignition sources; quantities greater than 1 liter should be stored in tightly sealed metal containers in areas separate from oxidizers in secondary containers.

Accidents

In the event of skin contact, immediately wash with soap and water and remove contaminated clothing. In case of eye contact, promptly wash with copious amounts of water for 15 min (lifting upper and lower lids occasionally) and obtain medical attention. If chloromethyl methyl ether is ingested, obtain medical attention immediately. If large amounts of this compound are inhaled, move the person to fresh air and seek medical attention at once.

In the event of a spill, remove all ignition sources, soak up the chloromethyl methyl ether with a spill pillow or absorbent material, place in an appropriate container, and dispose of properly. Respiratory protection may be necessary in the event of a large spill or release in a confined area.

Disposal

Excess chloromethyl methyl ether and waste material containing this substance should be placed in an appropriate container, clearly labeled, and handled according to your institution's waste disposal guidelines. For more information on disposal procedures, see Chapter 7 of this volume.

LABORATORY CHEMICAL SAFETY SUMMARY: CHROMIUM TRIOXIDE AND OTHER CHROMIUM(VI) SALTS

Substance

Chromium trioxide
(Chromic anhydride; chromic acid; chromium(VI) oxide; chromic trioxide; chromium oxide)
CAS 1333-82-0

Formula

CrO_3

Physical Properties

Dark red flakes or crystals
mp 196 °C, bp: decomposes at 250 °C
Very soluble in water (62 g/100 mL)

Flash Point

Noncombustible

Toxicity Data

LD_{50} oral (rat)	80 mg/kg
PEL (OSHA)	0.1 mg $(CrO_3)/m^3$ (ceiling)
TLV-TWA (ACGIH)	0.05 mg $(Cr)/m^3$

Major Hazards

Probable human carcinogen (OSHA "select carcinogen"); severely irritating to the skin and mucous membranes; very strong oxidizing agent.

Toxicity

Chromium trioxide and other chromium(VI) salts are moderately toxic substances by ingestion; 1 to 15 g may be a fatal dose in humans. Ingestion of nonlethal doses of these compounds can cause stomach, liver, and kidney damage; symptoms may include clammy, cyanotic skin, sore throat, gastric burning, vomiting, and diarrhea. Chromic acid is irritating to the skin, and prolonged contact can cause ulceration. Inhalation of chromate dust or chromic acid mist can result in severe irritation of the nose, throat, bronchial tubes, and lungs and may cause coughing, labored breathing, and swelling of the larynx. Eye contact with chromium trioxide and its solutions can cause severe burns and possible loss of vision.

Occupational exposure to chromium(VI) compounds has been related to an increased risk of lung cancer. Several hexavalent compounds of chromium, including chromium trioxide, are listed in IARC Group 1 ("carcinogenic to humans") and are classified as "select carcinogens" under the criteria of the OSHA Laboratory Standard. Long-term exposure to chromium trioxide or chromium(VI) salts may cause ulceration of the respiratory system and skin. Exposure to chromium trioxide by inhalation or skin contact may lead to sensitization. Chromium trioxide has exhibited teratogenic activity in animal tests.

Flammability and Explosibility

Chromium trioxide is not combustible but is a strong oxidizing agent and can accelerate the burning rate of combustible materials. Contact with easily oxidized organic or other combustible materials (including paper and oil) may result in ignition, violent combustion, or explosion. The use of dry chemical, carbon dioxide, Halon, or water spray extinguishers is recommended for fires involving chromium(VI) compounds.

(continued on facing page)

The information in this LCSS has been compiled by a committee of the National Research Council from literature sources and Material Safety Data Sheets and is believed to be accurate as of July 1994. This summary is intended for use by trained laboratory personnel in conjunction with the NRC report *Prudent Practices in the Laboratory: Handling and Disposal of Chemicals*. This LCSS presents a concise summary of safety information that should be adequate for most laboratory uses of the title substance, but in some cases it may be advisable to consult more comprehensive references. This information should not be used as a guide to the nonlaboratory use of this chemical.

Reactivity and Incompatibility

Chromium trioxide and certain other chromium(VI) compounds are useful as strong oxidizing agents in the laboratory, but appropriate precautionary measures should be taken when conducting these reactions. Chromium trioxide has been reported to react violently with a variety of substances, including readily oxidized organic compounds such as acetone, acetaldehyde, methanol, ethanol, diethyl ether, ethyl acetate, acetic acid, and DMF, and violent reactions may also occur on reaction with alkali metals, gaseous ammonia, phosphorus, and selenium.

Storage and Handling

Because of their carcinogenicity, chromium(VI) compounds should be handled using the "basic prudent practices" of Chapter 5.C, supplemented by the additional precautions for work with compounds of high toxicity (Chapter 5.D). In particular, chromium trioxide should be handled in a fume hood to avoid the inhalation of dust, and impermeable gloves should be worn at all times to prevent skin contact. The practice of using chromate solutions to clean glassware should be avoided. Chromium trioxide should be stored in areas separated from readily oxidized materials.

Accidents

In the event of skin contact, immediately wash with soap and water and remove contaminated clothing. In case of eye contact, promptly wash with copious amounts of water for 15 min (lifting upper and lower lids occasionally) and obtain medical attention. If chromium trioxide or other chromium compounds are ingested, give the person large amounts of water or milk and obtain medical attention immediately. If dust or aerosols of these compounds are inhaled, move the person to fresh air and seek medical attention at once.

In the event of a spill, remove all combustibles from the area, sweep up the chromium compounds, place in an appropriate container, and dispose of properly. In the event solutions containing chromium compounds are spilled, neutralize (if possible) with aqueous base, soak up with a spill pillow or appropriate noncombustible absorbent material, place in an appropriate container, and dispose of properly. Respiratory protection may be necessary in the event of a large spill of powder, particularly in a confined area.

Disposal

Excess chromium compounds and waste material containing these substances should be placed in an appropriate container, clearly labeled, and handled according to your institution's waste disposal guidelines. For more information on disposal procedures, see Chapter 7 of this volume.

LABORATORY CHEMICAL SAFETY SUMMARY: CYANOGEN BROMIDE

Substance	Cyanogen bromide (Bromine cyanide, bromocyanogen, cyanobromide) CAS 506-68-3
Formula	BrCN
Physical Properties	Colorless needles bp 62 °C, mp 52 °C Soluble in water
Odor	Penetrating odor
Vapor Pressure	100 mmHg at 22.6 °C
Toxicity Data	LC_{LO} inhal (human) 92 ppm (398 mg/m^3; 10 min) LC_{LO} inhal (mouse) 115 ppm (500 mg/m^3; 10 min)
Major Hazards	High acute toxicity; severely irritating.
Toxicity	The acute toxicity of cyanogen bromide is high. Toxic effects are similar to but not as severe as those of hydrogen cyanide. Toxic symptoms may include cyanosis, nausea, dizziness, headache, lung irritation, chest pain, and pulmonary edema, which may be fatal. Cyanogen bromide may cause chronic pulmonary edema.
Flammability and Explosibility	Cyanogen bromide is noncombustible. Impure material decomposes rapidly and can be explosive.
Reactivity and Incompatibility	Cyanogen bromide can react violently with large quantities of acid. It may decompose when exposed to heat, moist air, or water, producing toxic fumes of hydrogen cyanide and hydrogen bromide. Cyanogen bromide can polymerize violently on prolonged storage at ambient temperature.
Storage and Handling	Because of its high acute toxicity, cyanogen bromide should be handled using the "basic prudent practices" of Chapter 5.C, supplemented by the additional precautions for work with compounds of high toxicity (Chapter 5.D). In particular, work with BrCN should be conducted in a fume hood to prevent exposure by inhalation, and splash goggles and impermeable gloves should be worn at all times to prevent eye and skin contact. Containers of cyanogen bromide should be kept tightly sealed and stored under nitrogen in a secondary container in a refrigerator.
Accidents	In the event of skin contact, immediately wash with soap and water and remove contaminated clothing. In case of eye contact, promptly wash with copious amounts of water for 15 min (lifting upper and lower lids occasionally) and obtain medical attention. If cyanogen

(continued on facing page)

bromide is ingested, obtain medical attention immediately. If large amounts of this compound are inhaled, move the person to fresh air and seek medical attention at once.

In the event of a spill, sweep up cyanogen bromide, place in an appropriate container, and dispose of properly. Respiratory and appropriate impermeable protective gloves and clothing should be worn while conducting cleanup of this highly toxic substance.

Disposal   Excess cyanogen bromide and waste material containing this substance should be placed in an appropriate container, clearly labeled, and handled according to your institution's waste disposal guidelines. For more information on disposal procedures, see Chapter 7 of this volume.

LABORATORY CHEMICAL SAFETY SUMMARY: DIAZOMETHANE

Substance	Diazomethane (Diazirine, azimethylene) CAS 334-88-3
Formula	CH_2N_2
Physical Properties	Yellow gas bp -23 °C, mp -145 °C Reacts with water
Odor	Musty odor (no accepted threshold value)
Vapor Density	1.4 (air $= 1.0$)
Autoignition Temperature	150 °C; impure material explodes at lower temperature

Toxicity Data

LC_{LO} inhal (cat)	175 ppm (10 min)
PEL (OSHA)	0.2 ppm (0.4 mg/m³)
TLV-TWA (ACGIH)	0.2 ppm (0.4 mg/m³)

Major Hazards

Powerful allergen with high acute toxicity; extremely unstable; may explode on contact with alkali metals, calcium sulfate (Drierite), or rough edges such as those found on ground glass.

Toxicity

Diazomethane vapor causes severe irritation of the skin, eyes, mucous membranes, and lungs. It is considered to be a substance with poor warning properties, and the effects of exposure may be delayed in onset. Symptoms of exposure may include headache, chest pain, cough, fever, severe asthmatic attacks, and pulmonary edema, which can be fatal. Exposure of the skin and mucous membranes to diazomethane may cause serious burns.

Diazomethane is a powerful allergen. Prolonged or repeated exposure to diazomethane can lead to sensitization of the skin and lungs, in which case asthma-like symptoms or fever may occur as the result of exposure to concentrations of diazomethane that previously caused no symptoms. Chronic exposure to diazomethane has been reported to cause cancer in experimental animals, but this substance has not been identified as a human carcinogen.

Note that diazomethane is often prepared in situ from precursors that may themselves be highly toxic and/or carcinogenic.

Flammability and Explosibility

Pure diazomethane gas and liquid are readily flammable and can explode easily. A variety of conditions have been reported to cause explosions of diazomethane, including contact with rough surfaces such as ground-glass joints, etched or scratched flasks, and glass tubing that has not been carefully fire-polished. Direct sunlight and strong artificial light

(continued on facing page)

may also cause explosions of this substance. Violent reactions may occur on exposure of diazomethane to alkali metals.

Reactivity and Incompatibility

Explosions may occur on exposure of diazomethane to alkali metals and calcium sulfate (Drierite).

Storage and Handling

Because of its high toxicity and explosibility, diazomethane should be handled using the "basic prudent practices" of Chapter 5.C, supplemented by the additional precautions for work with compounds of high chronic toxicity (Chapter 5.D) and for work with reactive and explosive substances (Chapter 5.G). In particular, diazomethane should preferably be handled in solution using glassware specially designated for diazomethane (e.g., with Clear-Seal joints) and should be used as soon as possible after preparation. Storage of diazomethane solutions (even at low temperature) is not advisable. All work with diazomethane should be conducted in a fume hood behind a safety shield, and appropriate impermeable gloves, protective clothing, and safety goggles should be worn at all times.

Accidents

In the event of skin contact, immediately wash with soap and water and remove contaminated clothing. In case of eye contact, promptly wash with copious amounts of water for 15 min (lifting upper and lower lids occasionally) and obtain medical attention. If this compound is inhaled, move the person to fresh air and seek medical attention at once.

In the event of a spill, remove all ignition sources and close off the hood. Diazomethane solutions can be soaked up with a spill pillow or an absorbent material such as clay or vermiculite, placed in an appropriate container, and disposed of properly. Respiratory protection may be necessary in the event of a large spill or release in a confined area.

Disposal

Small amounts of excess diazomethane can be destroyed by carefully adding acetic acid dropwise to a dilute solution of the diazomethane in an inert solvent such as ether at 0 °C. Excess diazomethane solutions and waste material containing this substance should be placed in an appropriate container, clearly labeled, and handled according to your institution's waste disposal guidelines. For more information on disposal procedures, see Chapter 7 of this volume.

LABORATORY CHEMICAL SAFETY SUMMARY: DIBORANE

Substance	Diborane (Boroethane, boron hydride, diboron hexahydride) CAS 19287-45-7
Formula	B_2H_6
Physical Properties	Colorless gas bp -93 °C, mp -165 °C Rapidly decomposes in water to form hydrogen gas
Odor	Repulsive odor detectable at 1.8 to 3.5 ppm
Vapor Density	0.96 (air = 1.0)
Flash Point	-90 °C
Autoignition Temperature	38 to 52 °C

Toxicity Data

LC_{50} inhal (rat)	50 ppm (4 h)
PEL (OSHA)	0.1 ppm
TLV-TWA (ACGIH)	0.1 ppm

Major Hazards

Highly toxic, flammable, and reactive gas; contact with air or halogenated compounds results in fires and explosions.

Toxicity

Inhalation of diborane gas results in irritation of the respiratory tract and may result in headache, cough, nausea, difficulty in breathing, chills, fever, and weakness. The odor of diborane cannot be detected below the permissible exposure limit, so this substance is considered to have poor warning properties. Overexposure to diborane can cause damage to the central nervous system, liver, and kidneys. Death can result from pulmonary edema (fluid in the lungs) and/or from lack of oxygen. Exposure to diborane gas has not been found to have significant effects on the skin and mucous membranes, but high concentrations can cause eye irritation, and contact with the liquid can cause burns.

Chronic exposure to low concentrations of diborane may cause headache, lightheadedness, fatigue, weakness in the muscles, and tremors. Repeated exposure may produce chronic respiratory distress, particularly in susceptible individuals. An existing dermatitis may also be worsened by repeated exposure to the liquid. Diborane has not been shown to have carcinogenic or reproductive or developmental effects in humans.

Flammability and Explosibility

Diborane is a flammable gas that ignites spontaneously in moist air at room temperature and forms explosive mixtures with air from 0.8% up to 88% by volume. Diborane reacts with halogenated hydrocarbons, and fire extinguishing agents such as Halon or carbon

(continued on facing page)

tetrachloride are therefore not recommended. Carbon dioxide extinguishers should be used to fight diborane fires. Fires involving diborane sometimes release toxic gases such as boron oxide smoke.

Reactivity and Incompatibility

Explodes on contact with fluorine, chlorine, halogenated hydrocarbons (e.g., chloroform and carbon tetrachloride), fuming nitric acid, and nitrogen trifluoride. Diborane is a strong reducing agent that produces hydrogen upon heating or upon reaction with water. Contact with aluminum, lithium, and other active metals forms metal hydrides, which may ignite spontaneously. Diborane is incompatible with oxidizing agents, halogens, and halogenated compounds. Diborane will attack some forms of plastics, rubber, and coatings.

Storage and Handling

Diborane should be handled using the "basic prudent practices" of Chapter 5.C, supplemented by the additional precautions for work with reactive and explosive compounds described in Chapter 5.G. In particular, diborane should be used only in a fume hood free of ignition sources and should be stored in a cold, dry, well-ventilated area separated from incompatible substances and isolated from sources of sparks and open flames.

Accidents

In the event of skin contact, immediately wash with soap and water and remove contaminated clothing. In case of eye contact, promptly wash with copious amounts of water for 15 min (lifting upper and lower lids occasionally) and obtain medical attention. If this compound is inhaled, move the person to fresh air and seek medical attention at once.

In the event of a leak, remove all ignition sources and ventilate the area of the leak. Respiratory protection and protective clothing may be necessary in the event of a large spill or release in a confined area. If a cylinder is the source of the leak and the leak cannot be stopped, if possible remove the leaking cylinder to a fume hood or a safe place in the open air, and repair the leak or allow the cylinder to empty. If the leak has resulted in a fire, water spray can be used to cool the container and to reduce corrosive vapors, keeping in mind that if flames are extinguished, explosive re-ignition can occur.

Disposal

Excess diborane and waste material containing this substance should be placed in an appropriate container, clearly labeled, and handled according to your institution's waste disposal guidelines. For more information on disposal procedures, see Chapter 7 of this volume.

The information in this LCSS has been compiled by a committee of the National Research Council from literature sources and Material Safety Data Sheets and is believed to be accurate as of July 1994. This summary is intended for use by trained laboratory personnel in conjunction with the NRC report *Prudent Practices in the Laboratory: Handling and Disposal of Chemicals*. This LCSS presents a concise summary of safety information that should be adequate for most laboratory uses of the title substance, but in some cases it may be advisable to consult more comprehensive references. This information should not be used as a guide to the nonlaboratory use of this chemical.

LABORATORY CHEMICAL SAFETY SUMMARY: DICHLOROMETHANE

Substance	Dichloromethane (Methylene chloride; aerothene MM) CAS 75-09-2
Formula	CH_2Cl_2
Physical Properties	Colorless liquid bp 40 °C, mp -97 °C Slightly soluble in water (1.32 g/100 mL)
Odor	Odor threshold 160 to 230 ppm
Vapor Density	2.93 (air $= 1.0$)
Vapor Pressure	440 mmHg at 25 °C
Flash Point	Noncombustible
Autoignition Temperature	556 °C

Toxicity Data

LD_{50} oral (rat)	1600 mg/kg
LC_{50} inhal (rat)	88,000 mg/m^3; 30 min
PEL (OSHA)	500 ppm (8 h)
TLV-TWA (ACGIH)	50 ppm

Major Hazards Low acute toxicity; skin and eye irritant.

Toxicity Dichloromethane is classified as only slightly toxic by the oral and inhalation routes. Exposure to high concentrations of dichloromethane vapor (>500 ppm for 8 h) can lead to lightheadedness, fatigue, weakness, and nausea. Contact of the compound with the eyes causes painful irritation and can lead to conjunctivitis and corneal injury if not promptly removed by washing. Dichloromethane is a mild skin irritant, and upon prolonged contact (e.g., under the cover of clothing or shoes) can cause burns after 30 to 60 min exposure.

Dichloromethane is not teratogenic at levels up to 4500 ppm or embryotoxic in rats and mice at levels up to 1250 ppm.

Flammability and Explosibility Noncombustible. Dichloromethane vapor concentrated in a confined or poorly ventilated area can be ignited with a high-energy spark, flame, or high-intensity heat source.

Reactivity and Incompatibility Reacts violently with alkali metals, aluminum, magnesium powder, potassium *t*-butoxide, nitrogen tetroxide, and strong oxidizing agents.

(continued on facing page)

**Storage and
Handling**

This compound should be handled in the laboratory using the "basic prudent practices" described in Chapter 5.C.

Accidents

In the event of skin contact, immediately wash with soap and water and remove contaminated clothing. In case of eye contact, promptly wash with copious amounts of water for 15 min (lifting upper and lower lids occasionally) and obtain medical attention. If dichloromethane is ingested, obtain medical attention immediately. If large amounts of this compound are inhaled, move the person to fresh air and seek medical attention at once.

In the event of a spill, soak up dichloromethane with a spill pillow or absorbent material, place in an appropriate container, and dispose of properly. Respiratory protection may be necessary in the event of a large spill or release in a confined area.

Disposal

Excess dichloromethane and waste material containing this substance should be placed in an appropriate container, clearly labeled, and handled according to your institution's waste disposal guidelines. For more information on disposal procedures, see Chapter 7 of this volume.

LABORATORY CHEMICAL SAFETY SUMMARY: DIETHYL ETHER

Substance	Diethyl ether (Ethyl ether, ether) CAS 60-29-7
Formula	$(CH_3CH_2)_2O$
Physical Properties	Colorless liquid bp 35 °C, mp −116 °C Slightly soluble in water (8 g/100 mL)
Odor	Pungent odor detectable at 0.33 ppm
Vapor Density	2.6 (air = 1.0)
Vapor Pressure	442 mmHg at 20 °C
Flash Point	−45 °C
Autoignition Temperature	160 °C

Toxicity Data

LD_{50} oral (rat)	1215 mg/kg
LC_{50} inhal (rat)	73,000 ppm (2 h)
PEL (OSHA)	400 ppm
TLV-TWA (ACGIH)	400 ppm
STEL (ACGIH)	500 ppm

Major Hazards

Extremely flammable liquid and vapor; forms explosive peroxides upon storage in contact with air.

Toxicity

The acute toxicity of diethyl ether is low. Inhalation of high concentrations can cause sedation, unconsciousness, and respiratory paralysis. These effects are usually reversible upon cessation of exposure. Diethyl ether is mildly irritating to the eyes and skin, but does not generally cause irreversible damage. Repeated contact can cause dryness and cracking of the skin due to removal of skin oils. The liquid is not readily absorbed through the skin, in part because of its high volatility. Diethyl ether is slightly toxic by ingestion. Diethyl ether is regarded as having adequate warning properties.

There is no evidence for carcinogenicity of diethyl ether, and no reproductive effects have been reported. Chronic exposure to diethyl ether vapor may lead to loss of appetite, exhaustion, drowsiness, dizziness, and other central nervous system effects.

Flammability and Explosibility

Diethyl ether is extremely flammable (NFPA rating = 4) and is one of the most dangerous fire hazards commonly encountered in the laboratory, owing to its volatility and extremely low ignition temperature. Ether vapor may be ignited by hot surfaces such as hot plates

(continued on facing page)

The information in this LCSS has been compiled by a committee of the National Research Council from literature sources and Material Safety Data Sheets and is believed to be accurate as of July 1994. This summary is intended for use by trained laboratory personnel in conjunction with the NRC report *Prudent Practices in the Laboratory: Handling and Disposal of Chemicals*. This LCSS presents a concise summary of safety information that should be adequate for most laboratory uses of the title substance, but in some cases it may be advisable to consult more comprehensive references. This information should not be used as a guide to the nonlaboratory use of this chemical.

and static electricity discharges, and since the vapor is heavier than air, it may travel a considerable distance to an ignition source and flash back. Ether vapor forms explosive mixtures with air at concentrations of 1.9 to 36% (by volume). Carbon dioxide or dry chemical extinguishers should be used for ether fires. Diethyl ether forms unstable peroxides on exposure to air in a reaction that is promoted by light; the presence of these peroxides may lead to explosive residues upon distillation.

Reactivity and Incompatibility

Diethyl ether may react violently with halogens or strong oxidizing agents such as perchloric acid.

Storage and Handling

Diethyl ether should be handled in the laboratory using the "basic prudent practices" described in Chapter 5.C, supplemented by additional precautions for dealing with extremely flammable substances (Chapter 5.F). In particular, ether should be used only in areas free of ignition sources (including hot plates, incandescent lightbulbs, and steam baths), and this substance should be stored in tightly sealed metal containers in areas separate from oxidizers. Because of the tendency of diethyl ether to form peroxides on contact with air, containers should be dated upon receipt and at the time they are opened. Once opened, containers of diethyl ether should be tested periodically for the presence of peroxides according to the procedures described in Chapter 5. Diethyl ether is generally supplied with additives that inhibit peroxide formation; distillation removes these inhibitors and renders the liquid more prone to peroxide formation. Material found to contain peroxides should be treated to destroy the peroxides before use or disposed of properly.

Accidents

In the event of skin contact, immediately wash with soap and water and remove contaminated clothing. In case of eye contact, promptly wash with copious amounts of water for 15 min (lifting upper and lower lids occasionally) and obtain medical attention. If diethyl ether is ingested, obtain medical attention immediately. If large amounts of this compound are inhaled, move the person to fresh air and seek medical attention at once.

In the event of a spill of diethyl ether, exercise extreme caution because of its highly flammable nature. Remove all ignition sources, soak up the diethyl ether as quickly as possible with a spill pillow or absorbent material, place in an appropriate container, and dispose of properly. Respiratory protection may be necessary in the event of a large spill or release in a confined area.

Disposal

Excess diethyl ether and waste material containing this substance should be placed in an appropriate container, clearly labeled, and handled according to your institution's waste disposal guidelines. For more information on disposal procedures, see Chapter 7 of this volume.

The information in this LCSS has been compiled by a committee of the National Research Council from literature sources and Material Safety Data Sheets and is believed to be accurate as of July 1994. This summary is intended for use by trained laboratory personnel in conjunction with the NRC report *Prudent Practices in the Laboratory: Handling and Disposal of Chemicals*. This LCSS presents a concise summary of safety information that should be adequate for most laboratory uses of the title substance, but in some cases it may be advisable to consult more comprehensive references. This information should not be used as a guide to the nonlaboratory use of this chemical.

LABORATORY CHEMICAL SAFETY SUMMARY: DIETHYLNITROSAMINE (AND RELATED NITROSAMINES)

Substance
Diethylnitrosamine (and related nitrosamines)
(N-nitrosodiethylamine; N-ethyl-N-nitrosoethananime)
CAS 55-18-5

Formula
$(CH_3CH_2)_2N-NO$

Physical Properties
Yellow liquid
bp 177 °C
Soluble in water

Odor
Not available

Vapor Pressure
1.7 mmHg at 20 °C

Flash Point
61 °C

Autoignition Temperature
Not available

Toxicity Data
LD$_{50}$ oral (rat)   280 mg/kg

Major Hazards
Probable human carcinogen (OSHA "select carcinogen"); other nitrosamines should also be regarded as carcinogenic.

Toxicity
The acute toxicity of diethylnitrosamine is classified as moderate. Other nitrosamines of higher molecular weight are somewhat less toxic. Harmful exposure to nitrosamines can occur by inhalation and ingestion and may cause nausea, vomiting, and fever. This substance does not have adequate warning properties.

Chronic exposure to nitrosamines can cause severe liver damage. Diethylnitrosamine is listed in IARC Group 2A ("probable human carcinogen") and is classified as an OSHA "select carcinogen." Nitrosamines are suspected of causing cancers of the lung, nasal sinuses, brain, esophagus, stomach, liver, bladder, and kidney. Diethylnitrosamine is mutagenic and teratogenic.

Flammability and Explosibility
Volatilization during combustion produces hazardous vapors. Combustion products contain nitrogen oxides.

Reactivity and Incompatibility
Diethylnitrosamine is decomposed by strong acids, liberating nitrous acid. Nitrosamines are incompatible with strong oxidizing agents.

Storage and Handling
Because of its carcinogenicity, diethylnitrosamine should be handled using the "basic prudent practices" of Chapter 5.C, supplemented by the additional precautions for work with compounds of high chronic toxicity (Chapter 5.D). In particular, work with diethylni-

(continued on facing page)

trosamine should be conducted in a fume hood to prevent exposure by inhalation, and appropriate impermeable gloves and splash goggles should be worn at all times to prevent skin and eye contact.

Accidents

In the event of skin contact, immediately wash with copious amounts of water while removing contaminated clothing. Place contaminated items in a plastic bag. Seal the bag and dispose of it appropriately. In case of eye contact, promptly wash with copious amounts of water for 15 min (lifting upper and lower lids occasionally) and obtain medical attention. If diethylnitrosamine is ingested, obtain medical attention immediately. If diethylnitrosamine is inhaled, move the person to fresh air and seek medical attention at once.

In the event of a spill, immediately evacuate and isolate the area. Decontamination should be performed by trained people wearing self-contained breathing apparatus and impervious clothing. The diethylnitrosamine should be soaked up with absorbents and placed in closed containers for disposal. After pickup is complete, wash the spill site and ventilate the area.

Disposal

Excess diethylnitrosamine and waste material containing this substance should be placed in an appropriate container, clearly labeled, and handled according to your institution's waste disposal guidelines. For more information on disposal procedures, see Chapter 7 of this volume.

LABORATORY CHEMICAL SAFETY SUMMARY: DIMETHYL SULFATE

Substance	Dimethyl sulfate (Methyl sulfate; DMS) CAS 77-78-1
Formula	$(CH_3)_2SO_4$
Physical Properties	Colorless, oily liquid bp 189 °C, mp −32 °C Soluble in water (2.8 g/100 mL at 20 °C); reacts slowly with water to form sulfuric acid and methanol
Odor	Almost odorless
Vapor Density	4.3 (air = 1.0)
Vapor Pressure	0.5 mmHg at 20 °C
Flash Point	83 °C
Autoignition Temperature	495 °C

Toxicity Data

LD_{50} oral (rat)	205 mg/kg
LC_{50} inhal (rat)	9 ppm (45 mg/m^3; 4 h)
PEL (OSHA)	1 ppm (5 mg/m^3)—skin
TLV-TWA (ACGIH)	0.1 ppm (0.52 mg/m^3)—skin

Major Hazards   Liquid and vapor can cause severe burns to the skin, eyes, and respiratory tract; corrosive and moderately toxic by ingestion; probable human carcinogen (OSHA "select carcinogen").

Toxicity   Dimethyl sulfate is extremely hazardous because of its lack of warning properties and delayed toxic effects. The vapor of this compound is extremely irritating to the skin, eyes, and respiratory tract, and contact with the liquid can cause very severe burns to the eyes and skin. Ingestion of dimethyl sulfate causes burns to the mouth, throat, and gastrointestinal tract. The effects of overexposure to dimethyl sulfate vapor may be delayed. After a latent period of 10 hours or more, headache and severe pain to the eyes upon exposure to light may occur, followed by cough, tightness of the chest, shortness of breath, difficulty in swallowing and speaking, vomiting, diarrhea, and painful urination. Fatal pulmonary edema may develop. Systemic effects of dimethyl sulfate include damage to the liver and kidneys.

Dimethyl sulfate is listed by IARC in Group 2A ("probable human carcinogen") and is classified as a "select carcinogen" under the criteria of the OSHA Laboratory Standard. Data indicate that dimethyl sulfate does not specifically harm unborn animals; dimethyl

(continued on facing page)

The information in this LCSS has been compiled by a committee of the National Research Council from literature sources and Material Safety Data Sheets and is believed to be accurate as of July 1994. This summary is intended for use by trained laboratory personnel in conjunction with the NRC report *Prudent Practices in the Laboratory: Handling and Disposal of Chemicals.* This LCSS presents a concise summary of safety information that should be adequate for most laboratory uses of the title substance, but in some cases it may be advisable to consult more comprehensive references. This information should not be used as a guide to the nonlaboratory use of this chemical.

sulfate is not a developmental toxin. It is a strong alkylating agent and does produce genetic damage in animals and in bacterial and mammalian cell cultures.

Flammability and Explosibility

Dimethyl sulfate is a combustible liquid (NFPA rating = 2). Toxic dimethyl sulfate vapors are produced in a fire. Carbon dioxide or dry chemical extinguishers should be used to fight dimethyl sulfate fires.

Reactivity and Incompatibility

Dimethyl sulfate can react violently with ammonium hydroxide, sodium azide, and strong oxidizers.

Storage and Handling

Because of its carcinogenicity, dimethyl sulfate should be handled using the "basic prudent practices" of Chapter 5.C, supplemented by the additional precautions for work with compounds of high chronic toxicity (Chapter 5.D). In particular, work with dimethyl sulfate should be conducted in a fume hood to prevent exposure by inhalation, and appropriate impermeable gloves and safety goggles should be worn at all times to prevent skin and eye contact.

Accidents

In the event of skin contact, immediately wash with soap and water and remove contaminated clothing. In case of eye contact, promptly wash with copious amounts of water for 15 min (lifting upper and lower lids occasionally) and obtain medical attention. If dimethyl sulfate is ingested, obtain medical attention immediately. If inhaled, move the person to fresh air and seek medical attention at once.

In the event of a spill, remove all ignition sources, soak up the dimethyl sulfate with a spill pillow or absorbent material, place in an appropriate container, and dispose of properly. Respiratory protection may be necessary in the event of a large spill or release in a confined area.

Disposal

Excess dimethyl sulfate and waste material containing this substance should be placed in a covered metal container, clearly labeled, and handled according to your institution's waste disposal guidelines. For more information on disposal procedures, see Chapter 7 of this volume.

LABORATORY CHEMICAL SAFETY SUMMARY: DIMETHYL SULFOXIDE

Substance	Dimethyl sulfoxide (DMSO, methyl sulfoxide) CAS 67-68-5
Formula	$(CH_3)_2SO$
Physical Properties	Colorless liquid bp 189 °C (decomposes), mp 18.5 °C Miscible with water
Odor	Mild garlic odor
Vapor Pressure	0.37 mmHg at 20 °C
Flash Point	95 °C
Autoignition Temperature	215 °C

Toxicity Data

LD_{50} oral (rat)	14,500 mg/kg
LD_{50} skin (rabbit)	40,000 mg/kg
LC_{50} inhal (rat)	1600 mg/m^3 (4 h)

Major Hazards

Freely penetrates skin and may carry dissolved chemicals across the skin.

Toxicity

The acute toxicity of DMSO by all routes of exposure is very low. Inhalation of DMSO vapor can cause irritation of the respiratory tract, and at higher concentrations may cause vomiting, chills, headache, and dizziness. The material is only slightly toxic by ingestion and may cause vomiting, abdominal pain, and lethargy. Dimethyl sulfoxide is relatively nontoxic by skin absorption, but can cause itching, scaling, and a transient burning sensation. Dimethyl sulfoxide can increase the tendency for other chemicals to penetrate the skin and so increase their toxic effects. Contact of DMSO liquid with the eyes may cause irritation with redness, pain, and blurred vision.

Chronic exposure to dimethyl sulfoxide can cause damage to the cornea of the eye. Dimethyl sulfoxide has not been found to be carcinogenic or to show reproductive or developmental toxicity in humans.

Flammability and Explosibility

Combustible when exposed to heat or flame (NFPA rating = 1). Carbon dioxide or dry chemical extinguishers should be used to fight DMSO fires.

Reactivity and Incompatibility

DMSO reacts violently with strong oxidizers, many acyl halides, boron hydrides, and alkali metals. DMSO can form explosive mixtures with metal salts of oxoacids (sodium perchlorate, iron(III) nitrate).

(continued on facing page)

Storage and Handling

Dimethyl sulfoxide should be handled in the laboratory using the "basic prudent practices" described in Chapter 5.C.

Accidents

In the event of skin contact, immediately wash with soap and water and remove contaminated clothing. In case of eye contact, promptly wash with copious amounts of water for 15 min (lifting upper and lower lids occasionally) and obtain medical attention. If dimethyl sulfoxide is ingested, obtain medical attention immediately. If large amounts of this compound are inhaled, move the person to fresh air and seek medical attention at once.

In the event of a spill, remove all ignition sources, soak up the dimethyl sulfoxide with a spill pillow or absorbent material, place in an appropriate container, and dispose of properly. Respiratory protection may be necessary in the event of a large spill or release in a confined area.

Disposal

Excess dimethyl sulfoxide and waste material containing this substance should be placed in an appropriate container, clearly labeled, and handled according to your institution's waste disposal guidelines. For more information on disposal procedures, see Chapter 7 of this volume.

LABORATORY CHEMICAL SAFETY SUMMARY: DIMETHYLFORMAMIDE

Substance	Dimethylformamide (*N*,*N*-Dimethylformamide, DMF) CAS 68-12-2
Formula	$(CH_3)_2NCHO$
Physical Properties	Colorless, clear liquid bp 153 °C, mp –61 °C Miscible with water in all proportions
Odor	Faint, ammonia-like odor detectable at 100 ppm
Vapor Density	2.5 (air = 1.0)
Vapor Pressure	2.6 mmHg at 20 °C
Flash Point	58 °C
Autoignition Temperature	445 °C

Toxicity Data

LD_{50} oral (rat)	2800 mg/kg
LD_{50} skin (rabbit)	4720 mg/kg
LC_{50} inhal (mouse)	9400 mg/m^3; 2 h
PEL (OSHA)	10 ppm (30 mg/m^3—skin)
TLV-TWA (ACGIH)	10 ppm (30 mg/m^3—skin)

Major Hazards

Low acute toxicity; readily absorbed through the skin.

Toxicity

The acute toxicity of DMF is low by inhalation, ingestion, and skin contact. Contact with liquid DMF may cause eye and skin irritation. DMF is an excellent solvent for many toxic materials that are not ordinarily absorbed and can increase the hazard of these substances by skin contact. Exposure to high concentrations of DMF may lead to liver damage and other systemic effects.

Dimethylformamide is listed by IARC in Group 2B ("possible human carcinogen"). It is not classified as a "select carcinogen" according to the criteria of the OSHA Laboratory Standard. No significant reproductive effects have been observed in animal tests. Repeated exposure to DMF may result in damage to the liver, kidneys, and cardiovascular system.

Flammability and Explosibility

DMF is a combustible liquid (NFPA rating = 2). Vapors are heavier than air and may travel to source of ignition and flash back. DMF vapor forms explosive mixtures with air at concentrations of 2.2 to 15.2% (by volume). Carbon dioxide or dry chemical extinguishers should be used to fight DMF fires.

(continued on facing page)

The information in this LCSS has been compiled by a committee of the National Research Council from literature sources and Material Safety Data Sheets and is believed to be accurate as of July 1994. This summary is intended for use by trained laboratory personnel in conjunction with the NRC report *Prudent Practices in the Laboratory: Handling and Disposal of Chemicals*. This LCSS presents a concise summary of safety information that should be adequate for most laboratory uses of the title substance, but in some cases it may be advisable to consult more comprehensive references. This information should not be used as a guide to the nonlaboratory use of this chemical.

Reactivity and Incompatibility

Though stable at normal temperatures and storage conditions, DMF may react violently with halogens, acyl halides, strong oxidizers, and polyhalogenated compounds in the presence of iron. Decomposition products include toxic gases and vapors such as dimethylamine and carbon monoxide. DMF will attack some forms of plastics, rubber, and coatings.

Storage and Handling

DMF should be handled in the laboratory using the "basic prudent practices" described in Chapter 5.C. In particular, DMF should be used only in areas free of ignition sources, and quantities greater than 1 liter should be stored in tightly sealed metal containers in areas separate from oxidizers.

Accidents

In the event of skin contact, immediately wash with soap and water and remove contaminated clothing. Destroy contaminated shoes. In case of eye contact, promptly wash with copious amounts of water for 15 min (lifting upper and lower lids occasionally) and obtain medical attention. If DMF is ingested, do not induce vomiting. Obtain medical attention immediately. If large amounts of this compound are inhaled, move the person to fresh air and seek medical attention at once.

In the event of a spill, remove all ignition sources, soak up the DMF with a spill pillow or absorbent material, place in an appropriate container, and dispose of properly. Respiratory protection may be necessary in the event of a large spill or release in a confined area.

Disposal

Excess DMF and waste material containing this substance should be placed in an appropriate container, clearly labeled, and handled according to your institution's waste disposal guidelines. For more information on disposal procedures, see Chapter 7 of this volume.

LABORATORY CHEMICAL SAFETY SUMMARY: DIOXANE

Substance	Dioxane (1,4-Dioxane; *p*-dioxane; diethylene ether; 1,4-diethylene dioxide) CAS 123-91-1
Formula	$O(CH_2CH_2)_2O$
Physical Properties	Colorless liquid bp 101 °C, mp 12 °C Miscible with water
Odor	Mild ether-like odor detectable at 0.8 to 172 ppm (mean = 12 ppm)
Vapor Density	3 (air = 1.0)
Vapor Pressure	40 mmHg at 25 °C
Flash Point	12 °C
Autoignition Temperature	180 °C

Toxicity Data

LD_{50} oral (mouse)	5700 mg/kg
LC_{50} inhal (rat)	13,000 ppm (46,800 mg/m^3; 2 h)
LD_{50} skin (rabbit)	7600 mg/kg
PEL (OSHA)	100 ppm (360 mg/m^3)—skin
TLV-TWA (ACGIH)	25 ppm (90 mg/m^3)—skin

Major Hazards

Highly flammable; forms sensitive peroxides on exposure to air that may explode on concentration by distillation or drying.

Toxicity

The acute toxicity of 1,4-dioxane is low. Exposure to 200 to 300 ppm causes irritation of the eyes, nose, and throat. Inhalation of higher concentrations can result in damage to the kidneys and liver. Symptoms of overexposure may include upper respiratory tract irritation, coughing, drowsiness, vertigo, headache, stomach pains, nausea, and vomiting. Prolonged or repeated contact may produce drying and cracking of the skin. Ingestion of this substance will result in the effects of exposure by inhalation. The odor of dioxane is not unpleasant, and its irritating effects may be transitory; consequently, it is not regarded as a substance with adequate warning properties.

Dioxane shows carcinogenic effects in animal studies and is listed by IARC in Group 2B ("possible human carcinogen"). It is not classified as a "select carcinogen" according to the criteria of the OSHA Laboratory Standard. Prolonged or repeated exposure to this substance may result in liver and kidney injury. Dioxane has not been shown to be a reproductive or developmental toxin in humans.

(continued on facing page)

The information in this LCSS has been compiled by a committee of the National Research Council from literature sources and Material Safety Data Sheets and is believed to be accurate as of July 1994. This summary is intended for use by trained laboratory personnel in conjunction with the NRC report *Prudent Practices in the Laboratory: Handling and Disposal of Chemicals*. This LCSS presents a concise summary of safety information that should be adequate for most laboratory uses of the title substance, but in some cases it may be advisable to consult more comprehensive references. This information should not be used as a guide to the nonlaboratory use of this chemical.

Flammability and Explosibility

Dioxane is a highly flammable liquid (NFPA rating = 3). Its vapor is heavier than air and may travel a considerable distance to a source of ignition and flash back. Dioxane vapor forms explosive mixtures with air at concentrations of 2 to 22% (by volume). Fires involving dioxane should be extinguished with carbon dioxide or dry powder extinguishers.

Dioxane can form shock- and heat-sensitive peroxides that may explode on concentration by distillation or evaporation. Samples of this substance should always be tested for the presence of peroxides before distilling or allowing to evaporate. Dioxane should never be distilled to dryness.

Reactivity and Incompatibility

Dioxane can form potentially explosive peroxides upon long exposure to air. Dioxane may react violently with Raney nickel catalyst, nitric and perchloric acids, sulfur trioxide, and strong oxidizing reagents.

Storage and Handling

Dioxane should be handled in the laboratory using the "basic prudent practices" described in Chapter 5.C, supplemented by the additional precautions for dealing with extremely flammable substances (Chapter 5.F). In particular, dioxane should be used only in areas free of ignition sources, and quantities greater than 1 liter should be stored in tightly sealed metal containers in areas separate from oxidizers. Containers of dioxane should be dated when opened and tested periodically for the presence of peroxides.

Accidents

In the event of skin contact, immediately wash with soap and water and remove contaminated clothing. In case of eye contact, promptly wash with copious amounts of water for 15 min (lifting upper and lower lids occasionally) and obtain medical attention. If dioxane is ingested, obtain medical attention immediately. If large amounts of this compound are inhaled, move the person to fresh air and seek medical attention at once.

In the event of a spill, remove all ignition sources, soak up the dioxane with a spill pillow or absorbent material, place in an appropriate container, and dispose of properly. Respiratory protection may be necessary in the event of a large spill or release in a confined area.

Disposal

Excess dioxane and waste material containing this substance should be placed in an appropriate container, clearly labeled, and handled according to your institution's waste disposal guidelines. For more information on disposal procedures, see Chapter 7 of this volume.

LABORATORY CHEMICAL SAFETY SUMMARY: ETHANOL

Substance	Ethanol (Ethyl alcohol, alcohol, methylcarbinol) CAS 64-17-5
Formula	C_2H_5OH
Physical Properties	Colorless liquid bp 78 °C, mp −114 °C Miscible with water
Odor	Pleasant alcoholic odor detectable at 49 to 716 ppm (mean = 180 ppm)
Vapor Density	1.6 (air = 1.0)
Vapor Pressure	43 mmHg at 20 °C
Flash Point	13 °C
Autoignition Temperature	363 °C

Toxicity Data

LD_{50} oral (rat)	7060 mg/kg
LD_{50} skin (rabbit)	>20 mL/kg
LC_{50} inhal (rat)	20,000 ppm (10 h)
PEL (OSHA)	1000 ppm (1900 mg/m^3)
TLV-TWA (ACGIH)	1000 ppm (1900 mg/m^3)

Major Hazards — Flammable liquid

Toxicity

The acute toxicity of ethanol is very low. Ingestion of ethanol can cause temporary nervous system depression with anesthetic effects such as dizziness, headache, confusion, and loss of consciousness; large doses (250 to 500 mL) can be fatal in humans. High concentrations of ethanol vapor are irritating to the eyes and upper respiratory tract. Liquid ethanol does not significantly irritate the skin but is a moderate eye irritant. Exposure to high concentrations of ethanol by inhalation (over 1000 ppm) can cause central nervous system (CNS) effects, including dizziness, headache, and giddiness followed by depression, drowsiness, and fatigue. Ethanol is regarded as a substance with good warning properties.

Tests in some animals indicate that ethanol may have developmental and reproductive toxicity if ingested. There is no evidence that laboratory exposure to ethanol has carcinogenic effects.

To discourage deliberate ingestion, ethanol for laboratory use is often "denatured" by the addition of other chemicals; the toxicity of possible additives must also be considered when evaluating the risk of laboratory exposure to ethanol.

(continued on facing page)

Flammability and Explosibility

Ethanol is a flammable liquid (NFPA rating = 3), and its vapor can travel a considerable distance to an ignition source and "flash back." Ethanol vapor forms explosive mixtures with air at concentrations of 4.3 to 19% (by volume). Hazardous gases produced in ethanol fires include carbon monoxide and carbon dioxide. Carbon dioxide or dry chemical extinguishers should be used for ethanol fires.

Reactivity and Incompatibility

Contact of ethanol with strong oxidizers, peroxides, strong alkalis, and strong acids may cause fires and explosions.

Storage and Handling

Ethanol should be handled in the laboratory using the "basic prudent practices" described in Chapter 5.C, supplemented by the additional precautions for dealing with highly flammable substances (Chapter 5.F). In particular, ethanol should be used only in areas free of ignition sources, and quantities greater than 1 liter should be stored in tightly sealed metal containers in areas separate from oxidizers.

Accidents

In the event of skin contact, immediately wash with soap and water and remove contaminated clothing. In case of eye contact, promptly wash with copious amounts of water for 15 min (lifting upper and lower lids occasionally) and obtain medical attention. If ethanol is ingested, obtain medical attention immediately. If large amounts of this compound are inhaled, move the person to fresh air and seek medical attention at once.

In the event of a spill, remove all ignition sources, soak up the ethanol with a spill pillow or absorbent material, place in an appropriate container, and dispose of properly. Respiratory protection may be necessary in the event of a large spill or release in a confined area.

Disposal

Excess ethanol and waste material containing this substance should be placed in an appropriate container, clearly labeled, and handled according to your institution's waste disposal guidelines. For more information on disposal procedures, see Chapter 7 of this volume.

The information in this LCSS has been compiled by a committee of the National Research Council from literature sources and Material Safety Data Sheets and is believed to be accurate as of July 1994. This summary is intended for use by trained laboratory personnel in conjunction with the NRC report *Prudent Practices in the Laboratory: Handling and Disposal of Chemicals*. This LCSS presents a concise summary of safety information that should be adequate for most laboratory uses of the title substance, but in some cases it may be advisable to consult more comprehensive references. This information should not be used as a guide to the nonlaboratory use of this chemical.

LABORATORY CHEMICAL SAFETY SUMMARY: ETHIDIUM BROMIDE

Substance	Ethidium bromide (Dromilac, homidium bromide) CAS 1239-45-8
Formula	$C_{21}H_{20}BrN_3$
Physical Properties	Dark red crystals mp 260 to 262 °C Soluble in water (5 g/100 mL)
Odor	Odorless solid
Major Hazards	Potent mutagen

Toxicity

Acute toxic effects from exposure to ethidium bromide have not been thoroughly investigated. Ethidium bromide is irritating to the eyes, skin, mucous membranes, and upper respiratory tract.

Although there is no evidence for the carcinogenicity or teratogenicity of this substance in humans, ethidium bromide is strongly mutagenic and therefore should be regarded as a possible carcinogen and reproductive toxin.

Flammability and Explosibility

Ethidium bromide does not pose a flammability hazard (NFPA rating = 1).

Reactivity and Incompatibility

No incompatibilities are known.

Storage and Handling

Ethidium bromide should be handled in the laboratory using the "basic prudent practices" described in Chapter 5.C. Because of its mutagenicity, stock solutions of this compound should be prepared in a fume hood, and protective gloves should be worn at all times while handling this substance. Operations capable of generating ethidium bromide dust or aerosols of ethidium bromide solutions should be conducted in a fume hood to prevent exposure by inhalation.

Accidents

In the event of skin contact, immediately wash with soap and water and remove contaminated clothing. In case of eye contact, promptly wash with copious amounts of water for 15 min (lifting upper and lower lids occasionally) and obtain medical attention. If ethidium bromide is ingested, obtain medical attention immediately.

In the event of a spill, mix ethidium bromide with an absorbent material (avoid raising dust), place in an appropriate container, and dispose of properly. Soak up aqueous solutions with a spill pillow or absorbent material.

(continued on facing page)

Disposal   Excess ethidium bromide and waste material containing this substance should be placed in an appropriate container, clearly labeled, and handled according to your institution's waste disposal guidelines. For more information on disposal procedures, see Chapter 7 of this volume.

LABORATORY CHEMICAL SAFETY SUMMARY: ETHYL ACETATE

Substance	Ethyl acetate (Acetic acid ethyl ester, ethyl ethanoate, acetoxyethane) CAS 141-78-6
Formula	$CH_3COOC_2H_5$
Physical Properties	Colorless liquid bp 77 °C, mp −84 °C Moderately soluble in water (9 g/100 mL)
Odor	Pleasant fruity odor detectable at 7 to 50 ppm (mean = 18 ppm)
Vapor Density	3.0 (air = 1.0)
Vapor Pressure	76 mmHg at 20 °C
Flash Point	−4 °C
Autoignition Temperature	427 °C

Toxicity Data

LD_{50} oral (rat)	5620 mg/kg
LC_{50} inhal (rat)	1600 ppm (8 h)
PEL (OSHA)	400 ppm (1400 mg/m^3)
TLV-TWA (ACGIH)	400 ppm (1440 mg/m^3)

Major Hazards Flammable liquid and vapor

Toxicity The acute toxicity of ethyl acetate is low. Ethyl acetate vapor causes eye, skin, and respiratory tract irritation at concentrations above 400 ppm. Exposure to high concentrations may lead to headache, nausea, blurred vision, central nervous system depression, dizziness, drowsiness, and fatigue. Ingestion of ethyl acetate may cause gastrointestinal irritation and, with larger amounts, central nervous system depression. Eye contact with the liquid can produce temporary irritation and lacrimation. Skin contact produces irritation. Ethyl acetate is regarded as a substance with good warning properties.

No chronic systemic effects have been reported in humans, and ethyl acetate has not been shown to be a human carcinogen, reproductive, or developmental toxin.

Flammability and Explosibility Ethyl acetate is a flammable liquid (NFPA rating = 3), and its vapor can travel a considerable distance to an ignition source and "flash back." Ethyl acetate vapor forms explosive mixtures with air at concentrations of 2 to 11.5% (by volume). Hazardous gases produced in ethyl acetate fires include carbon monoxide and carbon dioxide. Carbon dioxide or dry chemical extinguishers should be used for ethyl acetate fires.

(continued on facing page)

Reactivity and Incompatibility

Contact with strong oxidizers, strong alkalis, and strong acids may cause fires and explosions.

Storage and Handling

Ethyl acetate should be handled in the laboratory using the "basic prudent practices" described in Chapter 5.C, supplemented by the additional precautions for dealing with highly flammable substances (Chapter 5.F). In particular, ethyl acetate should be used only in areas free of ignition sources, and quantities greater than 1 liter should be stored in tightly sealed metal containers in areas separate from oxidizers.

Accidents

In the event of skin contact, immediately wash with soap and water and remove contaminated clothing. In case of eye contact, promptly wash with copious amounts of water for 15 min (lifting upper and lower lids occasionally) and obtain medical attention. If ethyl acetate is ingested, obtain medical attention immediately. If large amounts of this compound are inhaled, move the person to fresh air and seek medical attention at once.

In the event of a spill, remove all ignition sources, soak up the ethyl acetate with a spill pillow or absorbent material, place in an appropriate container, and dispose of properly. Respiratory protection may be necessary in the event of a large spill or release in a confined area.

Disposal

Excess ethyl acetate and waste material containing this substance should be placed in an appropriate container, clearly labeled, and handled according to your institution's waste disposal guidelines. For more information on disposal procedures, see Chapter 7 of this volume.

LABORATORY CHEMICAL SAFETY SUMMARY: ETHYLENE DIBROMIDE

Substance	Ethylene dibromide (1,2-Dibromoethane, ethylene bromide, EDB) CAS 106-93-4
Formula	$BrCH_2CH_2Br$
Physical Properties	Colorless liquid bp 131 °C, mp 9 °C Slightly soluble in water (0.4 g/100 mL at 20 °C)
Odor	Mild, sweet odor detectable at 10 ppm
Vapor Density	6.5 (air = 1.0)
Vapor Pressure	12 mmHg at 25 °C
Flash Point	Noncombustible

Toxicity Data

LD_{50} oral (rat)	108 mg/kg
LD_{50} skin (rabbit)	300 mg/kg
LC_{50} inhal (rat)	14,300 mg/m^3 (30 min)
PEL (OSHA)	20 ppm (150 mg/m^3)

Major Hazards Suspected human carcinogen (OSHA "select carcinogen"); moderate acute toxicity; severe skin and eye irritant.

Toxicity Ethylene dibromide is moderately toxic by inhalation, ingestion, and skin contact and is a severe irritant of the skin, eyes, and mucous membranes. Symptoms of overexposure by inhalation may include depression of the central nervous system, respiratory tract irritation, and pulmonary edema. Oral intake of 5 to 10 mL can be fatal to humans owing to liver and kidney damage. Skin contact with EDB can produce severe irritation and blistering; serious skin injury can result from contact with clothing and shoes wet with EDB. This compound can be absorbed through the skin in toxic amounts. EDB vapors are severely irritating to the eyes, and contact with the liquid can damage vision.

EDB is listed in IARC Group 2A ("probable human carcinogen") and is classified as a "select carcinogen" under the criteria of the OSHA Laboratory Standard. Chronic inhalation may cause pulmonary, renal, and hepatic damage. EDB is a suspected reproductive toxin implicated in reduction in male fertility. Ethylene dibromide is considered to be a compound with poor warning properties due to potential chronic and carcinogenic effects.

Flammability and Explosibility Ethylene dibromide is a noncombustible substance (NFPA rating = 0).

(continued on facing page)

Reactivity and Incompatibility

EDB reacts vigorously with alkali metals, zinc, magnesium, aluminum, caustic alkalis, strong oxidizers, and liquid ammonia. Liquid EDB will attack some forms of plastics, rubber, and coatings.

Storage and Handling

Because of its carcinogenicity, EDB should be handled using the "basic prudent practices" of Chapter 5.C, supplemented by the additional precautions for work with compounds of high chronic toxicity (Chapter 5.D). In particular, work with EDB should be conducted in a fume hood to prevent exposure by inhalation, and appropriate impermeable gloves and safety goggles should be worn to prevent skin contact. Gloves and protective clothing should be changed immediately if EDB contamination occurs. Since EDB can penetrate neoprene and other plastics, protective apparel made of these materials does not provide adequate protection from contact with EDB.

Accidents

In the event of skin contact, immediately remove contaminated clothing and wash with soap and water. In case of eye contact, promptly wash with copious amounts of water for 15 min (lifting upper and lower lids occasionally) and obtain medical attention. If EDB is ingested, obtain medical attention immediately. If large amounts of this compound are inhaled, move the person to fresh air and seek medical attention at once.

Persons not wearing protective equipment and clothing should be restricted from areas of spill or leaks until cleanup has been completed. Soak up EDB with a spill pillow or absorbent material such as vermiculite or dry sand, place in an appropriate container, and dispose of properly. Evacuation and cleanup using respiratory protection may be necessary in the event of a large spill or release in a confined area.

Disposal

Excess EDB and waste material containing this substance should be placed in an appropriate container, clearly labeled, and handled according to your institution's waste disposal guidelines. For more information on disposal procedures, see Chapter 7 of this volume.

LABORATORY CHEMICAL SAFETY SUMMARY: ETHYLENE OXIDE

Substance	Ethylene oxide (1,2 Epoxyethane; oxacyclopropane; dimethylene oxide) CAS 75-21-8
Formula	C_2H_4O
Physical Properties	Colorless liquid or gas bp 10.7 °C, mp −111.3 °C Miscible with water
Odor	Sweet odor detectable at 257 to 690 ppm (mean = 420 ppm)
Vapor Density	1.5 at bp (air = 1.0)
Vapor Pressure	1095 mmHg at 20 °C
Flash Point	−20 °C
Autoignition Temperature	429 °C

Toxicity Data

LD_{50} oral (rat)	72 mg/kg
LC_{50} inhal (rat)	800 ppm (1600 mg/m^3)
PEL (OSHA)	1 ppm (2 mg/m^3)
TLV-TWA (ACGIH)	1 ppm (2 mg/m^3)

Major Hazards   OSHA "select carcinogen"; highly flammable; severe irritant.

Toxicity

Ethylene oxide is a severe irritant to the eyes, skin, and respiratory tract and exhibits moderate acute toxicity by all routes of exposure. Symptoms of overexposure by inhalation may be delayed and can include nausea, vomiting, headache, drowsiness, and difficulty breathing. Ethylene oxide can cause serious burns to the skin, which may only appear after a delay of 1 to 5 hours. This substance may also be absorbed through the skin to cause the systemic effects listed above. Eye contact can result in severe burns. Ethylene oxide is not considered to have adequate warning properties.

Ethylene oxide is listed by IARC in Group 2A ("probable human carcinogen") and is classified as a "select carcinogen" under the criteria of the OSHA Laboratory Standard. There is some evidence from animal studies that ethylene oxide may be a developmental and reproductive toxin in both males and females. Exposure to this substance may lead to sensitization.

Flammability and Explosibility

Ethylene oxide is an extremely flammable substance (NFPA rating = 4). Ethylene oxide vapor may be ignited by hot surfaces such as hot plates and static electricity discharges, and since the vapor is heavier than air, it may travel a considerable distance to an

(continued on facing page)

The information in this LCSS has been compiled by a committee of the National Research Council from literature sources and Material Safety Data Sheets and is believed to be accurate as of July 1994. This summary is intended for use by trained laboratory personnel in conjunction with the NRC report *Prudent Practices in the Laboratory: Handling and Disposal of Chemicals*. This LCSS presents a concise summary of safety information that should be adequate for most laboratory uses of the title substance, but in some cases it may be advisable to consult more comprehensive references. This information should not be used as a guide to the nonlaboratory use of this chemical.

ignition source and flash back. Ethylene oxide vapor forms explosive mixtures with air at concentrations of 3 to 100% (by volume). Carbon dioxide or dry chemical extinguishers should be used for ethylene oxide fires. Ethylene oxide may explode when heated in a closed vessel.

Reactivity and Incompatibility

Ethylene oxide can undergo violent polymerization, which can be initiated by contact with metal surfaces, strong acids or bases, alkali metals, iron oxide or chloride, and aluminum chloride.

Storage and Handling

Because of its carcinogenicity, flammability, and reactivity, ethylene oxide should be handled using the "basic prudent practices" of Chapter 5.C, supplemented by the additional precautions for work with compounds of high chronic toxicity (Chapter 5.D) and extremely flammable substances (Chapter 5.F). In particular, work with ethylene oxide should be conducted in a fume hood to prevent exposure by inhalation, and appropriate impermeable gloves and splash goggles should be worn at all times to prevent skin and eye contact. Ethylene oxide should be used only in areas free of ignition sources and should be stored in the cold in tightly sealed containers placed within a secondary container.

Accidents

In the event of skin contact, immediately wash with soap and water and remove contaminated clothing. In case of eye contact, promptly wash with copious amounts of water for 15 min (lifting upper and lower lids occasionally) and obtain medical attention. If ethylene oxide is ingested, obtain medical attention immediately. If large amounts of this compound are inhaled, move the person to fresh air and seek medical attention at once.

In the event of a spill of liquid ethylene oxide, remove all ignition sources, soak up the ethylene oxide with a spill pillow or absorbent material, place in an appropriate container, and dispose of properly. In the event of accidental release of ethylene oxide gas, evacuate the area and eliminate the source of the release, such as a leaking cylinder, if possible. Respiratory protection may be necessary in the event of a large spill or release in a confined area.

Disposal

Excess ethylene oxide and waste material containing this substance should be placed in an appropriate container, clearly labeled, and handled according to your institution's waste disposal guidelines. For more information on disposal procedures, see Chapter 7 of this volume.

Substance	Fluorides (inorganic) NaF: CAS 7681-49-4
Physical Properties	NaF: mp 993 °C, bp 1700 °C Slightly soluble to insoluble in water
Odor	Odorless

Toxicity Data

LD_{50} oral (rat)	245 mg/kg (potassium fluoride) 52 mg/kg (sodium fluoride) 377 mg/kg (stannous fluoride)
PEL (OSHA)	2.5 mg/m^3
TLV-TWA (ACGIH)	2.5 mg/m^3

Major Hazards

Moderate acute toxicity; irritating to eyes and respiratory tract.

Toxicity

The acute toxicity of fluorides is generally moderate. High exposures may cause irritation of the eyes and respiratory tract. Ingestion of fluoride may cause a salty or soapy taste, vomiting, abdominal pain, diarrhea, shortness of breath, difficulty in speaking, thirst, weak pulse, disturbed color vision, muscular weakness, convulsions, loss of consciousness, and death. In humans the approximate lethal dose of NaF by ingestion is 5 g.

Repeated inhalation of fluoride dust may cause excessive calcification of the bone and calcification of ligaments of the ribs, pelvis, and spinal column. Repeated skin contact may cause a rash.

Fluorides have not been shown to be carcinogenic or to show reproductive or developmental toxicity in humans.

Flammability and Explosibility

Fluorides are not combustible.

Reactivity and Incompatibility

Contact with strong acids may cause formation of highly toxic and corrosive hydrogen fluoride.

Storage and Handling

Fluorides should be handled in the laboratory using the "basic prudent practices" described in Chapter 5.C.

Accidents

In the event of skin contact, immediately wash with soap and water and remove contaminated clothing. In case of eye contact, promptly wash with copious amounts of water for 15 min (lifting upper and lower lids occasionally) and obtain medical attention. If fluorides are ingested, obtain medical attention immediately. If large amounts of fluorides are inhaled, move the person to fresh air and seek medical attention at once.

(continued on facing page)

The information in this LCSS has been compiled by a committee of the National Research Council from literature sources and Material Safety Data Sheets and is believed to be accurate as of July 1994. This summary is intended for use by trained laboratory personnel in conjunction with the NRC report *Prudent Practices in the Laboratory: Handling and Disposal of Chemicals*. This LCSS presents a concise summary of safety information that should be adequate for most laboratory uses of the title substance, but in some cases it may be advisable to consult more comprehensive references. This information should not be used as a guide to the nonlaboratory use of this chemical.

In the event of a spill, sweep up fluorides, place in an appropriate container, and dispose of properly. Respiratory protection may be necessary in the event of a large spill or release in a confined area.

Disposal   Excess fluorides and waste material containing this substance should be placed in an appropriate container, clearly labeled, and handled according to your institution's waste disposal guidelines. For more information on disposal procedures, see Chapter 7 of this volume.

LABORATORY CHEMICAL SAFETY SUMMARY: FLUORINE

Substance	Fluorine CAS 7782-41-4
Formula	F_2
Physical Properties	Pale yellow gas bp -188 °C, mp -219 °C Reacts with water
Odor	Strong ozone-like odor detectable at 0.1 to 0.2 ppm
Vapor Density	1.695 (air = 1.0)
Vapor Pressure	>760 mmHg at 20 °C

Toxicity Data

LC_{50} inhal (rat)	185 ppm (300 mg/m^3; 1 h)
PEL (OSHA)	0.1 ppm (0.2 mg/m^3)
TLV-TWA (ACGIH)	1 ppm (1.6 mg/m^3)
STEL (ACGIH)	2 ppm (3.1 mg/m^3)

Major Hazards

Dangerously reactive gas; contact with many materials results in ignition or violent reactions; highly irritating and corrosive to the eyes, skin, and mucous membranes.

Toxicity

The acute toxicity of fluorine is high. Even very low concentrations irritate the respiratory tract, and brief exposure to 50 ppm can be intolerable. High concentrations can cause severe damage to the respiratory system and can result in the delayed onset of pulmonary edema, which may be fatal. Fluorine is highly irritating to the eyes, and high concentrations cause severe injury and can lead to permanent damage and blindness. Fluorine is extremely corrosive to the skin, causing damage similar to second-degree thermal burns. Fluorine is not considered to have adequate warning properties.

Chronic toxicity is unlikely to occur due to the corrosive effects of fluorine exposure. Fluorine has not been found to be carcinogenic or to show reproductive or developmental toxicity in humans.

Flammability and Explosibility

Fluorine is not flammable, but is a very strong oxidizer, reacting vigorously with most oxidizable materials at room temperature, frequently with ignition. Water should not be used to fight fires involving fluorine.

Reactivity and Incompatibility

Fluorine is an extremely powerful oxidizing agent that reacts violently with a great many materials, including water, most organic substances (including greases, many plastics, rubbers, and coatings), silicon-containing compounds, and most metals. The reaction with water produces HF and ozone. Fluorine reacts explosively or forms explosive compounds, often at very low temperatures, with chemicals as diverse as graphite, sodium acetate, stainless steel, perchloric acid, and water or ice. Fluorine ignites in contact with ammonia,

(continued on facing page)

The information in this LCSS has been compiled by a committee of the National Research Council from literature sources and Material Safety Data Sheets and is believed to be accurate as of July 1994. This summary is intended for use by trained laboratory personnel in conjunction with the NRC report *Prudent Practices in the Laboratory: Handling and Disposal of Chemicals*. This LCSS presents a concise summary of safety information that should be adequate for most laboratory uses of the title substance, but in some cases it may be advisable to consult more comprehensive references. This information should not be used as a guide to the nonlaboratory use of this chemical.

ceramic materials, phosphorus, sulfur, copper wire, acetone, and many other organic and inorganic compounds. The literature on incompatibilities of fluorine should be carefully reviewed before attempting work with this substance.

Storage and Handling

Because of its extreme reactivity, toxicity, and gaseous nature, fluorine should be handled using the "basic prudent practices" of Chapter 5.C, supplemented by the additional precautions for work with reactive or explosive chemicals (Chapter 5.G) and work with compressed gases (Chapter 5.H). Work with fluorine requires special precautions and protective equipment and should be carried out only by specially trained personnel. Fluorine will react with many materials normally recommended for handling compressed gases.

Accidents

In the event of skin contact, immediately wash with soap and water and remove contaminated clothing. In case of eye contact, promptly wash with copious amounts of water for 15 min (lifting upper and lower lids occasionally) and obtain medical attention. If large amounts of fluorine are inhaled, move the person to fresh air and seek medical attention at once.

In the event of a small leak, stop flow of gas if possible, or move cylinder to a fume hood or to a safe location in the open air. Accidental releases of fluorine require evacuation of the affected area and should be handled only by trained personnel equipped with proper protective clothing and respiratory protection.

Disposal

Excess fluorine should be returned to the manufacturer if possible, according to your institution's waste disposal guidelines. For more information on disposal procedures, see Chapter 7 of this volume.

LABORATORY CHEMICAL SAFETY SUMMARY: FORMALDEHYDE

Substance	Formaldehyde (Methanal; 37% aqueous solution (usually containing 10 to 15% methanol) is called formalin; solid polymer is called paraformaldehyde) CAS 50-00-0
Formula	HCHO
Physical Properties	Clear, colorless liquid Formaldehyde: bp -19 °C, mp -92 °C Formalin: bp 96 °C, mp -15 °C Miscible with water
Odor	Pungent odor detectable at 1 ppm
Vapor Density	~1 (air = 1.0)
Vapor Pressure	Formaldehyde: 10 mmHg at -88 °C Formalin: 23 to 26 mmHg at 25 °C
Flash Point	50 °C for formalin containing 15% methanol
Autoignition Temperature	424 °C for formalin containing 15% methanol

Toxicity Data

LD_{50} oral (rat)	500 mg/kg
LD_{50} skin (rabbit)	270 mg/kg
LC_{50} inhal (rat)	203 mg/m^3 (2 h)
PEL (OSHA)	1 ppm (1.5 mg/m^3)
TLV-TWA (ACGIH)	0.3 ppm (ceiling)(0.37 mg/m^3)
STEL (OSHA)	2 ppm (2.5 mg/m^3)

Major Hazards   Probable human carcinogen (OSHA "select carcinogen"); moderate acute toxicity; skin sensitizer.

Toxicity   Formaldehyde is moderately toxic by skin contact and inhalation. Exposure to formaldehyde gas can cause irritation of the eyes and respiratory tract, coughing, dry throat, tightening of the chest, headache, a sensation of pressure in the head, and palpitations of the heart. Exposure to 0.1 to 5 ppm causes irritation of the eyes, nose, and throat; above 10 ppm severe lacrimation occurs, burning in the nose and throat is experienced, and breathing becomes difficult. Acute exposure to concentrations above 25 ppm can cause serious injury, including fatal pulmonary edema. Formaldehyde has low acute toxicity via the oral route. Ingestion can cause irritation of the mouth, throat, and stomach, nausea, vomiting, convulsions, and coma. An oral dose of 30 to 100 mL of 37% formalin can be fatal in humans. Formalin solutions can cause severe eye burns and loss of vision. Eye contact may lead to delayed effects that are not appreciably eased by eye washing.

(continued on facing page)

The information in this LCSS has been compiled by a committee of the National Research Council from literature sources and Material Safety Data Sheets and is believed to be accurate as of July 1994. This summary is intended for use by trained laboratory personnel in conjunction with the NRC report *Prudent Practices in the Laboratory: Handling and Disposal of Chemicals.* This LCSS presents a concise summary of safety information that should be adequate for most laboratory uses of the title substance, but in some cases it may be advisable to consult more comprehensive references. This information should not be used as a guide to the nonlaboratory use of this chemical.

Formaldehyde is regulated by OSHA as a carcinogen (Standard 1910.1048) and is listed in IARC Group 2A ("probable human carcinogen"). This substance is classified as a "select carcinogen" under the criteria of the OSHA Laboratory Standard. Prolonged or repeated exposure to formaldehyde can cause dermatitis and sensitization of the skin and respiratory tract. Following skin contact, a symptom-free period may occur in sensitized individuals. Subsequent exposures can then lead to itching, redness, and the formation of blisters.

Flammability and Explosibility

Formaldehyde gas is extremely flammable; formalin solution is a combustible liquid (NFPA rating = 2 for 37% formaldehyde (15% methanol), NFPA rating = 4 for 37% formaldehyde (methanol free)). Toxic vapors may be given off in a fire. Carbon dioxide or dry chemical extinguishers should be used to fight formaldehyde fires.

Reactivity and Incompatibility

Formaldehyde may react violently with strong oxidizing agents, ammonia and strong alkalis, isocyanates, peracids, anhydrides, and inorganic acids. Formaldehyde reacts with HCl to form the potent carcinogen, bis-chloromethyl ether.

Storage and Handling

Because of its carcinogenicity and flammability, formaldehyde should be handled using the "basic prudent practices" of Chapter 5.C, supplemented by the additional precautions for work with compounds of high chronic toxicity (Chapter 5.D) and extremely flammable substances (Chapter 5.F). In particular, work with formaldehyde should be conducted in a fume hood to prevent exposure by inhalation, and splash goggles and impermeable gloves should be worn at all times to prevent eye and skin contact. Formaldehyde should be used only in areas free of ignition sources. Containers of formaldehyde should be stored in secondary containers in areas separate from oxidizers and bases.

Accidents

In the event of skin contact, immediately wash with soap and water and remove contaminated clothing. In case of eye contact, promptly wash with copious amounts of water for 15 min (lifting upper and lower lids occasionally) and obtain medical attention. If formaldehyde is ingested, obtain medical attention immediately. If large amounts of this compound are inhaled, move the person to fresh air and seek medical attention at once.

In the event of a spill, remove all ignition sources, soak up the formaldehyde with a spill pillow or absorbent material, place in an appropriate container, and dispose of properly. Respiratory protection may be necessary in the event of a large spill or release in a confined area.

Disposal

Excess formaldehyde and waste material containing this substance should be placed in an appropriate container, clearly labeled, and handled according to your institution's waste disposal guidelines. For more information on disposal procedures, see Chapter 7 of this volume.

LABORATORY CHEMICAL SAFETY SUMMARY: HEXAMETHYLPHOSPHORAMIDE

Substance	Hexamethylphosphoramide (Hexamethylphosphoric triamide, HMPA, HMPT) CAS 680-31-9
Formula	$(Me_2N)_3P=O$
Physical Properties	Colorless liquid bp 233 °C, mp 6 °C Completely miscible with water
Odor	Spicy odor (no threshold data)
Vapor Density	6.2 (air = 1.0)
Vapor Pressure	0.07 mmHg at 25 °C
Flash Point	105 °C
Toxicity Data	LD_{50} oral (rat) 2525 mg/kg LD_{50} skin (rabbit) 2600 mg/kg
Major Hazards	Possible human carcinogen (OSHA "select carcinogen")
Toxicity	The acute toxicity of hexamethylphosphoramide is low. HMPA can cause irritation upon contact with the skin and eyes. Hexamethylphosphoramide has been found to cause cancer in laboratory animals exposed by inhalation and meets the criteria for classification as an OSHA "select carcinogen." Chronic exposure to HMPA can cause damage to the lungs and kidneys. Reproductive effects in male animals treated with hexamethylphosphoramide have been observed. HMPA should be regarded as a substance with poor warning properties.
Flammability and Explosibility	Combustible liquid. Its decomposition at high temperatures or in a fire can produce phosphine, phosphorus oxides, and oxides of nitrogen, which are extremely toxic. Carbon dioxide or dry chemical extinguishers should be used for HMPA fires.
Reactivity and Incompatibility	Incompatible with strong oxidizing agents and strong acids.
Storage and Handling	Because of its carcinogenicity, hexamethylphosphoramide should be handled using the "basic prudent practices" of Chapter 5.C, supplemented by the additional precautions for work with compounds of high chronic toxicity (Chapter 5.D). In particular, this compound should be handled only in a fume hood, using appropriate impermeable gloves and splash goggles to prevent skin and eye contact. Containers of this substance should be stored in secondary containers.

(continued on facing page)

The information in this LCSS has been compiled by a committee of the National Research Council from literature sources and Material Safety Data Sheets and is believed to be accurate as of July 1994. This summary is intended for use by trained laboratory personnel in conjunction with the NRC report *Prudent Practices in the Laboratory: Handling and Disposal of Chemicals*. This LCSS presents a concise summary of safety information that should be adequate for most laboratory uses of the title substance, but in some cases it may be advisable to consult more comprehensive references. This information should not be used as a guide to the nonlaboratory use of this chemical.

Accidents   In the event of skin contact, immediately wash with soap and water and remove contaminated clothing. In case of eye contact, promptly wash with copious amounts of water for 15 min (lifting upper and lower lids occasionally) and obtain medical attention. If hexamethylphosphoramide is ingested, obtain medical attention immediately. If large amounts of this compound are inhaled, move the person to fresh air and seek medical attention at once.

In the event of a spill, soak up the hexamethylphosphoramide with a spill pillow or absorbent material, place in an appropriate container, and dispose of properly. Respiratory protection may be necessary in the event of a large spill or release in a confined area.

Disposal   Excess hexamethylphosphoramide and waste material containing this substance should be placed in an appropriate container, clearly labeled, and handled according to your institution's waste disposal guidelines. For more information on disposal procedures, see Chapter 7 of this volume.

LABORATORY CHEMICAL SAFETY SUMMARY: HEXANE (AND RELATED HYDROCARBONS)

Substance	Hexane (and related aliphatic hydrocarbons) (Normal hexane, skellysolve B) CAS 110-54-3
Formula	C_6H_{14}
Physical Properties	Colorless liquid bp 69 °C, mp −95 °C Slightly soluble in water (0.014 g/100 mL)
Odor	Mild gasoline-like odor detectable at 65 to 248 ppm
Vapor Density	3.0 (air = 1.0)
Vapor Pressure	124 mmHg at 20 °C
Flash Point	−21.7 °C
Autoignition Temperature	225 °C

Toxicity Data

LD_{50} oral (rat)	28,700 mg/kg
PEL (OSHA)	500 ppm (1800 mg/m³)
TLV-TWA (ACGIH)	50 ppm

Major Hazards　　Highly flammable; chronic exposure may cause neurotoxic effects.

Toxicity

Hexane and related aliphatic hydrocarbons exhibit only slight acute toxicity by all routes of exposure. The liquid may cause irritation upon contact with skin or eyes. Hexane vapor (and the vapor of other volatile hydrocarbons) at high concentrations (>1000 ppm) is a narcotic, and inhalation may result in lightheadedness, giddiness, nausea, and headache. Ingestion of hexane or other hydrocarbons may lead to aspiration of the substance into the lungs, causing pneumonia. Prolonged skin exposure may cause irritation due to the ability of these solvents to remove fats from the skin. Hexane is regarded as a substance with good warning properties.

Chronic exposure to hexane or other aliphatic hydrocarbons may cause central nervous system toxicity. Hexane has not been found to be a carcinogen or reproductive toxin in humans.

Flammability and Explosibility

Hexane is extremely flammable (NFPA rating = 3), and its vapor can travel a considerable distance to an ignition source and "flash back." Hexane vapor forms explosive mixtures with air at concentrations of 1.1 to 7.5 % (by volume). Hydrocarbons of significantly higher molecular weight have correspondingly higher vapor pressures and therefore present a

(continued on facing page)

reduced flammability hazard. Carbon dioxide or dry chemical extinguishers should be used for hexane fires.

Reactivity and Incompatibility

Contact with strong oxidizing agents may cause explosions or fires.

Storage and Handling

Hexane and other aliphatic hydrocarbons should be handled in the laboratory using the "basic prudent practices" described in Chapter 5.C, supplemented by the additional precautions for dealing with extremely flammable substances (Chapter 5.F). In particular, hexane should be used only in areas free of ignition sources, and quantities greater than 1 liter should be stored in tightly sealed metal containers in areas separate from oxidizers.

Accidents

In the event of skin contact, immediately wash with soap and water and remove contaminated clothing. In case of eye contact, promptly wash with copious amounts of water for 15 min (lifting upper and lower lids occasionally) and obtain medical attention. If hexane is ingested, obtain medical attention immediately. If large amounts of this compound are inhaled, move the person to fresh air and seek medical attention at once.

In the event of a spill, remove all ignition sources, soak up the hexane with a spill pillow or absorbent material, place in an appropriate container, and dispose of properly. Respiratory protection may be necessary in the event of a large spill or release in a confined area.

Disposal

Excess hexane and waste material containing this substance should be placed in an appropriate container, clearly labeled, and handled according to your institution's waste disposal guidelines. For more information on disposal procedures, see Chapter 7 of this volume.

LABORATORY CHEMICAL SAFETY SUMMARY: HYDRAZINE

Substance	Hydrazine (Diamide, diamine) CAS 302-01-2
Formula	NH_2NH_2
Physical Properties	Colorless oily liquid that fumes in air bp 113.5 °C, mp 1.4 °C Miscible with water
Odor	Fishy or ammonia-like odor detectable at 3 to 4 ppm (mean = 3.7 ppm)
Vapor Density	1.04 (air = 1.0)
Vapor Pressure	14.4 mmHg at 25 °C
Flash Point	38 °C
Autoignition Temperature	24 °C on iron rust surface; 270 °C on glass surface

Toxicity Data

LD_{50} oral (rat)	60 mg/kg
LD_{50} skin (rabbit)	91 mg/kg
LC_{50} inhal (rat)	570 ppm (744 mg/m^3; 4 h)
PEL (OSHA)	1 ppm (1.3 mg/m^3)—skin
TLV-TWA (ACGIH)	0.1 ppm (0.13 mg/m^3)—skin—suspected human carcinogen
(proposed)	0.01 ppm (0.013 mg/m^3)

Major Hazards

Possible human carcinogen (OSHA "select carcinogen"); corrosive to eyes, skin, and mucous membranes; highly flammable and reactive.

Toxicity

Hydrazine is extremely destructive to the tissues of the mucous membranes and upper respiratory tract, eyes, and skin. Skin contact with the liquid can result in severe burns; hydrazine is readily absorbed through the skin, leading to systemic effects, which may include damage to the liver, kidney, nervous system, and red blood cells. Hydrazine vapor is irritating to the nose, throat, and respiratory tract, and inhalation of high concentrations may be fatal as a result of spasm, inflammation, chemical pneumonitis, and pulmonary edema. Symptoms of exposure may include a burning sensation, coughing, wheezing, laryngitis, shortness of breath, headache, nausea, and vomiting. Hydrazine vapor is extremely irritating to the eyes and can cause temporary blindness. Eye contact with the liquid can result in severe burns and permanent damage. Hydrazine is not considered to have adequate warning properties.

Hydrazine is listed by IARC in Group 2B "possible human carcinogen" and is classified as a "select carcinogen" according to the criteria of the OSHA Laboratory Standard.

(continued on facing page)

Chronic exposure to subacute levels of hydrazine can cause lethargy, vomiting, tremors, itching and burning of the eyes and skin, conjunctivitis, and contact dermatitis. Hydrazine has been found to exhibit reproductive and developmental toxicity in animal tests.

Flammability and Explosibility

Hydrazine is a flammable liquid (NFPA rating = 3) over a very broad range of vapor concentrations (4.7 to 100%). Hydrazine may undergo autoxidation and ignite spontaneously when brought in contact with porous substances such as rusty surfaces, earth, wood, or cloth. Fires should be extinguished with water spray, carbon dioxide, or dry chemical extinguishers.

Reactivity and Incompatibility

Hydrazine is a highly reactive reducing agent that forms shock-sensitive, explosive mixtures with many compounds. It explodes on contact with barium oxide, calcium oxide, chromate salts, and many other substances. On contact with metal catalysts (platinum black, Raney nickel, etc.), hydrazine decomposes to ammonia, hydrogen, and nitrogen gases, which may ignite or explode.

Storage and Handling

Because of its carcinogenicity, reactivity, and flammability, hydrazine should be handled using the "basic prudent practices" of Chapter 5.C, supplemented by the additional precautions for work with compounds of high chronic toxicity (Chapter 5.D), flammability (Chapter 5.F), and reactivity (Chapter 5.G). In particular, work with hydrazine should be conducted in a fume hood to prevent exposure by inhalation, and splash goggles and impermeable gloves should be worn at all times to prevent eye and skin contact. Hydrazine should be used only in areas free of ignition sources. Hydrazine should be stored under nitrogen in containers placed in secondary containers in areas separate from oxidizers and acids.

Accidents

In the event of skin contact, immediately wash with soap and water and remove contaminated clothing. In case of eye contact, promptly wash with copious amounts of water for 15 min (lifting upper and lower lids occasionally) and obtain medical attention. If hydrazine is ingested, obtain medical attention immediately. If significant quantities of this compound are inhaled, move the person to fresh air and seek medical attention at once.

In the event of a spill, remove all ignition sources, soak up the hydrazine with a spill pillow or absorbent material, place in an appropriate container, and dispose of properly. Evacuation and cleanup using respiratory protection may be necessary in the event of a large spill or release in a confined area.

Disposal

Excess hydrazine and waste material containing this substance should be placed in an appropriate container, clearly labeled, and handled according to your institution's waste disposal guidelines. For more information on disposal procedures, see Chapter 7 of this volume.

LABORATORY CHEMICAL SAFETY SUMMARY: HYDROBROMIC ACID AND HYDROGEN BROMIDE

Substance	Hydrobromic acid CAS 10035-10-6	Hydrogen bromide CAS 10035-10-6
Formula	Reagent grade conc HBr contains 48 wt % HBr in water	HBr
Physical Properties	bp 126 °C, mp −11 °C Miscible with water	bp −67° C, mp −87 °C Miscible with water
Odor	Sharp, irritating odor detectable at 2 ppm	Sharp, irritating odor detectable at 2 ppm
Vapor Density		2.71 (air = 1.0)

Toxicity Data

LD$_{50}$ oral (rabbit)	900 mg/kg
LC$_{50}$ inhal (rat)	2858 ppm/1 h
PEL (OSHA)	3 ppm (10 mg/m^3)
TLV (ACGIH)	3 ppm (10 mg/m^3; ceiling)

Major Hazards Highly corrosive; causes severe burns on eye and skin contact and upon inhalation of gas. .

Toxicity Hydrobromic acid and hydrogen bromide gas are highly corrosive substances that can cause severe burns upon contact with all body tissues. The aqueous acid and gas are strong eye irritants and lacrimators. Contact of concentrated hydrobromic acid or concentrated HBr vapor with the eyes may cause severe injury, resulting in permanent impairment of vision and possible blindness. Skin contact with the acid or HBr gas can produce severe burns. Ingestion can lead to severe burns of the mouth, throat, and gastrointestinal system and can be fatal. Inhalation of hydrogen bromide gas can cause extreme irritation and injury to the upper respiratory tract and lungs, and exposure to high concentrations may cause death. HBr gas is regarded as having adequate warning properties.

Hydrogen bromide has not been found to be carcinogenic or to show reproductive or developmental toxicity in humans.

Flammability and Explosibility Noncombustible, but contact with metals may produce highly flammable hydrogen gas.

Reactivity and Incompatibility Hydrobromic acid and hydrogen bromide react violently with many metals with the generation of highly flammable hydrogen gas, which may explode. Reaction with oxidizers such as permanganates, chlorates, chlorites, and hypochlorites may produce chlorine or bromine.

Storage and Handling Hydrobromic acid should be handled in the laboratory using the "basic prudent practices" described in Chapter 5.C. Splash goggles and rubber gloves should be worn when handling this acid, and containers of HBr should be stored in a well-ventilated location separated

(continued on facing page)

The information in this LCSS has been compiled by a committee of the National Research Council from literature sources and Material Safety Data Sheets and is believed to be accurate as of July 1994. This summary is intended for use by trained laboratory personnel in conjunction with the NRC report *Prudent Practices in the Laboratory: Handling and Disposal of Chemicals*. This LCSS presents a concise summary of safety information that should be adequate for most laboratory uses of the title substance, but in some cases it may be advisable to consult more comprehensive references. This information should not be used as a guide to the nonlaboratory use of this chemical.

from incompatible metals. Water should never be added to HBr because splattering may result; always add acid to water. Containers of hydrobromic acid should be stored in secondary plastic trays to avoid corrosion of metal storage shelves due to drips or spills.

Hydrogen bromide gas should be handled in the laboratory using the "basic prudent practices" described in Chapter 5.C, supplemented by the procedures described in Chapter 5.H for the handling of compressed gases. Cylinders of hydrogen bromide should be stored in cool, dry locations, separated from alkali metals and other incompatible substances.

Accidents

In the event of skin contact, remove contaminated clothing and immediately wash with flowing water for at least 15 min. In case of eye contact, immediately wash with copious amounts of water for at least 15 min while holding the eyelids open. Seek medical attention. In case of ingestion, do not induce vomiting. Give large amounts of water or milk if available and transport to medical facility. In case of inhalation, remove to fresh air and seek medical attention.

Carefully neutralize spills of hydrobromic acid with a suitable agent such as powdered sodium bicarbonate, further dilute with absorbent material, place in an appropriate container, and dispose of properly. Dilution with water before applying the solid adsorbent may be an effective means of reducing exposure to hydrogen bromide vapor. Respiratory protection may be necessary in the event of a large spill or release in a confined area.

Leaks of HBr gas are evident from the formation of dense white fumes on contact with the atmosphere. Small leaks can be detected by holding an open container of concentrated ammonium hydroxide near the site of the suspected leak; dense white fumes confirm a leak is present. In case of the accidental release of hydrogen bromide gas, such as from a leaking cylinder or associated apparatus, evacuate the area and eliminate the source of the leak if this can be done safely. Remove cylinder to a fume hood or remote area if it cannot be shut off. Full respiratory protection and protective clothing may be required to deal with a hydrogen bromide release.

Disposal

In many localities, hydrobromic acid or the residue from a spill may be disposed of down the drain after appropriate dilution and neutralization. Otherwise, hydrobromic acid and waste material containing this substance should be placed in an appropriate container, clearly labeled, and handled according to your institution's waste disposal guidelines. Excess hydrogen bromide in cylinders should be returned to the manufacturer. For more information on disposal procedures, see Chapter 7 of this volume.

LABORATORY CHEMICAL SAFETY SUMMARY: HYDROCHLORIC ACID AND HYDROGEN CHLORIDE

Substance	Hydrochloric acid (Muriatic acid) CAS 7647-01-0	Hydrogen chloride CAS 7647-01-0
Formula	Reagent grade conc HCl contains 37 wt % HCl in water; constant-boiling acid (an azeotrope with water) contains ~20% HCl	HCl
Physical Properties	Concentrated acid evolves HCl at 60 °C leading to the formation of an azeotrope of constant composition (20% HCl) bp 110 °C, mp −24 °C Miscible with water	bp −85 °C, mp −114 °C Miscible with water
Odor	Sharp, irritating odor detectable at 0.25 to 10 ppm	Sharp, irritating odor detectable at 0.25 to 10 ppm
Vapor Density		1.27 (air = 1.0)

Toxicity Data

LD_{50} oral (rabbit)	900 mg/kg
LC_{50} inhal (rat)	3124 ppm (1 h)
PEL (OSHA)	5 ppm (7 mg/m^3; ceiling)
TLV (ACGIH)	5 ppm (7.5 mg/m^3; ceiling)

Major Hazards

Highly corrosive; causes severe burns on eye and skin contact and upon inhalation of gas.

Toxicity

Hydrochloric acid and hydrogen chloride gas are highly corrosive substances that may cause severe burns upon contact with any body tissue. The aqueous acid and gas are strong eye irritants and lacrimators. Contact of conc hydrochloric acid or concentrated HCl vapor with the eyes may cause severe injury, resulting in permanent impairment of vision and possible blindness, and skin contact results in severe burns. Ingestion can cause severe burns of the mouth, throat, and gastrointestinal system and can be fatal. Inhalation of hydrogen chloride gas can cause severe irritation and injury to the upper respiratory tract and lungs, and exposure to high concentrations may cause death. HCl gas is regarded as having adequate warning properties.

Hydrogen chloride has not been found to be carcinogenic or to show reproductive or developmental toxicity in humans.

Flammability and Explosibility

Noncombustible, but contact with metals may produce highly flammable hydrogen gas.

(continued on facing page)

Reactivity and Incompatibility

Hydrochloric acid and hydrogen chloride react violently with many metals, with the generation of highly flammable hydrogen gas, which may explode. Reaction with oxidizers such as permanganates, chlorates, chlorites, and hypochlorites may produce chlorine or bromine.

Storage and Handling

Hydrochloric acid should be handled in the laboratory using the "basic prudent practices" described in Chapter 5.C. Splash goggles and rubber gloves should be worn when handling this acid, and containers of HCl should be stored in a well-ventilated location separated from incompatible metals. Water should never be added to HCl because splattering may result; always add acid to water. Containers of hydrochloric acid should be stored in secondary plastic trays to avoid corrosion of metal storage shelves due to drips or spills.

Hydrogen chloride gas should be handled in the laboratory using the "basic prudent practices" described in Chapter 5.C, supplemented by the procedures described in Chapter 5.H for the handling of compressed gases. Cylinders of hydrogen chloride should be stored in cool, dry locations separated from alkali metals and other incompatible substances.

Accidents

In the event of skin contact, remove contaminated clothing and immediately wash with flowing water for at least 15 min. In case of eye contact, immediately wash with copious amounts of water for at least 15 min while holding the eyelids open. Seek medical attention. In case of ingestion, do not induce vomiting. Give large amounts of water or milk if available and transport to medical facility. In case of inhalation, remove to fresh air and seek medical attention.

Carefully neutralize spills of hydrochloric acid with a suitable agent such as powdered sodium bicarbonate, further dilute with absorbent material, place in an appropriate container, and dispose of properly. Dilution with water before applying the solid adsorbent may be an effective means of reducing exposure to hydrogen chloride vapor. Respiratory protection may be necessary in the event of a large spill or release in a confined area.

Leaks of HCl gas are evident from the formation of dense white fumes on contact with the atmosphere. Small leaks can be detected by holding an open container of concentrated ammonium hydroxide near the site of the suspected leak; dense white fumes confirm that a leak is present. In case of accidental release of hydrogen chloride gas, such as from a leaking cylinder or associated apparatus, evacuate the area and eliminate the source of the leak if this can be done safely. Remove cylinder to a fume hood or remote area if it cannot be shut off. Full respiratory protection and protective clothing may be required to deal with a hydrogen chloride release.

Disposal

In many localities, hydrochloric acid or the residue from a spill may be disposed of down the drain after appropriate dilution and neutralization. Otherwise, hydrochloric acid and waste material containing this substance should be placed in an appropriate container, clearly labeled, and handled according to your institution's waste disposal guidelines. Excess hydrogen chloride in cylinders should be returned to the manufacturer. For more information on disposal procedures, see Chapter 7 of this volume.

LABORATORY CHEMICAL SAFETY SUMMARY: HYDROGEN

Substance	Hydrogen (Water gas) CAS 1333-74-0
Formula	H_2
Physical Properties	Colorless gas bp -252.8 °C, mp -259.2 °C Slightly soluble in water (0.17 mg/100 mL)
Odor	Odorless gas
Vapor Density	0.069 (air = 1.0)
Vapor Pressure	Critical temperature is -239.9 °C; noncondensible above this temperature
Autoignition Temperature	500 to 590 °C
Toxicity Data	TLV-TWA (ACGIH) None established; simple asphyxiant
Major Hazards	Highly flammable gas; explosion hazard in the presence of heat, flame, or oxidizing agents.
Toxicity	Hydrogen is practically nontoxic. In high concentrations this gas is a simple asphyxiant, and ultimate loss of consciousness may occur when oxygen concentrations fall below 18%. Skin contact with liquid hydrogen can cause frostbite.
Flammability and Explosibility	Hydrogen is a highly flammable gas that burns with an almost invisible flame and low heat radiation. Hydrogen forms explosive mixtures with air from 4 to 75% by volume. These explosive mixtures of hydrogen with air (or oxygen) can be ignited by a number of finely divided metals (such as common hydrogenation catalysts). In the event of fire, shut off the flow of gas and extinguish with carbon dioxide, dry chemical, or halon extinguishers. Warming of liquid hydrogen contained in an enclosed vessel to above its critical temperature can cause bursting of that container.
Reactivity and Incompatibility	Hydrogen is a reducing agent and reacts explosively with strong oxidizers such as halogens (fluorine, chlorine, bromine, iodine) and interhalogen compounds.
Storage and Handling	Because of its flammable and gaseous nature, hydrogen should be handled using the "basic prudent practices" of Chapter 5.C, supplemented by the additional precautions for work with flammable compounds (Chapter 5.F) and for work at high pressure (Chapter 5.H). In particular, hydrogen cylinders should be clamped or otherwise supported in place and used only in areas free of ignition sources and separate from oxidizers. Expansion of hydrogen released rapidly from a compressed cylinder will cause evolution of heat due to its negative Joule-Thompson coefficient.

(continued on facing page)

The information in this LCSS has been compiled by a committee of the National Research Council from literature sources and Material Safety Data Sheets and is believed to be accurate as of July 1994. This summary is intended for use by trained laboratory personnel in conjunction with the NRC report *Prudent Practices in the Laboratory: Handling and Disposal of Chemicals*. This LCSS presents a concise summary of safety information that should be adequate for most laboratory uses of the title substance, but in some cases it may be advisable to consult more comprehensive references. This information should not be used as a guide to the nonlaboratory use of this chemical.

Accidents   If large amounts of hydrogen are inhaled, move the person to fresh air and seek medical attention at once.

In the event of a leak, remove all ignition sources and allow the hydrogen to disperse with increased ventilation. Hydrogen disperses rapidly in normal open environments. Respiratory protection may be necessary in the event of a release in a confined area.

Disposal   Excess hydrogen cylinders should be returned to the vendor. Excess hydrogen gas present over reaction mixtures should be carefully vented to the atmosphere under conditions of good ventilation after all ignition sources have been removed. For more information on disposal procedures, see Chapter 7 of this volume.

LABORATORY CHEMICAL SAFETY SUMMARY: HYDROGEN CYANIDE

Substance	Hydrogen cyanide (Hydrocyanic acid; prussic acid) CAS 74-90-8
Formula	HCN
Physical Properties	Colorless or pale blue liquid or gas bp 26 °C, mp −13 °C Miscible in water in all proportions
Odor	Bitter almond odor detectable at 1 to 5 ppm; however, 20 to 60% of the population are reported to be unable to detect the odor of HCN
Vapor Pressure	750 mmHg at 25 °C
Flash Point	−18 °C
Autoignition Temperature	538 °C

Toxicity Data

Approx LD oral (rat)	10 mg/kg
Approx LD skin (rabbit)	~1500 mg/kg
LC_{50} inhal (rat)	63 ppm (40 min)
PEL (OSHA)	10 ppm (11 mg/m^3)—skin
TLV-TWA (ACGIH)	Ceiling 10 ppm (11 mg/m^3)—skin

Major Hazards High acute toxicity; inhalation, ingestion, or skin contact may be rapidly fatal.

Toxicity The acute toxicity of hydrogen cyanide is high, and exposure by inhalation, ingestion, or eye or skin contact can be rapidly fatal. Symptoms observed at low levels of exposure (e.g., inhalation of 18 to 36 ppm for several hours) include weakness, headache, confusion, nausea, and vomiting. Inhalation of 270 ppm can cause immediate death, and 100 to 200 ppm can be fatal in 30 to 60 min. Aqueous solutions of HCN are readily absorbed through the skin and eyes, and absorption of 50 mg can be fatal. In humans, ingestion of 50 to 100 mg of HCN can be fatal. Because there is wide variation in the ability of different individuals to detect the odor of HCN, this substance is regarded as having poor warning properties.

Effects of chronic exposure to hydrogen cyanide are nonspecific and rare.

Flammability and Explosibility Hydrogen cyanide is a highly flammable liquid. Liquid HCN contains a stabilizer (usually phosphoric acid), and old samples may explode if the acid stabilizer is not maintained at a sufficient concentration.

(continued on facing page)

The information in this LCSS has been compiled by a committee of the National Research Council from literature sources and Material Safety Data Sheets and is believed to be accurate as of July 1994. This summary is intended for use by trained laboratory personnel in conjunction with the NRC report *Prudent Practices in the Laboratory: Handling and Disposal of Chemicals.* This LCSS presents a concise summary of safety information that should be adequate for most laboratory uses of the title substance, but in some cases it may be advisable to consult more comprehensive references. This information should not be used as a guide to the nonlaboratory use of this chemical.

Reactivity and Incompatibility

HCN can polymerize explosively if heated above 50 °C or in the presence of trace amounts of alkali.

Storage and Handling

Because of its high acute toxicity, hydrogen cyanide should be handled using the "basic prudent practices" of Chapter 5.C, supplemented by the additional precautions for work with compounds of high toxicity (Chapter 5.D) and flammability (Chapter 5.F). In particular, work with HCN should be conducted in a fume hood to prevent exposure by inhalation, and splash goggles and impermeable gloves should be worn at all times to prevent eye and skin contact. Never work alone with hydrogen cyanide. HCN should be used only in areas free of ignition sources. Containers of HCN should be protected from physical damage and stored in areas separate from ignition sources and other materials. Hydrogen cyanide should not be stored for extended periods (>90 days) unless testing confirms product quality.

Accidents

In the event of skin contact, immediately wash with soap and water and remove contaminated clothing. In case of eye contact, promptly wash with copious amounts of water for 15 min (lifting upper and lower lids occasionally) and obtain medical attention. If hydrogen cyanide is ingested, obtain medical attention immediately. If HCN is inhaled, move the person to fresh air and seek medical attention at once. Specific medical procedures for treating cyanide exposure are available but usually must be administered by properly trained personnel. Consult your environmental safety office or its equivalent before beginning work with hydrogen cyanide.

In the event of a spill, remove all ignition sources. Cleanup should be conducted wearing appropriate chemical-resistant clothing and respiratory protection.

Disposal

Excess hydrogen cyanide and waste material containing this substance should be placed in an appropriate container, clearly labeled, and handled according to your institution's waste disposal guidelines. For more information on disposal procedures, see Chapter 7 of this volume.

LABORATORY CHEMICAL SAFETY SUMMARY: HYDROGEN FLUORIDE AND HYDROFLUORIC ACID

Substance	Hydrogen fluoride and hydrofluoric acid CAS 7664-39-3
Formula	HF
Physical Properties	Colorless, clear, fuming liquid Anhydrous HF: bp 20 °C, mp -83 °C Miscible with water
Odor	Acrid, irritating odor
Vapor Pressure	Anhydrous HF: 775 mmHg at 20 °C Hydrofluoric acid: 14 mmHg at 20 °C
Flash Point	Noncombustible

Toxicity Data

LC_{LO} inhal (humans)	50 ppm (0.5 h)
PEL (OSHA)	3 ppm (as fluoride)
TLV-TWA (ACGIH)	3 ppm (2.6 mg/m^3; ceiling as fluoride)

Major Hazards

Extremely corrosive liquid and vapor that can cause severe injury via skin and eye contact, inhalation, or ingestion.

Toxicity

Anhydrous hydrogen fluoride and hydrofluoric acid are extremely corrosive to all tissues of the body. Skin contact results in painful deep-seated burns that are slow to heal. Burns from dilute (<50%) HF solutions do not usually become apparent until several hours after exposure; more concentrated solutions and anhydrous HF cause immediate painful burns and tissue destruction. HF burns pose unique dangers distinct from other acids such as HCl and H_2SO_4: undissociated HF readily penetrates the skin, damaging underlying tissue; fluoride ion can then cause destruction of soft tissues and decalcification of the bones. Hydrofluoric acid and HF vapor can cause severe burns to the eyes, which may lead to permanent damage and blindness. At 10 to 15 ppm, HF vapor is irritating to the eyes, skin, and respiratory tract. Exposure to higher concentrations can result in serious damage to the lungs, and fatal pulmonary edema may develop after a delay of several hours. Brief exposure (5 min) to 50 to 250 ppm may be fatal to humans. Ingestion of HF can produce severe injury to the mouth, throat, and gastrointestinal tract and may be fatal.

HF has not been reported to be a human carcinogen. No acceptable animal test reports are available to define the developmental or reproductive toxicity of this substance.

Flammability and Explosibility

Hydrogen fluoride is not a combustible substance.

(continued on facing page)

The information in this LCSS has been compiled by a committee of the National Research Council from literature sources and Material Safety Data Sheets and is believed to be accurate as of July 1994. This summary is intended for use by trained laboratory personnel in conjunction with the NRC report *Prudent Practices in the Laboratory: Handling and Disposal of Chemicals*. This LCSS presents a concise summary of safety information that should be adequate for most laboratory uses of the title substance, but in some cases it may be advisable to consult more comprehensive references. This information should not be used as a guide to the nonlaboratory use of this chemical.

Reactivity and Incompatibility

HF reacts with glass, ceramics, and some metals. Reactions with metals may generate potentially explosive hydrogen gas.

Storage and Handling

Dilute solutions of hydrofluoric acid (<50%) should be handled using the "basic prudent practices" of Chapter 5.C. Because of its corrosivity and high acute toxicity, anhydrous hydrogen fluoride and concentrated solutions of HF should be handled using the "basic prudent practices" of Chapter 5.C, supplemented by the additional precautions for work with compounds of high toxicity (Chapter 5.D). All work with HF should be conducted in a fume hood to prevent exposure by inhalation, and splash goggles and neoprene gloves should be worn at all times to prevent eye and skin contact. Containers of HF should be stored in secondary containers made of polyethylene in areas separate from incompatible materials. Work with anhydrous HF should be undertaken using special equipment and only by well-trained personnel familiar with first aid procedures.

Accidents

Laboratory personnel should be familiar with first aid procedures before beginning work with HF; calcium gluconate gel should be readily accessible in areas where HF exposure potential exists.

First aid must be started within seconds in the event of contact of any form. In the event of skin contact, immediately wash with water for 15 min and remove contaminated clothing. If available, apply calcium gluconate gel. Obtain medical attention at once, and inform attending physician that injury involves HF rather than other acid. In case of eye contact, promptly wash with copious amounts of water for 5 min while holding the eyelids apart and seek medical attention at once. If HF is ingested, obtain medical attention immediately. If HF vapor is inhaled, move the person to fresh air and seek medical attention at once.

In the event of a spill of dilute hydrofluoric acid, soak up the acid with an HF-compatible spill pillow or neutralize with lime, transfer material to a polyethylene container, and dispose of properly. Respiratory protection may be necessary in the event of a large spill or release in a confined area. Releases of anhydrous HF require specially trained personnel.

Disposal

Excess hydrogen fluoride and waste material containing this substance should be placed in an appropriate container, clearly labeled, and handled according to your institution's waste disposal guidelines. For more information on disposal procedures, see Chapter 7 of this volume.

Substance	Hydrogen peroxide (Hydrogen dioxide) CAS 7722-84-1
Formula	HOOH
Physical Properties	Colorless liquid bp 150 °C, mp −0.4 °C Miscible in all proportions in water
Odor	Slightly pungent, irritating odor
Vapor Density	1.15 (air = 1.0)
Vapor Pressure	1 mm Hg at 15.3 °C 5 mm Hg at 30 °C
Flash Point	Noncombustible
Autoignition Temperature	None

Toxicity Data

LD_{50} oral (rat)	75 mg/kg (70%)
LD_{50} skin (rabbit)	700 mg/kg (90%)
LD_{50} skin (rabbit)	9200 mg/kg (70%)
LC_{50} inhal (rat)	>2000 ppm (90%)
PEL (OSHA)	1 ppm (1.4 mg/m^3) (90%)
TLV-TWA (ACGIH)	1 ppm (1.4 mg/m^3) (90%)

Major Hazards

Contact with certain metals and organic compounds can lead to fires and explosions; concentrated solutions can cause severe irritation or burns of the skin, eyes, and mucous membranes.

Toxicity

Contact with aqueous concentrations of less than 50% cause skin irritation, but more concentrated solutions of H_2O_2 are corrosive to the skin. At greater than 10% concentration, hydrogen peroxide is corrosive to the eyes and can cause severe irreversible damage and possibly blindness. Hydrogen peroxide is moderately toxic by ingestion and slightly toxic by inhalation. This substance is not considered to have adequate warning properties.

Hydrogen peroxide has not been found to be carcinogenic in humans. Repeated inhalation exposures produced nasal discharge, bleached hair, and respiratory tract congestion, with some deaths occurring in rats and mice exposed to concentrations greater than 67 ppm.

Flammability and Explosibility

Hydrogen peroxide is not flammable, but concentrated solutions may undergo violent decomposition in the presence of trace impurities or upon heating.

(continued on facing page)

The information in this LCSS has been compiled by a committee of the National Research Council from literature sources and Material Safety Data Sheets and is believed to be accurate as of July 1994. This summary is intended for use by trained laboratory personnel in conjunction with the NRC report *Prudent Practices in the Laboratory: Handling and Disposal of Chemicals*. This LCSS presents a concise summary of safety information that should be adequate for most laboratory uses of the title substance, but in some cases it may be advisable to consult more comprehensive references. This information should not be used as a guide to the nonlaboratory use of this chemical.

Reactivity and Incompatibility

Contact with many organic compounds can lead to immediate fires or violent explosions (consult Bretherick for references and examples). Hydrogen peroxide reacts with certain organic functional groups (ethers, acetals, etc.) to form peroxides, which may explode upon concentration. Reaction with acetone generates explosive cyclic dimeric and trimeric peroxides. Explosions may also occur on exposure of hydrogen peroxide to metals such as sodium, potassium, magnesium, copper, iron, and nickel.

Storage and Handling

Hydrogen peroxide should be handled in the laboratory using the "basic prudent practices" described in Chapter 5.C, supplemented by the procedures for work with reactive and explosive substances (Chapter 5.G). Use extreme care when carrying out reactions with hydrogen peroxide because of the fire and explosion potential (immediate or delayed). The use of safety shields is advisable, and is essential for experiments involving concentrated (>50%) solutions of hydrogen peroxide. Sealed containers of hydrogen peroxide can build up dangerous pressures of oxygen, owing to slow decomposition.

Accidents

In the event of skin contact, immediately wash with soap and water and remove contaminated clothing. In case of eye contact, promptly wash with copious amounts of water for 15 min (lifting upper and lower lids occasionally) and obtain medical attention. If hydrogen peroxide is ingested, obtain medical attention immediately. If large amounts of this compound are inhaled, move the person to fresh air and seek medical attention at once.

In the event of a spill, remove all ignition sources, soak up the hydrogen peroxide with a spill pillow or absorbent material, place in an appropriate container, and dispose of properly. Respiratory protection may be necessary in the event of a large spill or release in a confined area.

Disposal

Excess hydrogen peroxide and waste material containing this substance should be placed in an appropriate container, clearly labeled, and handled according to your institution's waste disposal guidelines. For more information on disposal procedures, see Chapter 7 of this volume.

LABORATORY CHEMICAL SAFETY SUMMARY: HYDROGEN SULFIDE

Substance	Hydrogen sulfide (Hydrosulfuric acid, sulfur hydride) CAS 7783-06-4
Formula	H_2S
Physical Properties	Colorless gas bp -61 °C, mp -83 °C Slightly soluble in water (2.9 g/100 mL at 20 °C)
Odor	Strong rotten egg odor detectable at 0.001 to 0.1 ppm (mean = 0.0094 ppm); olfactory fatigue occurs quickly at high concentrations
Vapor Density	1.189 (air = 1.0)
Vapor Pressure	20 atm at 25 °C
Flash Point	< -82.4 °C
Autoignition Temperature	260 °C

Toxicity Data		
	LC_{50} inhal (rat)	444 ppm (580 mg/m³)
	LC_{LO} inhal (human)	800 ppm (1110 mg/m³; 5 min)
	PEL (OSHA)	20 ppm (ceiling) (28 mg/m³)
	TLV-TWA (ACGIH)	10 ppm (14 mg/m³)
	STEL (ACGIH)	15 ppm (21 mg/m³)

Major Hazards Moderately toxic gas; inhalation of large concentrations can cause unconsciousness, respiratory paralysis, and death; highly flammable.

Toxicity The acute toxicity of hydrogen sulfide by inhalation is moderate. A 5-min exposure to 800 ppm has resulted in death. Inhalation of 1000 to 2000 ppm may cause coma after a single breath. Exposure to lower concentrations may cause headache, dizziness, and upset stomach. Low concentrations of H_2S (20 to 150 ppm) can cause eye irritation, which may be delayed in onset. Although the odor of hydrogen sulfide is detectable at very low concentrations, it rapidly causes olfactory fatigue at higher levels, and therefore is not considered to have adequate warning properties.

Hydrogen sulfide has not been shown to be carcinogenic or to have reproductive or developmental effects in humans.

Flammability and Explosibility Hydrogen sulfide is flammable in air in the range of 4.3 to 45.5% (NFPA rating = 4). Combustion products (sulfur oxides) are also toxic by inhalation. In the event of a hydrogen

(continued on facing page)

sulfide fire, stop the flow of gas if possible without risk of harmful exposure and let the fire burn itself out.

Reactivity and Incompatibility

Hydrogen sulfide is incompatible with strong oxidizers. It will attack many metals, forming sulfides. Liquid hydrogen sulfide will attack some forms of plastics, rubber, and coatings. H_2S reacts violently with a variety of metal oxides, including the oxides of chromium, mercury, silver, lead, nickel, and iron.

Storage and Handling

Because of its toxic, flammable, and gaseous nature, hydrogen sulfide should be handled using the "basic prudent practices" of Chapter 5.C, supplemented by the additional precautions for work with flammable compounds (Chapter 5.F) and for work at high pressure (Chapter 5.H). In particular, cylinders of hydrogen sulfide should be stored and used in a continuously ventilated gas cabinet or fume hood. Local fire codes should be reviewed for limitations on quantity and storage requirements.

Accidents

In the event of a release of hydrogen sulfide, the area should be evacuated immediately. Use appropriate respiratory protection to rescue an affected individual. Remove exposed individual to an uncontaminated area, and seek immediate emergency help. Keep victim warm, quiet, and at rest; provide assisted respiration if breathing has stopped.

In the event of skin contact, immediately wash with soap and water and remove contaminated clothing. In case of eye contact, promptly wash with copious amounts of water for 15 min (lifting upper and lower lids occasionally) and obtain medical attention.

To respond to a release, use appropriate protective equipment and clothing. Positive pressure air-supplied respiratory protection is required. Close cylinder valve and ventilate area. Remove cylinder to a fume hood or remote area if it cannot be shut off.

Disposal

Excess hydrogen sulfide should be returned to the manufacturer, according to your institution's waste disposal guidelines. For more information on disposal procedures, see Chapter 7 of this volume.

LABORATORY CHEMICAL SAFETY SUMMARY: IODINE

Substance	Iodine CAS 7553-56-2
Formula	I_2
Physical Properties	Blue-violet to black crystalline solid bp 185 °C, mp 114 °C Slightly soluble in water (0.03 g/100 mL at 20 °C)
Odor	Sharp, characteristic odor
Vapor Density	8.8 (air = 1.0)
Vapor Pressure	0.3 mmHg at 20 °C
Flash Point	Noncombustible

Toxicity Data

LD_{50} oral (rat)	14,000 mg/kg
LC_{LO} inhal (rat)	80 ppm (800 mg/m^3; 1 h)
PEL (OSHA)	0.1 ppm (ceiling, 1 mg/m^3)
TLV-TWA (ACGIH)	0.1 ppm (ceiling, 1 mg/m^3)

Major Hazards Iodine vapor is highly toxic and is a severe irritant to the eyes and respiratory tract.

Toxicity The acute toxicity of iodine by inhalation is high. Exposure may cause severe breathing difficulties, which may be delayed in onset; headache, tightness of the chest, and congestion of the lungs may also result. In an experimental investigation, four human subjects tolerated 0.57 ppm iodine vapor for 5 min without eye irritation, but all experienced eye irritation in 2 min at 1.63 ppm. Iodine in crystalline form or in concentrated solutions is a severe skin irritant; it is not easily removed from the skin, and the lesions resemble thermal burns. Iodine is more toxic by the oral route in humans than in experimental animals; ingestion of 2 to 3 g of the solid may be fatal in humans.

Iodine has not been found to be carcinogenic or to show reproductive or developmental toxicity in humans. Chronic absorption of iodine may cause insomnia, inflammation of the eyes and nose, bronchitis, tremor, rapid heartbeat, diarrhea, and weight loss.

Flammability and Explosibility Iodine is noncombustible and in itself represents a negligible fire hazard when exposed to heat or flame. However, when heated, it will increase the burning rate of combustible materials.

Reactivity and Incompatibility Iodine is stable under normal temperatures and pressures.

Iodine may react violently with acetylene, ammonia, acetaldehyde, formaldehyde, acrylonitrile, powdered antimony, tetraamine copper(II) sulfate, and liquid chlorine. Iodine

(continued on facing page)

can form sensitive, explosive mixtures with potassium, sodium, and oxygen difluoride; ammonium hydroxide reacts with iodine to produce nitrogen triiodide, which detonates on drying.

Storage and Handling

Iodine should be handled in the laboratory using the "basic prudent practices" described in Chapter 5.C. In particular, safety goggles and rubber gloves should be worn when handling iodine, and operations involving large quantities should be conducted in a fume hood to prevent exposure to iodine vapor or dusts by inhalation.

Accidents

In the event of skin contact, immediately wash with soap and water and remove contaminated clothing. In case of eye contact, promptly wash with copious amounts of water for 15 min (lifting upper and lower lids occasionally) and obtain medical attention. If iodine is ingested, obtain medical attention immediately. If large amounts of this compound are inhaled, move the person to fresh air and seek medical attention at once.

In the event of a spill, sweep up solid iodine, soak up liquid spills with absorbent material, place in an appropriate container, and dispose of properly. Respiratory protection may be necessary in the event of a large spill or release in a confined area.

Disposal

Excess iodine and waste material containing this substance should be placed in an appropriate container, clearly labeled, and handled according to your institution's waste disposal guidelines. For more information on disposal procedures, see Chapter 7 of this volume.

The information in this LCSS has been compiled by a committee of the National Research Council from literature sources and Material Safety Data Sheets and is believed to be accurate as of July 1994. This summary is intended for use by trained laboratory personnel in conjunction with the NRC report *Prudent Practices in the Laboratory: Handling and Disposal of Chemicals*. This LCSS presents a concise summary of safety information that should be adequate for most laboratory uses of the title substance, but in some cases it may be advisable to consult more comprehensive references. This information should not be used as a guide to the nonlaboratory use of this chemical.

LABORATORY CHEMICAL SAFETY SUMMARY: LEAD AND ITS INORGANIC COMPOUNDS

Substance	Lead and its inorganic compounds CAS 7439-92-1
Formula	Pb
Physical Properties	Bluish-white, silvery, or gray solid bp 1740 °C, mp 327 °C Insoluble (metal; solubility of lead salts varies)
Odor	Odorless

Toxicity Data

LD_{LO} oral (pigeon)	160 mg/kg
PEL (OSHA)	0.05 mg/m^3
PEL (action level)	0.03 mg/m^3
TLV-TWA (ACGIH)	0.05 mg/m^3
(PEL and TLV apply to lead and inorganic lead compounds)	

Major Hazards

Chronic toxin affecting the kidneys and central and peripheral nervous systems; reproductive and developmental toxin.

Toxicity

The acute toxicity of lead and inorganic lead compounds is moderate to low. Symptoms of exposure include decreased appetite, insomnia, headache, muscle and joint pain, colic, and constipation. Inorganic lead compounds are not significantly absorbed through the skin.

Chronic exposure to inorganic lead via inhalation or ingestion can result in damage to the peripheral and central nervous system, anemia, and chronic kidney disease. Lead can accumulate in the soft tissues and bones, with the highest accumulation in the liver and kidneys, and elimination is slow. Lead has shown developmental and reproductive toxicity in both male and female animals and humans. Lead is listed by IARC in Group 2B ("possible human carcinogen") and by NTP as "reasonably anticipated to be a carcinogen," but is not considered to be a "select carcinogen" under the criteria of the OSHA Laboratory Standard.

Flammability and Explosibility

Lead powder is combustible when exposed to heat or flame.

Reactivity and Incompatibility

Violent reactions of lead with sodium azide, zirconium, sodium acetylide, and chlorine trifluoride have been reported. Reactivity of lead compounds varies depending on structure.

(continued on facing page)

Storage and Handling

Lead should be handled in the laboratory using the "basic prudent practices" described in Chapter 5.C. In particular, work with lead dust, molten lead, and lead salts capable of forming dusts should be conducted in a fume hood to prevent exposure by inhalation.

Accidents

In the event of skin contact, immediately wash with soap and water and remove contaminated clothing. In case of eye contact, promptly wash with copious amounts of water for 15 min (lifting upper and lower lids occasionally) and obtain medical attention. If lead or lead compounds are ingested, obtain medical attention immediately. If large amounts of such substances are inhaled, move the person to fresh air and seek medical attention at once.

In the event of a spill, sweep up dry lead and its compounds, soak up solutions with a spill pillow or absorbent material, place in an appropriate container, and dispose of properly. Respiratory protection may be necessary in the event of a large spill or release causing significant airborne particulate levels.

Disposal

Excess lead and waste material containing this substance should be placed in an appropriate container, clearly labeled, and handled according to your institution's waste disposal guidelines. For more information on disposal procedures, see Chapter 7 of this volume.

LABORATORY CHEMICAL SAFETY SUMMARY: LITHIUM ALUMINUM HYDRIDE

Substance	Lithium aluminum hydride (LAH, lithium tetrahydroaluminate) CAS 16853-85-3
Formula	$LiAlH_4$
Physical Properties	White to gray crystalline solid Decomposes above 125° C Reacts vigorously with water
Odor	Odorless solid
Autoignition Temperature	Ignites in moist or heated air
Toxicity Data	TLV-TWA (ACGIH) 2 mg (Al)/m³
Major Hazards	Reacts violently with water, acids, and many oxygenated compounds; may ignite in moist air; corrosive to skin, eyes, and mucous membranes.
Toxicity	Lithium aluminum hydride is highly corrosive to the skin, eyes, and mucous membranes. Contact with moisture forms lithium hydroxide, which can cause severe burns. Powdered LAH forms dusts that can pose an inhalation hazard. Ingestion of this substance may cause aching muscles, nausea, vomiting, dizziness, and unconsciousness and may be fatal. Ingestion can result in gas embolism due to the formation of hydrogen. No chronic effects of lithium aluminum hydride have been identified.
Flammability and Explosibility	Lithium aluminum hydride is a highly flammable solid and may ignite in moist or heated air. Exposure to water results in the release of hydrogen, which can be ignited by the heat from the exothermic reaction. Lithium aluminum hydride should not be used as a drying agent for solvents because fires can easily result (LAH decomposes at about 125° C, a temperature easily reached at a flask's surface in a heating mantle). The decomposition products of LAH can be quite explosive, and the products of its reaction with carbon dioxide have been reported to be explosive. Use dry chemical powder or sand to extinguish fires involving lithium aluminum hydride. *Never use water or carbon dioxide extinguishers on an LAH fire.*
Reactivity and Incompatibility	Lithium aluminum hydride reacts violently with water, acids, oxidizers, alcohols, and many oxygenated organic compounds, including, in particular, peroxides, hydroperoxides, and peracids. LAH reacts with many metal halides to produce metal hydride products, which are flammable and toxic.
Storage and Handling	Lithium aluminum hydride should be handled in the laboratory using the "basic prudent practices" described in Chapter 5.C, supplemented by the additional precautions for work

(continued on facing page)

The information in this LCSS has been compiled by a committee of the National Research Council from literature sources and Material Safety Data Sheets and is believed to be accurate as of July 1994. This summary is intended for use by trained laboratory personnel in conjunction with the NRC report *Prudent Practices in the Laboratory: Handling and Disposal of Chemicals*. This LCSS presents a concise summary of safety information that should be adequate for most laboratory uses of the title substance, but in some cases it may be advisable to consult more comprehensive references. This information should not be used as a guide to the nonlaboratory use of this chemical.

with flammable (Chapter 5.F) and reactive (Chapter 5.G) substances. In particular, LAH should be handled in areas free of ignition sources under an inert atmosphere. Safety glasses, impermeable gloves, and a fire-retardant laboratory coat are required. A dry powder fire extinguisher or pail of sand (and shovel) must be available in areas where LAH is to be handled or stored. Work with large quantities of powdered LAH should be conducted in a fume hood under an inert gas such as nitrogen or argon. Lithium aluminum hydride should be stored in tightly sealed containers in a cool, dry area separate from combustible materials. Dry LAH powder should never be exposed to water or moist air. Lithium aluminum hydride can be a finely powdered reagent that produces a reactive dust on handling. The older practice of grinding lithium aluminum hydride prior to use can cause explosions and should not be employed.

Accidents

In the event of skin contact, immediately wash with soap and water and remove contaminated clothing. In case of eye contact, promptly wash with copious amounts of water for 15 min (lifting upper and lower lids occasionally) and obtain medical attention. If lithium aluminum hydride is ingested, obtain medical attention immediately. If large amounts of LAH dust are inhaled, move the person to fresh air and seek medical attention at once.

In the event of a spill, instruct others to maintain a safe distance; while wearing a face shield and goggles, laboratory coat, and butyl rubber gloves, cover the spilled material with sand. Scoop the resulting mixture into a container suitable for treatment or disposal as discussed below.

Disposal

Small amounts of excess LAH can be destroyed by forming a suspension or solution in an inert solvent such as diethyl ether or hexane, cooling in an ice bath, and slowly and carefully adding ethyl acetate dropwise with stirring. This is followed by the addition of a saturated aqueous solution of ammonium chloride.

Excess lithium aluminum hydride and the products of the treatment described above should be placed in an appropriate container, clearly labeled, and handled according to your institution's waste disposal guidelines. For more information on disposal procedures, see Chapter 7.

LABORATORY CHEMICAL SAFETY SUMMARY: MERCURY

Substance	Mercury (Quicksilver, hydrargyrum) CAS 7439-97-6
Formula	Hg
Physical Properties	Silvery, mobile liquid bp 357 °C, mp −39 °C Very slightly soluble in water (0.002 g/100 mL at 20 °C)
Odor	Odorless
Vapor Density	6.9 (air = 1.0)
Vapor Pressure	0.0012 mmHg at 20 °C
Flash Point	Noncombustible

Toxicity Data

LC_{LO} inhal (rabbit)	29 mg/m³ (30 h)
PEL (OSHA)	0.1 mg/m³ (ceiling)
TLV-TWA (ACGIH)	0.025 mg/m³—skin

Major Hazards

Repeated or prolonged exposure to mercury vapor is highly toxic to the central nervous system.

Toxicity

The acute toxicity of mercury varies significantly with the route of exposure. Ingestion is largely without effects. Inhalation of high concentrations of mercury causes severe respiratory irritation, digestive disturbances, and marked kidney damage. There are no warning properties for exposure to mercury vapor, which is colorless, odorless, and tasteless.

Toxicity caused by repeated or prolonged exposure to mercury vapor or liquid is characterized by emotional disturbances, inflammation of the mouth and gums, general fatigue, memory loss, headaches, tremors, anorexia, and weight loss. Skin absorption of mercury and mercury vapor adds to the toxic effects of vapor inhalation. At low levels the onset of symptoms is insidious; fine tremors of the hand, eyelids, lips, and tongue are often the presenting complaints. Mercury has been reported to be capable of causing sensitization dermatitis. Mercury has not been shown to be a human carcinogen or reproductive toxin.

Flammability and Explosibility

Mercury is not combustible.

Reactivity and Incompatibility

Mercury is a fairly unreactive metal that is highly resistant to corrosion. It can dissolve a number of metals, such as silver, gold, and tin, forming amalgams. Mercury can react violently with acetylene and ammonia.

(continued on facing page)

Storage and Handling

Mercury should be handled in the laboratory using the "basic prudent practices" described in Chapter 5.C. In particular, precautions should be taken to prevent spills of mercury because drops of the liquid metal can easily become lodged in floor cracks, behind cabinets, and equipment, etc., with the result that the mercury vapor concentration in the laboratory may then exceed the safe and allowable limits. Containers of mercury should be kept tightly sealed and stored in secondary containers (such as a plastic pan or tray) in a well-ventilated area. When breakage of instruments or apparatus containing significant quantities of Hg is possible, the equipment should be placed in a plastic tray or pan that is large enough to contain the mercury in the event of an accident. Transfers of mercury between containers should be carried out in a fume hood over a tray or pan to confine any spills.

Accidents

In the event of skin contact, immediately wash with soap and water and remove contaminated clothing. In case of eye contact, promptly wash with copious amounts of water for 15 min (lifting upper and lower lids occasionally) and obtain medical attention. If mercury is ingested, obtain medical attention immediately. If large amounts of this substance are inhaled, move the person to fresh air and seek medical attention at once.

In the event of a spill, collect the mercury using the procedures described in Chapter 5.C, place in an appropriate container, and dispose of properly. Respiratory protection will be necessary in the event of a large spill, release in a confined area, or spill under conditions of higher than normal temperatures.

Disposal

Excess mercury should be collected for recycling, and waste material containing mercury should be placed in an appropriate container, clearly labeled, and handled according to your institution's waste disposal guidelines. For more information on disposal procedures, see Chapter 7 of this volume.

LABORATORY CHEMICAL SAFETY SUMMARY: METHANOL

Substance	Methanol (Methyl alcohol, wood alcohol) CAS 67-56-1
Formula	CH_3OH
Physical Properties	Colorless liquid bp 65 °C, mp −98 °C Miscible with water in all proportions
Odor	Faint alcohol odor detectable at 4 to 6000 ppm (mean = 160 ppm)
Vapor Density	1.1 (air = 1.0)
Vapor Pressure	96 mmHg at 20 °C
Flash Point	11 °C
Autoignition Temperature	385 °C

Toxicity Data		
	LD_{50} oral (rat)	5628 mg/kg
	LD_{50} skin (rabbit)	15,840 mg/kg
	LC_{50} inhal (rat)	>145,000 ppm (1 h)
	PEL (OSHA)	200 ppm (260 mg/m^3)
	TLV-TWA (ACGIH)	200 ppm (260 mg/m^3)—skin
	STEL (ACGIH)	250 ppm (328 mg/m^3)

Major Hazards Highly flammable liquid; low acute toxicity.

Toxicity The acute toxicity of methanol by ingestion, inhalation, and skin contact is low. Ingestion of methanol or inhalation of high concentrations can produce headache, drowsiness, blurred vision, nausea, vomiting, blindness, and death. In humans, 60 to 250 mL is reported to be a lethal dose. Prolonged or repeated skin contact can cause irritation and inflammation; methanol can be absorbed through the skin in toxic amounts. Contact of methanol with the eyes can cause irritation and burns. Methanol is not considered to have adequate warning properties.

Methanol has not been found to be carcinogenic in humans. Information available is insufficient to characterize the reproductive hazard presented by methanol. In animal tests, the compound produced developmental effects only at levels that were maternally toxic; hence, it is not considered to be a highly significant hazard to the fetus. Tests in bacterial or mammalian cell cultures demonstrate no mutagenic activity.

(continued on facing page)

The information in this LCSS has been compiled by a committee of the National Research Council from literature sources and Material Safety Data Sheets and is believed to be accurate as of July 1994. This summary is intended for use by trained laboratory personnel in conjunction with the NRC report *Prudent Practices in the Laboratory: Handling and Disposal of Chemicals.* This LCSS presents a concise summary of safety information that should be adequate for most laboratory uses of the title substance, but in some cases it may be advisable to consult more comprehensive references. This information should not be used as a guide to the nonlaboratory use of this chemical.

Flammability and Explosibility

Methanol is a flammable liquid (NFPA rating = 3) that burns with an invisible flame in daylight; its vapor can travel a considerable distance to an ignition source and "flash back." Methanol-water mixtures will burn unless very dilute. Carbon dioxide or dry chemical extinguishers should be used for methanol fires.

Reactivity and Incompatibility

Methanol can react violently with strong oxidizing agents such as chromium trioxide, with strong mineral acids such as perchloric, sulfuric, and nitric acids, and with highly reactive metals such as potassium. Sodium and magnesium metal react vigorously with methanol.

Storage and Handling

Methanol should be handled in the laboratory using the "basic prudent practices" described in Chapter 5.C, supplemented by the additional precautions for dealing with extremely flammable substances (Chapter 5.F). In particular, methanol should be used only in areas free of ignition sources, and quantities greater than 1 liter should be stored in tightly sealed metal containers in areas separate from oxidizers.

Accidents

In the event of skin contact, immediately wash with soap and water and remove contaminated clothing. In case of eye contact, promptly wash with copious amounts of water for 15 min (lifting upper and lower lids occasionally) and obtain medical attention. If methanol is ingested, obtain medical attention immediately. If large amounts of this compound are inhaled, move the person to fresh air and seek medical attention at once.

In the event of a spill, remove all ignition sources, soak up the methanol with a spill pillow or absorbent material, place in an appropriate container, and dispose of properly. Respiratory protection may be necessary in the event of a large spill or release in a confined area.

Disposal

Excess methanol and waste material containing this substance should be placed in an appropriate container, clearly labeled, and handled according to your institution's waste disposal guidelines. For more information on disposal procedures, see Chapter 7 of this volume.

The information in this LCSS has been compiled by a committee of the National Research Council from literature sources and Material Safety Data Sheets and is believed to be accurate as of July 1994. This summary is intended for use by trained laboratory personnel in conjunction with the NRC report *Prudent Practices in the Laboratory: Handling and Disposal of Chemicals*. This LCSS presents a concise summary of safety information that should be adequate for most laboratory uses of the title substance, but in some cases it may be advisable to consult more comprehensive references. This information should not be used as a guide to the nonlaboratory use of this chemical.

Substance	Methyl ethyl ketone (2-Butanone, methyl acetone, MEK, butan-2-one) CAS 78-93-3
Formula	$CH_3COCH_2CH_3$
Physical Properties	Colorless liquid bp 80 °C, mp −86 °C Highly soluble in water (25.6 g/100 mL at 20 °C)
Odor	Sweet/sharp odor detectable at 2 to 85 ppm (mean = 16 ppm)
Vapor Density	2.5 (air = 1.0)
Vapor Pressure	71.2 mmHg at 20 °C
Flash Point	−9 °C
Autoignition Temperature	516 °C

Toxicity Data		
	LD_{50} oral (rat)	2737 mg/kg
	LD_{50} skin (rabbit)	6480 mg/kg
	LC_{50} inhal (rat)	23,500 mg/m^3 (8 h)
	PEL (OSHA)	200 ppm (590 mg/m^3)
	TLV-TWA (ACGIH)	200 ppm (590 mg/m^3)
	STEL (ACGIH)	300 ppm (885 mg/m^3)

Major Hazards   Highly flammable

Toxicity

The acute toxicity of methyl ethyl ketone is low. Exposure to high concentrations can cause headache, dizziness, drowsiness, vomiting, and numbness of the extremities. Irritation of the eyes, nose, and throat can also occur. Methyl ethyl ketone is considered to have adequate warning properties.

Repeated or prolonged skin exposure to methyl ethyl ketone can cause defatting of the skin, leading to cracking, secondary infection, and dermatitis. This compound has not been found to be carcinogenic or to show reproductive or developmental toxicity in humans. Methyl ethyl ketone has exhibited developmental toxicity in some animal tests.

Flammability and Explosibility

Methyl ethyl ketone is extremely flammable (NFPA rating = 3), and its vapor can travel a considerable distance to an ignition source and "flash back." MEK vapor forms explosive mixtures with air at concentrations of 1.9 to 11% (by volume). Carbon dioxide or dry chemical extinguishers should be used for MEK fires.

(continued on facing page)

Reactivity and Incompatibility

Fires and/or explosions may result from the reaction of methyl ethyl ketone with strong oxidizing agents and very strong bases.

Storage and Handling

Methyl ethyl ketone should be handled in the laboratory using the "basic prudent practices" described in Chapter 5.C, supplemented by the additional precautions for dealing with extremely flammable substances (Chapter 5.F). In particular, MEK should be used only in areas free of ignition sources, and quantities greater than 1 liter should be stored in tightly sealed metal containers in areas separate from oxidizers.

Accidents

In the event of skin contact, immediately wash with soap and water and remove contaminated clothing. In case of eye contact, promptly wash with copious amounts of water for 15 min (lifting upper and lower lids occasionally) and obtain medical attention. If methyl ethyl ketone is ingested, obtain medical attention immediately. If large amounts of this compound are inhaled, move the person to fresh air and seek medical attention at once.

In the event of a spill, remove all ignition sources, soak up the methyl ethyl ketone with a spill pillow or absorbent material, place in an appropriate container, and dispose of properly. Respiratory protection may be necessary in the event of a large spill or release in a confined area.

Disposal

Excess methyl ethyl ketone and waste material containing this substance should be placed in an appropriate container, clearly labeled, and handled according to your institution's waste disposal guidelines. For more information on disposal procedures, see Chapter 7 of this volume.

LABORATORY CHEMICAL SAFETY SUMMARY: METHYL IODIDE

Substance	Methyl iodide (Iodomethane) CAS 74-88-4
Formula	CH_3I
Physical Properties	Colorless liquid; may darken upon exposure to light bp 42 °C, mp −66 °C Slightly soluble in water (2 g/100 mL)
Odor	Sweet, ethereal odor (no threshold data available); inadequate warning properties
Vapor Density	4.9 (air = 1.0)
Vapor Pressure	400 mmHg at 25 °C
Flash Point	Noncombustible

Toxicity Data

LD_{LO} oral (rat)	150 mg/kg
LD_{LO} skin (rat)	800 mg/kg
LC_{50} inhal (rat)	1300 mg/m^3 (4 h)
PEL (OSHA)	5 ppm (28 mg/m^3)—skin
TLV-TWA (ACGIH)	2 ppm (11 mg/m^3)—skin

Major Hazards

Moderately toxic, volatile substance readily absorbed through skin.

Toxicity

The acute toxicity of methyl iodide is moderate by ingestion, inhalation, and skin contact. This substance is readily absorbed through the skin and may cause systemic toxicity as a result. Methyl iodide is moderately irritating upon contact with the skin and eyes. Methyl iodide is an acute neurotoxin. Symptoms of exposure (which may be delayed for several hours) can include nausea, vomiting, diarrhea, drowsiness, slurred speech, visual disturbances, and tremor. Massive overexposure may cause pulmonary edema, convulsions, coma, and death.

Chronic exposure to methyl iodide vapor may cause neurotoxic effects such as dizziness, drowsiness, and blurred vision. There is limited evidence for the carcinogenicity of methyl iodide to experimental animals; it is not classified as an OSHA "select carcinogen."

Flammability and Explosibility

Noncombustible. High vapor pressure may cause containers to burst at elevated temperatures.

Reactivity and Incompatibility

Methyl iodide may react vigorously with alkali metals and strong oxidizing agents.

(continued on facing page)

Storage and Handling

Methyl iodide should be handled using the "basic prudent practices" of Chapter 5.C.

Accidents

In the event of skin contact, immediately wash with soap and water and remove contaminated clothing. In case of eye contact, promptly wash with copious amounts of water for 15 min (lifting upper and lower lids occasionally) and obtain medical attention. If methyl iodide is ingested, obtain medical attention immediately. If large amounts of this compound are inhaled, move the person to fresh air and seek medical attention at once.

In the event of a spill, soak up methyl iodide with a spill pillow or absorbent material, place in an appropriate container, and dispose of properly. Respiratory protection may be necessary in the event of a large spill or release in a confined area.

Disposal

Excess methyl iodide and waste material containing this substance should be placed in an appropriate container, clearly labeled, and handled according to your institution's waste disposal guidelines. For more information on disposal procedures, see Chapter 7 of this volume.

LABORATORY CHEMICAL SAFETY SUMMARY: NICKEL CARBONYL

Substance	Nickel carbonyl (Tetracarbonyl nickel) CAS 13463-39-3
Formula	$Ni(CO)_4$
Physical Properties	Colorless liquid bp 43 °C, mp −25 °C Very slightly soluble in water (0.0018 g/100 mL at 20 °C)
Odor	Sooty odor detectable at 0.5 to 3 ppm
Vapor Density	5.89 (air = 1.0)
Vapor Pressure	321 mmHg at 20 °C
Flash Point	< −20 °C
Autoignition Temperature	Explodes above 60 °C

Toxicity Data

LC_{50} inhal (rat)	35 ppm (240 mg/m³; 30 min)
PEL (OSHA)	0.001 ppm (0.007 mg/m³)
TLV-TWA (ACGIH)	0.05 mg/m³

Major Hazards
High acute toxicity; possible human carcinogen (OSHA "select carcinogen"); highly flammable.

Toxicity
The acute toxicity of nickel carbonyl by inhalation is high. Acute toxic effects occur in two stages, immediate and delayed. Headache, dizziness, shortness of breath, vomiting, and nausea are the initial symptoms of overexposure; the delayed effects (10 to 36 h) consist of chest pain, coughing, shortness of breath, bluish discoloration of the skin, and in severe cases, delirium, convulsions, and death. Recovery is protracted and characterized by fatigue on slight exertion. Nickel carbonyl is not regarded as having adequate warning properties.

Repeated or prolonged exposure to nickel carbonyl has been associated with an increased incidence of cancer of the lungs and sinuses. Nickel carbonyl is listed by IARC in Group 2B ("possible human carcinogen"), is listed by NTP as "reasonably anticipated to be a carcinogen," and is classified as a "select carcinogen" under the criteria of the OSHA Laboratory Standard.

Flammability and Explosibility
Nickel carbonyl is a highly flammable liquid (NFPA rating = 3) that may ignite spontaneously and explodes when heated above 60 °C. Its lower flammable limit in air is 2% by

(continued on facing page)

volume; the upper limit has not been reported. Carbon dioxide, water, or dry chemical extinguishers should be used for nickel carbonyl fires.

Reactivity and Incompatibility

In the presence of air, nickel carbonyl forms a deposit that becomes peroxidized and may ignite. Nickel carbonyl is incompatible with mercury, nitric acid, chlorine, and other oxidizers, which may cause fires and explosions. Products of decomposition (nickel oxide and carbon monoxide) are less toxic that nickel carbonyl itself.

Storage and Handling

Because of its carcinogenicity and flammability, nickel carbonyl should be handled using the "basic prudent practices" of Chapter 5.C supplemented by the additional precautions for work with compounds of high chronic toxicity (Chapter 5.D) and extremely flammable substances (Chapter 5.F). In particular, work with nickel carbonyl should be conducted in a fume hood to prevent exposure by inhalation and splash goggles and impermeable gloves should be worn at all times to prevent eye and skin contact. Nickel carbonyl should only be used in areas free of ignition sources. Containers of nickel carbonyl should be stored in secondary containers in the dark in areas separate from oxidizers.

Accidents

In the event of skin contact, immediately wash with soap and water and remove contaminated clothing. In case of eye contact, promptly wash with copious amounts of water for 15 min (lifting upper and lower lids occasionally) and obtain medical attention. If nickel carbonyl is ingested, obtain medical attention immediately. If large amounts of this compound are inhaled, move the person to fresh air and seek medical attention at once.

In the event of a spill, remove all ignition sources, soak up the nickel carbonyl with a spill pillow or absorbent material, place in an appropriate container, and dispose of properly. Respiratory protection will be necessary in the event of a large spill or release in a confined area.

Disposal

Excess nickel carbonyl and waste material containing this substance should be placed in an appropriate container, clearly labeled, and handled according to your institution's waste disposal guidelines. For more information on disposal procedures, see Chapter 7 of this volume.

LABORATORY CHEMICAL SAFETY SUMMARY: NITRIC ACID

Substance	Nitric acid CAS 7697-37-2
Formula	HNO_3
Physical Properties	Colorless, yellowish, or reddish-brown fuming liquid "Concentrated nitric acid" (68 to 70% HNO_3 by wt): bp 122 °C "White fuming nitric acid" (97.5% HNO_3, 2% H_2O, <0.5% NO_x): bp 83 °C, mp −42 °C "Red fuming nitric acid" contains 85% HNO_3, <5% H_2O, and 6 to 15% NO_x Miscible with water in all proportions
Odor	Suffocating fumes detectable at <5.0 ppm
Vapor Density	>1 (air = 1.0)
Vapor Pressure	57 mmHg at 25 °C for white fuming nitric acid 49 mmHg at 20 °C for 70% nitric acid
Flash Point	Not flammable

Toxicity Data

LC_{50} inhal (rat)	2500 ppm (1 h)
PEL (OSHA)	2 ppm (5 mg/m³)
TLV-TWA (ACGIH)	2 ppm (5.2 mg/m³)
STEL (ACGIH)	4 ppm (10 mg/m³)

Major Hazards	Highly corrosive to the eyes, skin, and mucous membranes; powerful oxidizing agent that ignites on contact or reacts explosively with many organic and inorganic substances.
Toxicity	Concentrated nitric acid and its vapors are highly corrosive to the eyes, skin, and mucous membranes. Dilute solutions cause mild skin irritation and hardening of the epidermis. Contact with concentrated nitric acid stains the skin yellow and produces deep painful burns. Eye contact can cause severe burns and permanent damage. Inhalation of high concentrations can lead to severe respiratory irritation and delayed effects, including pulmonary edema, which may be fatal. Ingestion of nitric acid may result in burning and corrosion of the mouth, throat, and stomach. An oral dose of 10 mL can be fatal in humans. Tests in animals demonstrate no carcinogenic or developmental toxicity for nitric acid. Tests for mutagenic activity or for reproductive hazards have not been performed.
Flammability and Explosibility	Not a combustible substance, but a strong oxidizer. Contact with easily oxidizible materials including many organic substances may result in fires or explosions.
Reactivity and Incompatibility	Nitric acid is a powerful oxidizing agent and ignites on contact or reacts explosively with a variety of organic substances including acetic anhydride, acetone, acetonitrile, many alcohols, thiols, and amines, dichloromethane, DMSO, and certain aromatic compounds

(continued on facing page)

The information in this LCSS has been compiled by a committee of the National Research Council from literature sources and Material Safety Data Sheets and is believed to be accurate as of July 1994. This summary is intended for use by trained laboratory personnel in conjunction with the NRC report *Prudent Practices in the Laboratory: Handling and Disposal of Chemicals*. This LCSS presents a concise summary of safety information that should be adequate for most laboratory uses of the title substance, but in some cases it may be advisable to consult more comprehensive references. This information should not be used as a guide to the nonlaboratory use of this chemical.

including benzene. Nitric acid also reacts violently with a wide range of inorganic substances including many bases, reducing agents, alkali metals, copper, phosphorus, and ammonia. Nitric acid corrodes steel.

Storage and Handling

Nitric acid should be handled in the laboratory using the "basic prudent practices" described in Chapter 5.C. In particular, splash goggles and rubber gloves should be worn when handling this acid, and containers of nitric acid should be stored in a well ventilated location separated from organic substances and other combustible materials.

Accidents

In the event of skin contact, immediately wash with water and remove contaminated clothing. In case of eye contact, promptly wash with copious amounts of water for 15 min (lifting upper and lower lids occasionally) and obtain medical attention. If nitric acid is ingested, obtain medical attention immediately. If large amounts of this compound are inhaled, move the person to fresh air and seek medical attention at once.

In the event of a spill, soak up nitric acid with a spill pillow or absorbent material, place in an appropriate container, and dispose of properly. Respiratory protection may be necessary in the event of a large spill or release in a confined area.

Disposal

Excess nitric acid and waste material containing this substance should be placed in an appropriate container, clearly labeled, and handled according to your institution's waste disposal guidelines. For more information on disposal procedures, see Chapter 7 of this volume.

The information in this LCSS has been compiled by a committee of the National Research Council from literature sources and Material Safety Data Sheets and is believed to be accurate as of July 1994. This summary is intended for use by trained laboratory personnel in conjunction with the NRC report *Prudent Practices in the Laboratory: Handling and Disposal of Chemicals*. This LCSS presents a concise summary of safety information that should be adequate for most laboratory uses of the title substance, but in some cases it may be advisable to consult more comprehensive references. This information should not be used as a guide to the nonlaboratory use of this chemical.

LABORATORY CHEMICAL SAFETY SUMMARY: NITROGEN DIOXIDE

Substance	Nitrogen dioxide (Nitrogen peroxide) CAS 10102-44-0
Formula	NO_2
Physical Properties	Yellow-brown liquid to reddish brown gas; generally a mixture of NO_2 and N_2O_4 (at -11 °C liquid is 0.01% NO_2, at 21 °C liquid is 0.1% NO_2 and gas is 15.9% NO_2) bp 21 °C, mp -11 °C Miscible in all proportions with water reacting to form nitric and nitrous acids
Odor	Pungent, acrid odor detectable at 0.12 ppm
Vapor Density	1.58 (air = 1.0)
Vapor Pressure	720 mmHg at 20 °C
Flash Point	Noncombustible

Toxicity Data

LC_{50} inhal (rat)	88 ppm (4 h)
PEL (OSHA)	5 ppm (9 mg/m^3; ceiling)
TLV-TWA (ACGIH)	3 ppm (5.6 mg/m^3)
STEL (ACGIH)	5 ppm (9.4 mg/m^3)

Major Hazards   Highly toxic by inhalation; high concentrations of the gas and liquid NO_2-N_2O_4 are corrosive to the skin, eyes, and mucous membranes.

Toxicity   The acute toxicity of nitrogen dioxide by inhalation is high. Inhalation may cause shortness of breath and pulmonary edema progressing to respiratory illness, reduction in the blood's oxygen carrying capacity, chronic lung disorders and death; symptoms may be delayed for hours and may recur after several weeks. Toxic effects may occur after exposure to concentrations of 10 ppm for 10 min and include coughing, chest pain, frothy sputum, and difficulty in breathing. Brief exposure to 200 ppm can cause severe lung damage and delayed pulmonary edema, which may be fatal. Nitrogen dioxide at concentrations of 10 to 20 ppm is mildly irritating to the eyes; higher concentrations of the gas and liquid NO_2-N_2O_4 are highly corrosive to the skin, eyes, and mucous membranes. Nitrogen dioxide can be detected below the permissible exposure limit by its odor and irritant effects and is regarded as a substance with adequate warning properties.

Animal testing indicates that nitrogen dioxide does not have carcinogenic or reproductive effects. It does produce genetic damage in bacterial and mammalian cell cultures; however, most studies in animals indicate that it does not produce heritable genetic damage.

(continued on facing page)

The information in this LCSS has been compiled by a committee of the National Research Council from literature sources and Material Safety Data Sheets and is believed to be accurate as of July 1994. This summary is intended for use by trained laboratory personnel in conjunction with the NRC report *Prudent Practices in the Laboratory: Handling and Disposal of Chemicals*. This LCSS presents a concise summary of safety information that should be adequate for most laboratory uses of the title substance, but in some cases it may be advisable to consult more comprehensive references. This information should not be used as a guide to the nonlaboratory use of this chemical.

Flammability and Explosibility

Nitrogen dioxide is not combustible (NFPA rating = 0) but is a strong oxidizing agent and will support combustion. Cylinders of NO_2 gas exposed to fire or intense heat may vent rapidly or explode.

Reactivity and Incompatibility

Nitrogen dioxide-nitrogen tetroxide is a powerful oxidizer and can cause many organic substances to ignite. This substance may react violently with alcohols, aldehydes, acetonitrile, DMSO, certain hydrocarbons, and chlorinated hydrocarbons. Metals react vigorously and alkali metals ignite in NO_2.

Storage and Handling

Because of its high acute toxicity, nitrogen dioxide should be handled using the "basic prudent practices" of Chapter 5.C, supplemented by the additional precautions for work with compounds of high toxicity (Chapter 5.D) and compressed gases (Chapter 5.H). In particular, cylinders of nitrogen dioxide should be stored and used in a continuously ventilated gas cabinet or fume hood.

Accidents

If large amounts of this compound are inhaled, the person should be moved to fresh air and medical attention should be sought at once. In the event of skin contact, immediately wash with soap and water and remove contaminated clothing. In case of eye contact, wash promptly with copious amounts of water for 15 min (lifting upper and lower lids occasionally) and obtain medical attention. If nitrogen dioxide is ingested, obtain medical attention immediately.

In the event of a release of nitrogen dioxide, use appropriate protective equipment and clothing. Positive pressure air-supplied respiratory protection may be required in cases involving a large release of nitrogen dioxide gas. If a cylinder is the source of the leak and the leak cannot be stopped, remove the leaking cylinder to a fume hood or a safe place, if possible, in the open air, and repair the leak or allow the cylinder to empty.

Disposal

Excess nitrogen dioxide and waste material containing this substance should be placed in an appropriate container, clearly labeled, and handled according to your institution's waste disposal guidelines. For more information on disposal procedures, see Chapter 7 of this volume.

The information in this LCSS has been compiled by a committee of the National Research Council from literature sources and Material Safety Data Sheets and is believed to be accurate as of July 1994. This summary is intended for use by trained laboratory personnel in conjunction with the NRC report *Prudent Practices in the Laboratory: Handling and Disposal of Chemicals*. This LCSS presents a concise summary of safety information that should be adequate for most laboratory uses of the title substance, but in some cases it may be advisable to consult more comprehensive references. This information should not be used as a guide to the nonlaboratory use of this chemical.

LABORATORY CHEMICAL SAFETY SUMMARY: OSMIUM TETROXIDE

Substance	Osmium tetroxide (Osmic acid, perosmic oxide, osmium(IV) oxide) CAS 20816-12-0
Formula	OsO_4
Physical Properties	Colorless to pale yellow-green crystals bp 130 °C (but sublimes at lower temperature), mp 40 °C Moderately soluble in water (7 g/100 mL)
Odor	Acrid, chlorine-like odor detectable at 2 ppm (20 mg/m^3)
Vapor Density	8.8 (air = 1.0)
Vapor Pressure	7 mmHg at 20 °C

Toxicity Data

LD_{50} oral (rat)	14 mg/kg
LC_{LO} inhal (rat)	40 ppm (4 h)
PEL (OSHA)	0.0002 ppm (0.002 mg/m^3)
TLV-TWA (ACGIH)	0.0002 ppm (0.002 mg/m^3)
STEL (ACGIH)	0.0006 ppm (0.006 mg/m^3)

Major Hazards   High acute toxicity; severe irritant of the eyes and respiratory tract; vapor can cause serious eye damage.

Toxicity

The acute toxicity of osmium tetroxide is high, and it is a severe irritant of the eyes and respiratory tract. Exposure to osmium tetroxide vapor can damage the cornea of the eye. Irritation is generally the initial symptom of exposure to low concentrations of osmium tetroxide vapor, and lacrimation, a gritty feeling in the eyes, and the appearance of rings around lights may also be noted. In most cases, recovery occurs in a few days. Concentrations of vapor that do not cause immediate irritation can have an insidious cumulative effect; symptoms may not be noted until several hours after exposure. Contact of the eyes with concentrated solutions of this substance can cause severe damage and possible blindness. Inhalation can cause headache, coughing, dizziness, lung damage, and difficult breathing and may be fatal. Contact of the vapor with skin can cause dermatitis, and direct contact with the solid can lead to severe irritation and burns. Exposure to osmium tetroxide via inhalation, skin contact, or ingestion can lead to systemic toxic effects involving liver and kidney damage. Osmium tetroxide is regarded as a substance with poor warning properties.

Chronic exposure to osmium tetroxide can result in an accumulation of osmium compounds in the liver and kidney and damage to these organs. Osmium tetroxide has been reported to cause reproductive toxicity in animals; this substance has not been shown to be carcinogenic or to show reproductive or developmental toxicity in humans.

(continued on facing page)

The information in this LCSS has been compiled by a committee of the National Research Council from literature sources and Material Safety Data Sheets and is believed to be accurate as of July 1994. This summary is intended for use by trained laboratory personnel in conjunction with the NRC report *Prudent Practices in the Laboratory: Handling and Disposal of Chemicals*. This LCSS presents a concise summary of safety information that should be adequate for most laboratory uses of the title substance, but in some cases it may be advisable to consult more comprehensive references. This information should not be used as a guide to the nonlaboratory use of this chemical.

Flammability and Explosibility

Noncombustible

Reactivity and Incompatibility

Osmium tetroxide reacts with hydrochloric acid to form chlorine gas.

Storage and Handling

Because of its high acute toxicity, osmium tetroxide should be handled in the laboratory using the "basic prudent practices" of Chapter 5.C, supplemented by the additional precautions for work with compounds of high toxicity (Chapter 5.D). In particular, all work with osmium tetroxide should be conducted in a fume hood to prevent exposure by inhalation, and splash goggles and impermeable gloves should be worn at all times to prevent eye and skin contact. Osmium tetroxide as solid or solutions should be stored in tightly sealed containers, and these should be placed in secondary containers.

Accidents

In the event of skin contact, immediately wash with soap and water and remove contaminated clothing. In case of eye contact, promptly wash with copious amounts of water for 15 minutes (lifting upper and lower lids occasionally) and obtain medical attention. If osmium tetroxide is ingested, obtain medical attention immediately. If large amounts are inhaled, move the person to fresh air and seek medical attention at once.

In the event of a spill, mix osmium tetroxide with an absorbent material such as vermiculite or dry sand (avoid raising dust), place in an appropriate container, and dispose of properly. Evacuation and cleanup using respiratory protection may be necessary in the event of a large spill or release in a confined area.

Disposal

Excess osmium tetroxide solutions can be rendered safer by reaction with sodium sulfite to produce insoluble osmium dioxide. Ethanol will also react to produce the dioxide. Corn oil or sodium sulfide may also be used to deactivate osmium tetroxide. Osmium-containing waste should be placed in a tightly sealed, labeled container and handled according to your institution's waste disposal guidelines. For more information on disposal procedures, see Chapter 7 of this volume.

LABORATORY CHEMICAL SAFETY SUMMARY: OXYGEN

Substance	Oxygen (GOX, gas only; LOX, liquid only) CAS 7782-44-7
Formula	O_2
Physical Properties	Colorless gas bp -183 °C, mp -219 °C Slightly soluble in water (0.004 g/100 mL at 25°C)
Odor	Odorless gas
Vapor Density	1.11 (air $= 1.0$)
Vapor Pressure	>760 mmHg at 20 °C
Toxicity Data	OSHA recommends a minimum oxygen concentration of 19.5% for human occupancy.
Major Hazards	Powerful oxidizing agent; concentrations greater than 25% greatly enhance the combustion rate of many materials.
Toxicity	Oxygen is nontoxic under the usual conditions of laboratory use. Breathing pure oxygen at one atmosphere may produce cough and chest pains within 8 to 24 h, and concentrations of 60% may produce these symptoms in several days. Liquid oxygen can cause severe ''burns'' and tissue damage on contact with the skin due to extreme cold.
Flammability and Explosibility	Oxygen itself is nonflammable, but at concentrations greater than 25% supports and vigorously accelerates the combustion of flammable materials. Some materials (including metals) that are noncombustible in air will burn in the presence of oxygen.
Reactivity and Incompatibility	Oxygen is incompatible with combustible materials, including many lubricants and elastomers. Oil, greases, and other readily combustible substances should never be allowed to come in contact with O_2 cylinders, valves, regulators, and fittings. Contact of liquid oxygen with many organic substances can lead to an explosion.
Storage and Handling	Oxygen should be handled in the laboratory using the ''basic prudent practices'' described in Chapter 5.C, supplemented by the procedures for work with compressed gases found in Chapter 5.H.
Accidents	In the event of skin or eye contact with liquid oxygen, seek medical attention for cryogenic burns. Do not enter areas of high oxygen gas concentration, which can saturate clothing and increase its flammability. Ventilate area to evaporate and disperse oxygen.

(continued on facing page)

Disposal Excess liquid oxygen should be allowed to evaporate in a well-ventilated outdoor area. Vent oxygen gas to outside location. Locations should be remote from work areas, open flames, or sources of ignition and combustibles. Return empty and excess cylinders of oxygen to manufacturer. For more information on disposal procedures, see Chapter 7 of this volume.

LABORATORY CHEMICAL SAFETY SUMMARY: OZONE

Substance	Ozone CAS 10028-15-6
Formula	O_3
Physical Properties	Colorless to bluish gas bp -112 °C, mp -193 °C Almost insoluble in water (0.00003 g/100 mL at 20 °C)
Odor	Pungent odor, detectable at 0.01 to 0.04 ppm; sharp disagreeable odor at 1 ppm
Vapor Density	1.65 (air = 1.0)

Toxicity Data

LC_{50} inhal (rat)	4.8 ppm (4 h)
PEL (OSHA)	0.1 ppm (0.2 mg/m^3)
TLV-TWA (ACGIH)	0.1 ppm (0.2 mg/m^3)
STEL (ACGIH)	0.3 ppm (0.6 mg/m^3)

Major Hazards

Extremely irritating to the eyes and respiratory tract; high acute toxicity. Reacts violently with many oxidizable organic and inorganic substances; may form shock-sensitive and highly explosive reaction products.

Toxicity

Ozone is a highly toxic gas that is extremely irritating to the eyes, mucous membranes, and respiratory tract. The characteristic odor of ozone can be detected below the permissible exposure limit, and this compound is therefore regarded to have adequate warning properties. However, at higher concentrations the ability to smell ozone may decrease. Inhalation of 1 ppm ozone may cause headaches and irritation of the upper and lower respiratory tract. The first symptoms of exposure include irritation of the eyes, dryness of throat, and coughing; these symptoms disappear after exposure ceases. Exposure at higher levels may lead to lacrimation, vomiting, upset stomach, labored breathing, lowering of pulse rate and blood pressure, lung congestion, tightness in the chest, and pulmonary edema, which can be fatal. Exposure to 100 ppm of ozone for 1 hour can be lethal to humans.

Animal studies indicate that chronic exposure to ozone may result in pulmonary damage, leading to chronic lung impairment. Continual daily exposure to ozone can cause premature aging.

Flammability and Explosibility

Ozone by itself is not flammable. Liquid ozone and concentrated solutions are extremely hazardous and can explode on warming or when shocked.

Reactivity and Incompatibility

Ozone is a powerful oxidant and can react explosively with readily oxidizable substances and reducing agents. Explosions can occur when ozone is exposed to bromine, hydrogen bromide, hydrogen iodide, nitrogen oxides, lithium aluminum hydride, metal hydrides, hydrazine, alkyl metals, stilbene, ammonia, arsine, and phosphine. Ozone reacts with alkenes and other unsaturated organic compounds to form ozonides, many of which are

(continued on facing page)

highly unstable and explosive. Ozone combines with many aromatic compounds and ethers to form shock-sensitive and explosive products.

Storage and Handling

Because of its high degree of acute toxicity, ozone should be handled in the laboratory using the "basic prudent practices" described in Chapter 5.C, supplemented by the additional precautions for work with compounds of high toxicity (Chapter 5.D) and high reactivity (Chapter 5.G). In particular, work with ozone should be conducted in a fume hood to prevent exposure by inhalation. Ozone is usually produced in the laboratory with a ozone generator, and care should be taken to ensure adequate ventilation in the area where the ozone generation equipment is located. Because of the possibility of the generation of explosive ozonides, ozonolysis reactions should always be conducted in a fume hood behind a safety shield.

Accidents

An ozone leak can be easily detected by its characteristic pungent odor. If a large amount of ozone is inhaled, move the person to fresh air and seek medical attention at once. In the event of eye contact, promptly wash eyes with copious amounts of water for 15 min (lifting upper and lower lids occasionally) and obtain medical attention.

Respiratory protection may be necessary in the event of an accidental release of ozone.

Disposal

Ozone is usually produced on demand from a laboratory ozone generator, and a procedure for the treatment of excess ozone should be included in the experimental plan. Small to moderate amounts of excess ozone can be vented to the fume hood or other exhaust system. When large amounts of excess ozone are anticipated, the excess gas should be passed through a series of traps containing a 1 to 2% solution of potassium iodide or other reducing agent before venting to the fume hood.

LABORATORY CHEMICAL SAFETY SUMMARY: PALLADIUM ON CARBON

Substance	Palladium on carbon (Pd/C) CAS 7440-05-3 (palladium)
Formula	Pd/C
Physical Properties	Black powder mp 1555 °C, bp 3167 °C (palladium) Insoluble in water
Odor	Odorless
Toxicity Data	LD$_{50}$ oral (rat) 200 mg/kg (palladium chloride) LC$_{50}$ intratracheal (rat) 6 mg/kg (palladium chloride)
Major Hazards	May ignite on exposure to air, particularly when containing adsorbed hydrogen; readily causes ignition of flammable solvents in the presence of air.
Toxicity	Very little information is available on the toxicity of palladium and its compounds. There is some evidence that chronic exposure to palladium particles in dust can have toxic effects on the blood and respiratory systems. Finely divided carbon is irritating to mucous membranes and the upper respiratory tract.
Flammability and Explosibility	Palladium on carbon catalysts containing adsorbed hydrogen are pyrophoric, particularly when dry and at elevated temperatures. Palladium on carbon catalysts prepared by formaldehyde reduction are less pyrophoric than those reduced with hydrogen. Finely divided carbon, like most materials in powder form, is capable of creating a dust explosion.
Reactivity and Incompatibility	Catalysts prepared on high surface area supports are highly active and readily cause ignition of hydrogen/air and solvent/air mixtures. Methanol is notable for easy ignition because of its high volatility. Addition of catalyst to a tetrahydroborate solution may cause ignition of liberated hydrogen.
Storage and Handling	Because of its high potential for ignition, palladium on carbon should be handled using the "basic prudent practices" of Chapter 5.C, supplemented by the additional precautions for work with reactive and explosive chemicals (Chapter 5.G). In particular, palladium on carbon should always be handled under an inert atmosphere (preferably argon), and reaction vessels should be flushed with inert gas before the catalyst is added. Dry catalyst should never be added to an organic solvent in the presence of air. Palladium on carbon recovered from catalytic hydrogenation reactions by filtration requires careful handling because it is usually saturated with hydrogen and will ignite spontaneously on exposure to air. The filter cake should never be allowed to dry, and the moist material should be added to a large quantity of water and disposed of properly.

(continued on facing page)

The information in this LCSS has been compiled by a committee of the National Research Council from literature sources and Material Safety Data Sheets and is believed to be accurate as of July 1994. This summary is intended for use by trained laboratory personnel in conjunction with the NRC report *Prudent Practices in the Laboratory: Handling and Disposal of Chemicals*. This LCSS presents a concise summary of safety information that should be adequate for most laboratory uses of the title substance, but in some cases it may be advisable to consult more comprehensive references. This information should not be used as a guide to the nonlaboratory use of this chemical.

Accidents

In the event of skin contact, immediately wash with soap and water and remove contaminated clothing. In case of eye contact, promptly wash with copious amounts of water for 15 min (lifting upper and lower lids occasionally) and obtain medical attention. If palladium on carbon is ingested, obtain medical attention immediately. If large amounts of dust are inhaled, move the person to fresh air and seek medical attention at once.

In the event of a spill, remove all ignition sources, wet the palladium on carbon with water, place in an appropriate container, and dispose of properly. Respiratory protection may be necessary in the event of a large release in a confined area.

Disposal

Excess palladium on carbon and waste material containing this substance should be covered in water, placed in an appropriate container, clearly labeled, and handled according to your institution's waste disposal guidelines. For more information on disposal procedures, see Chapter 7 of this volume.

Substance	Peracetic acid (Peroxyacetic acid; acetyl hydroperoxide) CAS 79-21-0 Note: Although other percarboxylic acids have different physical properties, their reactivity and toxicology are similar to those of peracetic acid.
Formula	$CH_3C(O)OOH$
Physical Properties	Colorless liquid bp 105 °C (40% solution in acetic acid), mp 0.1 °C Miscible with water
Odor	Acrid odor
Vapor Pressure	Low
Flash Point	40.5 °C (open cup)
Autoignition Temperature	Explodes when heated to 110 °C
Toxicity Data	LD_{50} oral (rat) 1540 mg/kg LD_{50} skin (rabbit) 1410 mg/kg
Major Hazards	Severely irritating to the eyes, skin, and mucous membranes; can form explosive mixtures with easily oxidized substances.
Toxicity	The acute toxicity of peracetic acid is low. However, peracids are extremely irritating to the skin, eyes, and respiratory tract. Skin or eye contact with the 40% solution in acetic acid can cause serious burns. Inhalation of high concentrations of mists of peracetic acid solutions can lead to burning sensations, coughing, wheezing, and shortness of breath. Peracetic acid has not been found to be carcinogenic or to show reproductive or developmental toxicity in humans. There is some evidence that this compound is a weak carcinogen from animal studies (mice). Data on other peracids suggest peracetic acid may show the worst chronic and acute toxicity of this class of compounds. Other commonly available peracids, such as perbenzoic acid and *m*-chloroperbenzoic acid (MCPBA) are less toxic, less volatile, and more easily handled than the parent substance.
Flammability and Explosibility	Peracetic acid explodes when heated to 110 °C, and the pure compound is extremely shock sensitive. Virtually all peracids are strong oxidizing agents and decompose explosively on heating. Moreover, most peracids are highly flammable and can accelerate the combustion

(continued on facing page)

The information in this LCSS has been compiled by a committee of the National Research Council from literature sources and Material Safety Data Sheets and is believed to be accurate as of July 1994. This summary is intended for use by trained laboratory personnel in conjunction with the NRC report *Prudent Practices in the Laboratory: Handling and Disposal of Chemicals*. This LCSS presents a concise summary of safety information that should be adequate for most laboratory uses of the title substance, but in some cases it may be advisable to consult more comprehensive references. This information should not be used as a guide to the nonlaboratory use of this chemical.

of other flammable materials if present in a fire. Fires involving peracetic acid can be fought with water, dry chemical, or halon extinguishers. Containers of peracetic acid heated in a fire may explode.

Reactivity and Incompatibility

Peracids such as peracetic acid are strong oxidizing agents and react exothermically with easily oxidized substrates. In some cases the heat of reaction can be sufficient to induce ignition, at which point combustion is accelerated by the presence of the peracid. Violent reactions may potentially occur, for example, with ethers, metal chloride solutions, olefins, and some alcohols and ketones. Shock-sensitive peroxides may be generated by the action of peracids on these substances as well as on carboxylic anhydrides. Some metal ions, including iron, copper, cobalt, chromium, and manganese, may cause runaway peroxide decomposition. Peracetic acid is also reportedly sensitive to light.

Storage and Handling

Peracetic acid should be handled in the laboratory using the "basic prudent practices" described in Chapter 5.C, supplemented by the additional precautions for work with reactive and explosive substances (Chapter 5.G). Reactions involving large quantities of peracids should be carried out behind a safety shield. Peracetic acid should be used only in areas free of ignition sources and should be stored in tightly sealed containers in areas separate from oxidizable compounds and flammable substances. Other commonly available peracids, such as perbenzoic acid and m-chloroperbenzoic acid (MCPBA), are less toxic, less volatile, and more easily handled than peracetic acid.

Accidents

In the event of skin contact, immediately wash with soap and water and remove contaminated clothing. In case of eye contact, promptly wash with copious amounts of water for 15 min (lifting upper and lower lids occasionally) and obtain medical attention. If peracetic acid is ingested, obtain medical attention immediately. If large amounts of this compound are inhaled, move the person to fresh air and seek medical attention at once.

In the event of a spill, remove all ignition sources, soak up the peracetic acid solution with a spill pillow or a noncombustible absorbent material such as vermiculite, place in an appropriate container, and dispose of properly. Respiratory protection may be necessary in the event of a large spill or release in a confined area.

Disposal

Excess peracetic acid and waste material containing this substance should be placed in an appropriate container, clearly labeled, and handled according to your institution's waste disposal guidelines. Peracids may be incompatible with other flammable mixed chemical waste; for example, shock-sensitive peroxides can be generated by reaction with some ethers such as THF and diethyl ether. For more information on disposal procedures, see Chapter 7 of this volume.

LABORATORY CHEMICAL SAFETY SUMMARY: PERCHLORIC ACID (AND INORGANIC PERCHLORATES)

Substance	Perchloric acid (and inorganic perchlorates) CAS 7601-90-3
Formula	$HClO_4$ (maximum concentration commercially available is an aqueous solution containing about 72% $HClO_4$ by weight)
Physical Properties	Colorless liquid 72% $HClO_4$: bp 203 °C, mp -18 °C Miscible with water
Odor	Odorless
Vapor Pressure	6.8 mmHg at 25 °C

Toxicity Data

LD_{50} oral (rat)	1100 mg/kg
LD_{50} oral (dog)	400 mg/kg

Major Hazards

Highly corrosive to all tissues; reacts violently with many oxidizable substances; anhydrous form and certain salts are highly explosive.

Toxicity

Perchloric acid is a highly corrosive substance that causes severe burns on contact with the eyes, skin, and mucous membranes. The acute toxicity of perchloric acid is moderate. This substance is a severe irritant to the eyes, mucous membranes, and upper respiratory tract. Perchlorates are irritants to the body wherever they contact it.

Perchloric acid has not been shown to be carcinogenic or to show reproductive or developmental toxicity in humans.

Flammability and Explosibility

Perchloric acid is noncombustible. The anhydrous (dehydrated) acid presents a serious explosion hazard. It is unstable and can decompose explosively at ordinary temperatures or in contact with many organic compounds.

Many heavy metal perchlorates and organic perchlorate salts are extremely sensitive explosives; the ammonium, alkali metal, and alkali earth perchlorates are somewhat less hazardous. Mixtures of perchlorates with many oxidizable substances are explosive.

Reactivity and Incompatibility

Cold 70% perchloric acid is a strong acid but is not considered to be a strong oxidizing agent; however, more concentrated solutions are good oxidizers. Temperature increases the oxidizing power of perchloric acid, and hot concentrated solutions are very dangerous. Evaporation of a spill of the 70% solution may lead to the formation of more dangerous concentrations. Reaction of 70% perchloric acid with cellulose materials such as wood,

(continued on facing page)

The information in this LCSS has been compiled by a committee of the National Research Council from literature sources and Material Safety Data Sheets and is believed to be accurate as of July 1994. This summary is intended for use by trained laboratory personnel in conjunction with the NRC report *Prudent Practices in the Laboratory: Handling and Disposal of Chemicals*. This LCSS presents a concise summary of safety information that should be adequate for most laboratory uses of the title substance, but in some cases it may be advisable to consult more comprehensive references. This information should not be used as a guide to the nonlaboratory use of this chemical.

paper, and cotton can produce fires and explosions. Oxidizable organic compounds including alcohols, ketones, aldehydes, ethers, and dialkyl sulfoxides can react violently with concentrated perchloric acid.

All perchlorates are potentially hazardous when in contact with reducing agents.

Storage and Handling

Because of their extreme reactivity, perchloric acid and all organic and inorganic perchlorates should be handled using the "basic prudent practices" of Chapter 5.C, supplemented by the additional precautions for work with reactive and explosive compounds (Chapter 5.G). In particular, splash goggles and rubber gloves should be worn when handling perchloric acid, and containers of the acid should be stored in a well-ventilated location separated from organic substances and other combustible materials. Work with >85% perchloric acid requires special precautions and should be carried out only by specially trained personnel.

Accidents

In the event of skin contact, immediately wash with soap and water and remove contaminated clothing. In case of eye contact, promptly wash with copious amounts of water for 15 min (lifting upper and lower lids occasionally) and obtain medical attention. If perchloric acid is ingested, obtain medical attention immediately. If large amounts of this compound are inhaled, move the person to fresh air and seek medical attention at once.

In the event of a spill, dilute the perchloric acid with water to a concentration of <5%, absorbed with sand or vermiculite, place in an appropriate container, and dispose of properly. Organic absorbants must not be used. Respiratory protection may be necessary in the event of a large spill or release in a confined area.

Disposal

Excess perchloric acid and waste material containing this substance should be placed in an appropriate container, clearly labeled, and handled according to your institution's waste disposal guidelines. For more information on disposal procedures, see Chapter 7 of this volume.

LABORATORY CHEMICAL SAFETY SUMMARY: PHENOL

Substance	Phenol (Carbolic acid; hydroxybenzene) CAS 108-95-2
Formula	C_6H_5OH
Physical Properties	White crystalline solid bp 182 °C, mp 41 °C Slightly soluble in water (8.4 g/100 mL)
Odor	Sweet, medicinal odor detectable at 0.06 ppm
Vapor Density	3.24 at bp (air = 1.0)
Vapor Pressure	0.36 mmHg at 20 °C
Flash Point	79 °C
Autoignition Temperature	715 °C

Toxicity Data

LD_{50} oral (rat)	317 mg/kg
LD_{50} skin (rabbit)	850 mg/m^3
PEL (OSHA)	5 ppm (19 mg/m^3)—skin
TLV-TWA (ACGIH)	5 ppm (19 mg/m^3)—skin

Major Hazards Corrosive, moderately toxic substance readily absorbed through skin; can cause severe burns to the skin and eyes.

Toxicity Phenol is a corrosive and moderately toxic substance that affects the central nervous system and can cause damage to the liver and kidneys. Phenol is irritating to the skin but has a local anesthetic effect, so that no pain may be felt on initial contact. A whitening of the area of contact generally occurs, and later severe burns may develop. Phenol is rapidly absorbed through the skin, and toxic or even fatal amounts can be absorbed through relatively small areas. Exposure to phenol vapor can cause severe irritation of the eyes, nose, throat, and respiratory tract. Acute overexposure by any route may lead to nausea, vomiting, muscle weakness, and coma. Contact of phenol with the eyes may cause severe damage and possibly blindness. Ingestion of phenol leads to burning of the mouth and throat and rapid development of digestive disturbances and the systemic effects described above. As little as 1 g can be fatal to humans. Phenol is regarded as a substance with good warning properties.

Chronic exposure to phenol may cause vomiting, diarrhea, dizziness, difficulty in swallowing, headache, skin discoloration, and injury to the liver. Phenol has not been shown to

(continued on facing page)

be a carcinogen in humans. There is some evidence from animal studies that phenol may be a reproductive toxin.

Flammability and Explosibility

Phenol is a combustible solid (NFPA rating = 2). When heated, phenol produces flammable vapors that are explosive at concentrations of 3 to 10% in air. Carbon dioxide or dry chemical extinguishers should be used to fight phenol fires.

Reactivity and Incompatibility

Phenol may react violently with strong oxidizing agents.

Storage and Handling

Phenol should be handled in the laboratory using the "basic prudent practices" described in Chapter 5.C. Because of its corrosivity and ability to penetrate the skin, all work with phenol and its solutions should be conducted while wearing impermeable gloves, appropriate protective clothing, and splash goggles. Operations with the potential to produce dusts or aerosols of phenol or its solutions should be carried out in a fume hood.

Accidents

In the event of skin contact, immediately wash with soap and water and remove contaminated clothing. In case of eye contact, promptly wash with copious amounts of water for 15 min (lifting upper and lower lids occasionally) and obtain medical attention. If phenol is ingested, obtain medical attention immediately. If large amounts of this compound are inhaled, move the person to fresh air and seek medical attention at once.

In the event of a spill, remove all ignition sources, soak up the phenol with a spill pillow or absorbent material, place in an appropriate container, and dispose of properly. Respiratory protection may be necessary in the event of a large spill or release in a confined area. Care should be taken not to walk in spills of phenol or solutions of phenol because this substance can readily penetrate leather.

Disposal

Excess phenol and waste material containing this substance should be placed in an appropriate container, clearly labeled, and handled according to your institution's waste disposal guidelines. For more information on disposal procedures, see Chapter 7 of this volume.

The information in this LCSS has been compiled by a committee of the National Research Council from literature sources and Material Safety Data Sheets and is believed to be accurate as of July 1994. This summary is intended for use by trained laboratory personnel in conjunction with the NRC report *Prudent Practices in the Laboratory: Handling and Disposal of Chemicals*. This LCSS presents a concise summary of safety information that should be adequate for most laboratory uses of the title substance, but in some cases it may be advisable to consult more comprehensive references. This information should not be used as a guide to the nonlaboratory use of this chemical.

LABORATORY CHEMICAL SAFETY SUMMARY: PHOSGENE

Substance	Phosgene (Carbonyl chloride; chloroformyl chloride; carbon oxychloride) CAS 75-44-5
Formula	$COCl_2$
Physical Properties	Colorless gas bp 8.2 °C, mp -128 °C Decomposes in water with formation of HCl
Odor	Sweet, hay-like odor at lower levels, pungent at higher levels; detectable at 0.1 to 5.7 ppm
Vapor Density	3.4 at bp (air = 1.0)
Vapor Pressure	1180 mmHg at 20 °C

Toxicity Data

LC_{50} inhal (rat)	341 ppm (1364 mg/m^3; 30 min)
PEL (OSHA)	0.1 ppm (0.4 mg/m^3)
TLV-TWA (ACGIH)	0.1 ppm (0.4 mg/m^3)

Major Hazards

Highly toxic, irritating, and corrosive gas; inhalation can cause fatal respiratory damage.

Toxicity

Phosgene is severely irritating and corrosive to all body tissues. Irritation of the throat occurs immediately at 3 ppm, while 4 ppm causes immediate eye irritation. Exposure to 20 to 30 ppm for as little as 1 min may cause severe irritation of the upper and lower respiratory tract, with symptoms including burning throat, nausea, vomiting, chest pain, coughing, shortness of breath, and headache. Brief exposure to 50 ppm can be fatal within a few hours. Severe respiratory distress may not develop for 4 to 72 hours after exposure, at which point pulmonary edema progressing to pneumonia and cardiac failure may occur. Phosgene vapor is irritating to the eyes, and the liquid can cause severe burns to the eyes and skin. Phosgene is not regarded as a substance with adequate warning properties.

Phosgene has not been found to be carcinogenic or to show reproductive or developmental toxicity in humans.

Flammability and Explosibility

Noncombustible.

Reactivity and Incompatibility

Phosgene reacts with water to form HCl and carbon dioxide.

Storage and Handling

Because of its corrosivity and high acute toxicity, phosgene should be handled using the "basic prudent practices" of Chapter 5.C, supplemented by the additional precautions for work with compounds of high toxicity (Chapter 5.D). In particular, work with phosgene should be conducted in a fume hood to prevent exposure by inhalation, and splash goggles

(continued on facing page)

and impermeable gloves should be worn at all times to prevent eye and skin contact. Containers of phosgene solutions should be stored in secondary containers, and phosgene cylinders should be stored in a cool, well-ventilated area separate from incompatible materials.

Accidents

In the event of skin contact, immediately wash with soap and water and remove contaminated clothing. In case of eye contact, promptly wash with copious amounts of water for 15 min (lifting upper and lower lids occasionally) and obtain medical attention. If phosgene is ingested, obtain medical attention immediately. If phosgene is inhaled, move the person to fresh air and seek medical attention at once.

In case of the accidental release of phosgene gas, such as from a leaking cylinder or associated apparatus, evacuate the area and eliminate the source of the leak if this can be done safely. Remove cylinder to a fume hood or remote area if it cannot be shut off. In the event of a spill of a phosgene solution, soak up the solution with a spill pillow or absorbent material, place in an appropriate container, and dispose of properly. Full respiratory protection and protective clothing will be necessary in the event of a spill or release in a confined area.

Disposal

Excess phosgene and waste material containing this substance should be placed in an appropriate container, clearly labeled, and handled according to your institution's waste disposal guidelines. For more information on disposal procedures, see Chapter 7 of this volume.

The information in this LCSS has been compiled by a committee of the National Research Council from literature sources and Material Safety Data Sheets and is believed to be accurate as of July 1994. This summary is intended for use by trained laboratory personnel in conjunction with the NRC report *Prudent Practices in the Laboratory: Handling and Disposal of Chemicals*. This LCSS presents a concise summary of safety information that should be adequate for most laboratory uses of the title substance, but in some cases it may be advisable to consult more comprehensive references. This information should not be used as a guide to the nonlaboratory use of this chemical.

LABORATORY CHEMICAL SAFETY SUMMARY: PHOSPHORUS

Substance	Phosphorus (White phosphorus, yellow phosphorus) CAS 7723-14-0
Formula	P_4
Physical Properties	White to yellow, waxy soft solid bp 279 °C, mp 44 °C (Red phosphorus is an amorphous allotropic form that sublimes at 416 °C) Insoluble in water (0.0003 g/100 mL)
Odor	Acrid fumes when exposed to air
Vapor Density	4.4 at 279 °C (air = 1.0)
Vapor Pressure	0.03 mmHg at 20 °C
Flash Point	White phosphorus: 30 °C
Autoignition Temperature	White phosphorus: 29 °C Red phosphorus: 260 °C

Toxicity Data

LD_{50} oral (rat)	3 mg/kg
PEL (OSHA)	0.1 mg/m^3
TLV-TWA (ACGIH)	0.02 ppm (0.1 mg/m^3)

Major Hazards Spontaneously ignites in air; highly toxic by all routes of exposure.

Toxicity White phosphorus is a highly toxic substance by all routes of exposure. Contact of the solid with the skin produces deep painful burns, and eye contact can cause severe damage. Ingestion of phosphorus leads (after a delay of a few hours) to symptoms including nausea, vomiting, belching, and severe abdominal pain. Apparent recovery may be followed by a recurrence of symptoms. Death may occur after ingestion of 50 to 100 mg due to circulatory, liver, and kidney effects. Phosphorus ignites and burns spontaneously when exposed to air, and the resulting vapors are highly irritating to the eyes and respiratory tract.

Red phosphorus is much less toxic than the white allotrope; however, samples of red phosphorus may contain the white form as an impurity.

Early signs of chronic systemic poisoning by phosphorus are reported to include anemia, loss of appetite, gastrointestinal distress, chronic cough, a garlic-like odor to the breath, and pallor. A common response to severe chronic poisoning is damage of the jaw ("phossy jaw") and other bones. Phosphorus has not been reported to show carcinogenic effects in humans.

(continued on facing page)

The information in this LCSS has been compiled by a committee of the National Research Council from literature sources and Material Safety Data Sheets and is believed to be accurate as of July 1994. This summary is intended for use by trained laboratory personnel in conjunction with the NRC report *Prudent Practices in the Laboratory: Handling and Disposal of Chemicals*. This LCSS presents a concise summary of safety information that should be adequate for most laboratory uses of the title substance, but in some cases it may be advisable to consult more comprehensive references. This information should not be used as a guide to the nonlaboratory use of this chemical.

Flammability and Explosibility	White phosphorus ignites spontaneously upon contact with air, producing an irritating, dense white smoke of phosphorus oxides. Use water to extinguish phosphorus fires.
	Red phosphorus is a flammable solid but does not ignite spontaneously on exposure to air. At high temperatures (~300 °C), red phosphorus is converted to the white form.
Reactivity and Incompatibility	White phosphorus reacts with a number of substances to form explosive mixtures. For example, dangerous explosion hazards are produced upon reaction of phosphorus with many oxidizing agents, including chlorates, bromates, and many nitrates, with chlorine, bromine, peracids, organic peroxides, chromium trioxide, and potassium permanganate, with alkaline metal hydroxides (phosphine gas is liberated), and with sulfur, sulfuric acid, and many metals, including the alkali metals, copper, and iron.
	Red phosphorus is much less reactive than the white allotrope but may ignite or react explosively with strong oxidizing agents.
Storage and Handling	Because of its corrosivity, flammability, and high acute toxicity, white phosphorus should be handled using the "basic prudent practices" of Chapter 5.C, supplemented by the additional precautions for work with compounds of high toxicity (Chapter 5.D) and extremely flammable substances (Chapter 5.F). In particular, work with white phosphorus should be conducted in a fume hood to prevent exposure by inhalation, and splash goggles and impermeable gloves should be worn at all times to prevent eye and skin contact. Phosphorus should be stored under water in secondary containers in areas separate from oxidizing agents and other incompatible substances. The less dangerous red form of phosphorus can be handled using the "basic prudent practices" of Chapter 5.C.
Accidents	In the event of skin contact, immediately flush with water and remove contaminated clothing. Wet the skin until medical attention is obtained to prevent any remaining phosphorus from igniting. In case of eye contact, promptly wash with copious amounts of water for 15 min (lifting upper and lower lids occasionally) and obtain medical attention. If phosphorus is ingested, give the person (if conscious) large quantities of water to drink and obtain medical attention immediately. If large amounts of phosphorus or smoke and fumes from burning phosphorus are inhaled, move the person to fresh air and seek medical attention at once.
	In the event of a spill, douse with water and cover with wet sand or earth; collect material in a suitable container and dispose of properly. Respiratory protection may be necessary in the event of a spill or release in a confined area.
Disposal	Excess phosphorus and waste material containing this substance should be placed in an appropriate container, clearly labeled, and handled according to your institution's waste disposal guidelines. For more information on disposal procedures, see Chapter 7 of this volume.

LABORATORY CHEMICAL SAFETY SUMMARY: POTASSIUM

Substance	Potassium (Kalium) CAS 7440-09-7
Formula	K
Physical Properties	Silvery white metal that loses its luster on exposure to air or moisture bp 765.5 °C, mp 63 °C Explodes on contact with water
Odor	Odorless
Autoignition Temperature	25 °C or below in air or oxygen
Major Hazards	Ignites in air and reacts explosively with water; highly corrosive to the skin and eyes.
Toxicity	Potassium reacts with the moisture on skin and other tissues to form highly corrosive potassium hydroxide. Contact of metallic potassium with the skin, eyes, or mucous membranes causes severe burns; thermal burns may also occur due to ignition of the metal and liberated hydrogen.
Flammability and Explosibility	Potassium metal may ignite spontaneously on contact with air at room temperature. Potassium reacts explosively with water to form potassium hydroxide; the heat liberated generally ignites the hydrogen formed and can initiate the combustion of potassium metal itself. Potassium fires must be extinguished with a class D dry chemical extinguisher or by the use of sand, ground limestone, dry clay or graphite, or ''Met-L-X®'' type solids. *Water or CO_2 extinguishers must never be used on potassium fires.*
Reactivity and Incompatibility	Potassium is one of the most potent reducing reagents known. The metal reacts explosively with water, oxygen, and air to form potassium hydroxide and/or potassium oxide. Potassium reacts violently with many oxidizing agents and organic and inorganic halides and can form unstable and explosive mixtures with elemental halogens. Explosive mixtures form when potassium reacts with halogenated hydrocarbons such as carbon tetrachloride and upon reaction with carbon monoxide, carbon dioxide, and carbon disulfide. Potassium stored under mineral oil can form shock-sensitive peroxides if oxygen is present, so the metal must always be stored and handled under inert gases such as dry nitrogen or argon. It dissolves with such exothermicity in other metals such as mercury that the molten alloy may melt Pyrex glassware. Note that the reactivity of potassium is generally related to its surface area and the cleanliness of the surface at hand; chunks of potassium are less reactive than the very dangerous dispersions and sands.
Storage and Handling	Potassium should be handled in the laboratory using the ''basic prudent practices'' described in Chapter 5.C, supplemented by the additional precautions for work with flammable (Chapter 5.F) and reactive (Chapter 5.G) substances. Safety glasses, imperme-

(continued on facing page)

able gloves, and a fire-retardant laboratory coat should be worn at all times when working with potassium, and the metal should be handled under the surface of an inert liquid such as mineral oil, xylene, or toluene. Potassium should be used only in areas free of ignition sources and should be stored under mineral oil in tightly sealed metal containers under an inert gas such as argon. Potassium metal that has formed a yellow oxide coating should be disposed of immediately; do not attempt to cut such samples with a knife since the oxide coating may be explosive.

Accidents

In the event of skin contact, remove contaminated clothing and any metal particles and immediately wash with soap and water. In case of eye contact, promptly wash with copious amounts of water for 15 min (lifting upper and lower lids occasionally) and obtain medical attention. If potassium is ingested, obtain medical attention immediately.

In the event of a spill, remove all ignition sources, quench the resulting potassium fire with a dry chemical extinguishing medium, sweep up, place in an appropriate container under an inert atmosphere, and dispose of properly. Respiratory protection may be necessary in the event of a spill or release in a confined area.

Disposal

Excess potassium and waste material containing this substance should be placed in an appropriate container under an inert atmosphere, clearly labeled, and handled according to your institution's waste disposal guidelines. Experienced personnel can destroy small scraps of potassium by carefully adding *t*-butanol or *n*-butanol to a beaker containing the metal scraps covered in an inert solvent such as xylene or toluene. For more information on disposal procedures, see Chapter 7 of this volume.

LABORATORY CHEMICAL SAFETY SUMMARY: POTASSIUM HYDRIDE AND SODIUM HYDRIDE

Substance	Potassium hydride CAS 7693-26-7	Sodium hydride CAS 7646-69-7

(Commonly handled as dispersions in mineral oil)

Formula KH; NaH

Physical Properties White to brownish-gray crystalline powders (white-gray or white-beige dispersion in mineral oil)
NaH: mp 800 °C (decomposes)
Reacts violently with water

Autoignition Temperature Ignites spontaneously at room temperature in moist air

Major Hazards Reacts violently with water, liberating highly flammable hydrogen gas; causes severe burns on eye or skin contact.

Toxicity Sodium hydride and potassium hydride react with the moisture on skin and other tissues to form highly corrosive sodium and potassium hydroxide. Contact of these hydrides with the skin, eyes, or mucous membranes causes severe burns; thermal burns may also occur due to ignition of the liberated hydrogen gas.

Flammability and Explosibility Potassium hydride and sodium hydride are flammable solids that ignite on contact with moist air. Potassium hydride presents a more serious fire hazard than sodium hydride. The mineral oil dispersions do not ignite spontaneously on exposure to the atmosphere. Sodium hydride and potassium hydride fires must be extinguished with a class D dry chemical extinguisher or by the use of sand, ground limestone, dry clay or graphite, or "Met-L-X®" type solids. *Water or CO_2 extinguishers must never be used on sodium and potassium hydride fires.*

Reactivity and Incompatibility Potassium hydride and sodium hydride react violently with water, liberating hydrogen, which can ignite. Oil dispersions of these hydrides are much safer to handle because the mineral oil serves as a barrier to moisture and air. Potassium hydride may react violently with oxygen, CO, dimethyl sulfoxide, alcohols, and acids. Explosions can result from contact of these compounds with strong oxidizers. Potassium hydride is generally more reactive than sodium hydride.

Storage and Handling Sodium hydride and potassium hydride should be handled in the laboratory using the "basic prudent practices" described in Chapter 5.C, supplemented by the additional precautions for work with flammable (Chapter 5.F) and highly reactive (Chapter 5.G) substances. Safety glasses, impermeable gloves, and a fire-retardant laboratory coat should be worn at all times when working with these substances. These hydrides should be used only in areas free of ignition sources and should be stored preferably as mineral oil dispersions under an inert gas such as argon.

(continued on facing page)

The information in this LCSS has been compiled by a committee of the National Research Council from literature sources and Material Safety Data Sheets and is believed to be accurate as of July 1994. This summary is intended for use by trained laboratory personnel in conjunction with the NRC report *Prudent Practices in the Laboratory: Handling and Disposal of Chemicals*. This LCSS presents a concise summary of safety information that should be adequate for most laboratory uses of the title substance, but in some cases it may be advisable to consult more comprehensive references. This information should not be used as a guide to the nonlaboratory use of this chemical.

Accidents   In the event of skin contact, immediately wash with soap and water and remove contaminated clothing. In case of eye contact, promptly wash with copious amounts of water for 15 min (lifting upper and lower lids occasionally) and obtain medical attention. If potassium hydride or sodium hydride is ingested, obtain medical attention immediately. If sodium hydride dust is inhaled, move the person to fresh air and seek medical attention at once.

In the event of a spill, remove all ignition sources, quench the metal hydride, whether burning or not, with a dry chemical extinguishing medium, sweep up, place in an appropriate container under an inert atmosphere, and dispose of properly. Respiratory protection may be necessary in the event of a spill or release in a confined area.

Disposal   Excess potassium or sodium hydride and waste material containing these substances should be placed in an appropriate container under an inert atmosphere, clearly labeled, and handled according to your institution's waste disposal guidelines. Experienced personnel can destroy small quantities of sodium hydride and potassium hydride by the careful dropwise addition of *t*-butanol or *iso*-propanol to a suspension of the metal hydride in an inert solvent such as toluene under an inert atmosphere such as argon. Great care must be taken in the destruction of potassium hydride because of its greater reactivity. The resulting mixture of metal alkoxide should be placed in an appropriate container, clearly labeled, and handled according to your institution's waste disposal guidelines. For more information on disposal procedures, see Chapter 7 of this volume.

LABORATORY CHEMICAL SAFETY SUMMARY: PYRIDINE

Substance	Pyridine
	(Azabenzene; azine)
	CAS 110-86-1
Formula	C_5H_5N
Physical Properties	Colorless or pale yellow liquid
	bp 115 °C, mp −42 °C
	Miscible with water
Odor	Nauseating odor detectable at 0.23 to 1.9 ppm (mean = 0.66 ppm)
Vapor Density	2.72 at bp (air = 1.0)
Vapor Pressure	18 mmHg at 20 °C
Flash Point	20 °C
Autoignition Temperature	482 °C

Toxicity Data

LD_{50} oral (rat)	891 mg/kg
LD_{50} skin (rabbit)	1121 mg/m^3
PEL (OSHA)	5 ppm (15 mg/m^3)
TLV-TWA (ACGIH)	5 ppm (15 mg/m^3)

Major Hazards   Highly flammable liquid

Toxicity   The acute toxicity of pyridine is low. Inhalation causes irritation of the respiratory system and may affect the central nervous system, causing headache, nausea, vomiting, dizziness, and nervousness. Pyridine irritates the eyes and skin and is readily absorbed, leading to systemic effects. Ingestion of pyridine can result in liver and kidney damage. Pyridine causes olfactory fatigue, and its odor does not provide adequate warning of the presence of harmful concentrations.

Pyridine has not been found to be carcinogenic or to show reproductive or developmental toxicity in humans. Chronic exposure to pyridine can result in damage to the liver, kidneys, and central nervous system.

Flammability and Explosibility   Pyridine is a highly flammable liquid (NFPA rating = 3), and its vapor can travel a considerable distance and "flash back." Pyridine vapor forms explosive mixtures with air at concentrations of 1.8 to 12.4% (by volume). Carbon dioxide or dry chemical extinguishers should be used for pyridine fires.

(continued on facing page)

Reactivity and Incompatibility

Pyridine may react violently with dinitrogen tetroxide, acid chlorides and anhydrides, perchloric acid, and strong oxidizing agents.

Storage and Handling

Pyridine should be handled in the laboratory using the "basic prudent practices" described in Chapter 5.C. In particular, pyridine should be used only in areas free of ignition sources, and quantities greater than 1 liter should be stored in tightly sealed metal containers in areas separate from oxidizers.

Accidents

In the event of skin contact, immediately wash with soap and water and remove contaminated clothing. In case of eye contact, promptly wash with copious amounts of water for 15 min (lifting upper and lower lids occasionally) and obtain medical attention. If pyridine is ingested, obtain medical attention immediately. If large amounts of this compound are inhaled, move the person to fresh air and seek medical attention at once.

In the event of a spill, remove all ignition sources, soak up the pyridine with a spill pillow or absorbent material, place in an appropriate container, and dispose of properly. Respiratory protection may be necessary in the event of a large spill or release in a confined area.

Disposal

Excess pyridine and waste material containing this substance should be placed in an appropriate container, clearly labeled, and handled according to your institution's waste disposal guidelines. For more information on disposal procedures, see Chapter 7 of this volume.

LABORATORY CHEMICAL SAFETY SUMMARY: SILVER AND ITS COMPOUNDS

Substance	Silver and its compounds (Argentum) CAS 7440-22-4
Formula	Ag
Physical Properties	White metallic solid bp 2200 °C, mp 961 °C Insoluble in water
Odor	Odorless

Toxicity Data

PEL (OSHA)	0.01 mg/m^3
TLV-TWA (ACGIH)	0.1 mg/m^3 (silver metal)
TLV-TWA (ACGIH)	0.01 mg/m^3 (soluble silver compounds, as Ag)

Major Hazards

Exposure to silver metal or soluble silver compounds can cause discoloration or blue-gray darkening of the eyes, nose, throat, and skin.

Toxicity

The acute toxicity of silver metal is low. The acute toxicity of soluble silver compounds depends on the counterion and must be evaluated case by case. For example, silver nitrate is strongly corrosive and can cause burns and permanent damage to the eyes and skin.

Chronic exposure to silver or silver salts can cause a local or generalized darkening of the mucous membranes, skin, and eyes known as argyria. The other chronic effects of silver compounds must be evaluated individually.

Flammability and Explosibility

Silver and most soluble silver compounds are not combustible. However, silver nitrate and certain other silver compounds are oxidizers and can increase the flammability of combustible materials.

Silver acetylide, azide, fulminate, oxalate mixtures, styphnate, tartarate mixtures, and tetrazene are all explosives and must be handled as such.

Reactivity and Incompatibility

Contact of metallic silver and silver compounds with acetylene may cause formation of silver acetylide, which is a shock-sensitive explosive. Contact with ammonia may cause formation of compounds that are explosive when dry. Contact with strong hydrogen peroxide solutions causes violent decomposition with the formation of oxygen gas.

Many silver compounds are light sensitive, and many have significant reactivities or incompatibilities, which should be evaluated before use.

Storage and Handling

Silver and silver compounds should be handled in the laboratory using the "basic prudent practices" described in Chapter 5.C. Individual silver compounds should be evaluated on a case-by-case basis to determine whether additional handling procedures for high

(continued on facing page)

toxicity (Chapter 5.D) or reactivity and explosibility (Chapter 5.G) are appropriate. Most silver compounds should be protected from light during storage or while in use.

Accidents

In the event of skin contact, immediately wash with soap and water and remove contaminated clothing. In case of eye contact, promptly wash with copious amounts of water for 15 min (lifting upper and lower lids occasionally) and obtain medical attention. If silver or silver compounds are ingested, obtain medical attention immediately. If large amounts of silver dust or silver compounds are inhaled, move the person to fresh air and seek medical attention at once.

In the event of a spill, sweep up the silver or silver compounds or soak up with a nonreactive absorbent material or spill pillow, place in an appropriate container, and dispose of properly. Respiratory protection may be necessary in the event of a large spill or release in a confined area.

Disposal

Excess silver, silver compounds, and waste material containing these substances should be placed in an appropriate container, clearly labeled, and handled according to your institution's waste disposal guidelines. Collection for silver recovery should be considered. For more information on disposal procedures, see Chapter 7 of this volume.

LABORATORY CHEMICAL SAFETY SUMMARY: SODIUM

Substance	Sodium (Natrium) CAS 7440-23-5
Formula	Na
Physical Properties	Soft, silvery-white metal bp 881.4 °C, mp 97.8 °C Reacts violently with water
Vapor Pressure	1.2 mmHg at 400 °C
Autoignition Temperature	>115 °C in air
Major Hazards	Reacts violently with water, liberating highly flammable hydrogen gas; causes severe burns on eye or skin contact.
Toxicity	Sodium reacts with the moisture on skin and other tissues to form highly corrosive sodium hydroxide. Contact of metallic sodium with the skin, eyes, or mucous membranes causes severe burns; thermal burns may also occur due to ignition of the metal and liberated hydrogen.
Flammability and Explosibility	Sodium spontaneously ignites when heated above 115 °C in air that has even modest moisture content, and any sodium vapor generated is even more flammable. Sodium reacts violently on contact with water and often ignites or explodes the hydrogen formed. Sodium fires must be extinguished with a class D dry chemical extinguisher or by the use of sand, ground limestone, dry clay or graphite, or "Met-L-X®" type solids. *Water or CO_2 extinguishers must never be used on sodium fires.*
Reactivity and Incompatibility	Sodium is a potent reducing agent and reacts violently with water to form hydrogen and sodium hydroxide. It also reacts violently with mineral acids and halogens and reacts exothermically with oxidizing agents, organic and inorganic halides, and protic media. Shock-sensitive mixtures can form upon reaction of sodium with halogenated hydrocarbons such as carbon tetrachloride and chloroform. Sodium also reacts to generate shock-sensitive products with sulfur oxides and phosphorus, and reacts with incandescence with many metal oxides such as mercurous and lead oxides. Sodium dissolves in many other metals such as mercury and potassium with great evolution of heat. The reactivity of a sample of sodium is largely related to its surface area. Thus, reactions involving large pieces of sodium metal (especially those with some oxide or hydroxide coating) may be slow and controlled, but similar reactions involving clean, high-surface-area sodium dispersions may be vigorous or violent.
Storage and Handling	Sodium should be handled in the laboratory using the "basic prudent practices" described in Chapter 5.C, supplemented by the additional precautions for work with flammable

(continued on facing page)

The information in this LCSS has been compiled by a committee of the National Research Council from literature sources and Material Safety Data Sheets and is believed to be accurate as of July 1994. This summary is intended for use by trained laboratory personnel in conjunction with the NRC report *Prudent Practices in the Laboratory: Handling and Disposal of Chemicals*. This LCSS presents a concise summary of safety information that should be adequate for most laboratory uses of the title substance, but in some cases it may be advisable to consult more comprehensive references. This information should not be used as a guide to the nonlaboratory use of this chemical.

(Chapter 5.F) and reactive (Chapter 5.G) substances. Safety glasses, impermeable gloves, and a fire-retardant laboratory coat should be worn at all times when working with sodium, and the metal should be handled under the surface of an inert liquid such as mineral oil, xylene, or toluene. Sodium should be used only in areas free of ignition sources and should be stored under mineral oil in tightly sealed metal containers under an inert gas such as argon.

Accidents

In the event of skin contact, immediately remove contaminated clothing and any metal particles and wash with soap and water. In case of eye contact, promptly wash with copious amounts of water for 15 min (lifting upper and lower lids occasionally) and obtain medical attention. If sodium is ingested, obtain medical attention immediately.

In the event of a spill, remove all ignition sources, cover the sodium with a dry chemical extinguishing agent, sweep up, place in an appropriate container under an inert atmosphere, and dispose of properly. Respiratory protection may be necessary in the event of a spill or release in a confined area.

Disposal

Excess sodium and waste material containing this substance can be placed in an appropriate container under an inert atmosphere, clearly labeled, and handled according to your institution's waste disposal guidelines. Experienced personnel can destroy small scraps of sodium by carefully adding 95% ethanol to a beaker containing the metal scraps covered in an inert solvent such as xylene or toluene. The resulting mixture should then be placed in an appropriate container, clearly labeled, and handled according to your institution's waste disposal guidelines. For more information on disposal procedures, see Chapter 7 of this volume.

LABORATORY CHEMICAL SAFETY SUMMARY: SODIUM AZIDE

Substance	Sodium azide (Hydrazoic acid, sodium salt) CAS 26628-22-8
Formula	NaN_3
Physical Properties	Colorless crystalline solid mp >275 °C (decomposes) Readily soluble in water (41.7 g/100 mL at 17 °C)
Odor	Odorless solid

Toxicity Data

LD_{50} oral (rat)	27 mg/kg
LD_{50} skin (rabbit)	20 mg/kg
TLV-TWA (ACGIH)	0.29 mg/m^3 (ceiling)

Major Hazards   Highly toxic by inhalation, ingestion, or skin absorption.

Toxicity

The acute toxicity of sodium azide is high. Symptoms of exposure include lowered blood pressure, headache, hypothermia, and in the case of serious overexposure, convulsions and death. Ingestion of 100 to 200 mg in humans may result in headache, respiratory distress, and diarrhea. Target organs are primarily the central nervous system and brain. Sodium azide rapidly hydrolyzes in water to form hydrazoic acid, a highly toxic gas that can escape from solution, presenting a serious inhalation hazard. Symptoms of acute exposure to hydrazoic acid include eye irritation, headache, dramatic decrease in blood pressure, weakness, pulmonary edema, and collapse. Solutions of sodium azide can be absorbed through the skin.

Sodium azide has not been found to be carcinogenic in humans. Chronic, low-level exposure may cause nose irritation, episodes of falling blood pressure, dizziness, and bronchitis.

Flammability and Explosibility

Flammability hazard is low, but violent decomposition can occur when heated to 275 °C. Decomposition products include oxides of nitrogen and sodium oxide.

Reactivity and Incompatibility

Sodium azide should not be allowed to come into contact with heavy metals or their salts, because it may react to form heavy metal azides, which are notorious shock-sensitive explosives. Do not pour sodium azide solutions into a copper or lead drain. Sodium azide reacts violently with carbon disulfide, bromine, nitric acid, dimethyl sulfate, and a number of heavy metals, including copper and lead. Reaction with water and acids liberates highly toxic hydrazoic acid, which is a dangerous explosive. Sodium azide is reported to react with CH_2Cl_2 in the presence of DMSO to form explosive products.

Storage and Handling

Because of its high toxicity, sodium azide should be handled in the laboratory using the "basic prudent practices" of Chapter 5.C, supplemented by the additional precautions

(continued on facing page)

for work with compounds of high toxicity (Chapter 5.D). In particular, work with sodium azide should be conducted in a fume hood to prevent exposure by inhalation, and appropriate impermeable gloves and splash goggles should be worn at all times to prevent skin and eye contact. Containers of sodium azide should be stored in secondary containers in a cool, dry place separated from acids.

Accidents

In the event of skin contact, immediately wash with soap and water and remove contaminated clothing. In case of eye contact, promptly wash with copious amounts of water for 15 min (lifting upper and lower lids occasionally) and obtain medical attention. If sodium azide is ingested, obtain medical attention immediately. If large amounts of this compound are inhaled, move the person to fresh air and seek medical attention at once.

In the event of a spill, cover sodium azide with sand, sweep up, and place in a container for disposal. Soak up spilled solutions with a spill pillow or absorbent material, place in an appropriate container, and dispose of properly. Respiratory protection may be necessary in the event of a large spill or release in a confined area.

Disposal

Excess sodium azide and waste material containing this substance should be placed in an appropriate container, clearly labeled, and handled according to your institution's waste disposal guidelines. For more information on disposal procedures, see Chapter 7 of this volume.

The information in this LCSS has been compiled by a committee of the National Research Council from literature sources and Material Safety Data Sheets and is believed to be accurate as of July 1994. This summary is intended for use by trained laboratory personnel in conjunction with the NRC report *Prudent Practices in the Laboratory: Handling and Disposal of Chemicals*. This LCSS presents a concise summary of safety information that should be adequate for most laboratory uses of the title substance, but in some cases it may be advisable to consult more comprehensive references. This information should not be used as a guide to the nonlaboratory use of this chemical.

LABORATORY CHEMICAL SAFETY SUMMARY: SODIUM CYANIDE AND POTASSIUM CYANIDE

Substance	Sodium cyanide	Potassium cyanide
	CAS 143-33-9	CAS 151-50-8

Formula NaCN, KCN

Physical Properties
White solids
NaCN: bp 1496 °C, mp 564 °C
KCN: bp 1625 °C, mp 634 °C
Soluble in water (NaCN: 37 g/100 mL; KCN: 41 g/100 mL)

Odor
The dry salts are odorless, but reaction with atmospheric moisture produces HCN, whose bitter almond odor is detectable at 1 to 5 ppm; however, 20 to 60% of the population are reported to be unable to detect the odor of HCN.

Vapor Pressure Negligible

Flash Point Noncombustible

Toxicity Data

LD$_{50}$ oral (rat)	6.4 mg/kg (NaCN)
	5 mg/kg (KCN)
TLV-TWA (ACGIH)	5 mg/kg (KCN)—skin

Major Hazards
Highly toxic; exposure by eye or skin contact or ingestion can be rapidly fatal.

Toxicity
The acute toxicity of these metal cyanides is high. Ingestion of NaCN or KCN or exposure to the salts or their aqueous solutions by eye or skin contact can be fatal; exposure to as little as 50 to 150 mg can cause immediate collapse and death. Poisoning can occur by inhalation of mists of cyanide solutions and by inhalation of HCN produced by the reaction of metal cyanides with acids and with water. Symptoms of nonlethal exposure to cyanide include weakness, headache, dizziness, rapid breathing, nausea, and vomiting. These compounds are not regarded as having good warning properties.

Effects of chronic exposure to sodium cyanide or potassium cyanide are nonspecific and rare.

Flammability and Explosibility
Sodium cyanide and potassium cyanide are noncombustible solids. Reaction with acids liberates flammable HCN.

Reactivity and Incompatibility
Reaction with acid produces highly toxic and flammable hydrogen cyanide gas. Reaction with water can produce dangerous amounts of HCN in confined areas.

Storage and Handling
Sodium cyanide and potassium cyanide should be handled in the laboratory using the "basic prudent practices" described in Chapter 5.C, supplemented by the additional practices for work with compounds of high toxicity (Chapter 5.D). In particular, work with cyanides should be conducted in a fume hood to prevent exposure by inhalation,

(continued on facing page)

The information in this LCSS has been compiled by a committee of the National Research Council from literature sources and Material Safety Data Sheets and is believed to be accurate as of July 1994. This summary is intended for use by trained laboratory personnel in conjunction with the NRC report *Prudent Practices in the Laboratory: Handling and Disposal of Chemicals*. This LCSS presents a concise summary of safety information that should be adequate for most laboratory uses of the title substance, but in some cases it may be advisable to consult more comprehensive references. This information should not be used as a guide to the nonlaboratory use of this chemical.

and splash goggles and impermeable gloves should be worn at all times to prevent eye and skin contact. Cyanide salts should be stored in a cool, dry location, separated from acids.

Accidents

In the event of skin contact, immediately wash with soap and water and remove contaminated clothing. In case of eye contact, promptly wash with copious amounts of water for 15 min (lifting upper and lower lids occasionally) and obtain medical attention. If sodium or potassium cyanide is ingested, obtain medical attention immediately. If cyanide is inhaled, move the person to fresh air and seek medical attention at once. Specific medical procedures for treating cyanide exposure are available but usually must be administered by properly trained personnel. Consult your environmental safety office or its equivalent before beginning work with cyanides.

In the event of a spill, remove all ignition sources, soak up the sodium cyanide or potassium cyanide with a spill pillow or absorbent material, place in an appropriate container, and dispose of properly. Respiratory protection may be necessary in the event of a large spill or release in a confined area.

Disposal

Excess sodium or potassium cyanide and waste material containing this substance should be placed in an appropriate container, clearly labeled, and handled according to your institution's waste disposal guidelines. For more information on disposal procedures, see Chapter 7 of this volume.

LABORATORY CHEMICAL SAFETY SUMMARY: SODIUM HYDROXIDE AND POTASSIUM HYDROXIDE

Substance	Sodium hydroxide (Sodium hydrate, caustic soda, lye, caustic) CAS 1310-73-2	Potassium hydroxide (Potassium hydrate, caustic potash) CAS 1310-58-3
Formula	NaOH	KOH
Physical Properties	bp 1390 °C, mp 318 °C Highly soluble in water (109 g/100 mL)	bp 1320 °C, mp 360 °C Highly soluble in water
Odor	Odorless	Odorless

Toxicity Data

	Sodium hydroxide		Potassium hydroxide	
LD_{50} oral (rat)	140 to 340 mg/kg		LD_{50} oral (rat)	365 mg/kg
LD_{50} skin (rabbit)	1350 mg/kg		LD_{50} skin (rabbit)	1260 mg/kg
PEL (OSHA)	2 mg/m^3		PEL (OSHA)	2 mg/m^3
TLV (ACGIH)	2 mg/m^3; ceiling		TLV (ACGIH)	2 mg/m^3; ceiling

Major Hazards   Extremely corrosive; causes severe burns to skin, eyes, and mucous membranes.

Toxicity   The alkali metal hydroxides are highly corrosive substances; contact of solutions, dusts, or mists with the skin, eyes, and mucous membranes can lead to severe damage. Skin contact with the solid hydroxides or concentrated solutions can cause rapid tissue destruction and severe burns. In contrast to acids, hydroxides do not coagulate protein (which impedes penetration), and metal hydroxide burns may not be immediately painful while skin penetration occurs to produce severe and slow-healing burns. Potassium hydroxide is somewhat more corrosive than sodium hydroxide. Contact with even dilute solutions will also cause skin irritation and injury, the severity of which will depend on the duration of contact. Eye exposure to concentrated sodium hydroxide or potassium hydroxide solutions can cause severe eye damage and possibly blindness. Ingestion of concentrated solutions of sodium hydroxide or potassium hydroxide can cause severe abdominal pain, as well as serious damage to the mouth, throat, esophagus, and digestive tract. Inhalation of sodium/potassium hydroxide dust or mist can cause irritation and damage to the respiratory tract, depending on the concentration and duration of exposure. Exposure to high concentrations may result in delayed pulmonary edema.

Repeated or prolonged contact may cause dermatitis. Sodium hydroxide and potassium hydroxide have not been found to be carcinogenic or to show reproductive or developmental toxicity in humans.

Flammability and Explosibility   Sodium hydroxide and potassium hydroxide are not flammable as solids or aqueous solutions.

(continued on facing page)

The information in this LCSS has been compiled by a committee of the National Research Council from literature sources and Material Safety Data Sheets and is believed to be accurate as of July 1994. This summary is intended for use by trained laboratory personnel in conjunction with the NRC report *Prudent Practices in the Laboratory: Handling and Disposal of Chemicals*. This LCSS presents a concise summary of safety information that should be adequate for most laboratory uses of the title substance, but in some cases it may be advisable to consult more comprehensive references. This information should not be used as a guide to the nonlaboratory use of this chemical.

Reactivity and Incompatibility

Concentrated sodium hydroxide and potassium hydroxide react vigorously with acids with evolution of heat, and dissolution in water is highly exothermic. Reaction with aluminum and other metals may lead to evolution of hydrogen gas. The solids in prolonged contact with chloroform, trichloroethylene, and tetrachloroethanes can produce explosive products. Many organic compounds such as propylene oxide, allyl alcohol, glyoxal, acetaldehyde, acrolein, and acrylonitrile can violently polymerize on contact with concentrated base. Reaction with nitromethane and nitrophenols produces shock-sensitive explosive salts. Sodium hydroxide and potassium hydroxide as solids absorb moisture and carbon dioxide from the air to form the bicarbonates. Aqueous solutions also absorb carbon dioxide to form bicarbonate. Solutions stored in flasks with ground glass stoppers may leak air and freeze the stoppers, preventing removal.

Storage and Handling

Sodium hydroxide and potassium hydroxide should be handled in the laboratory using the "basic prudent practices" described in Chapter 5.C. In particular, splash goggles and impermeable gloves should be worn at all times when handling these substances to prevent eye and skin contact. Operations with metal hydroxide solutions that have the potential to create aerosols should be conducted in a fume hood to prevent exposure by inhalation. NaOH and KOH generate considerable heat when dissolved in water; when mixing with water, always add caustics slowly to the water and stir continuously. Never add water in limited quantities to solid hydroxides. Containers of hydroxides should be stored in a cool, dry location, separated from acids and incompatible substances.

Accidents

In cases of eye contact, immediate and continuous irrigation with flowing water for at least 15 min is imperative. Prompt medical consultation is essential. In case of skin contact, immediately remove contaminated clothing and flush affected area with large amounts of water for 15 min and obtain medical attention without delay. If sodium hydroxide or potassium hydroxide is ingested, do not induce vomiting; give large amounts of water and transport to medical facility immediately. If dusts or mists of these compounds are inhaled, move the person to fresh air and seek medical attention at once.

Disposal

In many localities, sodium/potassium hydroxide may be disposed of down the drain after appropriate dilution and neutralization. If neutralization and drain disposal is not permitted, excess hydroxide and waste material containing this substance should be placed in an appropriate container, clearly labeled, and handled according to your institution's waste disposal guidelines. For more information on disposal procedures, see Chapter 7 of this volume.

LABORATORY CHEMICAL SAFETY SUMMARY: SULFUR DIOXIDE

Substance	Sulfur dioxide (Sulfurous oxide, sulfur oxide, sulfurous anhydride) CAS 7446-09-5
Formula	SO_2
Physical Properties	Colorless gas or liquid under pressure bp -10.0 °C, mp -75.5 °C Soluble in water (10 g/100 mL at 20 °C)
Odor	Pungent odor detectable at 0.3 to 5 ppm
Vapor Density	2.26 (air = 1.0)
Vapor Pressure	1779 mmHg at 21 °C
Flash Point	Noncombustible

Toxicity Data

LC_{50} inhal (rat)	2520 ppm (6590 mg/m³; 1 h)
LC_{LO} inhal (human)	1000 ppm (2600 mg/m³; 10 min)
PEL (OSHA)	5 ppm (13 mg/m³)
TLV-TWA (ACGIH)	2 ppm (5.2 mg/m³)
STEL (ACGIH)	5 ppm (13 mg/m³)

Major Hazards   Intensely irritating to the skin, eyes, and respiratory tract; moderate acute toxicity.

The acute toxicity of sulfur dioxide is moderate. Inhalation of high concentrations may cause death as a result of respiratory paralysis and pulmonary edema. Exposure to 400 to 500 ppm is immediately dangerous, and 1000 ppm for 10 min is reported to have caused death in humans. Sulfur dioxide gas is a severe corrosive irritant of the eyes, mucous membranes, and skin. Its irritant properties are due to the rapidity with which it forms sulfurous acid on contact with moist membranes. When sulfur dioxide is inhaled, most of it is absorbed in the upper respiratory passages, where most of its effects then occur. Exposure to concentrations of 10 to 50 ppm for 5 to 15 min causes irritation of the eyes, nose, and throat, choking, and coughing. Some individuals are extremely sensitive to the effects of sulfur dioxide, while experienced workers may become adapted to its irritating properties. Sulfur dioxide is regarded as a substance with good warning properties except in the case of individuals with reactive respiratory tracts and asthmatics. Exposure of the eyes to liquid sulfur dioxide from pressurized containers can cause severe burns, resulting in the loss of vision. Liquid SO_2 on the skin produces skin burns from the freezing effect of rapid evaporation.

Sulfur dioxide has not been shown to be carcinogenic or to have reproductive or developmental effects in humans. Chronic exposure to low levels of sulfur dioxide has been shown to exacerbate pulmonary disease.

(continued on facing page)

The information in this LCSS has been compiled by a committee of the National Research Council from literature sources and Material Safety Data Sheets and is believed to be accurate as of July 1994. This summary is intended for use by trained laboratory personnel in conjunction with the NRC report *Prudent Practices in the Laboratory: Handling and Disposal of Chemicals*. This LCSS presents a concise summary of safety information that should be adequate for most laboratory uses of the title substance, but in some cases it may be advisable to consult more comprehensive references. This information should not be used as a guide to the nonlaboratory use of this chemical.

Flammability and Explosibility

Sulfur dioxide is a noncombustible substance (NFPA rating = 0).

Reactivity and Incompatibility

Contact with some powdered metals and with alkali metals such as sodium or potassium may cause fires and explosions. Liquid sulfur dioxide will attack some forms of plastics, rubber, and coatings.

Storage and Handling

Sulfur dioxide should be handled in the laboratory using the "basic prudent practices" described in Chapter 5.C, supplemented by the procedures for work with compressed gases (Chapter 5.H).

Accidents

In the event of skin contact, immediately wash with water and remove contaminated clothing. In case of eye contact, promptly wash with copious amounts of water for 15 min (lifting upper and lower lids occasionally) and obtain medical attention. If large amounts of this compound are inhaled, move the person to fresh air and seek medical attention at once.

Leaks of sulfur dioxide may be detected by passing a rag dampened with aqueous NH_3 over the suspected valve or fitting. White fumes indicate escaping SO_2 gas. To respond to a release, use appropriate protective equipment and clothing. Positive pressure air-supplied respiratory protection is required. Close cylinder valve and ventilate area. Remove cylinder to a fume hood or remote area if it cannot be shut off. If in liquid form, allow to vaporize.

Disposal

Excess sulfur dioxide should be returned to the manufacturer if possible, according to your institution's waste disposal guidelines. For more information on disposal procedures, see Chapter 7 of this volume.

LABORATORY CHEMICAL SAFETY SUMMARY: SULFURIC ACID

Substance	Sulfuric acid (Oil of vitriol) CAS 7664-93-9
Formula	H_2SO_4
Physical Properties	Clear, colorless, oily liquid bp 300 to 338 °C (loses SO_3 above 300 °C), mp 11 °C Miscible with water in all proportions
Odor	Odorless
Vapor Density	3.4 (air = 1.0)
Vapor Pressure	<0.3 mmHg at 25 °C
Flash Point	Noncombustible

Toxicity Data

LD_{50} oral (rat)	2140 mg/kg
LC_{50} inhal (rat)	347 mg/m^3 (1 h)
PEL (OSHA)	1 mg/m^3
TLV-TWA (ACGIH)	1 mg/m^3
STEL (ACGIH)	3 mg/m^3

Major Hazards

Highly corrosive; causes severe burns on eye and skin contact and upon inhalation of sulfuric acid mist; highly reactive, reacts violently with many organic and inorganic substances.

Toxicity

Concentrated sulfuric acid is a highly corrosive liquid that can cause severe, deep burns upon skin contact. The concentrated acid destroys tissue because of its dehydrating action, while dilute H_2SO_4 acts as a skin irritant because of its acid character. Eye contact with concentrated H_2SO_4 causes severe burns, which can result in permanent loss of vision; contact with dilute H_2SO_4 results in more transient effects from which recovery may be complete. Sulfuric acid mist severely irritates the eyes, respiratory tract, and skin. Because of its low vapor pressure, the principal inhalation hazard from sulfuric acid involves breathing in acid mists, which may result in irritation of the upper respiratory passages and erosion of dental surfaces. Higher inhalation exposures may lead to temporary lung irritation with difficulty breathing. Ingestion of sulfuric acid may cause severe burns to the mucous membranes of the mouth and esophagus.

Animal testing with sulfuric acid did not demonstrate carcinogenic, mutagenic, embryotoxic, or reproductive effects. Chronic exposure to sulfuric acid mist may lead to bronchitis, skin lesions, conjunctivitis, and erosion of teeth.

(continued on facing page)

Flammability and Explosibility

Sulfuric acid is noncombustible but can cause finely divided combustible substances to ignite. Sulfuric acid reacts with most metals, especially when dilute, to produce flammable and potentially explosive hydrogen gas.

Reactivity and Incompatibility

Concentrated sulfuric acid is stable, but may react violently with water and with many organic compounds because of its action as a powerful dehydrating, oxidizing, and sulfonating agent. Ignition or explosions may occur on contact of sulfuric acid with many metals, carbides, chlorates, perchlorates, permanganates, bases, and reducing agents. Sulfuric acid reacts with a number of substances to generate highly toxic products. Examples include the reaction of H_2SO_4 with formic or oxalic acid (CO formation), with cyanide salts (HCN formation), and sodium bromide (SO_2 and Br_2 formation).

Storage and Handling

Sulfuric acid should be handled in the laboratory using the "basic prudent practices" described in Chapter 5.C. Splash goggles and rubber gloves should be worn when handling this acid, and containers of sulfuric acid should be stored in a well-ventilated location, separated from organic substances and other combustible materials. Containers of sulfuric acid should be stored in secondary plastic trays to avoid corrosion of metal storage shelves due to drips or spills. Water should never be added to sulfuric acid because splattering may result; always add acid to water.

Accidents

In the event of skin contact, immediately wash with soap and water and remove contaminated clothing. In case of eye contact, promptly wash with copious amounts of water for 15 min (lifting upper and lower lids occasionally) and obtain medical attention. If sulfuric acid is ingested, obtain medical attention immediately. If large amounts of sulfuric acid mist are inhaled, move the person to fresh air and seek medical attention at once.

Carefully neutralize small spills of sulfuric acid with a suitable agent such as sodium carbonate, further dilute with absorbent material, place in an appropriate container, and dispose of properly. Respiratory protection may be necessary in the event of a large spill or release in a confined area.

Disposal

Excess sulfuric acid and waste material containing this substance should be placed in an appropriate container, clearly labeled, and handled according to your institution's waste disposal guidelines. For more information on disposal procedures, see Chapter 7 of this volume.

The information in this LCSS has been compiled by a committee of the National Research Council from literature sources and Material Safety Data Sheets and is believed to be accurate as of July 1994. This summary is intended for use by trained laboratory personnel in conjunction with the NRC report *Prudent Practices in the Laboratory: Handling and Disposal of Chemicals*. This LCSS presents a concise summary of safety information that should be adequate for most laboratory uses of the title substance, but in some cases it may be advisable to consult more comprehensive references. This information should not be used as a guide to the nonlaboratory use of this chemical.

Substance	Tetrahydrofuran (THF, oxacyclopentane, tetramethylene oxide) CAS 109-99-9
Formula	$(CH_2)_4O$
Physical Properties	Colorless liquid bp 66 °C, mp −108 °C Miscible with water
Odor	Ethereal, detectable at 2 to 50 ppm
Vapor Density	2.5 (air = 1.0)
Vapor Pressure	160 mmHg at 25 °C
Flash Point	−14 °C
Autoignition Temperature	321 °C

Toxicity Data		
LD_{50} oral (rat)	2880 mg/kg
LC_{50} inhal (rat)	21,000 ppm (3 h)
PEL (OSHA)	200 ppm (590 mg/m^3)
TLV-TWA (ACGIH)	200 ppm (590 mg/m^3)
STEL (ACGIH)	250 ppm (737 mg/m^3)

Major Hazards

Highly flammable; forms sensitive peroxides on exposure to air, which may explode on concentration by distillation or drying.

Toxicity

The acute toxicity of THF by inhalation and ingestion is low. Liquid THF is a severe eye irritant and a mild skin irritant, but is not a skin sensitizer. At vapor levels of 100 to 200 ppm, THF irritates the eyes and upper respiratory tract. At high concentrations (25,000 ppm), THF vapor can produce anesthetic effects. Since the odor threshold for THF is well below the permissible exposure limit, this substance is regarded as having good warning properties.

Limited animal testing indicates that THF is not carcinogenic and shows developmental effects only at exposure levels producing other toxic effects in adult animals. Bacterial and mammalian cell culture studies demonstrate no mutagenic activity with THF.

Flammability and Explosibility

THF is extremely flammable (NFPA rating = 3), and its vapor can travel a considerable distance to an ignition source and "flash back." A 5% solution of THF in water is flammable. THF vapor forms explosive mixtures with air at concentrations of 2 to 12% (by volume). Carbon dioxide or dry chemical extinguishers should be used for THF fires.

(continued on facing page)

The information in this LCSS has been compiled by a committee of the National Research Council from literature sources and Material Safety Data Sheets and is believed to be accurate as of July 1994. This summary is intended for use by trained laboratory personnel in conjunction with the NRC report *Prudent Practices in the Laboratory: Handling and Disposal of Chemicals*. This LCSS presents a concise summary of safety information that should be adequate for most laboratory uses of the title substance, but in some cases it may be advisable to consult more comprehensive references. This information should not be used as a guide to the nonlaboratory use of this chemical.

THF can form shock- and heat-sensitive peroxides, which may explode on concentration by distillation or evaporation. Always test samples of THF for the presence of peroxides before distilling or allowing to evaporate. THF should never be distilled to dryness.

Reactivity and Incompatibility

THF can form potentially explosive peroxides upon long exposure to air. THF may react violently with strong oxidizers and reacts vigorously with bromine and titanium tetrachloride. Polymerization can occur in the presence of cationic initiators such as certain Lewis acids and strong protic acids.

Storage and Handling

THF should be handled in the laboratory using the "basic prudent practices" described in Chapter 5.C, supplemented by the additional precautions for dealing with extremely flammable substances (Chapter 5.F). In particular, THF should be used only in areas free of ignition sources, and quantities greater than 1 liter should be stored in tightly sealed metal containers in areas separate from oxidizers. Containers of THF should be dated when opened and tested periodically for the presence of peroxides.

Accidents

In the event of skin contact, immediately wash with soap and water and remove contaminated clothing. In case of eye contact, promptly wash with copious amounts of water for 15 min (lifting upper and lower lids occasionally) and obtain medical attention. If THF is ingested, obtain medical attention immediately. If large amounts of this compound are inhaled, move the person to fresh air and seek medical attention at once.

In the event of a spill, remove all ignition sources, soak up the THF with a spill pillow or absorbent material, place in an appropriate container, and dispose of properly. Respiratory protection may be necessary in the event of a large spill or release in a confined area.

Disposal

Excess THF and waste material containing this substance should be placed in an appropriate container, clearly labeled, and handled according to your institution's waste disposal guidelines. For more information on disposal procedures, see Chapter 7 of this volume.

LABORATORY CHEMICAL SAFETY SUMMARY: TOLUENE

Substance	Toluene (Methylbenzene, toluol, phenylmethane) CAS 108-88-3
Formula	$C_6H_5CH_3$
Physical Properties	Colorless liquid bp 111 °C, mp −95 °C Poorly soluble in water (0.05 g/100 mL)
Odor	Aromatic, benzene-like odor detectable at 0.16 to 37 ppm (mean = 1.6 ppm)
Vapor Density	3.14 (air = 1.0)
Vapor Pressure	22 mmHg at 20 °C
Flash Point	4 °C
Autoignition Temperature	480 °C

Toxicity Data		
	LD_{50} oral (rat)	2650 to 7530 mg/kg
	LD_{50} skin (rabbit)	12,124 mg/kg
	LC_{50} inhal (rat)	26,700 ppm (1 h)
	PEL (OSHA)	200 ppm (750 mg/m^3)
	STEL (OSHA)	150 ppm (560 mg/m^3)
	TLV-TWA (ACGIH)	50 ppm (188 mg/m^3)—skin

Major Hazards

Highly flammable liquid and vapor.

Toxicity

The acute toxicity of toluene is low. Toluene may cause eye, skin, and respiratory tract irritation. Short-term exposure to high concentrations of toluene (e.g., 600 ppm) may produce fatigue, dizziness, headaches, loss of coordination, nausea, and stupor; 10,000 ppm may cause death from respiratory failure. Ingestion of toluene may cause nausea and vomiting and central nervous system depression. Contact of liquid toluene with the eyes causes temporary irritation. Toluene is a skin irritant and may cause redness and pain when trapped beneath clothing or shoes; prolonged or repeated contact with toluene may result in dry and cracked skin. Because of its odor and irritant effects, toluene is regarded as having good warning properties.

The chronic effects of exposure to toluene are much less severe than those of benzene. No carcinogenic effects were reported in animal studies. Equivocal results were obtained in studies to determine developmental effects in animals. Toluene was not observed to be mutagenic in standard studies.

(continued on facing page)

The information in this LCSS has been compiled by a committee of the National Research Council from literature sources and Material Safety Data Sheets and is believed to be accurate as of July 1994. This summary is intended for use by trained laboratory personnel in conjunction with the NRC report *Prudent Practices in the Laboratory: Handling and Disposal of Chemicals*. This LCSS presents a concise summary of safety information that should be adequate for most laboratory uses of the title substance, but in some cases it may be advisable to consult more comprehensive references. This information should not be used as a guide to the nonlaboratory use of this chemical.

Flammability and Explosibility

Toluene is a flammable liquid (NFPA rating = 3), and its vapor can travel a considerable distance to an ignition source and "flash back." Toluene vapor forms explosive mixtures with air at concentrations of 1.4 to 6.7% (by volume). Hazardous gases produced in fire include carbon monoxide and carbon dioxide. Carbon dioxide and dry chemical extinguishers should be used to fight toluene fires.

Reactivity and Incompatibility

Contact with strong oxidizers may cause fires and explosions.

Storage and Handling

Toluene should be handled in the laboratory using the "basic prudent practices" described in Chapter 5.C, supplemented by the additional precautions for dealing with highly flammable substances (Chapter 5.F). In particular, toluene should be used only in areas free of ignition sources, and quantities greater than 1 liter should be stored in tightly sealed metal containers in areas separate from oxidizers.

Accidents

In the event of skin contact, immediately wash with soap and water and remove contaminated clothing. In case of eye contact, promptly wash with copious amounts of water for 15 min (lifting upper and lower lids occasionally) and obtain medical attention. *If toluene is ingested, do not induce vomiting.* Obtain medical attention immediately. If large amounts of this compound are inhaled, move the person to fresh air and seek medical attention at once.

In the event of a spill, remove all ignition sources, soak up the toluene with a spill pillow or absorbent material, place in an appropriate container, and dispose of properly. Respiratory protection may be necessary in the event of a large spill or release in a confined area.

Disposal

Excess toluene and waste material containing this substance should be placed in an appropriate container, clearly labeled, and handled according to your institution's waste disposal guidelines. For more information on disposal procedures, see Chapter 7 of this volume.

Substance	Toluene diisocyanate (TDI; 2,4-toluene diisocyanate; 2,4-diisocyanato-1-methyl benzene) CAS 584-84-9
Formula	$C_9H_6N_2O_2$
Physical Properties	Colorless to pale yellow liquid bp 251 °C, mp 21 °C Insoluble in water (reacts exothermically)
Odor	Sharp, pungent odor detectable at 0.02 to 0.4 ppm
Vapor Density	6.0 (air = 1.0)
Vapor Pressure	0.05 mmHg at 25 °C
Flash Point	132 °C
Autoignition Temperature	>619 °C

Toxicity Data

LD_{50} oral (rat)	4130 mg/kg
LD_{50} skin (rabbit)	>10 g/kg
LC_{50} inhal (rat)	14 ppm/4 h (100 mg/m³; 4 h)
PEL (OSHA)	0.02 ppm (ceiling 0.14 mg/m³)
TLV-TWA (ACGIH)	0.005 ppm (0.036 mg/m³)
STEL (ACGIH)	0.02 ppm (0.14 mg/m³)

Major Hazards

Sensitizer by inhalation and skin contact; possible human carcinogen (OSHA "select carcinogen").

Toxicity

The acute toxicity of toluene diisocyanate by inhalation is high. Exposure to TDI can cause lung damage and decreased breathing capacity. Symptoms of exposure may include coughing, tightness of the chest, chest pain, nausea, vomiting, abdominal pain, headache, and insomnia. TDI irritates the skin, and eye contact can cause irritation with permanent damage if untreated. The oral acute toxicity of this substance is low. The odor of TDI does not provide an adequate warning to avoid overexposure.

Toluene diisocyanate has caused sensitization of the respiratory tract, manifested by acute asthmatic reaction upon return to work after a period of time away from exposure. Initial symptoms include coughing during the night, with difficult or labored breathing. Skin sensitization can also occur. Toluene diisocyanate is listed in IARC Group 2B ("possible human carcinogen"), is listed by NTP as "reasonably anticipated to be a carcinogen," and is classified as a "select carcinogen" under the criteria of the OSHA Laboratory Standard.

(continued on facing page)

The information in this LCSS has been compiled by a committee of the National Research Council from literature sources and Material Safety Data Sheets and is believed to be accurate as of July 1994. This summary is intended for use by trained laboratory personnel in conjunction with the NRC report *Prudent Practices in the Laboratory: Handling and Disposal of Chemicals*. This LCSS presents a concise summary of safety information that should be adequate for most laboratory uses of the title substance, but in some cases it may be advisable to consult more comprehensive references. This information should not be used as a guide to the nonlaboratory use of this chemical.

Flammability and Explosibility	TDI is a combustible liquid (NFPA rating = 1). Explosive limits in air are 0.9 to 9.5% by volume. Carbon dioxide or dry chemical extinguishers should be used for TDI fires.
Reactivity and Incompatibility	Contact with strong oxidizers may cause fires and explosions. Contact with water, acids, bases, and amines can lead to reactions that liberate heat and CO_2 and cause violent foaming and spattering. TDI will attack some forms of plastic, rubber, and coatings.
Storage and Handling	Because of its high toxicity, carcinogenicity, and ability to cause sensitization, toluene diisocyanate should be handled using the "basic prudent practices" of Chapter 5.C, supplemented by the additional precautions for work with compounds of high toxicity (Chapter 5.D). In particular, work with TDI should be conducted in a fume hood to prevent exposure by inhalation, and splash goggles and impermeable gloves should be worn at all times to prevent eye and skin contact.
Accidents	In the event of skin contact, immediately wash with soap and water and remove contaminated clothing. In case of eye contact, promptly wash with copious amounts of water for 15 min (lifting upper and lower lids occasionally) and obtain medical attention. If TDI is ingested, obtain medical attention immediately. If large amounts of this compound are inhaled, move the person to fresh air and seek medical attention at once.
	In the event of a spill, remove all ignition sources, soak up the TDI with a spill pillow or absorbent material, place in an appropriate container, and dispose of properly. Respiratory protection may be necessary in the event of a large spill or release in a confined area.
Disposal	Excess TDI and waste material containing this substance should be placed in an appropriate container, clearly labeled, and handled according to your institution's waste disposal guidelines. For more information on disposal procedures, see Chapter 7 of this volume.

The information in this LCSS has been compiled by a committee of the National Research Council from literature sources and Material Safety Data Sheets and is believed to be accurate as of July 1994. This summary is intended for use by trained laboratory personnel in conjunction with the NRC report *Prudent Practices in the Laboratory: Handling and Disposal of Chemicals*. This LCSS presents a concise summary of safety information that should be adequate for most laboratory uses of the title substance, but in some cases it may be advisable to consult more comprehensive references. This information should not be used as a guide to the nonlaboratory use of this chemical.

LABORATORY CHEMICAL SAFETY SUMMARY: TRIFLUOROACETIC ACID

Substance	Trifluoroacetic acid (TFA, trifluoroethanoic acid) CAS 76-05-1
Formula	CF_3COOH
Physical Properties	Colorless liquid bp 72 °C, mp -15 °C Miscible with water
Odor	Sharp, pungent odor
Vapor Density	3.9 (air = 1.0)
Vapor Pressure	107 mmHg at 25 °C
Flash Point	Noncombustible

Toxicity Data	LD_{50} oral (rat)	200 mg/kg
	LC_{50} inhal (rat)	2000 ppm (4 h)

Major Hazards Corrosive to the skin and eyes; vapor or mist is very irritating and can be destructive to the eyes and respiratory system; ingestion causes internal irritation and severe injury.

Toxicity Trifluoroacetic acid is a highly corrosive substance. Contact of the liquid with the skin, eyes, and mucous membranes can cause severe burns, and ingestion can result in serious damage to the digestive tract. TFA vapor is highly irritating of the eyes and respiratory tract, and inhalation of high concentrations can lead to severe destruction of the upper respiratory tract and may be fatal as a result of pulmonary edema. Symptoms of overexposure to TFA vapor include a burning feeling, coughing, headache, nausea, and vomiting.

Trifluoroacetic acid has not been found to be carcinogenic or to show reproductive or developmental toxicity in humans.

Flammability and Explosibility Trifluoroacetic acid is not combustible. Nevertheless, the presence of trifluoroacetic acid at the site of a fire would be of great concern because of its high vapor pressure and extreme corrosiveness.

Reactivity and Incompatibility Mixing trifluoroacetic acid and water evolves considerable heat.

Storage and Handling Trifluoroacetic acid should be handled in the laboratory using the "basic prudent practices" described in Chapter 5.C. In particular, trifluoroacetic acid should be stored in an acid cabinet away from other classes of compounds. Because of its high vapor pressure,

(continued on facing page)

fumes of trifluoroacetic acid can destroy labels on other bottles if the container is not tightly sealed.

Accidents

In the event of skin contact, the affected area should be flushed immediately with copious amounts of water. In case of eye contact, promptly wash with copious amounts of water for 15 min (lifting upper and lower lids occasionally). Medical attention should be obtained immediately in the event of contact with a large area of the skin or eye contact. If trifluoroacetic acid is ingested, obtain medical attention immediately. If large amounts of this compound are inhaled, move the person to fresh air and seek medical attention at once.

Carefully neutralize small spills of TFA with a suitable agent such as sodium carbonate, dilute with absorbent material, place in an appropriate container, and dispose of properly. Respiratory protection may be necessary in the event of a large spill or release in a confined area.

Disposal

Trifluoroacetic acid and waste material containing this substance should be placed in an appropriate container, clearly labeled, and handled according to your institution's waste disposal guidelines. For more information on disposal procedures, see Chapter 7 of this volume.

LABORATORY CHEMICAL SAFETY SUMMARY: TRIMETHYLALUMINUM (AND RELATED ORGANOALUMINUM COMPOUNDS)

Substance   Trimethylaluminum (and related organoaluminum compounds)
(Trimethylalane, trimethylaluminium)
CAS 75-24-1

Note: Although other alkylaluminum reagents may have different physical properties than trimethylaluminum, their toxicology and reactivity are similar.

Formula $(CH_3)_3Al$

Physical Properties   Colorless pyrophoric liquid
bp 125 to 126 °C, mp 15 °C
React explosively with water

Odor   Corrosive odor and "taste" may be detectable from trimethylaluminum fires

Vapor Density   Not available

Vapor Pressure   12 mmHg at 25 °C

Flash Point   −18 °C

Autoignition Temperature   Spontaneously ignites in air (even as a frozen solid)

Toxicity Data   TLV-TWA (ACGIH)   2 mg (Al)/m^3

Major Hazards   Highly reactive, pyrophoric substances; corrosive on contact with skin and eyes.

Toxicity   Trimethylaluminum and related alkylaluminum reagents are pyrophoric materials that can react explosively with the moisture in tissues, causing severe burns. The heat of reaction can also ignite the methane gas generated, resulting in thermal burns. Alkylaluminum reagents are corrosive substances, and contact is extremely destructive to the eyes, skin, and mucous membranes. Inhalation of trimethylaluminum and other volatile alkylaluminum compounds may cause severe damage to the respiratory tract and can lead to fatal pulmonary edema.

Flammability and Explosibility   Trimethylaluminum is pyrophoric and burns violently on contact with air or water. Other alkylaluminum reagents show similar behavior, although most are not as volatile as trimethylaluminum. *Water or CO_2 fire extinguishers must not be used to put out fires involving trialkylaluminum reagents.* Instead, dry chemical powders such as bicarbonate, Met-L-X®, or inert smothering agents such as sand or graphite should be used to extinguish fires involving trialkylaluminum compounds.

(continued on facing page)

Reactivity and Incompatibility

Trialkylaluminum reagents are highly reactive reducing and alkylating agents. They react violently with air, water, alcohols, halogenated hydrocarbons, and oxidizing agents. These reagents are often supplied as solutions in hydrocarbon solvents, which are less hazardous than the pure liquids.

Storage and Handling

Trialkylaluminum agents should be handled in the laboratory using the "basic prudent practices" described in Chapter 5.C, supplemented by the additional precautions for work with highly flammable (Chapter 5.F) and reactive (Chapter 5.G) substances. Safety glasses, impermeable gloves, and a fire-retardant laboratory coat should be worn at all times when working with these compounds. Trialkylaluminum reagents should be handled only under an inert atmosphere.

Accidents

In the event of skin contact, immediately wash with soap and water and remove contaminated clothing. In case of eye contact, promptly wash with copious amounts of water for 15 min (lifting upper and lower lids occasionally) and obtain medical attention. If trialkylaluminum compounds are ingested, obtain medical attention immediately. If any of these compounds are inhaled, move the person to fresh air and seek medical attention at once.

Any spill of trialkylaluminum will likely result in fire. Remove all ignition sources, put out the trialkylaluminum fire with a dry chemical extinguisher, sweep up the resulting solid, place in an appropriate container under an inert atmosphere, and dispose of properly. Respiratory protection may be necessary in the event of a large spill or release in a confined area.

Disposal

Excess trialkylaluminum reagents and waste material containing these substances should be placed in an appropriate container under an inert atmosphere, clearly labeled, and handled according to your institution's waste disposal guidelines. Alternately, small quantities of trialkylaluminum reagents can be destroyed in the laboratory by experienced personnel by slow addition of *t*-butanol to a solution of the reagent in an inert solvent such as toluene under an inert atmosphere such as argon. The resulting mixture should then be placed in an appropriate container, clearly labeled, and handled according to your institution's waste disposal guidelines. For more information on disposal procedures, see Chapter 7 of this volume.

The information in this LCSS has been compiled by a committee of the National Research Council from literature sources and Material Safety Data Sheets and is believed to be accurate as of July 1994. This summary is intended for use by trained laboratory personnel in conjunction with the NRC report *Prudent Practices in the Laboratory: Handling and Disposal of Chemicals*. This LCSS presents a concise summary of safety information that should be adequate for most laboratory uses of the title substance, but in some cases it may be advisable to consult more comprehensive references. This information should not be used as a guide to the nonlaboratory use of this chemical.

Substance	Trimethyltin chloride (chlorotrimethylstannane) CAS 1066-45-1 Other organotin compounds: tributyltin chloride, tributyltin hydride
Formula	C_3H_9ClSn
Physical Properties	White crystalline solid mp 37 to 39 °C Insoluble in water
Odor	Strong unpleasant stench; no threshold data available
Flash Point	97 °C

Toxicity Data

LD_{50} oral (rat)	12.6 mg/kg
PEL (OSHA)	0.1 mg tin/m^3

The ACGIH has established the following uniform exposure limits for all organotin compounds based on the concentration of tin in air:

TLV-TWA (ACGIH)	0.1 mg tin/m^3
STEL (ACGIH)	0.2 mg tin/m^3

Major Hazards

Trimethyltin chloride is highly toxic by all routes of exposure.

Toxicity

Trimethyltin chloride and other organotin compounds are highly toxic by ingestion, inhalation, or skin contact. Trimethyltin chloride can cause irritation and burns of the skin and eyes. Organotin compounds can affect the central nervous system. The degree of toxicity is greatest for compounds with three or four alkyl groups attached to tin. Di- and monoalkyltin compounds are moderately toxic. The toxicity diminishes as the size of the alkyl groups increases. Thus, the oral LD_{50}s in rats are as follows: dimethyltin dichloride, 74 to 237 mg/kg; tributyltin chloride, 122 to 349 mg/kg; dibutyltin oxide, 487 to 520 mg/kg; trioctyltin chloride, >4000 mg/kg.

Organotin compounds have been shown to cause reproductive effects in laboratory animals.

Flammability and Explosibility

Not a significant fire hazard. Emits toxic fumes in fire.

Reactivity and Incompatibility

Trimethyltin chloride and other organotin halides react with water to produce hydrogen halides. Organotin hydrides react with water to produce hydrogen gas, which is flammable and explosive.

(continued on facing page)

Storage and Handling

Because of its high acute toxicity, trimethyltin chloride should be handled using the "basic prudent practices" of Chapter 5.C, supplemented by additional precautions for work with compounds of high acute toxicity (Chapter 5.D). Other alkyltin compounds should be handled using the "basic prudent practices" of Chapter 5.C.

Accidents

In the event of skin contact, immediately wash with soap and water and remove contaminated clothing. In case of eye contact, promptly wash with copious amounts of water for 15 min (lifting upper and lower lids occasionally) and obtain medical attention. If trimethyltin chloride or another organotin compound is ingested, obtain medical attention immediately. If large amounts of this compound are inhaled, move the person to fresh air and seek medical attention at once.

In the event of a spill, sweep up the organotin compound or soak up with a spill pillow or absorbent material, place in an appropriate container, and dispose of properly. Respiratory protection may be necessary in the event of a large spill or release in a confined area.

Disposal

Excess trimethyltin chloride or other organotin compound and waste material containing this substance should be placed in an appropriate container, clearly labeled, and handled according to your institution's waste disposal guidelines. For more information on disposal procedures, see Chapter 7 of this volume.

Index

NOTE: *Page numbers in* **boldface** *under chemical names indicate a Laboratory Chemical Safety Summary for that chemical. These summaries contain information regarding a substance's chemistry, toxicity, and handling in a laboratory setting.*